Illustrated Guide to the 1999 National Electrical Code®

John E. Traister — Revised & Updated by Bradford Maher

Craftsman

Craftsman Book Company
6058 Corte del Cedro / P.O. Box 6500 / Carlsbad, CA 92018

Acknowledgements

A deep and grateful bow must be made in the direction of several individuals and organizations who helped in the preparation and development of this book. Without their help, there would be no book.

National Fire Protection Association, Inc.

Square D Company

John Karns, Art Work

Keeler Chapman, Art Work

Ruby Updike, Typist and Organizer

Library of Congress Cataloging-in-Publication Data
Traister, John E.
Illustrated guide to the 1999 National electrical code / by John E. Traister ; revised and edited by Bradford Maher.
p. c.m.
ISBN 1-57218-075-7
1. Electric engineering--Insurance requirements--United States. 2. National Fire Protection Association. National Electrical Code (1999) I. Maher, Bradford. II. Title. III. Title: National electrical code, 1999. IV. Title: Illustrated guide to the National electrical code, 1999.
TK260.T72 1999
621.319'24'0218--dc21 99-28540
CIP

Contents

Chapter 10
Transformers 241

Chapter 11
Capacitors, Resistors and Reactors 275

Chapter 12
Special Occupancies 283

Chapter 1
About the *NEC*®

All electrical installations must conform to applicable National Electrical Code® (NEC) requirements, local ordinances, and instructions provided by equipment and component manufacturers. This chapter explains the key terms and basic layout of the NEC.

PURPOSE AND HISTORY OF THE *NEC*

Owing to the potential fire and explosion hazards caused by the improper handling and installation of electrical wiring, certain rules in the selection of materials, quality of workmanship, and precautions for safety must be followed. To standardize and simplify these rules and provide a reliable guide for electrical construction, the *National Electrical Code* (*NEC*) was developed. The *NEC*, originally prepared in 1897, is frequently revised to meet changing conditions, improved equipment and materials, and new fire hazards. It is the result of the best efforts of electrical engineers, manufacturers of electrical equipment, insurance underwriters, fire fighters, and other concerned experts throughout the country.

The *NEC* is now published by the National Fire Protection Association (NFPA), Batterymarch Park, Quincy, Massachusetts 02269. It contains specific rules and regulations intended to help in the practical safeguarding of persons and property from hazards arising from the use of electricity.

The *NEC* states, "This Code is not intended as a design specification, but includes such provisions where considered necessary for safety. It is not intended as an instruction manual for untrained persons." However, it does provide a sound basis for the study of electrical installation procedures — under the proper guidance. The probable reason for the *NEC*'s self-analysis is that the Code also states, "This Code contains provisions considered necessary for safety. Compliance therewith and proper maintenance will result in an installation essentially free from hazard, but not necessarily efficient, convenient, or adequate for good service or future expansion of electrical use."

The NEC, however, has become the bible of the electrical construction industry, and anyone involved in electrical work, in any capacity, should obtain an up-to-date copy, keep it handy at all times, and refer to it frequently.

NEC TERMINOLOGY

There are two basic types of rules in the NEC: mandatory rules and advisory rules. Here is how to recognize the two types of rules and how they relate to all types of electrical systems.

- Mandatory rules — All mandatory rules have the word *shall* in them. The word "shall" means must. If a rule is mandatory, you must comply with it.
- Advisory rules — All advisory rules have the word *should* in them. The word "should" in this case means recommended but not necessarily required. If a rule is advisory, compliance is discretionary.

Be alert to local amendments to the NEC. Local ordinances may amend the language of the NEC, changing it from *should* to *shall*. This means that you must do in that county or city what may only be recommended in some other area. The office that issues building permits will either sell you a copy of the code that's enforced in that area or tell you where the code is sold. In rare instances, the electrical inspector having jurisdiction may issue these regulations verbally.

There are a few other "landmarks" that you will encounter while looking through the NEC. These are summarized in Figure 1-1, and a brief explanation of each follows:

Explanatory material: Explanatory material in the form of Fine Print Notes is designated (FPN). Where these appear, the FPNs normally apply to the NEC Section or paragraph immediately preceding the FPN.

Change bar: A change bar in the margins indicates that a change in the NEC has been made since the last edition. When becoming familiar with each new edition of the NEC, always review these changes. There are also several illustrated publications on the market that point out changes in the NEC with detailed explanations of each. Such publications make excellent reference material.

Bullets: A filled-in circle called a "bullet" indicates that something has been deleted from the last edition of the NEC. Although not absolutely necessary, many electricians like to compare the previous NEC edition to the most recent one when these bullets are encountered, just to see what has been omitted from the latest edition. The most probable reasons for the deletions are errors in the previous edition, or obsolete items.

Extracted text: Material identified by the superscript letter x includes text extracted from other NFPA documents as identified in Appendix A of the NEC. For example, ". . . 516-6.^{X} This section shall apply to processes in which combustible dry powders . . . ."

NEC Text Formats

As you open the NEC book, you will notice several different styles of text used. Here is an explanation of each.

1. *Normal black letters:* Basic definitions and explanations of the NEC requirements.
2. *Bold black letters:* Used for Article, Section and Subsection headings.
3. *Exceptions:* These explain the situations when a specific rule does not apply. Exceptions are written in italics under the Section or paragraph to which they apply.
4. *Tables:* Tables are often included when there is more than one possible application of a requirement.
5. *Diagrams:* A few diagrams are scattered throughout the NEC to illustrate certain NEC applications.

LEARNING THE NEC LAYOUT

The *NEC* is divided into the Introduction (Article 90) and nine chapters. Chapters 1, 2, 3, and 4 apply generally; Chapters 5, 6, and 7 apply to special occupancies, special equipment, or other special conditions. These latter chapters supplement or modify the general rules. Chapters 1 through 4 apply except as amended by Chapters 5, 6, and 7 for the particular conditions.

While looking through these *NEC* chapters, if you should encounter a word or term that is unfamiliar, look in Chapter 1, Article 100 — Definitions. Chances are, the term will be found here. If not, look in the Index for the word and the NEC page number. Many terms are included in Article 100, but others are scattered throughout the book.

For definitions of terms not found in the NEC, check the glossary in the back of this book or obtain a copy of *Illustrated Dictionary for Electrical Workers*, available from Delmar Publishers, Inc., Albany, New York.

Chapter 8 of the NEC covers communications systems and is independent of the other chapters except where they are specifically referenced therein.

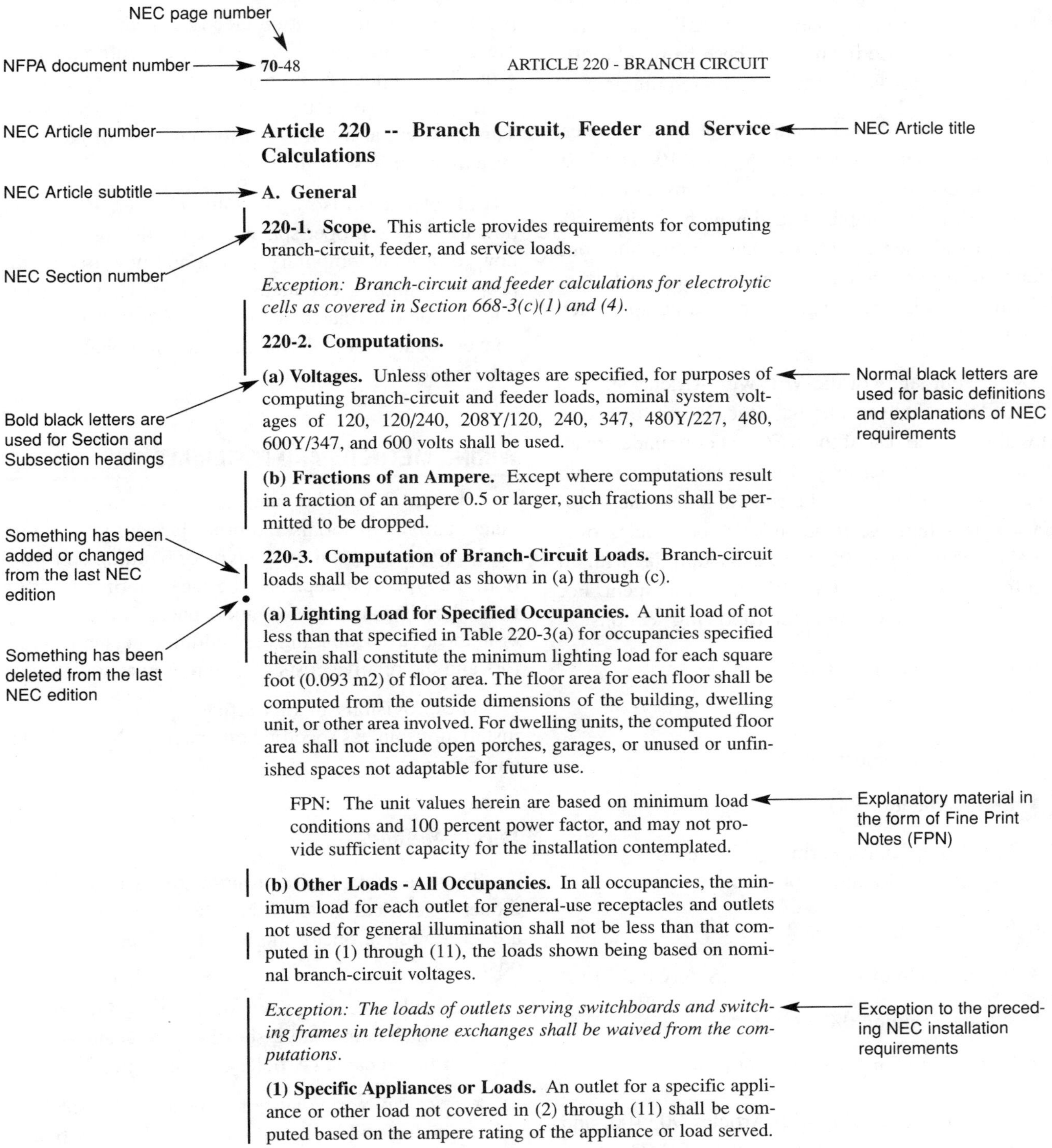

70-48 ARTICLE 220 - BRANCH CIRCUIT

Article 220 -- Branch Circuit, Feeder and Service Calculations

A. General

220-1. Scope. This article provides requirements for computing branch-circuit, feeder, and service loads.

Exception: Branch-circuit and feeder calculations for electrolytic cells as covered in Section 668-3(c)(1) and (4).

220-2. Computations.

(a) Voltages. Unless other voltages are specified, for purposes of computing branch-circuit and feeder loads, nominal system voltages of 120, 120/240, 208Y/120, 240, 347, 480Y/227, 480, 600Y/347, and 600 volts shall be used.

(b) Fractions of an Ampere. Except where computations result in a fraction of an ampere 0.5 or larger, such fractions shall be permitted to be dropped.

220-3. Computation of Branch-Circuit Loads. Branch-circuit loads shall be computed as shown in (a) through (c).

(a) Lighting Load for Specified Occupancies. A unit load of not less than that specified in Table 220-3(a) for occupancies specified therein shall constitute the minimum lighting load for each square foot (0.093 m2) of floor area. The floor area for each floor shall be computed from the outside dimensions of the building, dwelling unit, or other area involved. For dwelling units, the computed floor area shall not include open porches, garages, or unused or unfinished spaces not adaptable for future use.

FPN: The unit values herein are based on minimum load conditions and 100 percent power factor, and may not provide sufficient capacity for the installation contemplated.

(b) Other Loads - All Occupancies. In all occupancies, the minimum load for each outlet for general-use receptacles and outlets not used for general illumination shall not be less than that computed in (1) through (11), the loads shown being based on nominal branch-circuit voltages.

Exception: The loads of outlets serving switchboards and switching frames in telephone exchanges shall be waived from the computations.

(1) Specific Appliances or Loads. An outlet for a specific appliance or other load not covered in (2) through (11) shall be computed based on the ampere rating of the appliance or load served.

NEC page number
NFPA document number
NEC Article number
NEC Article title
NEC Article subtitle
NEC Section number
Normal black letters are used for basic definitions and explanations of NEC requirements
Bold black letters are used for Section and Subsection headings
Something has been added or changed from the last NEC edition
Something has been deleted from the last NEC edition
Explanatory material in the form of Fine Print Notes (FPN)
Exception to the preceding NEC installation requirements

Figure 1-1: *Layout of a typical NEC page*

Chapter 9 consists of tables and examples.

There is also the *NEC* Contents at the beginning of the book and a comprehensive index at the back. You will find frequent use for both of these helpful "tools" when searching for various installation requirements.

Each chapter is divided into one or more Articles. For example, Chapter 1 contains Articles 100 and 110. These Articles are subdivided into Sections. For example, Article 110 of Chapter 1 begins with Section 110-2 — Approval. Some sections may contain only one sentence or a paragraph, while others may be further subdivided into lettered or numbered paragraphs such as (a), (1), (2), and so on.

Begin your study of the *NEC* with Articles 90, 100 and 110. These three articles have the basic information that will make the rest of the *NEC* easier to understand. Article 100 defines terms you will need to understand the code. Article 110 gives the general requirements for electrical installations. Read these three articles over several times until you are thoroughly familiar with all the information they contain. It's time well spent. For example, Article 90 contains the following sections:

- Purpose (90-1)
- Scope (90-2)
- Code Arrangement (90-3)
- Enforcement (90-4)
- Mandatory Rules, Permissive Rules and Explanatory Material (90-5)
- Formal Interpretations (90-6)
- Examination of Equipment for Safety (90-7)
- Wiring Planning (90-8)
- Metric Units of Measurement (90-9)

Once you are familiar with Articles 90, 100, and 110 you can move on to the rest of the *NEC*. There are several key sections you will use often while installing and servicing electrical systems. Let's discuss each of these important sections.

WIRING DESIGN AND PROTECTION

Chapter 2 of the *NEC* discusses wiring design and protection, the information electrical technicians need most often. It covers the use and identification of grounded conductors, branch circuits, feeders, calculations, services, overcurrent protection and grounding. This is essential information for any type of electrical system, regardless of the type.

Chapter 2 is also a "how-to" chapter. It explains how to provide proper spacing for conductor supports, how to provide temporary wiring and how to size the proper grounding conductor or electrode. If you run into a problem related to the design/installation of a conventional electrical system, you can probably find a solution for it in this chapter.

WIRING METHODS AND MATERIALS

Chapter 3 has the rules on wiring methods and materials. The materials and procedures used on a particular system depend on the type of building construction, the type of occupancy, the location of the wiring in the building, the type of atmosphere in the building or in the area surrounding the building, mechanical factors and the relative costs of different wiring methods.

The provisions of this article apply to all wiring installations unless specified otherwise in *NEC* Articles and Sections.

Wiring Methods

There are four basic wiring methods used in most modern electrical systems. Nearly all wiring methods are a variation of one or more of these four basic methods:

- Sheathed cables of two or more conductors, such as nonmetallic-sheathed cable and armored cable (Articles 330 through 339)
- Raceway wiring systems, such as rigid steel conduit and electrical metallic tubing (Articles 342 to 358)
- Busways (Article 364)
- Cabletray (Article 318)

Electrical Conductors

Article 310 in Chapter 3 gives a complete description of all types of electrical conductors. Electrical conductors come in a wide range of sizes and forms. Be sure to check the working drawings and specifications to see what sizes and types of conductors are required for a specific job. If conductor type and size are not specified, choose the most appropriate type and size meeting standard *NEC* requirements.

When workers have the choice of selecting the wiring method to use, most will select the least expensive method allowed by the *NEC*. However, in some cases, what appears to be the least expensive method may not hold true in the final results. For example, when rewiring existing buildings where much "fishing" of cable is necessary, workers have found that Type AC armored cable (MC) is usually easier to fish in concealed partitions than Type NM (Romex) cable. Although MC cable is more expensive, the savings in labor usually offset the cost.

Boxes, Cabinets, And Enclosures

Articles 318 through 384 give rules for raceways, boxes, cabinets and raceway fittings. Outlet boxes vary in size and shape, depending on their use, the size of the raceway, the number of conductors entering the box, the type of building construction and atmospheric conditions of the areas. Chapter 3 should answer most questions on the selection and use of these items.

The *NEC* does not describe in detail all types and sizes of outlet boxes. But manufacturers of outlet boxes have excellent catalogs showing all of their products. Collect these catalogs. They are essential to your work.

Wiring Devices And Switchgear

Article 380 covers the switches, pushbuttons, pilot lamps, receptacles and power outlets. Again, get the manufacturers' catalogs on these items. They will provide you with detailed descriptions of each.

Article 384 covers switchboards and panelboards, including their location, installation methods, clearances, grounding and overcurrent protection.

EQUIPMENT FOR GENERAL USE

Chapter 4 of the *NEC* begins with the use and installation of flexible cords and cables, including the trade name, type letter, wire size, number of conductors, conductor insulation, outer covering and use of each. The chapter also includes fixture wires, again giving the trade name, type letter and other important details.

Article 410 on lighting fixtures is especially important. It gives installation procedures for fixtures in specific locations. For example, it covers fixtures near combustible material and fixtures in closets. The *NEC* does not describe how many fixtures will be needed in a given area to provide a certain amount of illumination.

Article 430 covers electric motors, including mounting the motor and making electrical connections to it. Motor controls and overload protection are also covered.

Articles 440 through 460 cover air conditioning and heating equipment, transformers and capacitors.

Article 480 gives most requirements related to battery-operated electrical systems. Storage batteries are seldom thought of as part of a conventional electrical system, but they often provide standby emergency lighting service. They may also supply power to security systems that are separate from the main ac electrical system.

SPECIAL OCCUPANCIES

Chapter 5 of the *NEC* covers special occupancy areas. These are areas where the sparks generated by electrical equipment may cause an explosion or fire. The hazard may be due to the atmosphere of the area or just the presence of a volatile material in the area. Commercial garages, aircraft hangers and service stations are typical special occupancy locations.

Articles 500 - 501 cover the different types of special occupancy atmospheres that are considered to be hazardous areas. The atmospheric groups were established to make it easy to test and approve equipment for various types of uses.

Articles 501-4, 502-4, and 503-3 cover the installation of wiring in hazardous (Classified) locations. Wiring in these areas must be designed to prevent the ignition of a surrounding explosive atmosphere when arcing occurs within the electrical system.

There are three main classes of special occupancy locations:

- Class I (Article 501): Areas containing flammable gases or vapors in the air. Class I areas include paint spray booths, dyeing plants where hazardous liquids are used and gas generator rooms.
- Class II (Article 502): Areas where combustible dust is present, such as grain handling and storage plants, dust and stock collector areas and sugar pulverizing plants. These are areas where, under normal operating conditions, there may be enough combustible dust in the air to produce explosive or ignitable mixtures.
- Class III (Article 503): Areas that are hazardous because of the presence of easily ignitable fibers or flyings in the air, although not in large enough quantity to produce ignitable mixtures. Class III locations include cotton mills, rayon mills, and clothing manufacturing plants.

Articles 511 and 514 regulate garages and similar locations where volatile or flammable liquids are used. While these areas are not always considered critically hazardous locations, there may be enough danger to require special precautions in the electrical installation. In these areas, the *NEC* requires that volatile gases be confined to an area not more than 4 ft above the floor. So in most cases, conventional raceway systems are permitted above this level. If the area is judged critically hazardous, explosionproof wiring (including seal-offs) may be required.

Article 520 regulates theaters and similar occupancies where fire and panic can cause hazards to life and property. Drive-in theaters do not present the same hazards as enclosed auditoriums. But the projection rooms and adjacent areas must be properly ventilated and wired for the protection of operating personnel and others using the area.

Chapter 5 also covers residential storage garages, aircraft hangars, service stations, bulk storage plants, health care facilities, mobile homes and parks, and recreation vehicles and parks.

When security technicians are installing systems in hazardous locations, extreme caution must be used. You may be working with only 12 or 24 V, but a spark caused by, say, an improper connection can set off a violent explosion. You may have already witnessed a low-voltage explosion in the common automotive battery. Although only 12 V dc are present, if a spark occurs near the battery and battery gases are leaking through the battery housing, chances are the battery will explode with a report similar to a shotgun firing.

When installing security systems in Class I, Division 1 locations, explosionproof fittings are required and most electrical wiring must be enclosed in rigid steel conduit (pipe).

SPECIAL EQUIPMENT

Article 600 covers electric signs and outline lighting. Article 610 applies to cranes and hoists. Article 620 covers the majority of the electrical work involved in the installation and operation of elevators, dumbwaiters, escalators and moving walks. The manufacturer is responsible for most of this work. The electrician usually just furnishes a feeder terminating in a disconnect means in the bottom of the elevator shaft. The electrician may also be responsible for a lighting circuit to a junction box midway in the elevator shaft for connecting the elevator cage lighting cable and exhaust fans. Articles in Chapter 6 of the *NEC* give most of the requirements for these installations.

Article 630 regulates electric welding equipment. It is normally treated as a piece of industrial power equipment requiring a special power outlet. But there are special conditions that apply to the circuits supplying welding equipment. These are outlined in detail in Chapter 6 of the *NEC*.

Article 640 covers wiring for sound-recording and similar equipment. This type of equipment normally requires low-voltage wiring. Special outlet boxes or cabinets are usually provided with the equipment. But some items may be mounted in or on standard outlet boxes. Some sound-recording electrical systems require direct current, supplied from rectifying equipment, batteries or motor generators. Low-voltage alternating current comes from relatively small transformers connected on the primary side to a 120-V circuit within the building.

Other items covered in Chapter 6 of the *NEC* include: X-ray equipment (Article 660), induction and dielectric heat-generating equipment (Article 665) and machine tools (Article 670).

If you ever have work that involves Chapter 6, study the chapter before work begins. That can save a lot of installation time. Here is another way to cut down on labor hours and prevent installation errors. Get a set of rough-in drawings of the equipment being installed. It is easy to install the wrong outlet box or to install the right box in the wrong place. Having a set of rough-in drawings can prevent those simple but costly errors.

SPECIAL CONDITIONS

In most commercial buildings, the *NEC* and local ordinances require a means of lighting public rooms, halls, stairways and entrances. There must be enough light to allow the occupants to exit from the building if the general building lighting is interrupted. Exit doors must be clearly indicated by illuminated exit signs.

Chapter 7 of the *NEC* covers the installation of emergency lighting systems. These circuits should be arranged so that they can automatically transfer to an alternate source of current, usually storage batteries or gasoline-driven generators. As an alternative in some types of occupancies, you can connect them to the supply side of the main service so disconnecting the main service switch would not disconnect the emergency circuits. See Article 700. *NEC* Chapter 7 also covers a variety of other equipment, systems and conditions that are not easily categorized elsewhere in the *NEC*.

Chapter 8 is a special category for wiring associated with electronic communications systems including telephone and telegraph, radio and TV, fire and burglar alarms, and community antenna systems.

Once you become familiar with the *NEC* through repeated usage, you will generally know where to look for a particular topic. While this chapter provides you with an initial familiarization of the *NEC* layout, much additional usage experience will be needed for you to feel comfortable with the *NEC*'s content.

The *NEC* is not an easy book to read and understand at first. In fact, seasoned electrical workers and technicians sometimes find it confusing. Basically, it is a reference book written in a legal, contract-type language and its content does assume prior knowledge of most subjects listed. Consequently, you will sometimes find the *NEC* frustrating to use because terms aren't always defined, or because of some unknown prerequisite knowledge.

DEFINITIONS

Many definitions of terms dealing with the *NEC* may be found in *NEC* Article 100. However, other definitions are scattered throughout the *NEC* under their appropriate category. For example the term lighting track is not listed in Article 100. The term is listed under *NEC* Section 410-100 and reads as follows:

> *Lighting track is a manufactured assembly designed to support and energize lighting fixtures that are capable of being readily repositioned on the track. Its length may be altered by the addition or subtraction of sections of track.*

Regardless of where the definition may be located — in Article 100 or under the appropriate *NEC* Section elsewhere in the book — the best way to learn and remember these definitions is to form a mental picture of each item or device as you read the definition. For example, turn to page 70-25 of the 1996 *NEC* and under Article 100 — Definitions, scan down the page until you come to the term "Attachment Plug (Plug Cap) (Plug)." After reading the definition, you will probably have already formed a mental picture of attachment plugs.

Once again, scan through the definitions until the term "Appliance" is found. Read the definition and then try to form a mental picture of what appliances look like. They should be familiar to everyone.

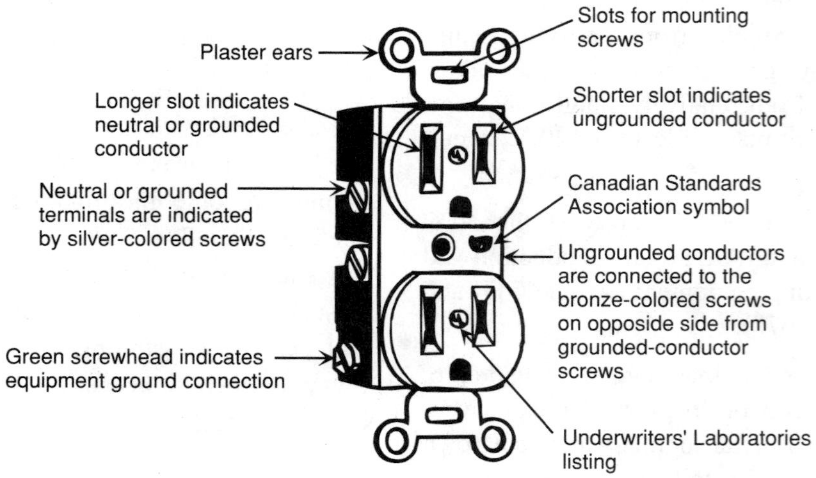
Plaster ears
Longer slot indicates neutral or grounded conductor
Neutral or grounded terminals are indicated by silver-colored screws
Green screwhead indicates equipment ground connection
Slots for mounting screws
Shorter slot indicates ungrounded conductor
Canadian Standards Association symbol
Ungrounded conductors are connected to the bronze-colored screws on opposite side from grounded-conductor screws
Underwriters' Laboratories listing

Figure 1-2: *Description of required markings and other components of a duplex receptacle*

Each and every term listed in the *NEC* should be understood. Know what the item looks like and how it is used on the job. If a term is unfamiliar, try other reference books such as manufacturers' catalogs for an illustration of the item. Then research the item further to determine its purpose in electrical systems. Once you are familiar with all the common terms and definitions found in the *NEC*, navigating through the *NEC* (and understanding what you read) will be much easier.

There are many definitions included in Article 100. You should become familiar with the definitions. Since a copy of the latest *NEC* is compulsory for any type of electrical wiring, there is no need to duplicate them here. However, here are two definitions that you should become especially familiar with:

- Labeled — Equipment or materials to which has been attached a label, symbol or other identifying mark of an organization acceptable to the authority having jurisdiction and concerned with product evaluation, that maintains periodic inspection of production of labeled equipment or materials, and by whose labeling the manufacturer indicates compliance with appropriate standards or performance in a specified manner.
- Listed — Equipment or materials included in a list published by an organization acceptable to the authority having jurisdiction and concerned with product evaluation, that maintains periodic inspection of production of listed equipment or materials, and whose listing states either that the equipment or material meets appropriate designated standards or has been tested and found suitable for use in a specified manner.

TESTING LABORATORIES

Besides installation rules, you will also have to be concerned with the type and quality of materials that are used in electrical wiring systems. Nationally recognized testing laboratories (Underwriters' Laboratories, Inc. is one) are product safety-certification laboratories. They establish and operate product safety certification programs to make sure that items produced under the service are safeguarded against reasonable foreseeable risks. Some of these organizations maintain a worldwide network of field representatives who make unannounced visits to manufacturing facilities to countercheck products bearing their "seal of approval." See Figure 1-2.

However, proper selection, overall functional performance and reliability of a product are factors that are not within the basic scope of UL activities.

To fully understand the *NEC*, it is important to understand the organizations that govern it. The following organizations will frequently be encountered and associated with materials and equipment used on almost every electrical installation.

Nationally Recognized Testing Laboratory (NRTL)

Nationally Recognized Testing Laboratories are product safety certification laboratories. They establish and operate product safety certification programs to make sure that items produced under the service are safeguarded against reasonable foreseeable risks. An approved item, however, does not mean that the item is approved for all uses; it is safe only for the purpose for which it is intended. NRTL maintains a worldwide network of field representatives who make unannounced visits to factories to countercheck products bearing the safety mark.

National Electrical Manufacturers Association (NEMA)

The National Electrical Manufacturers Association was founded in 1926. It is made up of companies that manufacture equipment used for generation, transmission, distribution, control, and utilization of electric power. The objectives of NEMA are to maintain and improve the quality and reliability of products; to ensure safety standards in the manufacture and use of products; to develop product standards covering such matters as naming, ratings, performance, testing, and dimensions. NEMA participates in developing the *NEC* and the National Electrical Safety Code and advocates their acceptance by state and local authorities.

National Fire Protection Association (NFPA)

The NFPA was founded in 1896. Its membership is drawn from the fire service, business and industry, health care, educational and other institutions, and individuals in the fields of insurance, government, architecture, and engineering. The duties of the NFPA include:

- Developing, publishing, and distributing standards prepared by approximately 175 technical committees. These standards are intended to minimize the possibility and effects of fire and explosion.
- Conducting fire safety education programs for the general public.
- Providing information on fire protection, prevention, and suppression.
- Compiling annual statistics on causes and occupancies of fires, large-loss fires (over $1 million), fire deaths, and fire fighter casualties.
- Providing field service by specialists on electricity, flammable liquids and gases, and marine fire problems.
- Conducting research projects that apply statistical methods and operations research to develop computer modes and data management systems.

The Role Of Testing Laboratories

Testing laboratories are an integral part of the development of the code. The NFPA, NEMA, and NRTL all provide testing laboratories to conduct research into electrical equipment and its safety. These laboratories perform extensive testing of new products to make sure they are built to code standards for electrical and fire safety. These organizations receive statistics and reports from agencies all over the United States concerning electrical shocks and fires and their causes. Upon seeing trends developing concerning association of certain equipment and dangerous situations or circumstances, this equipment will be specifically targeted for research.

MAJOR CHANGES TO THE 1999 *NEC*

The following summarize the major areas of change to the 1999 *NEC*. As mentioned previously, when using the *NEC* book itself, a change is readily identified by a vertical line (change bar) in the margin. A bullet in the margin represents a deletion.

Article 90 — Introduction

- Section 90-1. Purpose
- Section 90-2. Scope

Article 100 — Definitions

- Bonding (Bonded)
- Feeder
- Festoon Lighting
- Service Conductors
- Service Equipment

Article 110 — Requirements for Electrical Installations

- Section 110-22. Identification of Disconnecting Means

Part B. Over 600 volts, Nominal, or Less

- Section 110-31. Enclosures for Electrical Installations
- Section 110-34. Work Space and Guarding
- Section 110-36. Circuit Conductors

Part D. Tunnel Installations Over 600 Volts, Nominal

- Section 110-51. General
- Section 110-52. Overcurrent Protection
- Section 110-53. Conductors
- Section 110-54. Bonding and Equipment Grounding Conductors
- Section 110-55. Transformers, Switches, and Electric Equipment
- Section 110-56. Energized Parts
- Section 110-57. Ventilation System Controls
- Section 110-58. Disconnecting Means
- Section 110-59. Enclosures

Article 200 — Use and Identification of Grounded Conductors

- Section 200-2. General
- Section 200-6. Means of Identifying Grounded Conductors
- Section 200-7. Use of Insulation of a White or Natural Gray Color or with Three Continuous White Stripes
- Section 200-10. Identification of Terminals

Article 210 — Branch Circuits

Part A. General Provisions

- Section 210-2. Other Articles for Specific-Purpose Branch Circuits
- Section 210-4. Multiwire Branch Circuits
- Section 210-5. Identification for Branch Circuits
- Section 210-8. Ground-Fault Circuit-Interrupter Protection for Personnel
- Section 210-11. Branch Circuits Required

Part B. Branch-Circuit Ratings

- Section 210-19. Conductors — Minimum Ampacity and Size
- Section 210-20. Overcurrent Protection
- Section 210-23. Permissible Loads
- Section 210-24. Branch-Circuit Requirements — Summary

Part C. Required Outlets

- Section 210-52. Dwelling Unit Receptacle Outlets
- Section 210-60. Guest Rooms
- Section 210-70. Lighting Outlets Required

Article 215 — Feeders

- Section 215-1. Scope
- Section 215-2. Minimum Size and Rating
- Section 215-3. Overcurrent Protection

- Section 215-10. Ground-Fault Protection of Equipment

Article 220 — Branch-Circuit, Feeder, and Service Calculations

Part A. General

- Section 220-2. Computations
- Section 220-3. Computation of Branch Circuit Loads

Part B. Feeders and Services

- Section 220-10. General
- Section 220-12. Show-Window and Track Lighting
- Section 220-13. Receptacle Loads — Nondwelling Units
- Section 220-15. Fixed Electric Space Heating
- Section 220-16. Small Appliance and Laundry Loads — Dwelling Unit
- Section 220-18. Electric Clothes Dryers — Dwelling Unit(s)

Part D. Method for Computing Farm Loads

- Section 220-40. Farm Loads — Buildings and Other Loads

Article 225 — Outside Branch Circuits and Feeders

- Section 225-6. Conductor Size and Support
- Section 225-9. Overcurrent Protection
- Section 225-19 Clearances from Buildings for Conductors of Not Over 600 Volts, Nominal

Article 230 — Services

Part B. Overhead Service-Drop Conductors

Part D. Service-Entrance Conductors

- Section 230-40. — Number of Service Entrance Conductor Sets
- Section 230-42. — Minimum Size and Rating
- Section 230-46. — Spliced Conductors

Part G. Service Equipment — Overcurrent Protection

- Section 230-91. Location
- Section 230-95. Ground-Fault Protection of Equipment

Part H. Services Exceeding 600 Volts, Nominal

- Section 230-204. Isolating Switches
- Section 230-205. Disconnecting Means

Article 240 — Overcurrent Protection

Part A. General

- Section 240-3. Protection of Conductors
- Section 240-4. Protection of Flexible Cords and Fixture Wires
- Section 240-6. Standard Ampere Ratings
- Section 240-8. Fuses or Circuit Breakers in Parallel
- Section 240-13. Ground-Fault Protection of Equipment

Part B. Location

- Section 240-20. Ungrounded Conductors
- Section 240-21. Location in Circuit
- Section 240-24. Location in or on Premises

Part C. Enclosures

- Section 240-33. Vertical Position

Part F. Cartridge Fuses and Fuseholders

- Section 240-60 General

Part G. Circuit Breakers

- Section 240-83. Marking
- Section 240-86. Series Ratings

Part H. Supervised Industrial Installations

Part I. Overcurrent Protection Over 600 Volts, Nominal

Article 250 — Grounding (Entirely rewritten)

Article 300 — Wiring Methods

Part A. General Requirements

- Section 300-3. Conductors
- Section 300-5. Underground Installations
- Section 300-11. Securing and Supporting
- Section 300-14. Length of Free Conductors at Outlets, Junctions and Switch Points
- Section 300-15. Boxes, Conduit Bodies, or Fittings — Where Required

Part B. Requirements for Over 600 Volts, Nominal

- Section 300-32. — Conductors of Different Systems
- Section 300-37. — Aboveground Wiring Methods
- Section 300-39. — Braid-Covered Insulated Conductors — Open Installation
- Section 300-40. — Insulation Shielding
- Section 300-42. — Moisture or Mechanical Protection for Metal-Sheathed Cables
- Section 300-50. — Underground Installations
- Table 300-50. — Minimum Cover Requirements

Article 305 — Temporary Wiring

- Section 305-4. General
- Section 305-6. Ground-Fault Protection for Personnel

Article 310 — Conductors for General Wiring

- Section 310-8. Locations
- Section 310-11. Marking
- Section 310-12. Conductor Identification
- Section 310-15. Ampacities for Conductors Rated 0-2000 Volts
- Table 310-15(b)(2)(a). Adjustment Factors for More Than Three Current-Carrying Conductors in a Raceway or Cable
- Table 310-15(b)(6). Conductor Types and Sizes for 120/240-Volt, 3-Wire, Single-Phase Dwelling Services and Feeders
- Table 310-20. Ampacities of Two or Three Single-Insulated Conductors, Rated 0 through 2000 Volts, Supported on a Messenger, Based on Ambient Air Temperature of 40 degrees C (104 degrees F)
- Table 310-21. Ampacities for Bare or Covered Conductors, Based on 40 degrees C (104 degrees F) Ambient, 80 degrees C (176 degrees F) Total Conductor Temperature, 2 ft/sec (610 mm/sec) Wind Velocity
- Section 310-60. Conductors Rated 2001 to 35,000 Volts

Article 318 — Cable Trays

- Section 318-6. Installation

Article 320 — Open Wiring on Insulators

- Section 320-5. Conductors
- Section 320-15. Unfinished Attics and Roof Spaces

Article 333 — Armored Cable: Type AC

- Section 333-11. Exposed Work

Article 336 — Nonmetallic-Sheathed Cable: Types NM, NMC, and NMS

Article 362 — Metal Wireways and Nonmetallic Wireways

Part A. Metal Wireways

- Section 362-2. Uses
- Section 362-6. Deflected Insulated Conductors

Part B. Nonmetallic Wireways

- Section 362-19. Number of Conductors

Article 364 — Busways

Part A. General Requirements

- Section 364-10. Rating of Overcurrent Protection — Feeders

Article 365 — Cablebus

- Section 365-5. Overcurrent Protection

Article 370 — Outlet, Device, Pull and Junction Boxes, Conduit Bodies and Fittings

Part B. Installation

- Section 370-23. Supports

Part D. Manholes and Other Electric Enclosures Intended for Personnel Entry

Article 373 — Cabinets, Cutout Boxes and Meter Socket Enclosures

- Section 373-5. Cabinets, Cutout Boxes, and Meter Socket Enclosures

Article 384 — Switchboards and Panelboards

Part C. Panelboards

- Section 384-14. Classification of Panelboards
- Section 384-(16)(b). Power Panelboard Protection

Article 410 — Lighting Fixtures, Lampholders, Lamps and Receptacles

Part G. Construction of Fixtures

- Section 410-42. Portable Lamps

Part L. Receptacles, Cord Connectors, and Attachment Plugs (Caps)

- Section 410-56(g). Attachment Plugs
- Section 410-58(a). Grounding Poles

Part M. Special Provisions for Flush and Recessed Fixtures

- Section 410-66(a). Clearance

Part P. Special Provisions for Electric Discharge Lighting Systems of 1000 Volt or Less

- Section 410-73(f). High-Intensity Discharge Fixtures

Part R. Lighting Track

- Section 410-101(c). Locations Not Permitted

Article 422 — Appliances

Part A. General

- Section 422-4. Live Parts

Part B. Installation

- Section 422-10. Branch-Circuit Rating
- Section 422-11. Overcurrent Protection
- Section 422-12. Central Heating Equipment
- Section 422-13. Storage-Type Water Heaters
- Section 422-14. Infrared Lamp Industrial Heating Appliances
- Section 422-15. Central Vacuum Outlet Assemblies
- Section 422-16. Flexible Cords
- Section 422-17. Protection of Combustible Materials
- Section 422-18. Support of Ceiling-Suspended (Paddle) Fans
- Section 422-20. Other Installation Methods

Part C. Disconnecting Means

- Section 422-30. General
- Section 422-31. Disconnection of Permanently Connected Appliances
- Section 422-32. Disconnection of Cord- and Plug-Connected Appliances
- Section 422-33. Unit Switch(es) as Disconnecting Means
- Section 422-34. Switch and Circuit Breaker to Be Indicating
- Section 422-35. Disconnecting Means for Motor-Driven Appliance

Part D. Construction

- Section 422-40. Polarity in Cord- and Plug-Connected Appliances
- Section 422-41. Cord- and Plug-Connected Appliances Subject to Immersion
- Section 422-42. Signals for Heated Appliances
- Section 422-43. Flexible Cords
- Section 422-44. Cord- and Plug-Connected Immersion Heaters
- Section 422-45. Stands for Cord- and Plug-Connected Appliances
- Section 422-46. Flatirons
- Section 422-47. Water Heater Controls
- Section 422-48. Infrared Lamp Industrial Heating Appliances
- Section 422-49. High-Pressure Spray Washers
- Section 422-50. Cord- and Plug-Connected Pipe Heating Assemblies

Part E. Marking

- Section 422-60. Nameplate
- Section 422-61. Marking of Heating Elements
- Section 422-62. Appliances Consisting of Motors and Other Loads

Article 424 — Fixed Electric Space-Heating Equipment

Part E. Electric Space-Heating Cables

- Section 424-44(g). Ground-Fault Circuit-Interrupter Protection for Conductive Heated Floors of Bathrooms, Hydromassage Bathtub, Spa, and Hot Tub Locations

Article 430 — Motors, Motor Circuits, and Controllers

Part A. General

- Section 430-6. Ampacity and Motor Rating Determination

Part B. Motor Circuit Conductors

- Section 430-22(b). Other than Continuous Duty

Part D. Motor Branch-Circuit Short-Circuit and Ground-Fault Protection

- Section 430-52. Rating or Setting for Individual Motor Circuit

Part G. Motor Controllers

- Section 430-83. Ratings

Part J. Disconnecting Means

- Section 430-109. Type

Article 445 — Generators

- Section 445-9. Generator Terminal Housings
- Section 445-10. Disconnecting Means Required for Generators

Article 490 — Equipment, Over 600 Volts, Nominal (New)

Article 500 — Hazardous (Classified) Locations, Classes I, II, and III, Divisions 1 and 2 (Entirely rewritten)

Article 513 — Aircraft Hangers

- Section 513-1. Scope
- Section 513-2. Definitions

  Mobile Equipment

  Portable Equipment

Article 514 — Gasoline Dispensing and Service Stations

- Section 514-5. Circuit Disconnects

Article 517 — Health Care Facilities

Part C. Essential Electrical Systems

- Section 517-30(b)(6). Hospital power sources to serve essential electrical systems

Part D. Inhalation Anesthetizing Locations

- Section 517-63(a). Battery-Powered Emergency Lighting Units

Part G. Isolated Power Systems

- Section 517-160(a)(5). Identification of isolated circuit conductors

Article 518 — Places of Assembly

- Section 518-2. General Classifications
- Section 518-4. Wiring Methods

Article 520 — Theaters, Audience Areas of Motion Picture and Television Studios, and Similar Locations

Part A. General

- Section 520-5. Wiring Methods

Part D. Portable Switchboards on Stage

- Section 520-53(h). Supply Conductors

Part E. Portable Stage Equipment Other than Switchboards

- Section 520-68. Conductors for Portables

Part F. Dressing Rooms

- Section 520-73. Switches Required

Article 525 — Carnivals, Circuses, Fairs and Similar Events

Part A. General Requirements

- Section 525-3(c). Audio Signal Processing, Amplification and Reproduction Equipment

Part B. Installation

- Section 525-10(b). Services
- Section 525-13(g). Inside Tents and Concessions
- Section 525-18. Ground-Fault Circuit-Interrupter Protection for Personnel

Part D. Disconnecting Means

- Section 530-30. Type and Location

Part E. Attractions Utilizing Pools, Fountains and Similar Installations with Contained Volumes of Water

- Section 525-40. Wiring and Equipment

Article 530 — Motion Picture and Television Studios and Similar Locations

Part B. Stage or Set

- Section 530-12. Portable Wiring

Article 547 — Agricultural Building

- Section 547-4. Wiring methods
- Section 547-8. Service Equipment, Separately Derived Systems, Feeders, Disconnecting Means, and Grounding
- Section 547-9. Bonding and Equipotential Plane

Article 550 — Mobile Homes, Manufactured Homes and Mobile Home Parks

Part B. Mobile Homes

- Section 550-8(g) Pipe Heating Cable Outlet

Article 551 — Recreational Vehicles and Recreational Vehicle Parks

Part G. Recreational Vehicle Parks

- Section 551-72. Distribution System

Article 552 — Park Trailers

Part D. Nominal 120-Volt or 120/240 Volt Systems

- Section 552-41(d). Pipe Heating Cable Outlet
- Section 552-43. Power Supply

Article 555 — Marinas and Boatyards

- Section 555-6. Feeders and Services

Article 600 — Electric Signs and Outline Lighting

Part A. General

- Section 600-6. Disconnects

Part B. Field Installed Skeleton Tubing

- Section 600-32(j). Length of High Voltage Cable

Article 604 — Manufactured Wiring Systems

- Section 604-5. Uses Not Permitted

Article 610 — Cranes and Hoists

Part A. General

- Section 610-2(b). Combustible Materials

Part B. Wiring

- Section 610-13(d) Conductors for Class 1, Class 2, and Class 3 installation

Part E. Overcurrent Protection

- Section 610-41(b). More Than One Feeder Circuit

Article 620 — Elevators, Dumbwaiters, Escalators, Moving Walks, Wheelchair Lifts, and Stairway Chair Lifts

Part C. Wiring

- Section 620-22(a). Car Light Source

Article 625 — Electric Vehicle Charging System Equipment

Part A. General

- Section 625-2. Definitions

Electric Vehicle Coupler

Part B. Wiring and Methods

- Section 625-9. Electric Vehicle Coupler

Part C. Equipment Construction

- Section 625-13. Electric Vehicle Supply Equipment
- Section 625-15. Markings

Part E. Electric Vehicle Supply Equipment Location

- Section 625-29. Indoor Sites

Article 668 — Electrocytic Cells

- Section 668-2. Definitions

  Electrically Connected

Article 680 — Swimming Pools, Fountains and Similar Installations

Part A. General

- Section 680-1. Scope
- Section 680-4. Definitions

  Packaged Therapeutic Tub or Hydrotherapeutic Tank Equipment Assembly

  Self-Contained Therapeutic Tubs or Hydrotherapeutic Tanks

- Section 680-6. Receptacles, Lighting Fixtures, Lighting Outlets, Switching Devices and Ceiling Suspended (Paddle) Fans

Part B. Permanently Installed Pools

- Section 680-22. Bonding
- Section 680-25(c). Motors

Part D. Spas and Hot Tubs

- Section 680-38. Emergency Switch for Spas and Hot Tubs
- Section 680-40. Outdoor Installations

Part E. Fountains

1 Section 680-57. Signs

Part G. Hydromassage Bathtubs

1 Section 680-72. Accessibility

Article 690 — Solar Photovoltaic Systems

Part A. General

- Table 690-1(a). Identification of solar photovoltaic system components
- Table 690-1(b). Identification of solar photovoltaic system components in common configurations
- Section 690-2. Definitions

  Alternating-Current Module (AC Photovoltaic Module)

  Charge Controller

  Electrical Production and Distribution Network

  Hybrid System

  Interactive System

  System Voltage

- Section 690-3. Other Articles

- Section 690-4(c). Module Connection Arrangement
- Section 690-5. Ground-Fault Protection

Part B. Circuit Requirements

- Section 690-7. Maximum Voltage
- Table 690-7. Voltage Correction Factors for Crystalline and Multi-Crystalline Silicon Modules
- Section 690-8. Circuit Sizing and Current

Part H. Storage Batteries

- Section 690-72. Charge Control

Part I. Systems Over 600 Volts

Article 695 — Fire Pumps

- Section 695-6. Power Wiring

Article 700 — Emergency Systems

Part B. Circuit Wiring

- Section 700-9. Wiring, Emergency System

Part C. Sources of Power

- Section 700-12(c). Uninterruptible Power Supplies
- Section 700-12(d). Separate Service

Part F. Overcurrent Protection

Article 725. Class 1, Class 2, and Class 3 Remote-Control, Signaling and Power-Limited Circuits

Part C. Class 2 and Class 3 Circuits

- Section 725-71(g). Class 3 Single Conductors

Article 727 — Instrumentation Tray Cable: Type ITC

Article 760 — Fire Alarm Systems

Part A. General

- Section 760-2. Definitions

Fire Alarm Circuit Integrity (CI) Cable

Part B. Nonpower-Limited Fire Alarm (NPLFA) Circuits

- Section 760-24. NPLFA Circuit Overcurrent Device Location (Exception #3)
- Section 760-31. Listing and Marking of NPLFA Cables

Part C. Power Limited Fire Alarm (PLFA) Circuits

- Section 760-61. Applications of Listed PLFA Cables

Article 770 — Optical Fiber Cables and Raceways

Part A. General

- Section 770-2. Definitions

Article 800 — Communications Circuits

Part A. General

- Section 800-2. Definitions

  Exposed

  Premises

Part E. Communications Wires and Cables Within Buildings

- Section 800-51(j). Plenum Communications Raceway
- Section 800-51(k). Riser Communications Raceway
- Section 800-51(l). General-Purpose Communications Raceway
- Section 800-53(f). Cable Substitutions
- Table 800-53. Cable Uses and Permitted Substitutions

Article 810 — Radio and Television Equipment

- Section 810-1. Scope

Article 820 — Community Antenna Television and Radio Distribution Systems

- Section 820-2. Definitions

  Exposed

  Premises

- Section 820-3. Locations and Other Articles

Article 830 — Networked-Powered Broadband Communications Systems

Appendix D

SUMMARY

The *NEC* specifies the minimum provisions necessary for protecting people and property from hazards arising from the use of electricity and electrical equipment. Anyone involved in any phase of the electrical industry must be aware of how to use and apply the code on the job. Using the NEC will help you to safely install and maintain the electrical security equipment and systems that you come into contact with.

The *NEC* is composed of the following components:

Appendix: Appendix A includes material extracted from other NFPA documents. Appendix B is not part of the requirements of the *NEC* and contains additional material for informational purposes only. Appendix C includes tables for the amount and type of cables allowed in all the standard size conduits and tubings available. Appendix D includes examples illustrating load computations for various types of dwelling and electrical systems. These appendices are located at the back of the code book.

Article: Beginning with Article 90 — Introduction, and ending with Article 830 — Network Powered Broadband Communication Systems, the *NEC* Articles are the main topics in the code book.

Chapter: The *NEC* includes nine chapters. Chapter 1 — General, Chapter 2 — Wiring and Protection, Chapter 3 — Wiring Methods and Materials, Chapter 4 — Equipment for General Use, Chapter 5 — Special Occupancies, Chapter 6 — Special Equipment, Chapter 7 — Special Conditions, Chapter 8 — Communications Systems and Chapter 9 — Tables and Examples. The Chapters form the broad structure of the *NEC*.

Contents: Located among the first pages of the code book, the contents section provides a complete outline of the Chapters, Articles, Parts, Tables, and Examples. The contents section, used with the index, provides excellent direction for locating answers to electrical problems and questions.

Diagrams and Figures: Diagrams and Figures appear in the *NEC* to illustrate the relationship of Articles and Parts of the *NEC*. For example, Diagram 230-1, Services, shows the relationship of Articles and Parts relating to the installation of electric services.

Examples: Service and feeder calculations for various types of buildings.

Exceptions: Exceptions follow code sections and allow alternative methods, to be used under specific conditions, to the rule stated in the section.

FPN (Fine Print Note): A Fine Print Note is defined in *NEC* Section 110-1; that is, explanatory material is in the form of Fine Print Notes (FPN).

Notes: Notes typically follow tables and are used to provide additional information to the tables or clarification of tables.

Part: Certain Articles in the *NEC* are divided into Parts. Article 220 — Branch Circuit, Feeder and Service Calculations is divided into Part A, B, C, and D.

Section: Parts and Articles are divided into Sections. A reference to a section will look like the following:

> **300-19. Supporting Conductors in Vertical Raceways.**

NEC Sections provide more detailed information within *NEC* Articles.

Tables: Tables are located within Chapters to provide more detailed information explaining code content. For example, Table 310-16 lists ampacities for insulated conductors for copper, aluminum, and copper-clad aluminum conductors with insulation types, sizes, temperature ratings, and ampacity correction factors.

See Figure 1-3 for a summary of *NEC* installation requirements for various occupancies.

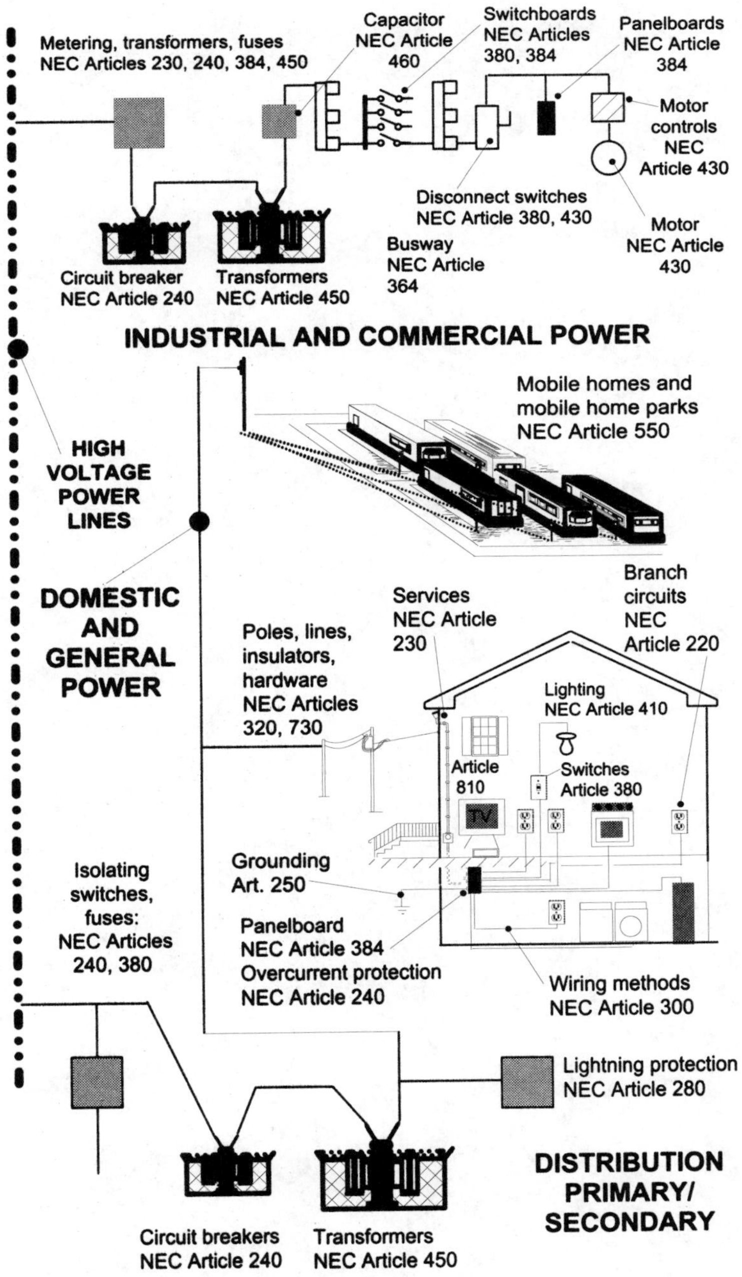
Metering, transformers, fuses
NEC Articles 230, 240, 384, 450
Capacitor
NEC Article
460
Switchboards
NEC Articles
380, 384
Panelboards
NEC Article
384
Motor
controls
NEC
Article 430
Disconnect switches
NEC Article 380, 430
Motor
NEC Article
430
Busway
NEC Article
364
Circuit breaker
NEC Article 240
Transformers
NEC Article 450
INDUSTRIAL AND COMMERCIAL POWER
Mobile homes and
mobile home parks
NEC Article 550
HIGH
VOLTAGE
POWER
LINES
DOMESTIC
AND
GENERAL
POWER
Services
NEC Article
230
Branch
circuits
NEC
Article 220
Poles, lines,
insulators,
hardware
NEC Articles
320, 730
Lighting
NEC Article 410
Article
810
Switches
Article 380
TV
Grounding
Art. 250
Isolating
switches,
fuses:
NEC Articles
240, 380
Panelboard
NEC Article 384
Overcurrent protection
NEC Article 240
Wiring methods
NEC Article 300
Lightning protection
NEC Article 280
DISTRIBUTION
PRIMARY/
SECONDARY
Circuit breakers
NEC Article 240
Transformers
NEC Article 450

Figure 1-3: *Summary of NEC installation requirements for various occupancies*

Chapter 2
Electric Services

This chapter presents NEC installation requirements and related calculations for determining sizes and ratings of the various conductors and equipment to be included in each electric service installation — including the feeders and branch-circuit loads.

All buildings containing equipment that utilizes electricity require an electric service. An electric service will enable the passage of electrical energy from the power company's lines to points of use within the buildings. Figure 2-1 on the next page shows the basic sections of a residential electric service. In this illustration, note that the high-voltage lines terminate on a power pole near the building that is being served. A transformer is mounted on the pole to reduce the voltage to a usable level (120/240 volts in this case). The remaining sections are described as follows:

- *Service drop:* The overhead service conductors from the last pole or other aerial support to and including the splices, if any, connecting to the service-entrance conductors at the building or other structure.
- *Service entrance:* All components between the point of termination of the overhead service drop or underground service lateral and the building main disconnecting device, except for the power company's metering equipment.
- *Service-entrance conductors, overhead system:* The service conductors between the terminals of the service equipment and a point usually outside the building, clear of the building walls, where joined by tap or splice to the service drop.
- *Service-entrance conductors, underground system:* The service conductors between the terminals of the service equipment and the point of connection to the service lateral.
- *Service-entrance equipment:* The necessary equipment, usually consisting of a circuit breaker(s) or switch(es) and fuse(s) and their accessories, connected to the load end of service conductors to a building or other structure, or an otherwise designated area, and intended to constitute the main control and cut-off of the supply.

When the service conductors to the building are routed underground, as shown in Figure 2-2, these conductors are known as the service lateral, defined as follows:

- *Service lateral:* The underground service conductors between the street main, including any risers at a pole and other structure or from transformers, and the first point of connection to the service-entrance conductors in a terminal box or meter or other enclosure, inside or outside the building wall. Where there is no terminal box, meter, or other enclosure, the point of connection shall be considered to be the point of entrance of the service conductors into the building.

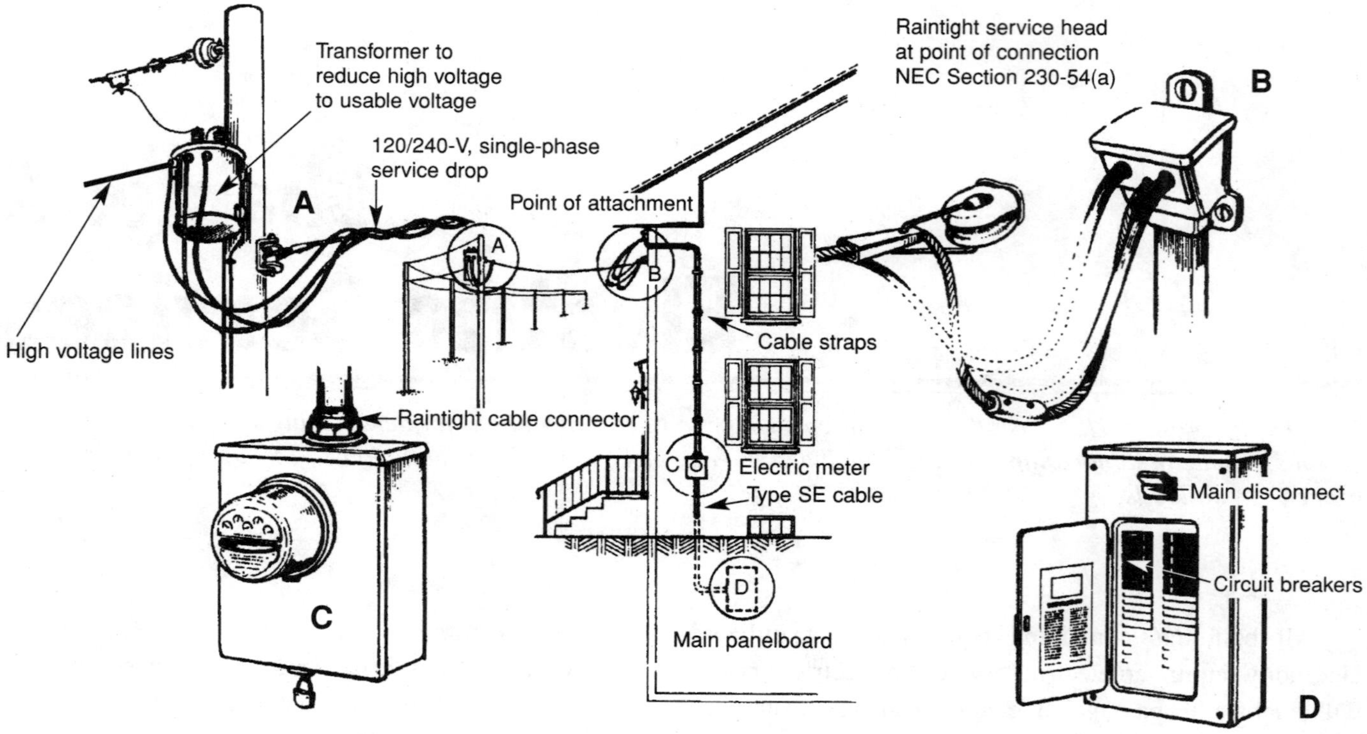
Transformer to
reduce high voltage
to usable voltage
120/240-V, single-phase
service drop
Raintight service head
at point of connection
NEC Section 230-54(a)
B
A
Point of attachment
A
B
High voltage lines
Cable straps
Raintight cable connector
C
Electric meter
Type SE cable
Main disconnect
D
Circuit breakers
C
Main panelboard
D

Figure 2-1: *Components of a typical residential single-phase electric service*

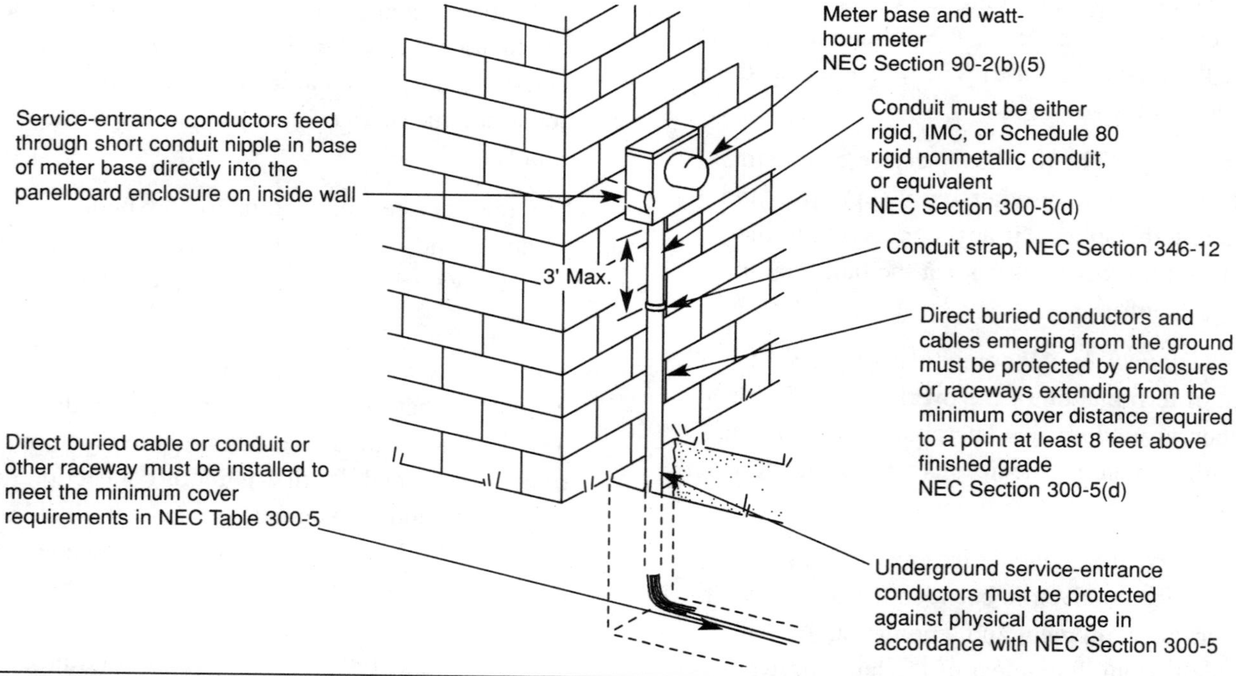
Meter base and watt-
hour meter
NEC Section 90-2(b)(5)
Service-entrance conductors feed
through short conduit nipple in base
of meter base directly into the
panelboard enclosure on inside wall
Conduit must be either
rigid, IMC, or Schedule 80
rigid nonmetallic conduit,
or equivalent
NEC Section 300-5(d)
Conduit strap, NEC Section 346-12
3' Max.
Direct buried conductors and
cables emerging from the ground
must be protected by enclosures
or raceways extending from the
minimum cover distance required
to a point at least 8 feet above
finished grade
NEC Section 300-5(d)
Direct buried cable or conduit or
other raceway must be installed to
meet the minimum cover
requirements in NEC Table 300-5
Underground service-entrance
conductors must be protected
against physical damage in
accordance with NEC Section 300-5

Figure 2-2: *Underground service lateral*

The service-entrance conductors are normally enclosed in a service raceway such as conduit, but sometimes consist of open conductors or Type SE cable. The conductors continue through a metering device and finally terminate at the service disconnecting means, which is usually a switch or circuit breaker. The service or main overcurrent protection is a set of fuses or a circuit breaker that protects the service-entrance conductors. Each metallic part of the service must be bonded together — either by means of a metallic raceway or an equipment bonding jumper. If one conductor of the circuit, such as the neutral conductor, is grounded, a grounding electrode conductor that connects the grounded conductor to two or more grounding electrodes is required. Each feeder circuit may require an equipment grounding conductor to ground the noncurrent-carrying metal parts of equipment.

Each conductor, the disconnecting means, and the overcurrent protective device must satisfy *NEC* installation requirements which specify the size or rating as appropriate for the load to be served. These requirements form the basis for the electrical load calculation of services, feeders, and branch circuits.

SIZING THE SERVICE

Sometimes it is confusing just which comes first: the layout of the outlets, or the sizing of the service. In many cases, the service size (size of main disconnect, panelboard, service conductors, etc.) can be sized using *NEC* procedures before the outlets are actually laid out. In other cases, the outlets and equipment connections will have to laid out first. However, in either case, the service-entrance and panelboard size will have to be calculated and located before the circuits may be designed or installed.

Traditionally, consulting engineers and electrical designers locate all outlets on the working drawings prior to sizing the service entrance. Once the total connected load has been determined, demand factors and continuous-load factors are applied to size the branch-circuit requirements. Sub-panels are then placed at strategic locations throughout the building, and feeders are then sized according to *NEC* requirements. The sum of these feeder circuits (allowing appropriate demand factors) determines the size of the electric service. The main distribution panel, number and size of disconnect switches and overcurrent protection is then determined to finalize the load calculation. Such procedures for determining the load of any project normally surpass *NEC* requirements, but this is quite acceptable since the *NEC* specifies minimum requirements. In fact, in *NEC* Section 90-1(b), the *NEC* itself states:

> *"This Code contains provisions considered necessary for safety. Compliance therewith and proper maintenance will result in an installation essentially free from hazard but not necessarily efficient, convenient, or adequate for good service or future expansion of electrical use."*

Consequently, many electrical installations are designed for conditions that surpass *NEC* requirements to obtain a more efficient and convenient electrical system. However, since most electrical examinations are based on the latest *NEC* installation requirements, procedures contained in this chapter are strictly based on *NEC* regulations. To illustrate why, an electrician's examination may ask the trade size of conduit that is necessary to contain a certain number of conductors. The *NEC* may specify, say, 2-inch conduit as a minimum size for this condition. You can use, say, 2½-inch conduit on the job which will surpass *NEC* requirements. In doing so, no electrical inspector is going to reject the installation because it surpasses *NEC* requirements. However, on most electrician's examinations, those taking the exam must give exact answers in order to get the questions right. Therefore, even though 2½-inch conduit may work on the job, your answer will be wrong if the *NEC* minimum requirement calls for 2-inch conduit. This is the main reason for sticking with *NEC* methods of load calculations throughout this chapter.

Steps For Calculating Service Loads

In general, the type of occupancy (dwelling, store, bank, etc.) is first determined and categorized before the load calculations for the service begin. The basic steps proceed as follows:

Step 1. Determine the area of the building using outside dimensions, less any areas such as garages, porches, or any unfinished spaces not adaptable for future use.

Step 2. Multiply the resulting area by the load per square foot amount listed in *NEC* Table 220-3(b) for the type of occupancy. See Figure 2-3.

Step 3. Determine the volt-amperes (va) of continuous loads (if any) and multiply the va for these loads by a factor of 1.25 (125 percent) as per *NEC* Section 220-3(a).

Step 4. Apply demand factors to any qualifying loads.

Step 5. Calculate the total adjusted general lighting load.

Step 6. Determine the type and va ratings of any other loads, such as appliance circuits, circuits for track lighting, show windows in stores and the like.

Step 7. Add these "other loads" to the total adjusted general lighting load to obtain the total load (in volt-amperes) for the project.

In actual situations, plans and specifications that provide ample space in raceways, spare raceways, and additional spaces in panelboards will allow for future increases in the use of electricity. Distribution centers located in readily accessible locations will provide convenience and safety of operation.

Although the *NEC* does not specifically state the exact amount of additional space to allow for future expansion, electrical inspectors normally require a minimum of 20 percent. Therefore, if the required service for an electrical installation is, say, exactly 100 amperes, electrical inspectors may require that the service size be increased to 125 amperes — the next higher standard overcurrent protective-device rating.

Furthermore, if a residential panelboard requires 20 spaces for circuit breakers to protect branch-circuit loads, a panelboard or load center that has at least 24 spaces (excluding any main circuit breaker or disconnect) should be considered minimum; a 30-space panelboard would be better.

The flowchart in Figure 2-4 gives an overview of the basic steps for determining electrical loads for all types of occupancies. Details concerning the use of this chart should become apparent as the following practical examples are reviewed.

Multiply Area By ⇨	Volt-amperes ¼	½	1	1½	2	3	3½
Armories and auditoriums			●				
Assembly halls			●				
Banks (add 1 volt-ampere/sq ft for outlets)							●
Barber shops						●	
Beauty parlors						●	
Churches			●				
Clubs					●		
Court rooms					●		
Dwelling units Additional load not required for general purpose outlets						●	
Garages — commercial		●					
Hospitals					●		
Hotels and motels					●		

Multiply Area By ⇨	Volt-amperes ¼	½	1	1½	2	3	3½
Industrial commercial (loft buildings)					●		
Lodge rooms				●			
Office buildings (add 1 volt-ampere/sq ft for outlets)							●
Restaurants					●		
Schools						●	
Stores						●	
Warehouses (storage)	●						
The following applies to all occupancies listed except dwellings:							
Halls, corridors		●					
Closets, stairways		●					
Storage spaces	●						

Figure 2-3: *Summary of NEC Table 220-3(b)*

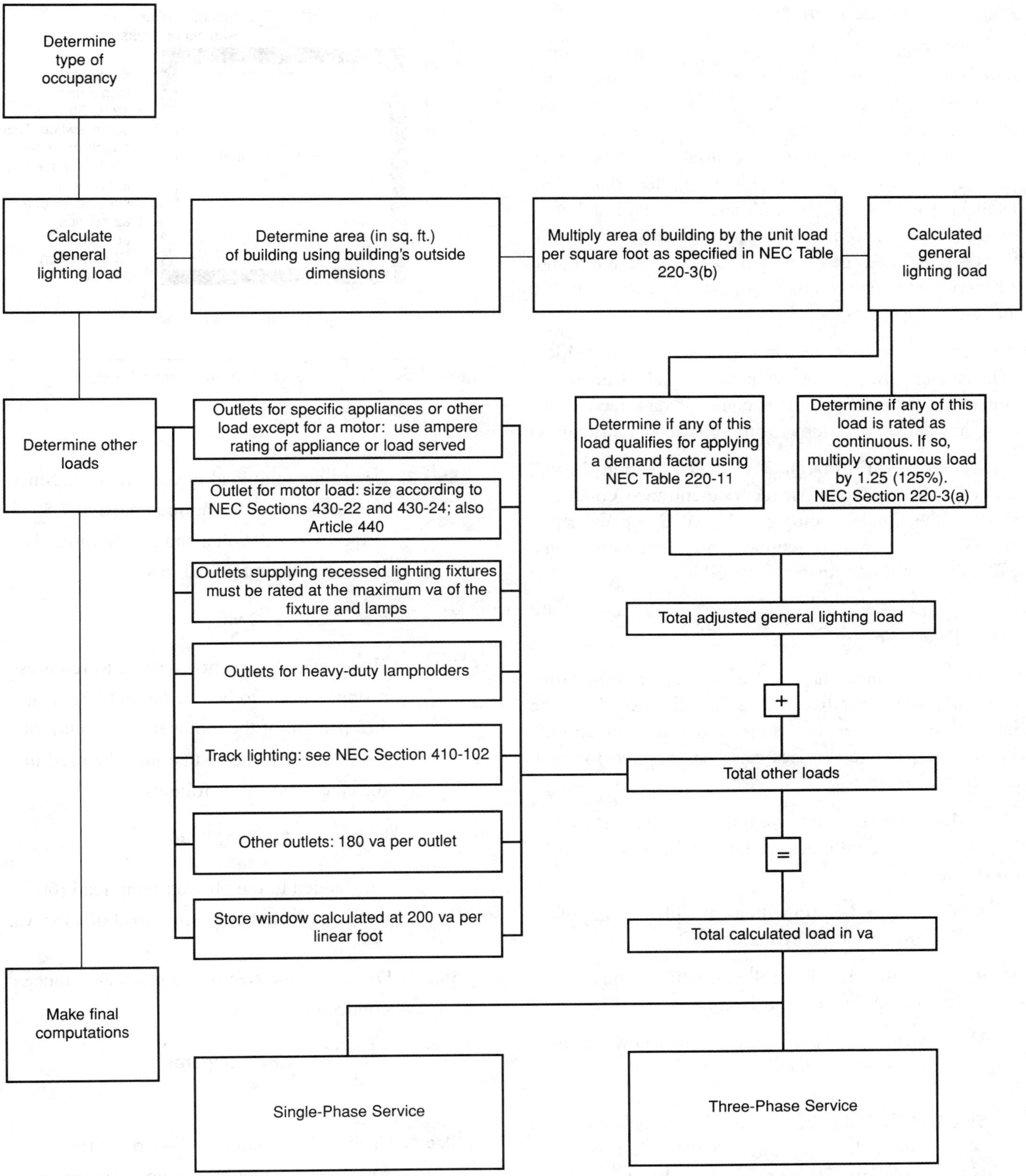
Determine type of occupancy
Calculate general lighting load
Determine area (in sq. ft.) of building using building's outside dimensions
Multiply area of building by the unit load per square foot as specified in NEC Table 220-3(b)
Calculated general lighting load
Determine other loads
Outlets for specific appliances or other load except for a motor: use ampere rating of appliance or load served
Outlet for motor load: size according to NEC Sections 430-22 and 430-24; also Article 440
Outlets supplying recessed lighting fixtures must be rated at the maximum va of the fixture and lamps
Outlets for heavy-duty lampholders
Track lighting: see NEC Section 410-102
Other outlets: 180 va per outlet
Store window calculated at 200 va per linear foot
Determine if any of this load qualifies for applying a demand factor using NEC Table 220-11
Determine if any of this load is rated as continuous. If so, multiply continuous load by 1.25 (125%). NEC Section 220-3(a)
Total adjusted general lighting load
+
Total other loads
=
Total calculated load in va
Make final computations
Single-Phase Service
Three-Phase Service

Figure 2-4: *Basic steps for sizing electric services for most occupancies*

Minimum Service Rating

NEC Section 230-42(a) through (b) specifies minimum sizes and ratings for ungrounded service-entrance conductors. In general, service-entrance conductors must be of sufficient size to carry the computed load. The minimum rating for residential service-entrance conductors is 60 amperes provided that the total calculated load is less than 10 kVA and the total number of branch circuits is 5 or less. If either of these two conditions are not met, the minimum rating of the ungrounded service-entrance conductors must not be less than 100 amperes.

Few, if any, new electric services will be rated less than 100 amperes. However, in some small structures, a smaller service size may be adequate. Consequently, the *NEC* has made provisions for these specialty structures.

Exceptions 1 through 3 of *NEC* Section 230-42 allow further reduction in service-entrance conductor sizes — the smallest being No. 12 AWG copper or No. 10 AWG aluminum or copper-clad aluminum. Some examples of these mini-services follow.

Rural Pump House

Small rural pump houses are required on some farms to supply water for livestock some distance from the farmhouse or other buildings containing electricity. Therefore, separate services must be supplied for such structures.

A floor plan of a small rural pump house is shown in Figure 2-5. The total loads for this facility consist of the following:

- Shallow well pump with a nameplate rating of 7.5 amperes at 120 volts.
- One wall-switch controlled lighting fixture containing one 60-watt lamp.

Using strict *NEC* procedures, here's how the load and service size is calculated.

Step 1. *NEC* Table 220-3(b) does not list a small pump house. Therefore, we must use *NEC* Section 220-3(c) — "Other Loads" — for our calculation.

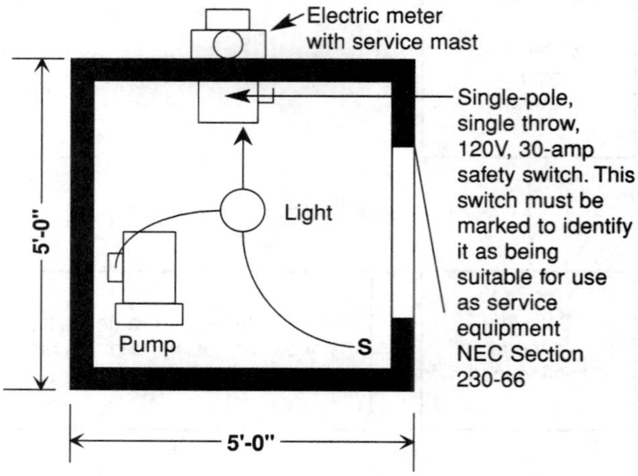
Electric meter
with service mast
Single-pole, single throw, 120V, 30-amp safety switch. This switch must be marked to identify it as being suitable for use as service equipment NEC Section 230-66
Light
Pump
S
5'-0"
5'-0"

Figure 2-5: *Floor plan of a small pump house.*

Step 2. Determine total load in volt-amperes. Since the full-load of the pump is rated at 7.5 amperes at 120 volts, and no HP rating is specified, the total load in va is:

$$7.5 \times 120 = 900 \text{ va}$$

However, *NEC* Section 430-22(a) requires a single motor to be rated at not less than 125 percent of the motor full-load current. Consequently, the va that must be used in the calculation is as follows:

$$900 \text{ va} \times 1.25 = 1125 \text{ va}$$

This added to the 60-watt lamp load (60 va) gives a total connected load of 1185 va.

Step 3. Determine the size of the service-entrance conductors.

$$\frac{1185 \text{ va}}{120\text{V}} = 9.875 \text{ amperes}$$

Step 4. Check *NEC* Section 230-42(b)(3) for service size. In doing so, note Exception No. 3 which states the following:

"Exception No. 3: For limited loads of a single branch circuit, No. 12 copper or No. 10 aluminum or copper-clad aluminum, but in no case smaller than the branch-circuit conductors."

Since the total connected load for the pump house is less than 10 amperes, and the single branch circuit feeding the pump and light fixture need only be No. 14 AWG copper (15-amperes), a No. 12 AWG copper or No. 10 aluminum conductor will qualify for the service-entrance conductors under *NEC* Section 230-42(b)(3), Exception 3.

Since 30 amperes is the smallest standard size safety switch, a 120-volt, single-pole, single-throw safety switch, with solid neutral and rated at 30 amperes, will be selected for the disconnect. Overcurrent protection will be provided by one 15-ampere Type S plug fuse. A time-delay fuse is recommended to compensate for the starting current of the motor.

Some inspection jurisdictions and utility companies now require a fault-current study prior to issuance of a permit. The fault-current study may result in a requirement to utilize a conductor size larger than the minimum *NEC* requirement.

Roadside Vegetable Stand

Another practical application of *NEC* Section 230-42(b)(3) is shown in Figure 2-6. This is a floor plan of a typical roadside vegetable stand. Again, *NEC* Table 230-3(b) does not list this facility; we must again use *NEC* Section 220-3(c),"Other Loads."

Step 1: Determine the lighting load. Two fluorescent fixtures are used to illuminate the 9- x 12-foot area. Each fixture contains two 40-watt fluorescent lamps, and the ballast is rated at 0.083 amperes which translates to approximately 10 va. The total connected load for each fixture is therefore 10 va, or a total of 20 va for both fixtures. However, since this stand will be open late at night during the season, this is considered to be a continuous load, and as per *NEC* Section 210-22(c), a factor of 1.25 (125 percent) must be applied as follows:

20 va x 1.25 = 25 va

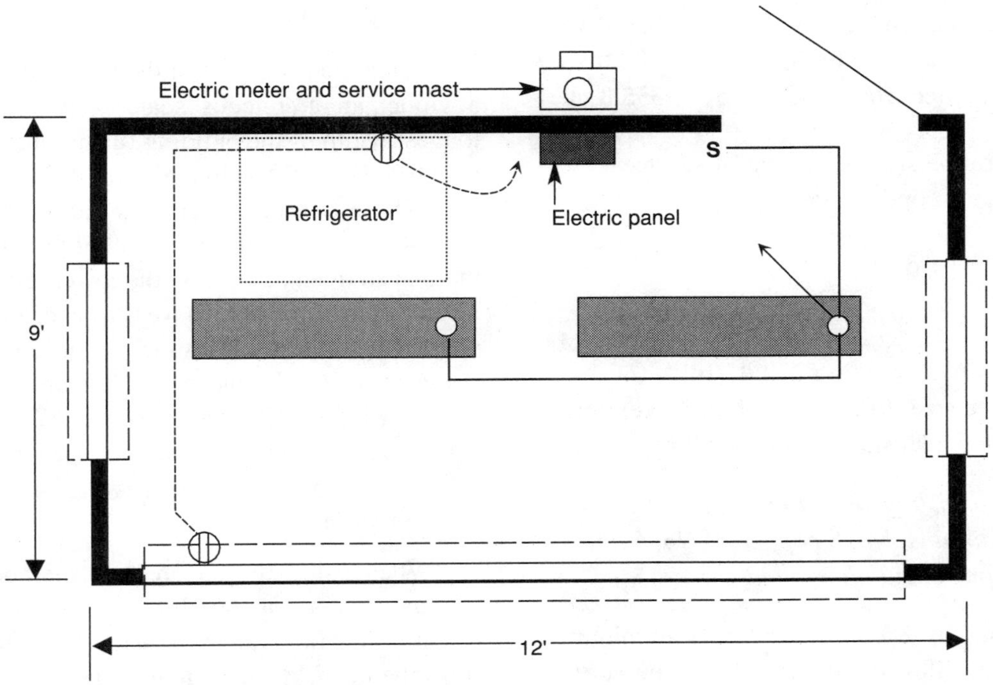
Electric meter and service mast
S
Refrigerator
Electric panel
9'
12'

Figure 2-6: *Floor plan of a typical roadside vegetable stand*

Step 2. Determine remaining loads. The only other electrical outlets in the stand consist of two receptacles: one furnishes power to a refrigerator with a nameplate full-load rating of 12.2 amperes; the other furnishes power for an electric cash register rated at 300 va and an electronic calculator rated at 200 va.

Step 3. Determine if any of the receptacle loads are continuous. Since the refrigerator will more than likely operate for more than three hours during hot summer months, this load will be rated as continuous. The cash register and electronic calculator, however, will operate intermittently, and are not continuous loads.

Refrigerator load = 12.2 x 120
= 1464 va x 1.25
= 1830 va

Step 4. Determine total connected load with appropriate continuous-load factors applied.

Fluorescent fixtures	25.0 va
Receptacle for refrigerator	1830.0 va
Receptacle for other loads	500.0 va
Total calculated load	**2355.0 va**

Step 5. Determine size and rating of service-entrance conductors.

$$\frac{2355 \text{ va}}{240\text{V}} = 9.81 \text{ amperes}$$

Step 6. Check *NEC* Section 230-42(b)(3) for service size. In doing so, note Exception No. 1 which states the following:

"For loads consisting of not more than two 2-wire branch circuits, No. 8 copper or No. 6 aluminum or copper-clad aluminum."

Consequently, No. 8 copper or No. 6 aluminum is the minimum size allowed by the *NEC* for service-entrance conductors on this project.

SINGLE AND MULTIFAMILY DWELLINGS

A floor plan of a small residence is shown in Figure 2-7 on the next page. This building is constructed on a concrete slab with no basement or crawl space. There is an unfinished attic above the living area, and an open carport just outside the kitchen entrance. Appliances include a 12 kVA (12,000 volt-amperes or 12 kilovolt-amperes) electric range and a 4.5 kVA water heater. There is also a washer/dryer (rated at 5.5 kVA) in the utility room. Gas heaters are installed in each room with no electrical requirements. What size service-entrance should be provided for this residence if no other information is specified?

General Lighting Loads

General lighting loads are calculated on the basis of *NEC* Table 220-3(b). For all residential occupancies, 3 volt-amperes (watts) per square foot of living space is the figure to use. This includes nonappliance duplex receptacles into which table lights, television, etc. may be connected. Therefore, the area of the building must be calculated first. If the building is under construction, the dimensions can be determined by scaling the working drawings used by the builder. If the residence is an existing building, with no drawings, actual measurement will have to be made on the site.

Using the floor plan of the residence in Figure 2-7 as a guide, an architect's scale is used to measure the longest width of the building (using outside dimensions) which is 33 feet. The longest length of the building is 48 feet. These two measurements multiplied together give (33 x 48 feet =) 1584 square feet of living area. However, there is an open carport on the lower left of the drawing. This carport area will have to be calculated and then deducted from the 1584 square-foot figure above to give a true amount of living space. This open area (carport) is 12 feet wide by 19.5 feet long. So, 12 x 19.5 feet = 234 square feet. Consequently, the carport area deducted from 1584 square feet leaves (1584 − 234 =) 1350 square feet of living area.

When using the square-foot method to determine lighting loads for buildings, *NEC* Section 220-3(b) requires the floor area for each floor to be computed from the *outside* dimensions. When calculating lighting loads for residences, the computed floor area must not

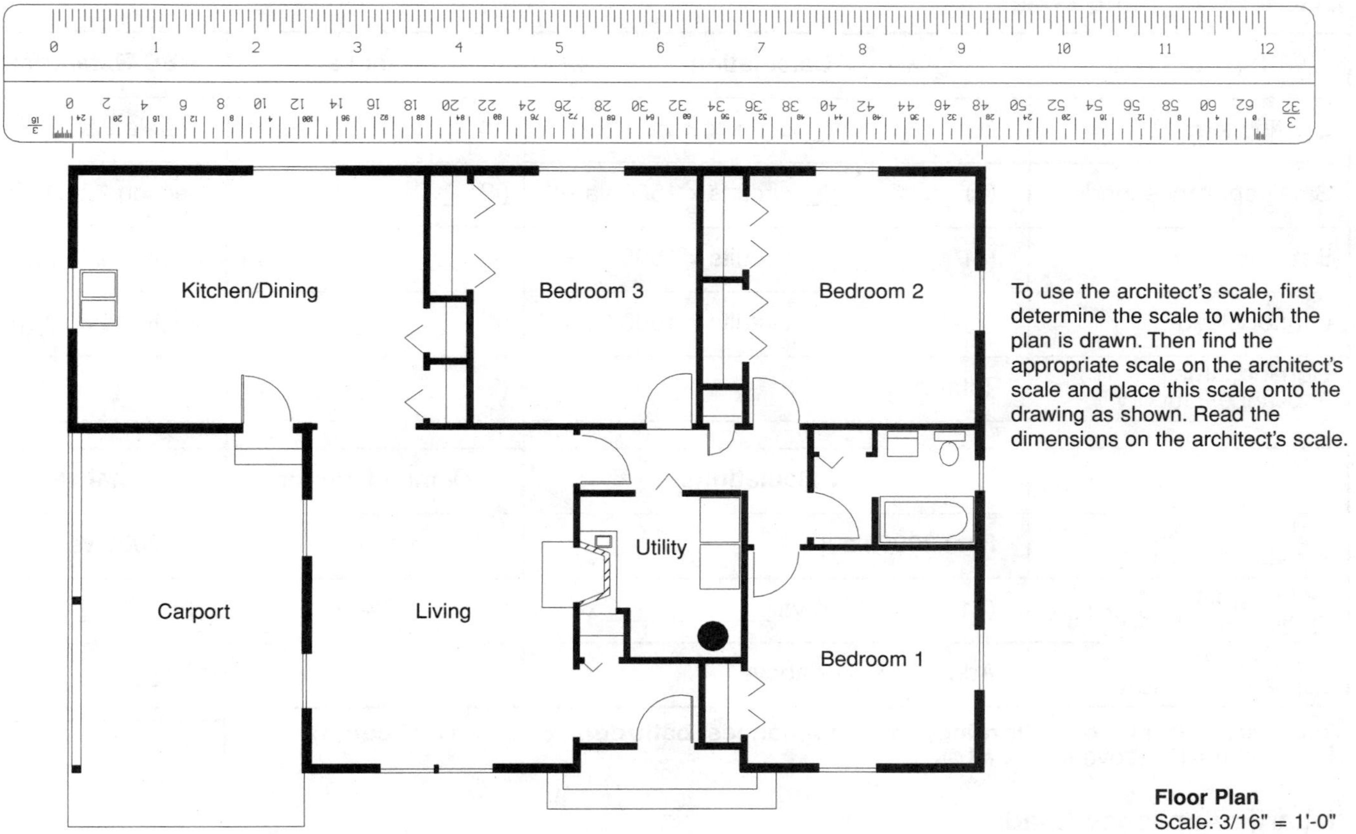
Kitchen/Dining
Bedroom 3
Bedroom 2
Utility
Carport
Living
Bedroom 1
To use the architect's scale, first determine the scale to which the plan is drawn. Then find the appropriate scale on the architect's scale and place this scale onto the drawing as shown. Read the dimensions on the architect's scale.
Floor Plan
Scale: 3/16" = 1'-0"

Figure 2-7: *Floor plan of a small residence*

include open porches, carports, garages, or unused or unfinished spaces not adaptable for future use.

Calculating The Electric Load

Figure 2-8 shows a blank calculation worksheet for a single-family dwelling. Figure 2-9 shows the completed form for the calculation we will be working out. Using this worksheet as a guide, we have determined the total area of our sample dwelling to be 1350 square feet of living space. This figure is entered in the appropriate space (1) on the form and multiplied by 3 va for a total general lighting load of 4050 volt-amperes (2).

Small Appliance Loads

NEC Section 220-4(b) requires at least two, 120-volt, 20-ampere small appliance circuits to be installed for small appliance loads in the kitchen, dining area, breakfast nook, and similar areas where toasters, coffee makers, etc. will be used. *NEC* Section 220-16 gives further requirements for residential small appliance circuits; that is, each must be rated at 1500 volt-amperes. Since two such circuits are used in our sample residence, the number "2" is entered in the appropriate space (3) and then multiplied for a total small appliance load of 3000 volt-amperes (4).

Bathroom Loads

NEC Section 215-52(d) requires that an additional 20 amp circuit be provided for the exclusive use of the dwelling's bathroom duplexes (5). This circuit must not have any other outlets or *loads* (such as the bathroom lighting). Therefore, enter 1500 (volt-amperes) in space (6) on the form.

I. General Lighting Loads

Type of Load	Calculation	Total va	NEC Reference
Lighting load	(1) ________ sq. ft. x 3 va =	(2) ____________ va	Table 220-3(b)
Small appliance loads	(3) ________ circuits x 1500 va =	(4) ____________ va	Section 220-16(a)
Bathroom load	(5) ________ circuits x 1500 va =	(6) ____________ va	Section 215-52(d)
Laundry load	(7) ________ circuits x 1500 va =	(8) ____________ va	Section 220-16(b)
Lighting, small appliance, laundry	Total va =	(9) ____________ va	
Load	**Calculations**	**Demand Factor**	**Total va**
Lighting, small appliance, laundry	First 3000 va x	100% =	(10) 3000 va
Lighting, small appliance, laundry	(11) Remaining va ________ x	35% =	(12) ________va
Lighting, small appliance, laundry	Add #10 & #12 above	=	(13) ________va

Total calculated load for lighting, small appliances, bathroom, & laundry circuit(s)
(Enter item #13 above in box #14)

14

II. Large Appliance Loads

Type of Load	Nameplate Rating	Demand Factor	Total va	NEC Reference
Electric range	Not over 12 kva	Use 8 kva	(15) 8,000 va	Table 220-19
Clothes dryer	(16) ________va	100%	(17) ________va	Table 220-19
Water heater	(18) ________va	100%	(19) ________va	
Other appliances	(20) ________va	100%	(21) ________va	

Total calculated load for large appliances (Enter items #17, #19 and #21 above).
Enter total in box #22

22

Total calculated load (Add boxes #14 & #22) Enter total in box #23

23

III. Convert va to Amperes

$$\frac{\textit{Total va in Box \#23}}{\textit{240 volts}} = \textit{amperes} \qquad \frac{\rule{3cm}{0.4pt}\ \text{va}}{\text{240 volts}} = (\#24)\ \rule{2cm}{0.4pt}\ \text{amperes}$$

IV. Ungrounded and grounded (neutral) conductor size — NEC Article 310

Notes to Ampacity Tables: No. 3 *(see Figure 2-9 on next page)*

Figure 2-8: *Calculation worksheet for residential service requirements*

I. General Lighting Loads

Type of Load	Calculation	Total va	NEC Reference
Lighting load	(1) 1350 sq. ft. x 3 va =	(2) 4050 va	Table 220-3(b)
Small appliance loads	(3) 2 circuits x 1500 va =	(4) 3000 va	Section 220-16(a)
Bathroom load	(5) 1 circuits x 1500 va =	(6) 1500 va	Section 215-52(d)
Laundry load	(7) 1 circuits x 1500 va =	(8) 1500 va	Section 220-16(b)
Lighting, small appliance, laundry	Total va =	(9) 10050 va	
Load	**Calculation**	**Demand Factor**	**Total va**
Lighting, small appliance, laundry	First 3000 va x	100% =	(10) 3000 va
Lighting, small appliance, laundry	(11) Remaining va 7050 x	35% =	(12) 2467.5 va
Lighting, small appliance, laundry	Add #10 & #12 above	=	(13) 5467.5 va

Total calculated load for lighting, small appliances, bathroom, & laundry circuit(s)
(Enter item #13 above in box #14)

14	5,467.5

II. Large Appliance Loads

Type of Load	Nameplate Rating	Demand Factor	Total va	NEC Reference
Electric range	Not over 12 kva	Use 8 kva	(15) 8,000 va	Table 220-19
Clothes dryer	(16) 5500 va	100%	(17) 5500 va	Table 220-19
Water heater	(18) 4500 va	100%	(19) 4500 va	
Other appliances	(20) ______ va	100%	(21) ______ va	

Total calculated load for large appliances (Enter items #17, #19 and #21 above).
Enter total in box #22

22	18,000 va

Total calculated load (Add boxes #14 & #22) Enter total in box #23

23	23,467.5 va

III. Convert va to Amperes

$$\frac{\text{Total va in Box \#23}}{240 \text{ volts}} = \text{amperes} \qquad \frac{23{,}467.5 \text{ va}}{240 \text{ volts}} = (\#24)\ \underline{97.8}\ \text{amperes}$$

IV. Ungrounded and grounded (neutral) conductor size — NEC Article 310

Notes to Ampacity Tables: No. 3

Figure 2-9: *Completed calculation worksheet for residential service requirements*

Minimum Ungrounded Conductor Size			Minimum Grounded (Neutral) Conductor Size		
Rating in Amperes	Copper AWG Size	Aluminum or Copper-Clad AL	Rating in Amperes	Copper AWG Size	Aluminum or Copper-Clad AL
100	4	2	100	8	6
110	3	1	110	8	6
125	2	1/0	125	8	6
150	1	2/0	150	6	4
175	1/0	3/0	175	6	4
200	2/0	4/0	200	4	2

Figure 2-10: *Minimum size neutral conductors allowed provided the requirements of NEC Sections 215-2, 220-22, and 230-42 are met*

Laundry Circuit

NEC Section 220-4(c) requires that an additional 20-ampere branch circuit be provided for the exclusive use of the laundry area (7). This circuit must not have any other outlets connected except for the laundry receptacle(s) as required by *NEC* Section 210-52(f). Therefore, enter 1500 (volt-amperes) in space (8) in the form.

Thus far we have enough information to complete the first portion of the service-calculation form. See Figure 2-10.

General Lighting	4050.0 va (2)
Small Appliance Load	3000.0 va (4)
Bathroom Load	1500.0 va (6)
Laundry Load	1500.0 va (8)
Total General Lighting & Appliance Loads	***10,050.0 va (9)***

Demand Factors

All residential electrical outlets are never used at one time. There may be a rare instance where all the lighting could be on for a short time every night, but if so, all the small appliances, all burners on the electric range, water heater, furnace, dryer, washer, and the numerous receptacles throughout the house will never be used simultaneously. Knowing this, the *NEC* allows a diversity or demand factor in sizing electric services. Our calculation continues:

First 3000 va are rated at 100% = 3000 va (10)

The remaining 7050 watts (va) (11) may be rated at 35% (the allowable demand factor). Therefore:

7050 x .35 = 2467.5 va (12)

Therefore, the net General Lighting & Small Appliance Load equals 5467.5 va (13). Also enter this number in Box 14 on the form.

The electric range, water heater, and clothes dryer must now be considered in the service calculation. Although we previously learned that the nameplate rating of the electric range is 12 kVA, seldom will every burner be on high at once. Nor will the oven remain on all the time during cooking. When the oven reaches the temperature set on the oven controls, the thermostat shuts off the power until it cools down. Again, the *NEC* allows a diversity or demand factor.

When one electric range is installed and the nameplate rating is not over 12 kVA, *NEC* Table 220-19 allows a demand factor resulting in a total rating of 8 kVA. Therefore, 8 kVA may be used in the service calculations instead of the nameplate rating of 12 kVA. The electric clothes dryer and water heater, however, must be calculated at 100% when using this method to calculate residential electric services. The total large appliance load is entered in Box 22 on the form. Here is the service calculation thus far:

Net computed small appliance load	5467.5 va (13)(14)
Electric range (using demand factor)	8000.0 va (15)

Clothes dryer	5500.0 va (16)(17)
Water heater	4500.0 va (18)(19)
Total load	**23,467.5 va (23)**

Required Service Size

The conventional electric service for residential use is 120/240-volt, 3-wire, single-phase. Services are sized in amperes and when the volt-amperes are known on single-phase services, amperes may be found by dividing the (highest) voltage into the total volt-amperes.

$$\frac{23467.5 \text{ va}}{240\text{V}} = 97.8 \text{ amperes (24)}$$

SIZING NEUTRAL CONDUCTORS

The neutral conductor in a 3-wire, single-phase service carries only the unbalanced load between the two "hot" legs. Since there are several 240-volt loads in the above calculations, these 240-volt loads will be balanced and therefore reduce the load on the service neutral conductor. Consequently, in most cases, the service neutral does not have to be as large as the ungrounded ("hot") conductors.

In the above example, the water heater does not have to be included in the neutral-conductor calculation, since it is strictly 240 volts with no 120-volt loads. The clothes dryer and electric range, however, have 120-volt lights that will unbalance the current between phases. The *NEC* allows a demand factor of 70 percent for these two appliances. Using this information, the neutral conductor may be sized accordingly:

General Lighting and Appliance Load	5467.5 va
Electric Range (8000 va x .70)	5600.0 va
Clothes Dryer (5500 va x .70)	3850.0 va
Total	**14,917.5 va**

To find the total phase-to-phase amperes, divide the total volt-amperes by the voltage between phases.

$$14{,}917.5 \div 240 \text{ V} = 62.2 \text{ amperes}$$

The service-entrance conductors have now been calculated and must be rated at 125 amperes with a neutral conductor rated for at least 70 amperes.

In *NEC* "Notes to Ampacity Tables (No. 3) . . ." that follow *NEC* Table 310-19, special consideration is given 120/240 volts, single-phase residential services. Conductor sizes are shown in the table that follows these *NEC* notes. Reference to this table shows that the *NEC* allows a No. 1 AWG copper or No. 2/0 AWG aluminum or copper-clad aluminum for 125 amperes. Furthermore, the *NEC* states that the grounded or neutral conductor may be smaller than the ungrounded conductors provided *NEC* Sections 215-2, 220-22, and 230-42 are met, along with other applicable Sections.

Let's take another single-family dwelling and calculate the required service size. This dwelling consists of 1625 square feet of living area. The following loads are to be accounted for:

- 1 — 12 kW range
- 1 — 5,500 va dryer
- 1 — 1,250 va dishwasher
- 1 — 750 va disposal
- 1 — 1,440 va attic fan

The home is heated with a natural gas furnace using a 1/3 HP, 828 va blower motor.

General Lighting = 1625 sq. ft. x 3 va = 4,875 va
Small Appliance Load = 1500 va x 2 = 3,000 va
Bathroom Load = 1,500 va

Load Calculation

General lighting	4,875 va
Small appliance load	3,000 va
Bathroom load	1,500 va
Laundry load	1,500 va
Total general lighting and small appliance load	**10,875 va**

Apply Demand Factors

3,000 va @100%	3,000 va

10,875 va – 3,000 va = 7,875 va @35% = 2,756 va

Net general lighting and small appliance load	5,756 va

Other Loads

Range load (*NEC* Table 220-19)	12,000 kVA
Dryer load (*NEC* Table 220-18)	5,500 va
Dishwasher	1,500 va
Hot water heater	6,000 va
Disposal	750 va
Attic fan	1,440 va
Blower motor	828 va

Total Fixed Appliance Loads (*NEC* 220-17)

28,018 va x 75% = 21,014 va

25% of largest motor

1440 va x 25% = 360 va

Total load	27,130 va

The size of this 120/240 volt single-phase service may be determined as follows:

27,130 va/240 volts = 113 amperes

Per *NEC* 230-42(b), the load is larger than 10 kVA and the service size is required to be at least 100 amps. However, the calculation comes to 113 amperes and the service conductors and service equipment will have to be sized accordingly. The calculated load of 113 amperes is not a standard size overcurrent protective device. According to *NEC* 240-6, the next standard size overcurrent protective device for this service is 125 amperes, so this is the size that will be used.

SIZING THE LOAD CENTER

Each circuit in the building must be provided with overload protection, either in the form of fuses or circuit breakers. If more than six such devices are needed, you must also provide a means of disconnecting the entire service — either with a main disconnect switch or a main circuit breaker. To calculate the number of fuse holders or circuit breakers required in the residence above, let's look at the general lighting load first. Since we have a total general lighting load of 4875 va, this can be divided by 120 volts (Ohm's Law states Amperes = W/V) which give us 40.625 amperes. You now have a choice of using either 15-ampere circuits or 20-ampere circuits for the lighting load. Two 20-ampere circuits (2 x 20) gives us a total amperage of 40, so three 20-ampere circuits would be adequate for the lighting. However, three 15-ampere circuits which total 45 amperes would also be satisfactory.

Many electricians performing residential wiring prefer to use 15-ampere circuits (No. 14 Type NM Cable with ground wire) for their lighting because the smaller size wire makes it easier to make up the connections at the lighting fixtures and switches. However, to cut down on voltage drop throughout the house, there are other electricians who prefer the higher-quality 20-ampere circuits (No. 12 Type NM Cable with ground wire) for all branch circuits throughout the building.

You will need a minimum of two 20-ampere appliance circuits for the small appliance load, one 20-ampere circuit for the bathrooms, one 20-ampere circuit for the laundry, and another 20-ampere circuit for the dishwasher. Thus far, we can count the following 120-volt branch circuits which will require 20-ampere, one-pole circuit breakers for each:

General Lighting Load	3
Small Appliance Loads	2
Bathroom Load	1
Laundry Load	1
Dishwasher	1
Total 20A, 1-pole breakers	**8**

Now let's take the 15-ampere, one-pole circuit breakers. A total of three will be needed for the disposal, attic fan and gas-fired furnace blower motor.

Most load centers and panelboards come with an even number of circuit breaker or fuse-holder connectors; that is, 4, 6, 8, 10, etc. The load center for this residence should therefore have either 8 single-pole circuit breakers or 8 single-pole fuse holders for the above loads.

Don't forget about the 240-volt appliances. The electric range will require a 2-pole circuit breaker, as will the clothes dryer. These two appliances will therefore require an additional (2 poles x 2 appliances =) 4 spaces in the load center. We could get by with a 14-space circuit-breaker load center with a 125-ampere main circuit breaker. However, it is always better to provide a few extra spaces for future use. Many designers prefer to add 20% extra space in the load center for additional circuits that may be added later. The circuit breakers themselves do not have to be installed, but space should be provided for them. This would put the load center for the house in question up to 16 or 18 spaces. You would need to order a 125-ampere load center with a 125-ampere main circuit breaker. Furthermore, you will need no less than three 15-ampere, single-pole circuit breakers, eight 20-ampere, single-pole circuit breakers, one double-pole circuit breaker for the electric range and one double-pole circuit breaker for the clothes dryer. The most common sizes of panelboards and load centers are as follows:

Single-Phase, Three-Wire Load Centers

Ampere Rating	Maximum Capacity
100 amperes	6 poles
100 amperes	8 poles
100 amperes	10 poles
100 amperes	12 poles
100 amperes	14 poles
100 amperes	16 poles
100 amperes	18 poles
100 amperes	20 poles
125 amperes	6 poles
125 amperes	8 poles
125 amperes	10 poles
125 amperes	12 poles
125 amperes	14 poles
125 amperes	16 poles
125 amperes	18 poles
125 amperes	20 poles
125 amperes	24 poles
150 amperes	20 poles
150 amperes	24 poles
150 amperes	28 poles
150 amperes	30 poles
200 amperes	20 poles
200 amperes	24 poles
200 amperes	28 poles
200 amperes	30 poles
200 amperes	36 poles
200 amperes	40 poles

Three-Phase, Three-Wire Load Centers

Ampere Rating	Maximum Capacity
100 amperes	10 poles
100 amperes	12 poles
125 amperes	12 poles
125 amperes	18 poles
150 amperes	24 poles
150 amperes	30 poles
200 amperes	30 poles
200 amperes	42 poles

Three-Phase, Four-Wire Panelboards

Ampere Rating	Maximum Capacity
50 amperes	8 poles
50 amperes	10 poles
50 amperes	12 poles
100 amperes	16 poles
100 amperes	18 poles
100 amperes	20 poles
100 amperes	22 poles
100 amperes	24 poles
100 amperes	26 poles
100 amperes	28 poles
100 amperes	30 poles
125 amperes	20 poles
125 amperes	22 poles
125 amperes	24 poles
225 amperes	22 poles
225 amperes	24 poles

Three-Phase, Four-Wire Panelboards (cont.)

Ampere Rating	Maximum Capacity
225 amperes	26 poles
225 amperes	28 poles
225 amperes	30 poles
225 amperes	32 poles
225 amperes	34 poles
225 amperes	36 poles
225 amperes	38 poles
225 amperes	40 poles
225 amperes	42 poles
400 amperes	30 poles
400 amperes	42 poles

OPTIONAL CALCULATIONS FOR SINGLE-FAMILY DWELLINGS

Now let's use the preceding residential example to illustrate the *NEC* optional calculation for single-family dwelling per *NEC* Section 220-30, *NEC* Table 220-30 and the example in *NEC* Chapter 9.

This single-family dwelling consists of 1625 square feet of living area. The following loads are to be accounted for:

- 1 – 12 kW range
- 1 – 5,500 va dryer
- 1 – 1,250 va dishwasher
- 1 – 750 va disposer
- 1 – 1,440 va attic fan

The home is heated with a natural gas furnace using a 1/3 hp, 828 va blower motor.

General Lighting = 1625 sq. ft. x 3 va =	4,875 va
Small Appliance Load = 1500 va x 2 =	3,000 va
Bathroom Load	1,500 va
Laundry Load	1,500 va
Range (Nameplate Rating)	12,000 va
Dryer Load	5,500 va
Dishwasher	1,250 va
Disposer	750 va
Attic Fan	1,440 va
Blower Motor	828 va
Total Load:	**32,643 va**

Apply Demand Factors (Table 220-30):

First 10 kVA @ 100% =	10,000 va
Remainder @ 40% (22,643 va x 40%) =	9,057 va
Total Load:	**19,057 va**

The service size for a 120/240 volt single-phase service is calculated as follows:

$$\frac{19{,}057\ \text{va}}{240\text{V}} = 79\ \text{amperes}$$

Per *NEC* 230-42(b), the load is larger than 10 kVA and the service size is required to be at least 100 amps. Per *NEC* 240-6, the next standard size overcurrent-protective device for this service would be 100 amps.

MULTIFAMILY DWELLING CALCULATIONS

The service load for multifamily dwellings is not simply the sum of the individual dwelling unit loads because of demand factors that may be applied when either the standard calculation or the optional calculation is used to compute the service load.

When the standard calculation is used to compute the service load, the total lighting, small appliance, and laundry loads as well as the total load from all electric ranges, clothes dryers, etc. are subject to the application of demand factors. In addition, further demand factors may be applied to the portion of the neutral load contributed by electric ranges and the portion of the total neutral load greater than 200 amperes.

When the optional calculation is used, the total connected load is subject to the application of a demand factor that varies according to the number of individual units in the structure.

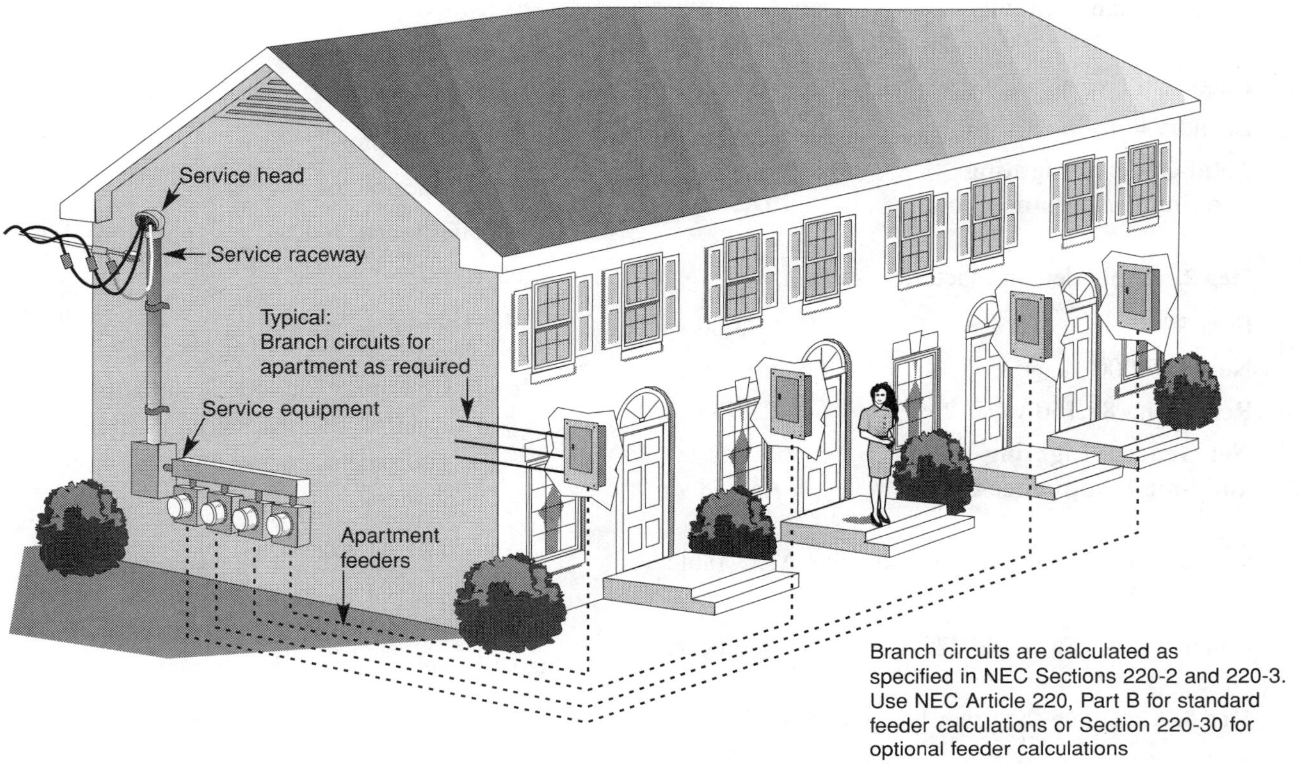
Service head
Service raceway
Typical:
Branch circuits for
apartment as required
Service equipment
Apartment
feeders
Branch circuits are calculated as
specified in NEC Sections 220-2 and 220-3.
Use NEC Article 220, Part B for standard
feeder calculations or Section 220-30 for
optional feeder calculations

Figure 2-11: *Summary of NEC installation requirements for multifamily dwellings*

A summary of the calculation methods for designing wiring systems in multifamily dwellings and the applicable *NEC* references are shown in Figure 2-11. The selection of a calculation method for computing the service load is not affected by the method used to design the feeders to the individual dwelling units.

The rules for computing the service load are also used for computing a main feeder load when the wiring system consists of a service that supplies main feeders which, in turn, supply a number of subfeeders to individual dwelling units.

Multifamily Dwelling Sample Calculation

A 30 unit apartment building consists of 18 one-bedroom units of 650 sq. ft. size, 6 two-bedroom units of 775 sq. ft. size, and 6 three-bedroom units of 950 sq. ft. size. The kitchen in each unit contains the following loads:

- 7.5 kW electric range
- 1250 va dishwasher
- 750 va garbage disposer

Each unit has an air-conditioning unit rated at 3600 va. The entire building is heated by a boiler that utilizes two electric pump motors: a 2 hp squirrel-cage motor and a 2½ hp squirrel-cage induction motor — both 240-volt, single-phase, two-wire. There is a community laundry room in the building with four washers and four clothes dryers; the latter rated at 4,500 va each. The building is to be furnished with a 120/240-volt, single-phase, 3-wire service. What size service is required?

Step 1. Calculate the total general lighting and small appliance load.

General Lighting Load:

650 sq. ft. x 3 = 1950 va x 18 units =	35,100 va
775 sq. ft. x 3 = 2325 va x 6 units =	13,950 va
950 sq. ft. x 3 = 2850 va x 6 units =	17,100 va

Small appliance circuits = 1500 va x 2 x 30 =	90,000 va
Bathroom circuit = 1500 va x 30 =	45,000 va
Laundry = 1500 va x 4 =	6,000 va
Total General Lighting and Small-Appliance Load:	**207,150 va**

Step 2. Apply demand factors.

First 3000 va @ 100% =	3,000 va
Next 117,000 va @ 35% =	40,950 va
Remaining 87,150 va @ 25% =	21,788 va
Net General Lighting and Small-Appliance Load:	**65,738 va**

Step 3. Calculate the range load using *NEC* Table 220-19.

Range load (*NEC* Table 220-19) =	45,000 va

Step 4. Calculate the dryer load using *NEC* Table 220-18.

Step 5. Calculate the fixed appliances using *NEC* Section 220-17 as a guide.

Dishwashers: 30 x 1250 x .75 =	28,125 va
Disposers: 30 x 750 x 75% (.75) =	16,875 va
2 hp boiler motor =	2,760 va
2½ hp boiler motor =	2,254 va

Step 6. Calculate the air-conditioning load as per *NEC* Article 220-21.

3600 va x 30 = 108,000 va

Step 7. Determine the largest motor and multiply the full-load amps (FLA) or va by 25% (.25).

2760 x .25 = 690 va

Step 8. Determine the total calculated load by adding all summaries in Steps 2 through 7.

Net General Lighting and Small-Appliance Loads	65,738 va
Range load	45,000 va
Dryer load	22,000 va
Dishwashers	28,125 va
Garbage disposers	16,875 va
2 hp boiler motor	2,760 va
2½ hp boiler motor	2,254 va
Air-conditioning loads	108,000 va
25% of largest motors	690 va
Total Load	**291,442 va**

Step 9. Determine the total load in amperes in order to select service conductors, equipment, and overcurrent protective device.

$$\frac{291{,}442 \text{ va}}{240\text{V}} = 1{,}214 \text{ amperes}$$

Step 10. Select the service-entrance conductors, service equipment, and overcurrent protection.

NOTE: Since the exact calculated load is not a standard size, the rating of the service for this apartment building would be 1600 amperes; this is the next largest standard size; this is permitted by NEC regulations.

COMMERCIAL OCCUPANCY CALCULATIONS

Calculating load requirements for commercial occupancies is based on specific *NEC* requirements that relate to the loads present. The basic approach is to separate the loads into the following:

- Lighting
- Receptacles
- Motors
- Appliances
- Other special loads

In general, all loads for commercial occupancies should be considered continuous unless specific information is available to the contrary.

Smaller commercial establishments will utilize single-phase, three-wire services, while the larger projects will almost always use a three-phase, four-wire service. Furthermore, it is not uncommon to have secondary feeders supplying panelboards which, in turn, supply branch circuits operating at different voltages. In this case, the calculation of the feeder and branch circuits for each voltage is considered separately. The rating of the main service is based on the total load with the load values transformed according to the various circuit voltages if necessary.

Demand factors are also applicable to some commercial establishments. For example, the lighting load in hospitals, hotels and motels, and warehouses, is subject to the application of demand factors. In restaurants and similar establishments, the load of electric cooking equipment is subject to a demand factor if there are more than three cooking units. Optional calculation methods to determine feeder or service loads for schools and similar occupancies are also provided in the *NEC*.

Special occupancies, such as mobile homes and recreational vehicles, require the feeder or service load to be calculated in accordance with specific *NEC* requirements. The service for mobile home parks and recreational vehicle parks is also designed based on specific *NEC* requirements that apply only to those locations. The feeder or service load for receptacles supplying shore power for boats in marinas and boatyards is also specified in the *NEC*.

When transformers are not involved, a relatively simple calculation involving only one voltage results. If step-down transformers are used, the transformer itself must be protected by an overcurrent device which may also protect the circuit conductors in most cases.

Switches and panelboards used for the distribution of electricity within a commercial building are also subject to *NEC* rules. In general, a lighting and appliance panelboard cannot have more than 42 overcurrent protective devices to protect the branch circuits originating at the panelboard. This requirement could affect the number of feeders required when a large number of lighting or appliance circuits are needed.

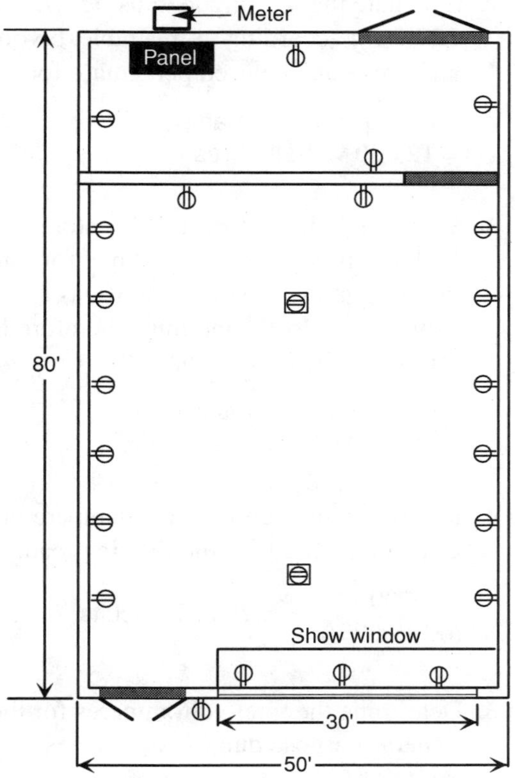
Meter
Panel
80'
Show window
30'
50'

Figure 2-12: *Floor plan of a retail store with a show window*

RETAIL STORES WITH SHOW WINDOWS

The drawing in Figure 2-12 shows a small store building with a show window in front. Note that the storage area has four general-purpose duplex receptacles, while the retail area has 14 wall-mounted duplex receptacles and two floor-mounted receptacles for a total of 16 in this area. These combined with the storage-area receptacles bring the total to 20 general-purpose duplex receptacles that do not supply a continuous load. What are the conductor sizes for the service-entrance if a 120/240-volt single-phase service will be used? How many branch circuits are required if 20-ampere circuits are used throughout?

Step 1. Determine the total square feet in the building by multiplying length x width.

50 x 80 = 4000 sq. ft.

Step 2. Calculate the lighting load using *NEC* Table 220-3(b); according to this table, 3 volt-amperes must be used per square foot.

4000 sq. ft. x 3 volt-amperes
= 12,000 volt-amperes

Step 3. Since the lighting load for this store building is expected to continue for three hours or more, this will qualify as a "continuous load" and must therefore be multiplied by a factor of 1.25.

12,000 volt-amperes x 1.25
= 15,000 volt-amperes

Step 4. Determine the number of 20-ampere branch circuits required for the lighting load.

$$\frac{15000 \text{ va}}{120\text{V} \times 20\text{A}} = 6.25 \text{ or } 7 \text{ circuits}$$

Step 5. Determine the total volt-amperes for the 20 general-purpose duplex receptacles.

20 x 180 va = 3600 volt-amperes

Step 6. Determine the number of 20-ampere branch circuits required for the general-purpose receptacles.

$$\frac{3600 \text{ va}}{120\text{V} \times 20\text{A}} = 1.5 \text{ or } 2 \text{ circuits}$$

Therefore, the branch circuits for lighting and the 20 general-purpose receptacles require a total of nine 20-amperes circuits. These will be installed with conductors with a current-carrying capacity of 20 amperes and protected with an overcurrent device — either fuse or circuit breaker — rated at 20 amperes.

Step 7. Calculate the load for the 30-foot show windows on the basis of 200 volt-amperes per linear foot which is considered a continuous load value.

30 x 200 = 6000 volt-amperes

Step 8. Determine the number of 20-ampere branch circuits required to feed the show-window outlets.

$$\frac{6000 \text{ va}}{120\text{V} \times 20\text{A}} = 2.5 \text{ or } 3 \text{ circuits}$$

Step 9. Allow one additional 20-ampere circuit for the outside outlet for sign or outline lighting if the store is on the ground floor.

The feeder or service load in this example is simply the sum of the branch-circuit loads. If it is assumed that the sign circuit is to be continuously loaded, its maximum load is calculated as follows:

.8 x 120 volts x 20 amperes = 1920 volt-amperes

However, if the actual load of this circuit is not known, a 1200-volt-ampere load may be used in the calculation.

Step 10. Calculate the total load in volt-amperes.

Lighting load	15,000 volt-amperes
Receptacle load	3,600 volt-amperes
Show window	6,000 volt-amperes
Sign circuit	1,200 volt-amperes (min.)
Total calculated load	**25,800 volt-amperes**

Step 11. Calculate the service size in amperes.

$$\frac{25{,}800 \text{ va}}{240\text{V}} = 107.5 \text{ circuits}$$

Consequently, the service-entrance conductors must be rated for no less than 107.5 amperes at 240 volts. The standard 110-ampere overcurrent protective device would be used. In actual practice, most contractors would install a 125-ampere service with disconnects, panelboard, overcurrent protection, and conductors rated for 125 amperes.

OFFICE BUILDING

A 20,000 square foot office building is served by a 480Y/277-volt, three-phase service. The building contains the following loads:

- 10,000 va, 208-volt, three-phase sign
- 100 duplex receptacles supplying continuous loads

- 40-foot long show window
- 12-kVA, 208/120-volt, three-phase electric range
- 10-kVA, 208/120-V, three-phase electric oven
- 20-kVA, 480-volt, three-phase water heater
- Seventy-five 150-watt, 120-volt incandescent lighting fixtures
- Two hundred 200-watt, 277-volt fluorescent lighting fixtures
- 7.5 hp, 480-volt, three-phase motor for fan-coil unit
- 40-kVA, 480-volt, three-phase electric heating unit
- 60-ampere, 480-volt, three-phase air-conditioning unit

The ratings of the service equipment, transformers, feeders, and branch circuits are to be determined, along with the required size of service grounding conductor. Circuit breakers are used to protect each circuit and THHN copper conductors are used throughout the electrical system.

A one-line diagram of the electrical system is shown in Figure 2-13. Note that the incoming 3-phase, 4-wire, 480Y/277V main service terminates into a main distribution panel containing six overcurrent protective devices. Since there are only six circuit breakers in this enclosure, no main circuit breaker or disconnect is required as allowed by *NEC* Section 230-71. Five of these circuit breakers protect feeders and branch circuits to 480/277-volt equipment, while the sixth circuit breaker protects the feeder to a 480 — 208Y/120-volt transformer. The secondary side of this transformer feeds a 208/120-volt lighting panel with all 120-volt loads balanced and all loads on this panel are continuous. Now let's start at the loads connected to the 208/120-volt panel and perform the required calculations.

Step 1. Calculate the load for the 100 receptacles, remembering that all of these are rated as a continuous load.

100 x 180 x 1.25 = 22,500 va

Step 2. Calculate the load for the show window using 200 va per linear foot.

200 va x 40 feet = 8,000 va

Step 3. Calculate the load for the outside lighting.

75 lamps x 150 va x 1.25 = 14,062.5 or 14,063 va

Step 4. Calculate the load for the 10 kW sign.

10,000 x 1.25 = 12,500 va

Step 5. Calculate the load for the 12 kW range.

12,000 x 1.25 = 15,000 va

Step 6. Calculate the load for the 10 kW oven.

10,000 x 1.25 = 12,500 va

Step 7. Determine the sum of the loads on the 208/120-volt lighting panel.

Receptacle load	22,500
Show window	8,000
Outside lighting	14,063
Electric sign	12,500
Electric range	15,000
Electric oven	12,500
Total connected load on subpanel	**84,563**

Step 8. Determine the feeder rating for the subpanel.

$$\frac{84563}{\sqrt{3} \times 208V} = 235 \text{ amperes}$$

Step 9. Refer to *NEC* Table 310-16 and find that 4/0 THHN conductor (rated at 260 amperes) is the closest conductor size that will handle the load. Since a 235-ampere circuit breaker is not standard, one rated at 250-amperes is the next standard size, and is permitted by the *NEC*.

The 208Y/120-volt circuit is a separately derived system from the transformer and is grounded by means of a grounding electrode conductor that must be at least a No. 2 copper conductor based on the No. 4/0 copper feeder conductors.

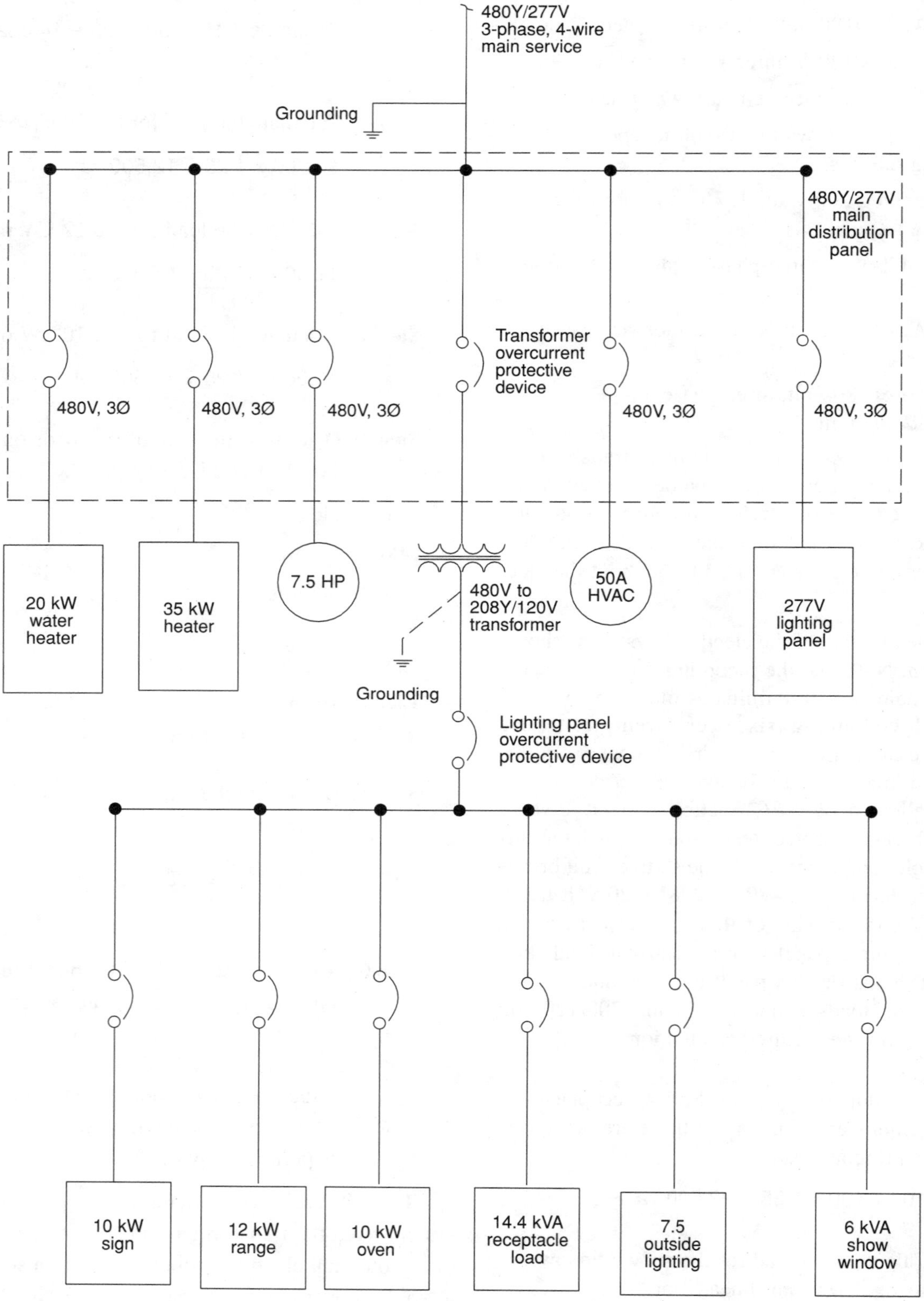
480Y/277V
3-phase, 4-wire
main service
Grounding
480Y/277V
main
distribution
panel
Transformer
overcurrent
protective
device
480V, 3Ø
480V, 3Ø
480V, 3Ø
480V, 3Ø
480V, 3Ø
20 kW
water
heater
35 kW
heater
7.5 HP
480V to
208Y/120V
transformer
50A
HVAC
277V
lighting
panel
Grounding
Lighting panel
overcurrent
protective device
10 kW
sign
12 kW
range
10 kW
oven
14.4 kVA
receptacle
load
7.5
outside
lighting
6 kVA
show
window

Figure 2-13: *One-line diagram of electrical systems for office building*

Step 10. Size the transformer and the transformer overcurrent protective device. Since the transformer supplying the subpanel is rated for a continuous load, the kVA rating of the transformer need only be the actual connected load. The 1.25 continuous-rating factor does not have to be applied. Therefore, the kVA rating of the transformer need only be 69.3 kVA. A commercially available 75-kVA transformer would be selected.

Step 11. Select the overcurrent protective device for the transformer.

$$\frac{75{,}000 \text{ va}}{\sqrt{3} \times 480\text{V}} = 90.21 \text{ amperes}$$

The maximum setting of the transformer overcurrent protective device is then 1.25 x 90.21 = 112.76 amperes; that is, it may not exceed this rating. Since 110 amperes is a standard fuse and circuit-breaker rating, this would be the maximum size normally used. No. 4 THHN copper conductors may be used for the transformer feeder.

Calculations For The Primary Feeder

Calculations for the primary feeder and other 480/277-volt circuits are based on the assumption that all of these loads are continuous, including the water heater for the building. Let's take a look at the lighting first. *NEC* Table 220-3(b) requires a minimum general lighting load for office buildings to be based on 3.5 va per square foot. Since the building has already been sized at 20,000 sq. ft., the minimum va for lighting may be determined as follows:

$$20{,}000 \times 3.5 = 70{,}000 \text{ va}$$

Since this load is continuous, the 70 kVA figure will have to be multiplied by a factor of 1.25 to obtain the total calculated load:

$$70{,}000 \times 1.25 = 87{,}500$$

The actual connected load, however, is as follows:

$$200 \text{ light fixtures} \times 200 \text{ va each} \times 1.25 = 50{,}000 \text{ va}$$

Since this connected load is less than the *NEC* requirements, it is neglected in the calculation and the 3.5 va per-square-foot figure is used. Therefore, the total load on the 277-volt lighting panel may be determined as follows:

$$\frac{1.25 \times 70000}{\sqrt{3} \times 480\text{V}} = 105.25 \text{ amperes}$$

No. 3 THHN conductors will be used for the feeder supplying the 277-volt lighting panel, and the overcurrent protective device will be rated for 110 amperes — the closest standard fuse or circuit breaker size.

NEC Section 422-14 requires that the feeder or branch circuit supplying a fixed storage-type water heater having a capacity of 120 gallons or less be sized not less than 125% of the nameplate rating of the water heater. Consequently, the load for the water heater may be calculated as follows:

$$1.25 \times \frac{20{,}000 \text{ va}}{\sqrt{3} \times 480\text{V}} = 30.01 \text{ amperes}$$

The feeder for the water heater will therefore require a No. 8 AWG THHN conductor with an overcurrent protective device of either 35 or 40 amperes.

The electric-heating load is sized in a similar way. *NEC* Section 424-3(b) requires that this load be sized at not less than 125 percent of the total load of the motors and the heaters. Therefore, the load for the electric heating may be calculated as follows:

$$1.25 \times \frac{40{,}000 \text{ va}}{\sqrt{3} \times 480\text{V}} = 60.1 \text{ amperes}$$

This load requires a conductor size of No. 6 AWG, THHN, with an overcurrent protective device rated at 70 amperes.

An instantaneous trip circuit breaker is selected to protect the 7.5 hp motor circuit. This arrangement is allowed if the breaker is adjustable and is part of an approved controller. A full-load current of 11 amperes requires No. 14 conductors protected by a breaker set as high as (13 x 11 =) 143 amperes. However, in this case, the circuit breaker will probably be set at approximately 70 to 80 amperes.

Type of Load	Load in Amperes	Neutral	NEC References
208Y/120-volt system	101.83A	0	
277-volt lighting panel	105.37A	104	
Water heater	30.1A	0	
Electric heat (neglected)	0	0	Section 220-21
7.5 hp motor	11.0	0	
Air conditioner	75	0	
25% of largest motor = .25 x 60A	15	0	Section 430-25
Service Load	**338.3**	**104**	

Figure 2-14: *Main service load calculation*

The air-conditioning circuit must have an ampacity of:

1.25 x 60 amperes = 75 amperes

for which No. 6 THHN conductors are selected from *NEC* Table 310-16. An overcurrent protective device for this circuit cannot be set higher than (1.75 x 60 =) 105 amperes.

Main Service Calculations

When performing the calculations for the main service, assume that all loads are balanced and may be computed in terms of volt-amperes, but are more than likely computed in terms of amperes. Calculation of the loads in amperes also simplifies the selection of the main overcurrent protective device and service conductors. A summary of this calculation is shown in Figure 2-14.

The 84,563 va, 208Y/120 lighting panel load represents a line load of 101.83 amperes at 480 volts. The 480/277-volt loads are then added to this for a total of 338.3 amperes. The conductors are then selected from *NEC* Table 310-16 which indicates that 350 kcmil THHN conductors have a current-carrying capacity of 350 amperes. An overcurrent protective device of the same amperage may also be selected. The load on the neutral is only 104 amperes, but the neutral conductor cannot be smaller than the grounding electrode conductor. Referring to *NEC* Table 250-66, the grounding electrode conductor must be 1/0 copper. Consequently, the neutral cannot be smaller than 1/0 THHN copper conductor.

In actual practice, the service size for this building would probably be increased to 400 amperes to allow for some future development.

Let's take another office building with some different equipment and determine the service size.

A 100- x 80-foot office building has a 120/208V 3Ø service requirement with the following loads:

- 5 – 180 va outside lighting outlets (120-volt continuous duty)
- 1 – sign outlet for an 1800 va, 120-volt continuous load
- 1 – 2000 va sign (120-volt continuous duty)
- 165 receptacles (120-volt noncontinuous duty)
- 15 – 120-volt receptacles (continuous duty)
- 24 feet of multioutlet assembly (120-volt continuous duty)
- 35,000 va heating unit
- 22,000 va air-conditioning unit
- 2 – 1560 va copy machines
- 2 – ½ hp exhaust fans (120-volt, 1176 va)

General lighting load	
80 x 100 = 8,000 sq. ft x 3.5 va =	35,000 va
Outside lighting load	
180 va x 5 x 125% =	1,125 va
Sign lighting	
1800 va x 125% =	2,250 va
2000 va x 125% =	2,500 va
Total Lighting Loads:	40,875 va
Receptacle Loads	
165 x 180 va =	29,700 va
Demand: First 10,000 va @ 100% =	10,000 va
Remaining 19,700 va @ 50% =	9,850 va
15 x 180 va x 125% =	3,375 va
24' multioutlet assembly x 180 va =	4,320 va
Total receptacle load =	27,545 va
Special Loads:	
2 copy machines @ 1560 va =	3,120 va
Heating or Air-Conditioning Load	

NOTE: Both heating and air-conditioning loads do not have to be included in the load calculation; only the larger of the two. In this case the 35,000 va heating load is larger than the 22,000 va air-conditioning load. Therefore, only the 35,000 va heating load is used in the calculation.

Heating load =	35,000 va
Motor Loads	
2 — ½ hp motors @1,176 va =	2,352 va
Largest motor load = 1,176 va x 25% =	294 va
Total Load for building =	**106,686 va**

To determine the total service-conductor size in amperes, use the following equation:

$$\text{Amperes} = \frac{\text{Total va}}{\text{Voltage} \times \sqrt{3}}$$

Therefore,

$$\frac{109{,}186}{208 \times \sqrt{3}} = 303.43 \text{ amperes}$$

Since electrical equipment is manufactured in standard sizes, the rating of the service size for this office building would be 400 amperes. Depending upon the type of conductor insulation, *NEC* Table 310-16 is used to select the conductor size.

MOBILE HOME AND RV PARKS

During the past couple of decades, mobile home and trailer parks have increased in number to the point where the NFPA found it necessary to include detailed wiring instructions covering such operations. *NEC* Articles 550 and 551 cover the electrical conductors and the equipment installed within or on mobile homes and recreation vehicles, as well as the means of connecting the units to an electrical supply.

Sizing Electrical Services For Mobile Homes

The electrical service for a mobile home may be installed either underground or overhead, but the point of attachment must be a pole or power pedestal located adjacent to but not mounted on or in the mobile home. The power supply to the mobile home itself is provided by a feeder assembly consisting of not more than one mobile home power cord, rated at 50 amperes, or a permanently installed feeder.

The *NEC* gives specific instructions for determining the size of the supply-cord and distribution-panel load for each feeder assembly for each mobile home. The calculations are based on the size of the mobile home, the small-appliance circuits, and other electrical equipment that will be connected to the service.

Lighting loads are computed on the basis of the mobile home's area: width times length (outside dimensions exclusive of coupler) times 3 watts per square foot.

Length x width x 3 = lighting va

Small-appliance loads are computed on the basis of the number of circuits times 1500 watts for each 20-ampere appliance receptacle circuit.

Number of circuits x 1500 = small-appliance watts

Nameplate Rating (va or Watts)	Use
10,000 or less	80% of rating
10,001 - 12,500	8,000 watts
12,501 - 13,500	8,400 watts
13,501 - 14,500	8,800 watts
14,501 - 15,500	9,200 watts
15,501 - 16,500	9,600 watts

Figure 2-15: *Power demand factors for free-standing electric ranges in mobile homes*

The sum of the two loads gives the total load in watts. However, there is a diversity (demand) factor that may be applied to this total in sizing the service and power cord. The first 3000 watts (obtained from the previous calculation) is rated at 100 percent. The remaining watts should be multiplied by a demand factor of 0.35 (35 percent). The total wattage so obtained is divided by the feeder voltage to obtain the service size in amperes.

If other electrical loads are to be used in the mobile home, the nameplate rating of each must be determined and entered in the summation. Therefore, to determine the total load for a mobile home power supply, perform the following calculations:

Step 1. Lighting and small appliance load, as discussed previously.

Step 2. Nameplate amperes for motors and heater loads, including exhaust fans, air conditioners, and electric heaters. Since air conditioners and heaters will not operate simultaneously, only the larger of the two needs to be included in the total load figures. Multiply the largest motor nameplate rating by 1.25 and add the answer in the calculations.

Step 3. Total of nameplate amperes for any garbage disposals, dishwashers, electric water heaters, clothes dryers, cooking units, and the like. Where there are more than three of these appliances, use 75 percent of the total load.

Step 4. The amperes for free-standing ranges (as distinguished from separate ovens and cooking units) by dividing the values shown in table in Figure 2-15 by the voltage between phases.

Step 5. The anticipated load if outlets or circuits are provided for other than factory-installed appliances.

To illustrate this procedure for determining the size of the electrical service and power cord for a mobile home, assume that a mobile home is 70 feet by 10 feet; has two portable appliance circuits; a laundry circuit; a 1200-watt air conditioner; a 200-watt, 120-volt exhaust fan; a 1000-watt water heater; and a 6000-watt electric range. The load is calculated as shown in Figure 2-16.

Lighting and Small-Appliance Load	Volt-Amperes
Lighting load: 70 x 10 x 3 W/ft^2	2100
Small appliance load: 1500 va x 2	3000
Laundry	1500
First 3000 va at 100%	3000
Remainder (6600 - 3000) at 35%	1260
Total lighting and small appliance load	4260

$$\frac{4260\text{ va}}{240\text{V}} = 17.75 \text{ amperes per phase}$$

Summary	Amperes Phase A	Amperes Phase B
Lighting and appliances	17.75	17.75
Water heater, 1000 va / 240	4.2	4.2
Fan, 200 va / 120	1.7	—
Air conditioner, 1200 va / 240	5.0	5.0
Range, 6000 va x .8/240	20.0	20.0
Total	48.65	46.95

Figure 2-16: *Summary of mobile home calculations*

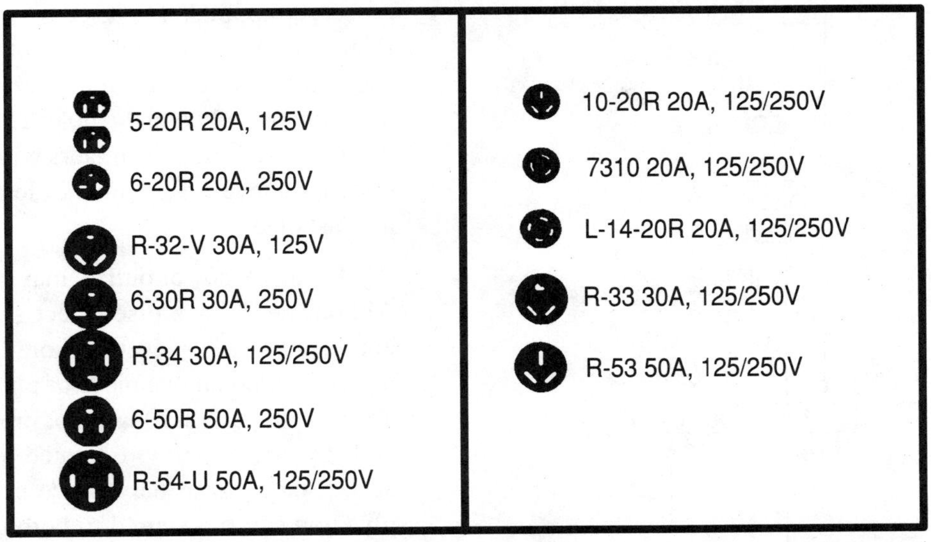
5-20R 20A, 125V
6-20R 20A, 250V
R-32-V 30A, 125V
6-30R 30A, 250V
R-34 30A, 125/250V
6-50R 50A, 250V
R-54-U 50A, 125/250V
10-20R 20A, 125/250V
7310 20A, 125/250V
L-14-20R 20A, 125/250V
R-33 30A, 125/250V
R-53 50A, 125/250V

Figure 2-17: *Receptacle configurations*

Based on the higher current for either phase, a 50-ampere power cord should be used to furnish electric power for the mobile home. The service should be rated for a minimum of 50 amperes and must be provided with overcurrent protection accordingly.

Types Of Equipment

Weatherproof electrical equipment for mobile homes, mobile home parks, and similar outdoor applications are available from many sources. The trainee should obtain catalogs and study the many types of mobile home utility power outlets and service equipment that are available. Any electrical equipment supplier should have these on hand.

Receptacle configurations used in mobile home and recreation vehicle applications are shown in Figure 2-17. Receptacles 1, 2, 4, and 8 are the most commonly used in mobile home and recreation vehicle parks.

Power units for use in mobile homes and recreation vehicles vary in design with fuse or circuit-breaker projection, attached meter sockets, mounting and junction posts, and special corrosion-resistant finish for ocean-side areas. The latter are used for boats where the outlets are located along the docks and power is leased by the dock owners on a daily basis. In addition to the standard units, manufacturers build equipment to meet special requirements.

Where more than one mobile home is to be fed, a power outlet and service equipment mounting cubicle (section shown in Figure 2-18) is ideal. The busbars in this unit accommodate wire sizes to 600 kcmil for a 400-ampere capacity. Therefore, either two 200-ampere units or four 100-ampere units may be mounted on the cubicle for feeding mobile home units underground. The main service should also enter from underground.

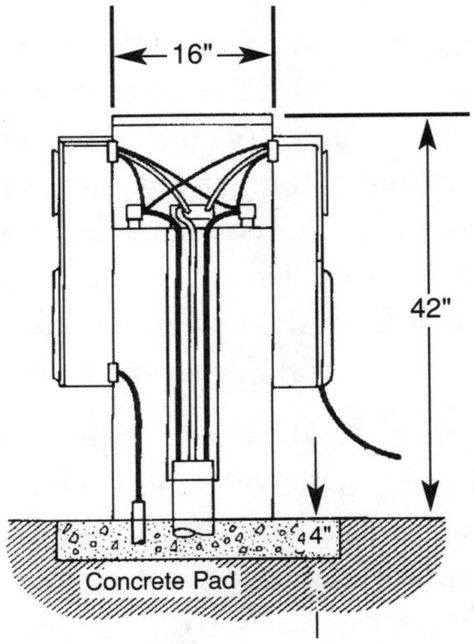
16"
42"
4"
Concrete Pad

Figure 2-18: *Mounting cubicle for mobile homes*

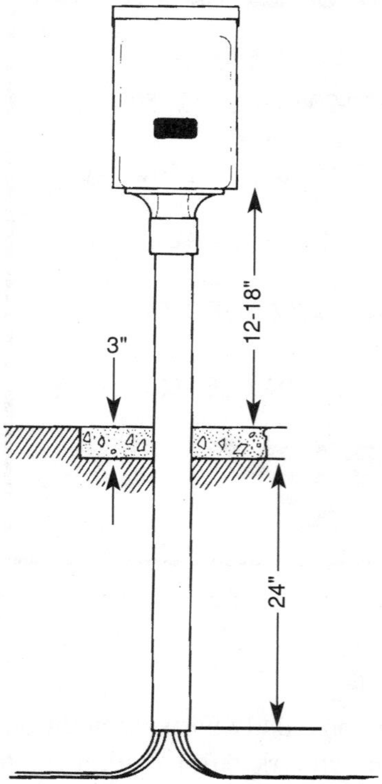
3"
12-18"
24"

Figure 2-19: *Power-outlet mounting post*

A power outlet mounting post is very popular for travel trailer parks and marinas. A cast-aluminum mounting base is provided to mount the power outlet on. The installer provides a length of 2-inch rigid conduit of the desired length. This is for underground installations. Note the conductors feeding in and out of the bottom of the conduit. See Figure 2-19.

Before beginning an installation, consult with the local power company for the method of serving the mobile home park and for the location of service entrance poles. Power company regulations vary from area to area, but the power company will furnish and set the pole in most cases. However, the electrician must obtain permission from the power company before performing any work on facilities on the poles.

Once a definite plan has been settled on, the electrician should obtain a piece of ¾-inch outdoor-rated plywood of sufficient size to hold the service equipment (i.e., a wire trough, meter bases, and weatherproof power outlets as shown in Figure 2-20). This piece of plywood should be primed with paint, and a final coat of wood preserver applied. Two pieces of 2- by 4-inch pressure-treated timbers are spiked or otherwise secured to the sheet of plywood for reinforcement before the entire assembly is spiked to the pole. The wood backing should be arranged so that the meters will be no more than 5 feet 6 inches above the ground nor less than 4 feet when they are installed.

Up to six power outlets may be fed from one service without need for a disconnect switch to shut down the entire service. However, if more than four power outlets are assembled on one piece of plywood, the arrangement shown in Figure 2-20 will not provide adequate support. A 4- by 4-inch pressure-treated post should be installed at a distance from the service pole so that the sheet of plywood can be secured to both poles (Figure 2-21) for added support.

With the plywood backing secured in place, a wire trough, sized according to *NEC* Article 374-5 should be installed at the very top of the board. The wire trough (auxiliary gutter) should not contain more than 30 current-carrying conductors nor should the sum of the

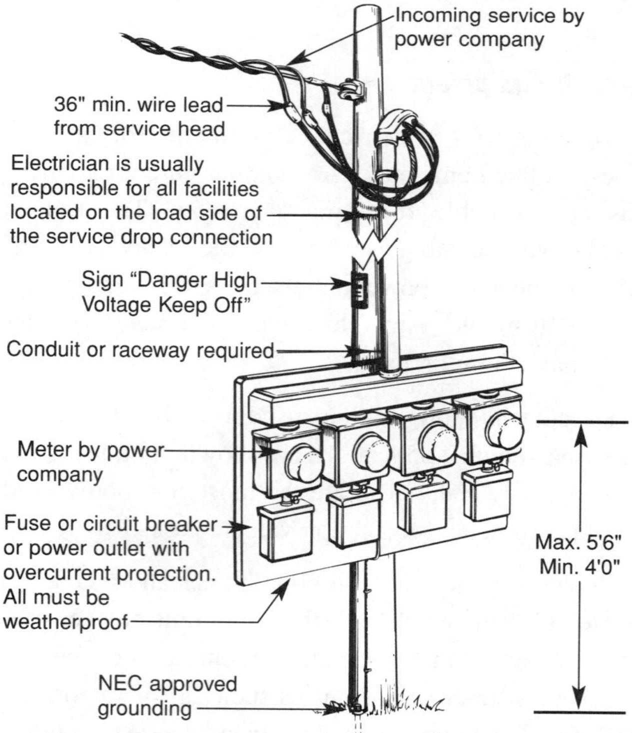
Incoming service by
power company
36" min. wire lead
from service head
Electrician is usually
responsible for all facilities
located on the load side of
the service drop connection
Sign "Danger High
Voltage Keep Off"
Conduit or raceway required
Meter by power
company
Fuse or circuit breaker
or power outlet with
overcurrent protection.
All must be
weatherproof
Max. 5'6"
Min. 4'0"
NEC approved
grounding

Figure 2-20: *Typical mobile home park service*

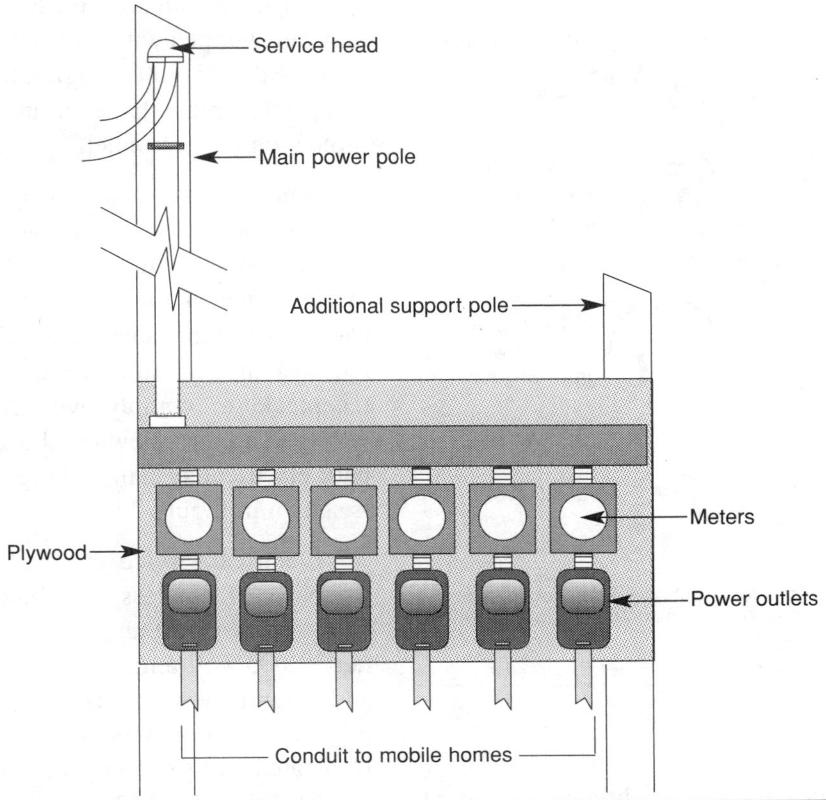
Service head
Main power pole
Additional support pole
Meters
Plywood
Power outlets
Conduit to mobile homes

Figure 2-21: *Additional pole for added support*

cross-sectional areas of all contained conductors at any cross section exceed 20 percent of the interior cross-sectional area of the gutter. The auxiliary gutter should be approved for outdoor use.

Once the entire installation is complete and has been inspected, the power company will install the meters. The connections of the meter bases to the wire trough are made with short, rigid conduit nipples using locknuts and bushings. Although straight nipples are often used for these connections, an offset nipple usually does a better job.

Weatherproof fuse or circuit-breaker disconnects are installed directly under the meter bases, again by means of conduit nipples. A weatherproof, 50-ampere mobile home power outlet with overcurrent protection is also used quite often.

The service mast comes next and should consist either of rigid metallic conduit or of EMT with weatherproof fittings. Once installed and secured to the pole with pipe straps, the service-entrance conductors may be pulled into the conduit and out into the wire trough. An approved weatherhead is then installed on top of the mast, and at least three feet of service conductors should be left for the power company to make their connections.

With the service-entrance conductors in place, meter taps are made to the service conductors in the trough. All such splices and taps made and insulated by approved methods may be located within the gutter when the taps are accessible by means of removable covers or doors on the wire trough. The conductors, including splices and taps, must not fill more than 75 percent of the gutter area (*NEC* Article 374-8a). These taps must leave the gutter opposite their terminal connections, and conductors must not be brought in contact with uninsulated current-carrying parts of opposite polarity.

The taps in the auxiliary gutter go directly to the line side of the meter bases. Once secured, the load side of the meter bases is connected to the disconnects or power outlets. All wiring should be sized according to the *NEC*.

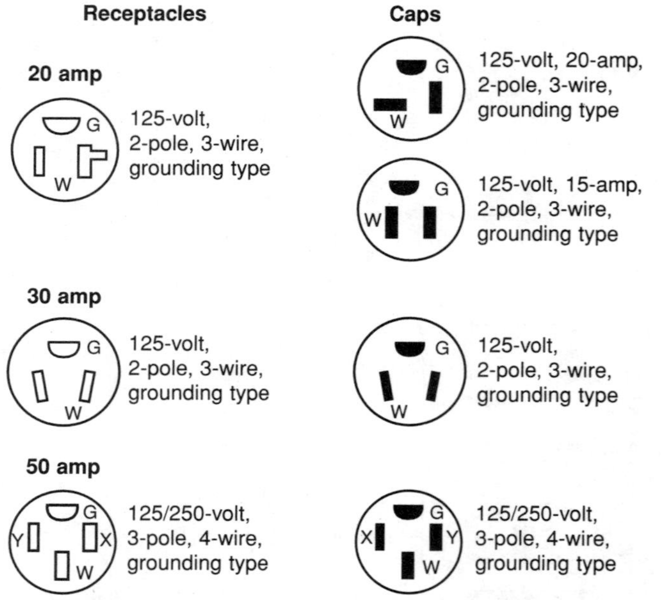
Receptacles
Caps
20 amp
125-volt,
2-pole, 3-wire,
grounding type
125-volt, 20-amp,
2-pole, 3-wire,
grounding type
125-volt, 15-amp,
2-pole, 3-wire,
grounding type
30 amp
125-volt,
2-pole, 3-wire,
grounding type
125-volt,
2-pole, 3-wire,
grounding type
50 amp
125/250-volt,
3-pole, 4-wire,
grounding type
125/250-volt,
3-pole, 4-wire,
grounding type
G
W
X
Y

Figure 551-46(c). Configurations for Grounding-Type Receptacles and Attachment Plug Caps Used for Recreational Vehicle Supply Cords and Recreational Vehicle Lots.

Figure 2-22: *Receptacles and caps*

Most water supplies for mobile homes consist of PVC (plastic) water pipe and, therefore, cannot provide an adequate ground for the service equipment. In cases like these, a grounding electrode, such as a ¾-inch by 8-foot ground rod driven in the ground near the service equipment, is used. A piece of bare copper ground wire is connected to the ground rod on one end with an approved ground clamp, and the other end is connected to the neutral wire in the auxiliary gutter. This wire must be sized according to *NEC* Table 250-94.

When all the work is complete, the service installation should be inspected by the local electrical inspector. The power company should then be notified to provide final connection of their lines.

Sizing Electrical Services And Feeders For Parks

A minimum of 70 percent of all recreation vehicle park lots with electrical service equipment must be equipped with both a 20-ampere, 125-volt receptacle and a 30-ampere, 125-volt receptacle. A minimum of 5 percent of all sites must be equipped with a 50-ampere, 125/250-volt receptacle conforming to the configuration as identified in *NEC* Figure 551-46(c) shown in Figure 2-22. The remainder with electrical service equipment may be equipped with only a 20-ampere, 125-volt receptacle. All 125-volt, single-phase 15- and 20-ampere receptacles must have ground-fault circuit-interrupter protection for personnel.

Since most travel trailers and recreation vehicles built recently are equipped with 30-ampere receptacles, an acceptable arrangement is to install a power pedestal in the corner of four lots so that four different vehicles can utilize the same pedestal. Such an arrangement requires three 30-ampere receptacles and one 20-ampere receptacle to comply with *NEC* Section 551-44. A wiring diagram showing the distribution system of a park electrical system serving 20 recreation vehicle lots is shown in Figure 2-23.

Electrical service and feeders must be calculated on the basis of not less than 9,600 volt-amperes per site equipped with 50-ampere, 120/240-volt supply facilities; 3,600 volt-amperes per site with both 20-ampere and 30-ampere supply facilities; and 2,400 volt-amperes per site equipped with only 20-ampere supply facilities. Dedicated tent sites equipped with only 20-ampere supply facilities should be calculated on the basis of 600 volt-amperes per site. The demand factors set forth in Table 551-73 are the minimum allowable demand factors permitted in calculating loads for service and feeders. Where the electrical supply for a recreational vehicle site has more than one receptacle, compute the load only for the highest rated receptacle.

Example:

Park area A has a capacity of 20 lots served by electricity; park B has 44. Find:

- The diversity (demand) factor of area A
- The diversity (demand) factor of area B
- The total demand of area A
- The total demand of area B

Step 1. The diversity factor is 45 percent, read directly from Table 551-73 of the *NEC*.

Step 2. The diversity factor is 41 percent, read directly from Table 551-73 of the *NEC*.

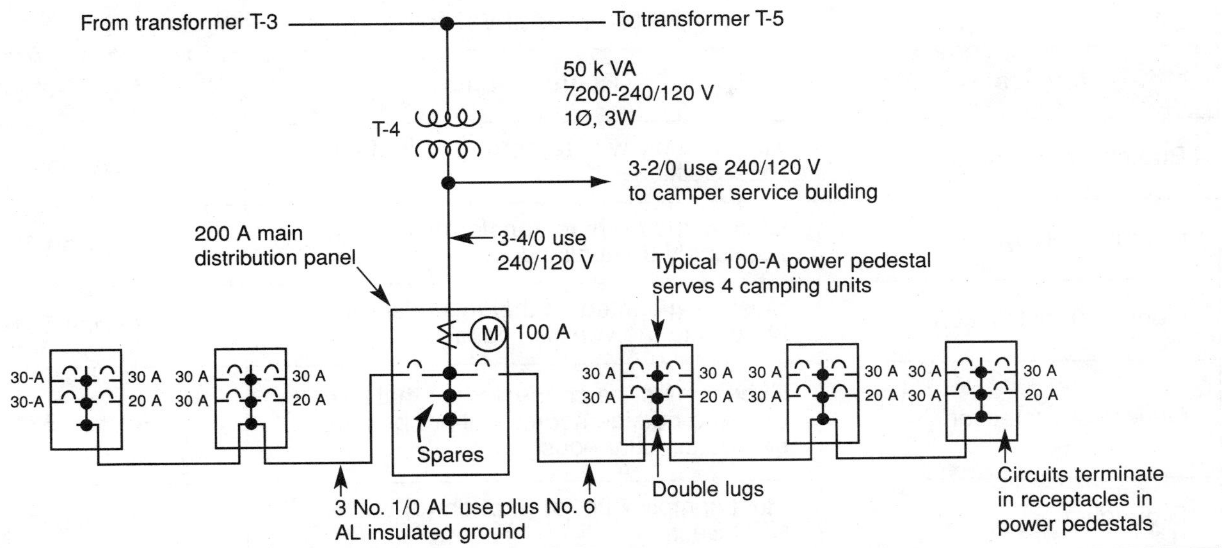
From transformer T-3
To transformer T-5
50 k VA
7200-240/120 V
1Ø, 3W
T-4
3-2/0 use 240/120 V
to camper service building
200 A main
distribution panel
3-4/0 use
240/120 V
Typical 100-A power pedestal
serves 4 camping units
M
100 A
30-A
30-A
30 A
20 A
30 A
30 A
Spares
Double lugs
3 No. 1/0 AL use plus No. 6
AL insulated ground
Circuits terminate
in receptacles in
power pedestals

Figure 2-23: *Wiring diagram of an RV trailer park distribution system*

Step 3. Since 5 percent of all sites having power must be equipped with a 50-ampere supply rated at 9,600 volt-amperes, and this park has 20 sites, one site must be rated at 9,600 volt-amperes. The remaining 19 may be rated at 3,600 volt-amperes.

1 x 9,600 va = 9,600
19 x 3,600 va = 68,400
Total va = 78,000
x .45 (diversity factor)
Total demand of A = 35,100 volt-amperes

Step 4. Since 5 percent of all sites having power must be equipped with a 50-ampere supply rated at 9,600 volt-amperes, and this park has 44 sites, two sites must be rated at 9,600 volt-amperes. The remaining 42 may be rated at 3,600 volt-amperes.

2 x 9,600 va = 19,200
42 x 3,600 va = 151,200
Total va = 170,400
x .41 (diversity factor)
Total demand of B = 69,864 volt-amperes

NEC installation requirements for RV parks are summarized in Figures 2-24, 2-25 and 2-26.

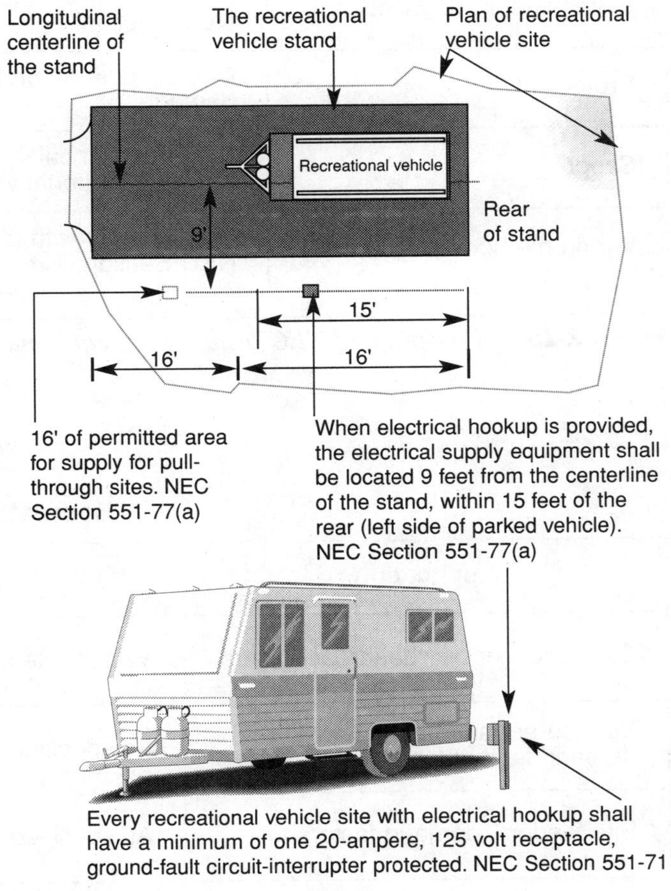
Longitudinal
centerline of
the stand
The recreational
vehicle stand
Plan of recreational
vehicle site
Recreational vehicle
Rear
of stand
9'
15'
16'
16'
16' of permitted area
for supply for pull-
through sites. NEC
Section 551-77(a)
When electrical hookup is provided,
the electrical supply equipment shall
be located 9 feet from the centerline
of the stand, within 15 feet of the
rear (left side of parked vehicle).
NEC Section 551-77(a)
Every recreational vehicle site with electrical hookup shall
have a minimum of one 20-ampere, 125 volt receptacle,
ground-fault circuit-interrupter protected. NEC Section 551-71

Figure 2-24: *NEC installation requirements for RV and trailer parks*

Recreational Vehicles

Application	Requirements	NEC Reference
Branch circuits required	Must comply with (a) through (d) of this NEC Section.	Section 551-42
Combination systems	Circuits fed from ac transformation must not supply dc appliances.	Section 551-20
Generator installations	Must be mounted so that it is effectively bonded to the vehicle chassis.	Section 551-30(a)
Generator protection	Circuits must be connected so that generator and outside source will not be connected simultaneously.	Section 551-30(b)
Grounding	Must comply with (a) through (d) of this NEC Section.	Section 551-76
Power supply assembly	Must comply with (a) through (d) of this NEC Section.	Section 551-44
Power supply, connection to	Must comply with (a) through (e) of this NEC Section.	Section 551-46
Protection	Cables must be protected against physical damage.	Section 300-4
Scope	Covers electrical systems installed within or on recreational vehicles.	Section 551-1
Wiring methods	Must comply with (a) through (r) of this NEC Section.	Section 551-47

Figure 2-25: *Description of NEC installation requirements for recreational vehicles*

Recreational Areas

Application	Requirements	NEC Reference
Clearance for overhead conductors	Vertical clearance of not less than 18 ft.	Section 551-79
Service-entrance conductors, demand factors for	As specified.	Section 551-73
Site feeders, demand factors for	As specified.	Section 551-73

Figure 2-26: *Description of NEC installation requirements for recreational vehicle area*

RESTAURANTS

The load of three or more cooking appliances and other equipment for a commercial kitchen may be reduced in accordance with *NEC* demand factors. This provision would apply to restaurants, bakeries, and similar locations.

For example, a small restaurant is supplied by a 240/120-volt, four-wire, three-phase service. The restaurant has the following loads:

- 1000-square foot area lighted by 120-volt lamps
- Ten duplex receptacles
- 20-ampere, 240-volt, three-phase motor-compressor
- 5-horsepower, 240-volt, three-phase roof ventilation fan protected by an inverse time circuit breaker
- More than six units of kitchen equipment with a total connected load of 80 kilovolt amperes. All units are 240-volt, three-phase equipment
- Two 20-ampere sign circuits

The main service uses type THHN copper conductors and is wired as shown in Figure 2-27. Lighting and receptacle loads contribute 27.9 amperes to phases A and C and 25.8 amperes to the neutral. The 80-kVA kitchen equipment load is subject to the application of a 65% demand factor which reduces it to a demand load of .65 x 80 kVA = 52 kVA. This load requires a minimum ampacity of 125 amperes per phase at 240 volts. The load of the three-phase motors and 25 percent of the largest motor load bring the service load total to 193.1 amperes for phases A and C and 165.2 amperes for phase B.

If the phase conductors are three No. 3/0 type THHN copper conductors, the grounding electrode conductor and the neutral conductor must each be at least a No. 4 copper conductor.

The fuses are selected in accordance with the *NEC* rules for motor-feeder protection. The ungrounded conductors, therefore, are protected at 200 amperes each.

SERVICES FOR HOTELS AND MOTELS

The portion of the feeder or service load contributed by general lighting in hotels and motels without provisions for cooking by tenants is subject to the application of demand factors. In addition, the receptacle load in the guest rooms is included in the general lighting load at 2 watts per square foot. The demand factors, however, do not apply to any area where the entire lighting is likely to be used at one time, such as the dining room or a ballroom. All other loads for hotels or motels are calculated as shown previously.

For example, let's determine the 120/240-volt feeder load contributed by general lighting in a 100-unit motel. Each guest room is 240 square feet in area. The general lighting load is:

$$2 \text{ va/ft}^2 \times 240 \text{ ft}^2\text{/unit} \times 100 \text{ units} = 48{,}000 \text{ va}$$

but the reduced lighting load is:

First 20,000 at 50% =	10,000 va
Remainder (48,000 – 20,000) at 40% =	11,200 va
Total =	**21,200 va**

This load would be added to any other loads on the feeder or service to compute the total capacity required.

OPTIONAL CALCULATION FOR SCHOOLS

The *NEC* provides an optional method for determining the feeder or service load of a school equipped with electric space heating or air conditioning, or both. This optional method applies to the building load, not to feeders within the building.

The optional method for schools basically involves determining the total connected load in volt-amperes, converting the load to volt-amperes/square foot, and applying the demand factors from the *NEC* table. If both air-conditioning and electric space-heating loads are present, only the larger of the loads is to be included in the calculation.

Let's take one example. A school building has 200,000 square feet of floor area. The electrical loads are as follows:

- Interior lighting at 3 va per square foot
- 300 kVA power load

Service Loads

Load	Calculation	Line A,C	Neutral	Line B
A. 240/120-volt loads				
Lighting	$= \frac{1.25 \times 2\text{VA per sq. ft.} \times 1000 \text{ sq. ft.}}{240\text{V}}$ = 10.4 amperes	10.4	8.3	-0-
Receptacles	$= \frac{180 \text{ VA} \times 10}{240\text{V}}$ = 7.5 amperes	7.5	7.5	-0-
B. Three-phase loads Kitchen equipment (6 or more units)	$= \frac{80{,}000\text{W} \times .65}{\sqrt{3} \times 240\text{V}}$ = 125 amperes	125	-0-	125
20-ampere three-phase motor compressor	breaker setting = 1.75 x 20A = 35 amperes	20	-0-	20
5-hp, three-phase motor	breaker setting = 2.5 x 15.2A = 38 amperes use 40-ampere std. size	15.2	-0-	15.2
C. 25% of largest motor load	= .25 x 20A = 5 amperes	5	-0-	5
D. Sign circuit	$= \frac{1200\text{va}}{120\text{V}}$ Each	10	10	-0-
Service load		193.1 amperes	25.8 amperes	165.2 amperes

Table 310-16, Section 250-23(b)

1. Conductors: Use No. 3/0 THHN copper for ungrounded conductors; use No. 4 THHN copper conductor for neutral (neutral based on size of grounding eletrode conductor)

Table 430-63, Section 240-6

2. Overcurrent protective device:
   Phases A and C = 40A (largest motor device) + 10.4A + 7.5A + 125A + 20A + 10A = 212.9 amperes.
   Use standard size 200-ampere fuses.
   Phase B = 40A + 125A + 20A = 185 amperes

Table 250-94

3. Grounding electrode conductor required to be No. 4 copper.

Figure 2-27: *Service specifications for a small restaurant*

- 100 kVA water heating load
- 100 kVA cooking load
- 100 kVA miscellaneous loads
- 200 kVA air-conditioning load
- 300 kVA heating load

The service load in volt-amperes is to be determined by the optional calculation method for schools.

As shown in Figure 2-28, the combined connected load is 1500 kVA. Based on the 200,000 square feet of floor area, the load per square foot is:

$$\frac{1{,}500{,}000\text{va}}{200{,}000 \text{ sq. ft.}} = 7.5 \text{ va per sq. ft.}$$

The demand factor for the portion of the load up to and including 3 volt-amperes/square foot is 100 percent. The remaining 4.5 volt-amperes/square foot in the example is added at a 75% demand factor for a total load of 1,275,000 volt-amperes.

Service Load Calculation For School

Lighting = 3 va per sq. ft x 200,000 sq. ft	600,000 va
Power Load	300,000 va
Water-Heating Load	100,000 va
Cooking Load	100,000 va
Miscellaneous Load	100,000 va
Heating Load (Neglect air conditioning)	300,000 va
Connected Load	**1,500,000 va**

The following equation may be used to determine the connected load per square foot:

$$\frac{1{,}500{,}000\text{va}}{200{,}000 \text{ sq. ft.}} = 7.5 \text{ va per sq. ft.}$$

Application of Demand Factor

3 va per sq. ft. x 200,000 sq. ft. @ 100%	600,000 va
4.5 va per sq. ft x 200,000 sq. ft. @ 75%	675,000 va
Service Load	**1,275,000 va**

Figure 2-28: *Optional calculation for schools*

SHORE POWER CIRCUITS FOR MARINAS AND BOATYARDS

The wiring system for marinas and boatyards is designed by using the same *NEC* rules as for other commercial occupancies except for the application of several special rules dealing primarily with the design of circuits supplying power to boats.

The smallest sized receptacle that may be used to provide shore power for boats is 20 amperes. Each single receptacle that supplies power to boats must be supplied by an individual branch circuit with a rating corresponding to the rating of the receptacle.

The feeder or service ampacity required to supply the receptacles depends on the number of receptacles and their rating, but demand factors may be applied that will reduce the load of five or more receptacles. For example, a feeder supplying ten 30-ampere shore power receptacles in a marina requires a minimum ampacity of:

$$10 \times 30\text{A} \times .8 = 240 \text{ amperes}$$

Although this computed feeder ampacity might seem rather large, this is the minimum required by *NEC* Section 555-5.

FARM LOAD CALCULATIONS

Electricity on the farm has progressed in a little more than 50 years from 32-volt dc battery systems that powered mostly lighting, to ultramodern 120/240-volt ac systems used to operate numerous time-saving devices around the farm. Quartz and HID lighting fixtures have now replaced the inefficient incandescent lamps that once were sparingly placed in essential locations, allowing much work to be done after daylight hours. In fact, farm wiring has become so specialized that some electrical contractors work only on farm electrical installations, and have their hands full at that!

Adequate wiring, properly planned and installed, provides safe lighting for every task, reduces eyestrain and fatigue, and reduces the time required for household, farm work and chores. Electricity provides convenient power for milking machines, milk coolers, feed grinders,

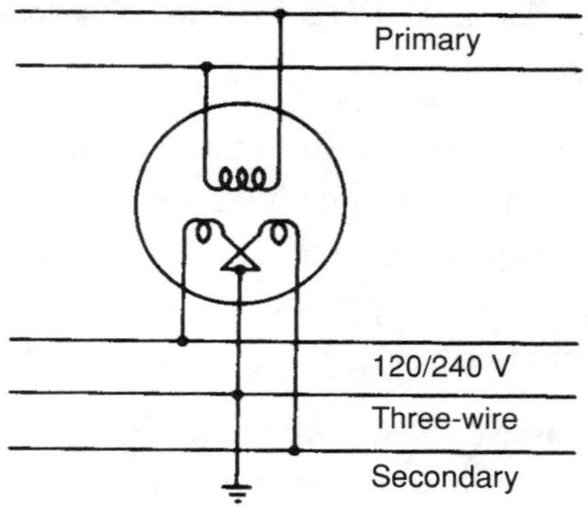
Primary
120/240 V
Three-wire
Secondary

Figure 2-29: *Single-phase transformer*

silo loaders, chicken brooders, and grain elevators, just to name a few. But these conveniences and comforts are not provided without adequate planning and wiring.

Farm Service Entrances

The service for the farm begins at the secondary terminals of the transformers connected to the utility company's lines. Two types are usually readily available: single phase and three phase. Of these two, single phase is by far the most common type found on farms throughout the United States and Canada. It will provide satisfactory power for lighting circuits and convenience outlets and for single-phase motors rated up to about 7½ horsepower. When motors larger than 7½ horsepower are to be used, a three-phase delta service is almost always provided.

The single-phase, three-wire service is provided by a transformer as shown in the wiring diagram in Figure 2-29. Note that two single-phase primary conductors are connected to the primary winding of the transformer through fuse cutouts; lightning arresters are also provided. The secondary windings have three conductors connected to the transformer terminals, two "hot" wires and a neutral, the latter of which is always grounded. The voltage between the two hot wires is 240 volts, while the voltage between the neutral and any one hot wire is 120 volts.

When possible, the transformer should be located near the center of the electrical load to reduce the size and length of feeder wires and thus reduce the cost of the installation. A typical farm electric service is shown in Figure 2-30.

Three-phase service is quite common when electric motors are used that are in excess of 7½ horsepower. Motors driving large irrigation pumps, feed grinders, crop-drying fans, and the like often require 10 horsepower or more. Voltage regulation with three-phase service is better, and the cost of three-phase motors is generally less than for single-phase motors of the same horsepower rating.

Two types of three-phase service are available in most areas: four-wire wye and four-wire delta. Of the two, the four-wire delta seems to be used the most, mainly because 240-volt single-phase motors and appliances may also be operated satisfactorily on this service, whereas the wye type supplies only 208 volts between phases. However, caution must be employed with the four-wire delta system to make sure the "high leg," which supplies about 208 volts, is not mistakenly connected to a 120-volt outlet. Also, 208-volt appliances and motors are available for use on the wye system, and where a number of 120-volt motors and appliances are in use, the wye is easier to balance than the delta system. There is also less chance of an improper connection that might be made by an inexperienced person.

NEC Article 220, Part D provides a separate method for computing farm loads other than the dwelling. Tables of demand factors are provided for use in computing the feeder loads of individual buildings as well as the service load of the entire farm. See *NEC* Sections 220-40 and 220-41.

The demand factors may be applied to the 120/240-volt feeders for any building or load (other than the dwelling) that is supplied by two or more branch circuits. All loads that operate without diversity, that is, the entire load is on at one time, must be included in the calculation at 100 percent of connected load. All "other" loads may be included at reduced demands. The load to be included at 100 percent demand, however, cannot be less than 125 percent of the largest motor and not less than the first 60 amperes of the total load. In other words, if the nondiverse and largest motor load is less than 60 amperes, a portion of the "other" loads will have to be included at 100 percent in order to reach the 60-ampere minimum.

After the loads from individual buildings are computed, it may be possible to reduce the total farm load further by applying additional demand factors.

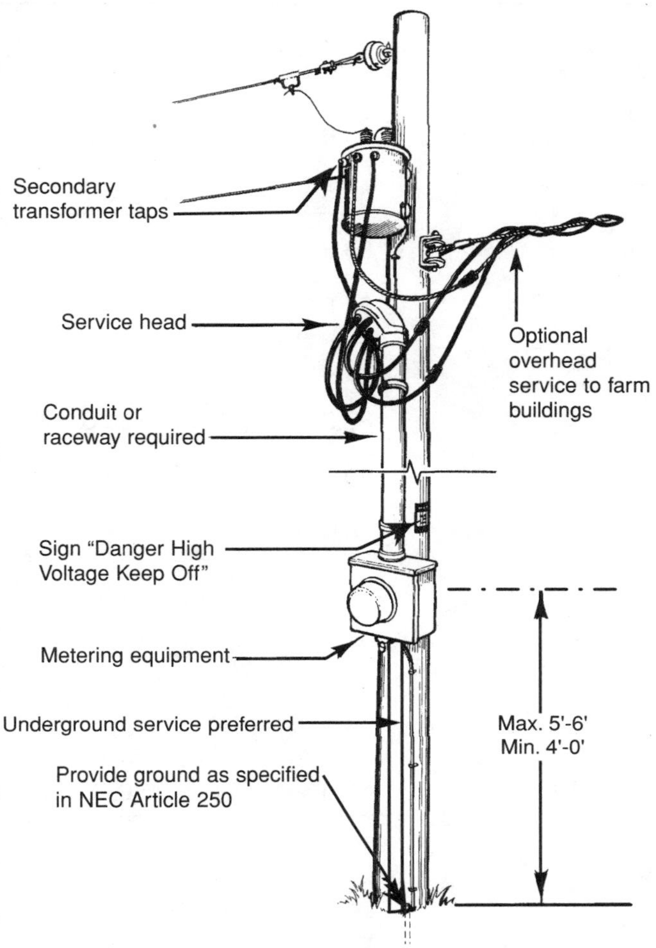
Secondary
transformer taps
Service head
Conduit or
raceway required
Sign "Danger High
Voltage Keep Off"
Metering equipment
Underground service preferred
Provide ground as specified
in NEC Article 250
Optional
overhead
service to farm
buildings
Max. 5'-6'
Min. 4'-0'

Figure 2-30: *Typical farm service*

For example, a farm has a dwelling and two other buildings supplied by the same 120/240-volt service. The electrical loads are as follows:

- Dwelling — 100-ampere load as computed by the calculation method for dwellings
- Building No. 1 — 5-kilovolt-ampere continuous lighting load operated by a single switch, 10-horsepower, 240-volt motor, 21 kilovolt-amperes of other loads
- Building No. 2 — 2-kilovolt-ampere continuous load operated by a single switch and 15 kilovolt-amperes of other loads

Determine the individual building loads and the total farm load as illustrated in Figure 2-31. The nondiverse load for building No. 1 consists of the 5-kilovolt ampere lighting load and the 10-horsepower motor for a total of 83.3 amperes. This value is included in the calculation at the 100 percent demand factor. Since the requirement for adding at least the first 60 amperes of load at the 100 percent demand factor has been satisfied, the next 60 amperes of the 87.5 amperes from all other loads are added at a 50 percent demand factor, and the remaining (87.5 − 60 =) 27.5 amperes are added at a 25 percent demand factor.

In the case of building No. 2, the nondiverse load is only 8.3 amperes; therefore, 51.7 amperes of "other" loads must be added at the 100 percent demand factor in order to meet the 60-ampere minimum.

Using the method given for computing the total farm load, we see that the service load is:

Largest load at 100% =	120.2 A (building No. 1)
Second largest at 75% =	49.1 A (building No. 2)
Total for both buildings =	169.3 A
Dwelling =	100.0 A
Total load in amperes	**269.3 A**

The total service load of 269 amperes requires the ungrounded service-entrance conductors to be at least 300 kcmil THW copper conductors. The neutral load of the dwelling is assumed to be 100 amperes which brought the total farm neutral load to 207 amperes. See Figure 2-31.

INDUSTRIAL MOTOR-CONTROL CENTER

Electricians working on industrial projects are frequently required to size the feeder load for motor-control centers, as well as the motor branch-circuit conductors and overcurrent devices. In general, when sizing feeders for motor-control centers — serving several motors — the feeder conductors must have an ampacity at least equal to the sum of the full-load current rating of all the motors, plus 25 percent of the highest rated motor in the group, plus the ampere rating of any other loads determined in accordance with *NEC* Article 220 and other *NEC* applicable sections.

A typical motor-control center (MCC) is shown in Figure 2-32, while its related motor-control schedule is shown in Figure 2-33. Note that this MCC contains

Building No. 1 Feeder Load		
Lighting (5-kva nondiverse load) = 5000va/240V	20.8	20.8
10hp motor = 1.25 x 50A	62.5	-0-
Total motor and nondiverse load	83.3	20.8
"Other loads" = 21000va/240V	87.5	87.5
Application of demand factors		
Motor and nondiverse loads @ 100%	83.3	20.8
Next 60A of "other" loads @ 50%	30.0	30.0
Remainder of "other" loads (87.5 - 60) @ 25%	6.9	6.9
Feeder load	**120.2A**	**57.7A**
Building No. 2 Feeder Load		
Lighting (2-kva nondiverse load) = 2000va/240V	8.3	8.3
"Other" loads = 15,000kva/240V	62.5	62.5
Application of demand factors		
Nondiverse load @100%	8.3	8.3
Remainder of first 60 (60 - 8.3) @ 100%	51.7	51.7
Remainder of "other load" (62.5 - 51.7) @ 50%	5.4	5.4
Feeder load	**65.4A**	**65.4A**
Total Farm Load		
Application of demand factors		
Largest load (Bld. No. 1) @ 100%	120.2	57.7
Next largest load (Bld. No. 2) @ 75%	49.1	49.1
Farm load (less dwelling)	169.3	106.8
Farm dwelling load	100.0	100.0
Total farm load	**269.3A**	**206.8A**

Figure 2-31: *Summary of farm calculations*

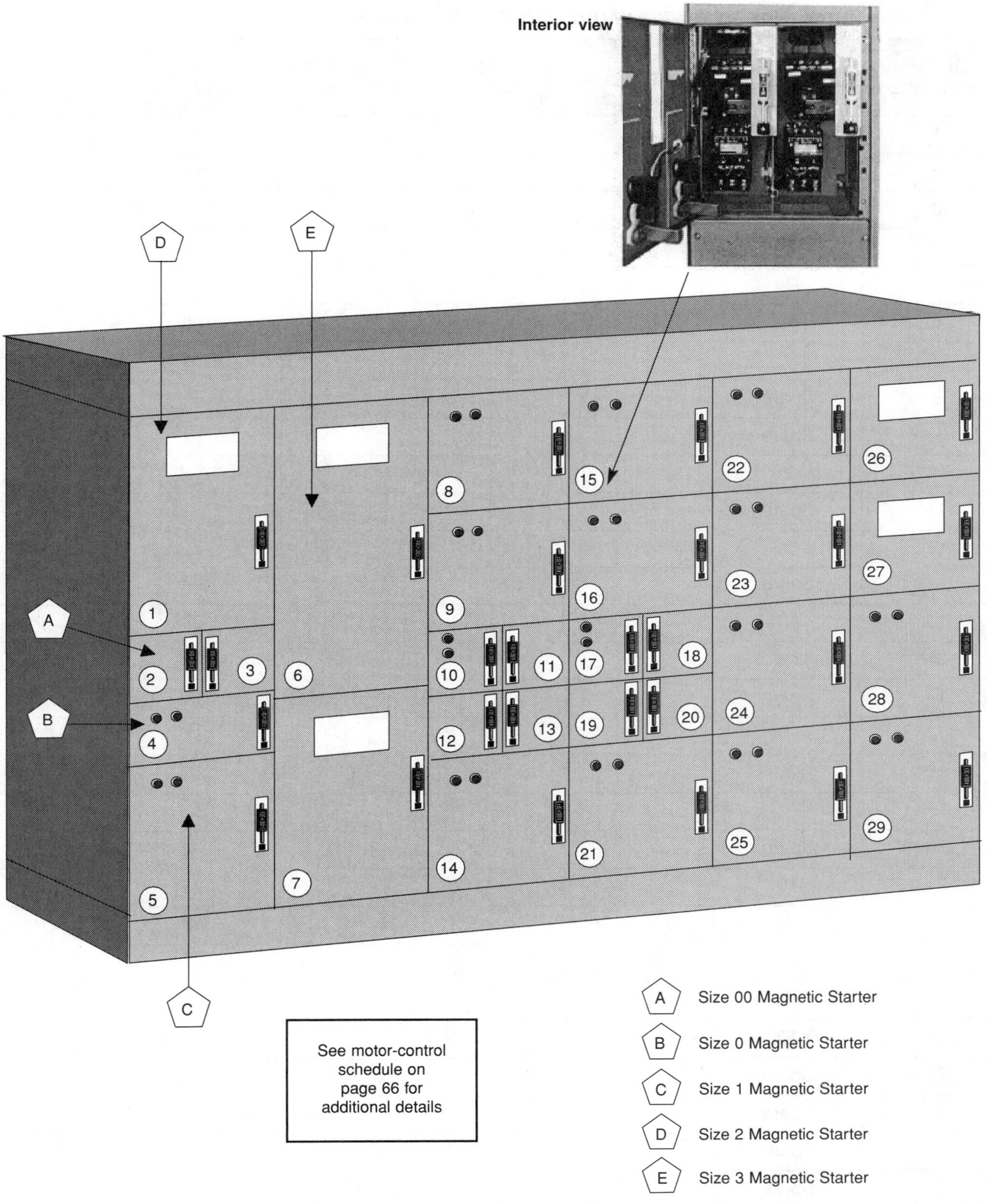
Interior view
D
E
A
B
C
1
2
3
4
5
6
7
8
9
10
11
12
13
14
15
16
17
18
19
20
21
22
23
24
25
26
27
28
29
See motor-control schedule on page 66 for additional details
A Size 00 Magnetic Starter
B Size 0 Magnetic Starter
C Size 1 Magnetic Starter
D Size 2 Magnetic Starter
E Size 3 Magnetic Starter

Figure 2-32: *Typical motor-control center*

Motor-Control Schedule

Circuit Number	Motor Horsepower	Temp °C	Code Letter	Service Factor	NEC FLA	NEMA Size
1	20	40	B	1.15	54	2
2	1½				5.2	00
3	1½				5.2	00
4	5				15.2	0
5	7½				22	1
6	50				130	3
7	20				54	2
8	Spare				—	1
9	Spare				—	1
10	2				6.8	00
11	Spare				—	00
12	1½				5.2	00
13	Spare				—	00
14	10				28	1
15	7½				22	1
16	Spare				—	1
17	1				3.6	00
18	¾				2.8	00
19	¾				2.8	00
20	Spare				—	00
21	10				28	1
22	10				28	1
23	7½				22	1
24	7½				22	1
25	10				28	1
26	Spare				—	1
27	Spare				—	1
28	Spare				—	1
29	Spare				—	1

Figure 2-33: *Motor-control schedule for motor control center in Figure 2-32*

spaces for 29 controllers ranging in size from NEMA size 00 to NEMA size 3. A total of 19 motors are currently connected with 10 spaces remaining as "spares" for future extensions.

Let's size the feeder for this MCC using connected loads only and assuming a power factor of 90 percent.

Step 1. Find the full-load current rating of all motors in the group: Since the *NEC* full-load amperes have already been determined in the motor-control schedule shown in Figure 2-33, the sum of all motor FLA is first determined. The total FLA of all motors = 484.8 amperes.

Step 2. Calculate 25 percent of the largest motor's FLA: The 50 hp is the largest motor in the group with a FLA of 130 amperes. Therefore,

130 x .25 = 32.5 amperes

Step 3. Add the results obtained from the last calculation to the sum of all motor FLA obtained previously; that is,

484.8 + 32.5 = 517.3 amperes

Step 4. The results obtained thus far would normally be the number of amperes to determine the feeder size. However, since the power factor of this electrical system is 90 percent, the footnotes to *NEC* Table 430-150 require that the load determined previously must be multiplied by a factor of 1.1. Therefore, the following equation must be used to determine the demand amperes for the MCC with a 90 percent power factor:

517.3 x 1.1 = 569.03 amperes

Step 5. Refer to *NEC* Table 310-16 to find a conductor, say, THHN, that will carry no less than 569.03 amperes. In doing so, it is found that 900 kcmil is the closest size obtainable to the load; that is, this conductor is rated at 585 amperes.

Summary Of Service Load Calculations

Load calculations are necessary for determining sizes and ratings of conductors, equipment, and overcurrent protection required by the *NEC* to be included in each electrical installation — including the service, feeders and branch-circuit loads. These calculations are necessary in every electrical installation from the smallest roadside vegetable stand to the largest industrial establishment. Therefore, electricians must know how to calculate services, feeders and branch circuits for any given electrical installation, and also know what *NEC* requirements apply.

When making any type of electrical calculation, the job will go smoother and faster if preprinted calculation forms are used. Such forms also ensure that nothing will be overlooked while making the calculation. Preprinted calculation forms are available from some of the electrical trade organizations such as the National Electrical Contractor's Association, and the many regional electrical contractor associations. However, most electrical technicians like to devise their own; forms that suit their own individual needs exactly. Once the form is fine-tuned and neatly arranged, it may be printed on a computer printer or the basic form may be taken to a job printer for typesetting and printing multiple copies.

UNDERGROUND SYSTEMS

There are several methods used to install underground wiring, but the most common include direct-burial cables and the use of duct lines or duct banks.

The method used depends on the type of wiring, soil conditions, allotted budget for the work, etc.

Direct-burial installations will range from small, single-conductor wires to multiconductor cables for power or communications or alarm systems. In any case, the conductors are installed in the ground by placing them in an excavated trench that will be backfilled later, or by burying them with a directional boring machine. Or, they can be installed by burying them directly with some form of cable plow that opens a furrow, feeds the conductors into the furrow, and then closes the furrow over the conductor.

Sometimes it becomes necessary to use lengths of conduit in conjunction with direct-burial installations, especially where the cables emerge on the surface of the ground or terminate at an outlet or junction box. Also, where the cables cross a roadway or concrete pavement, it is best to install a length of conduit under these areas in case the cable must be removed at a later date. By doing so, the road or concrete pad will not have to be disturbed.

Figure 2-34 shows a cross section of a trench with direct-burial cable installed. Note the sand base on which the conductors lie to protect them from sharp stones and such. A treated board is placed over the conductors in the trench to offer protection during any digging that might occur in the future. Also, a continuous warning ribbon is laid in the trench, some distance above the board, to warn future diggers that electrical conductors are present in the area.

Conductor Types

Type USE (Underground service-entrance cable): This type of cable is approved for underground use since it has a moisture-resistant covering.

Cabled single-conductor Type USE constructions may have a bare copper conductor cabled with the assembly. Type USE single, parallel, or cabled conductor assemblies may have a bare copper concentric conductor applied. These constructions do not require an outer overall covering.

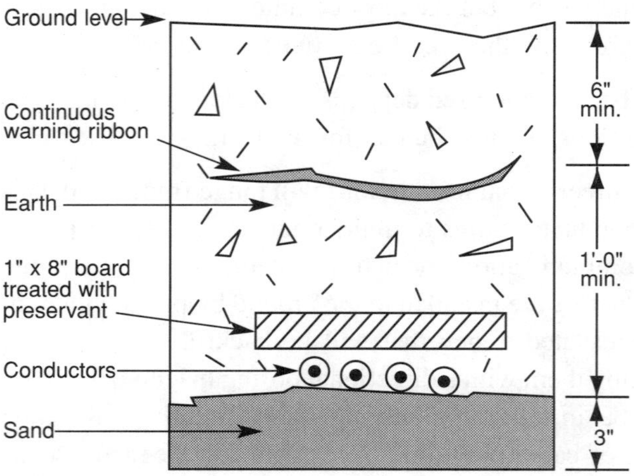
Ground level
Continuous warning ribbon
Earth
1" x 8" board treated with preservant
Conductors
Sand
6" min.
1'-0" min.
3"

Figure 2-34: *Cross section of direct-burial cable*

Type USE cable may be used for underground services, feeders, and branch circuits.

Type UF (Underground feeder and branch-circuit cable): This type of cable is manufactured in sizes from No. 14 AWG copper through No. 4/0. In general, the overall covering of Type UF cable is flame-retardant, moisture-, fungus-, and corrosion-resistant, and is suitable for direct burial in the ground.

Type UF cable may be used for direct-burial, underground installations as feeders or branch circuits when provided with overcurrent protection of the rated ampacity as required by the *NEC*.

Where single-conductor cables are installed, all cables of the feeder circuit, subfeeder circuit, or branch circuit — including the neutral and equipment grounding conductor (if any) — must be run together in the same trench or raceway.

Nonmetallic-armored cable: This type of cable is also used for underground installations. The interlocking armor consists of a single strip of interlocking tape that extends for the length of the cable. The surface of the cable is rounded, which allows it to deflect blows from picks and shovels much better than flat-bend armor. The cable must have an outer covering that will not corrode or rot. An asphalt-jute finish may be placed over the cable if it is to be subjected to particularly harsh corrosive environments.

Minimum Cover Requirements

The *NEC* specifies minimum cover requirements for direct buried cable (Figure 2-35). Furthermore, all underground installations must be grounded and bonded in accordance with *NEC* Article 250.

Where direct buried cables emerge from the ground, they must be protected by enclosures or raceways extending from the minimum cover distance specified in the table in Figure 2-35. However, in no cases will the protection be required to exceed 18 inches below the finished grade.

	Direct Burial	Rigid metal conduit or IMC	Rigid nonmetallic conduit approved without encasement or raceway	Residential branch circuit 120 volts or less; GFCI; 20 amp	Landscape lighting or irrigation circuit 30 volts or less
In trench below 2 inches of concrete	18 inches	6 inches	12 inches	6 inches	6 inches
Under a building (no clearance required)	Raceway required	—	—	Raceway required	Raceway required
In trench below 4 inches of concrete extending 6 inches beyond the installation	18 inches	4 inches	4 inches	6 inches (4 inches if in raceway)	6 inches (4 inches if in raceway)
Under streets, alleys, etc.	24-inch clearance required	24-inch clearance required	24-inch clearance required	24-inch clearance required	24-inch clearance required
Parking areas; drives, used for one- and two-family dwellings	18 inches	18 inches	18 inches	12 inches	18 inches
Airport runways	18-inch clearance required	18-inch clearance required	18-inch clearance required	18-inch clearance required	18-inch clearance required
In solid rock and covered by at least 2 inches of concrete (all cases)	Raceway required	—	—	Raceway required	Raceway required

Figure 2-35: *Minimum cover requirement for direct-burial cable, 0-600 volts*

Practical Application

Methods of installing direct-burial underground cable vary according to the length of the installation, the size of cable being installed, and the soil conditions. For short runs, from, say, a residential basement to a garage located 20 or so feet away, the excavation is often done by hand. For longer runs, power equipment is almost always used.

In general, the trench is opened to the correct depth with an entrenching tractor or backhoe. All sharp rocks, roots, and similar items are then removed from the trench to prevent these objects from damaging the direct-burial cable. If soil conditions dictate, a 3-inch layer of clean sand is poured into the bottom of the trench to further protect the direct-burial cable.

Underground cable will almost always come from the manufacturer on either metal or wooden reels. In the case of relatively short runs of the smaller sizes of multiconductor cable, the reel containing the cable is set up at one end of the trench and — using two reel jacks and a length of conduit through the center hole in the reel to allow the reel to rotate — the cable is "paid off." See Figure 2-36.

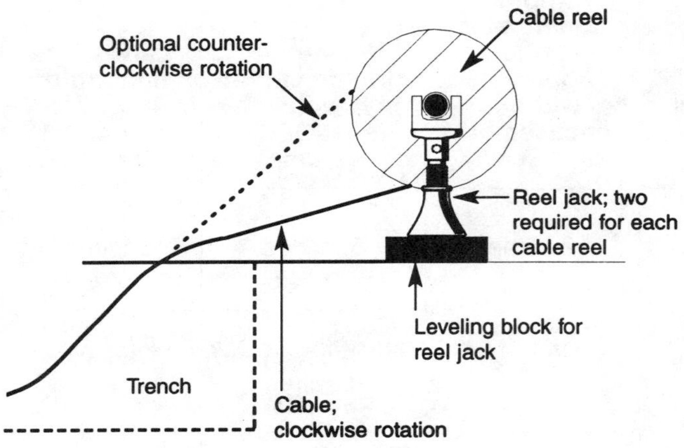
Cable reel
Optional counter-
clockwise rotation
Reel jack; two
required for each
cable reel
Leveling block for
reel jack
Trench
Cable;
clockwise rotation

Figure 2-36: *Method of setting up cable reel for underground pull*

Where several single conductors are installed, more than one reel is set up in a manner described above. Then cable is pulled from all the reels simultaneously as shown in Figure 2-37.

For longer runs, or where the larger sizes of cable are installed, the weight of the cable dictates a different method. In this case, the end of the cable is secured at one end of the trench, while the reel is attached to a tractor or backhoe and arranged so that the reel will rotate freely. The tractor or backhoe then runs along the trench — either at the trench edge or straddles it — and the cable is paid out by workers and allowed to fall into the trench.

Once installed, another layer of sand may be placed over the cable for protection against sharp rocks; the trench is then backfilled. A treated wooden plank may also be used for cable protection; a yellow warning ribbon — designed for the purpose — is also a good idea; both are shown in Figure 2-34.

Duct Systems

By definition, a duct is a single enclosed raceway through which conductors or cables are pulled. One or more ducts in a single trench is usually referred to as a duct bank. A duct system provides a safe passageway for power lines, communication cables, or both.

Depending upon the wiring system and the soil conditions, a duct bank may be placed in a trench and covered with earth or enclosed in concrete. Underground duct systems also include manholes, handholes, transformer vaults, and risers.

Manholes are set at various intervals in an underground duct system to facilitate pulling conductors or cables when first installed, and to allow for testing and maintenance later on. Access to manholes is provided through *throats* extending from the manhole compartment to the surface (ground level). At ground level, a manhole cover closes off the manhole area tightly.

In general, underground cable runs normally terminate at a manhole, where they are spliced to another length of cable. Manholes are sometimes constructed of brick and concrete. Most, however, are prefabricated, reinforced concrete, made in two parts — the base and the throat — for quicker installation. Their design provides room for workers to carry out all appropriate activities inside them, and they are also provided with a means for drainage. See Figure 2-38.

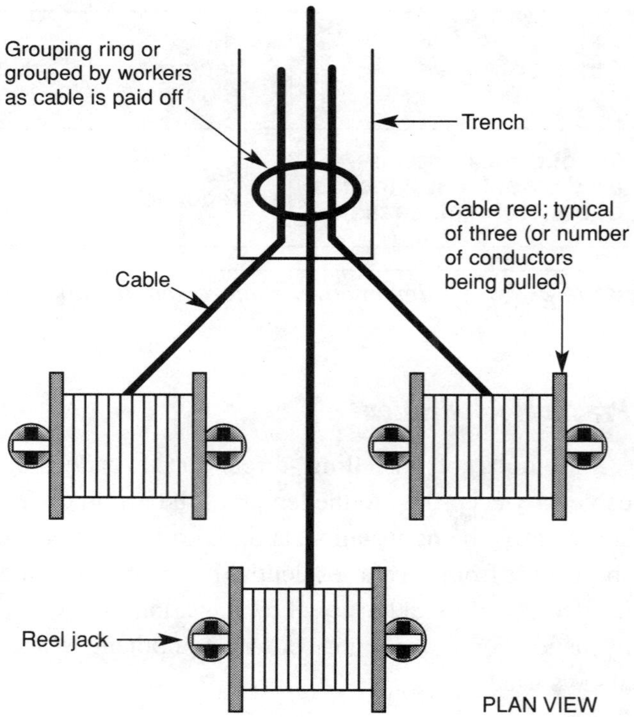
Grouping ring or
grouped by workers
as cable is paid off
Trench
Cable reel; typical
of three (or number
of conductors
being pulled)
Cable
Reel jack
PLAN VIEW

Figure 2-37: *Method of pulling three conductors at once for direct burial in an open trench*

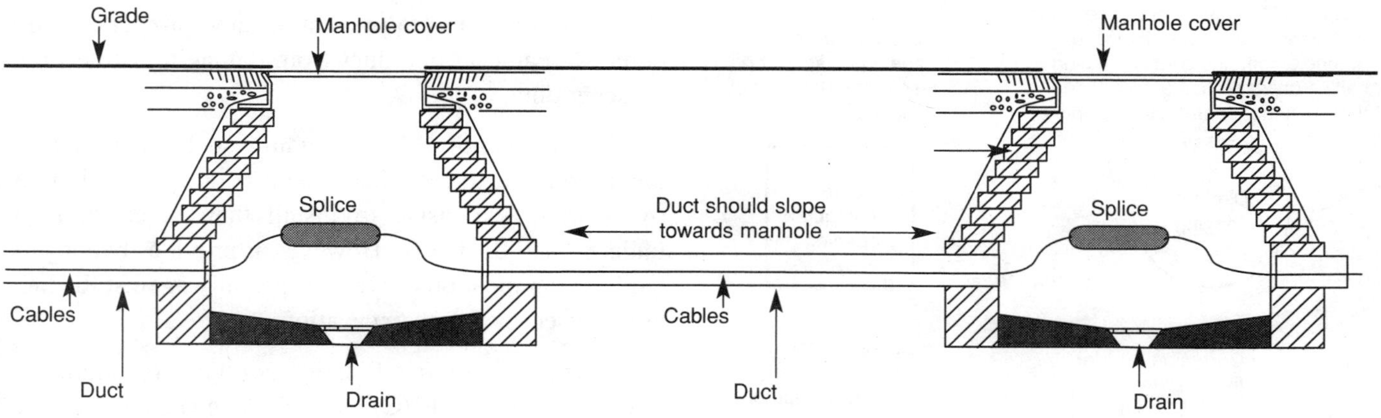
Grade
Manhole cover
Manhole cover
Splice
Duct should slope towards manhole
Splice
Cables
Cables
Duct
Drain
Duct
Drain

Figure 2-38: *Cross section of typical underground duct system running between two manholes*

There are three basic designs of manholes: two-way, three-way, and four-way. In a two-way manhole, ducts and cables enter and leave in two directions. A three-way manhole is similar to a two-way manhole, except that one additional duct/cable run leaves the manhole. Four duct/cable runs are installed in a four-way manhole. See Figure 2-39. Also see Figure 2-40 for specifications of a typical manhole.

Transformer vaults house power transformers, voltage regulators, network protectors, meters and circuit breakers. Other cables end at a substation or terminate as risers — connecting to overhead lines by means of a pothead. See Chapter 10 of this book.

Types Of Ducts

Ducts for use in underground electrical systems are made of fiber, vitrified tile, metal conduit, plastic or poured concrete. In some existing installations, the worker may find that asbestos/cement ducts have been used. In most areas, a contractor must be certified before removing or disturbing asbestos ductwork, and then extreme caution must be practiced at all times.

The inside diameter of ducts for specific installations is determined by the size of the cable that the ducts will house. Sizes from 2 to 6 inches are common.

Fiber duct: Fiber duct is made with wood pulp and various chemicals to provide a lightweight raceway that will resist rotting. It can be used enclosed in a concrete

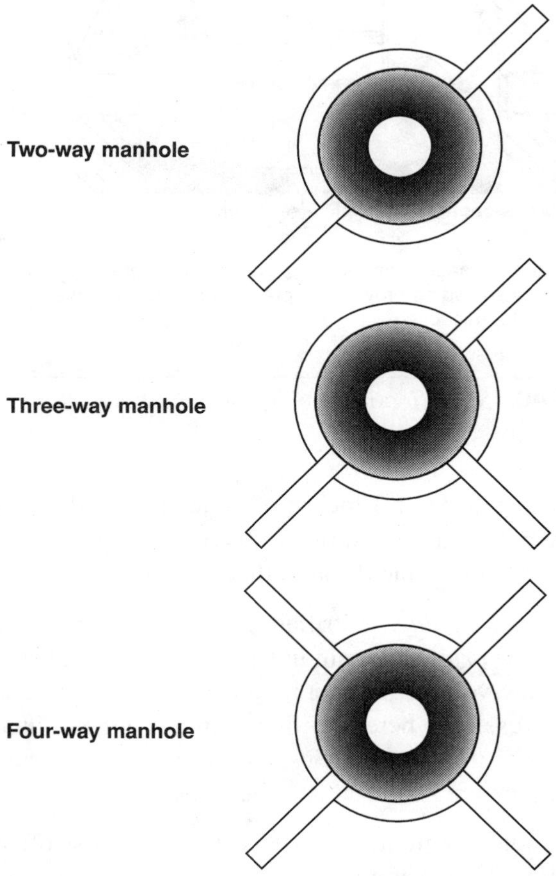
Two-way manhole
Three-way manhole
Four-way manhole

Figure 2-39: *Plan view of two-, three-, and four-way manholes*

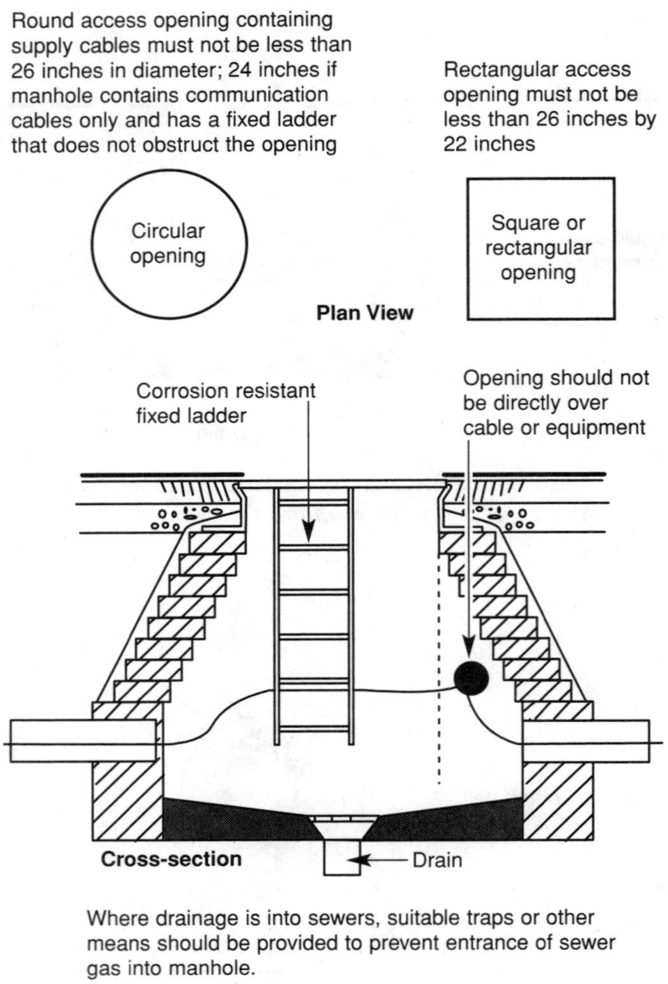
Round access opening containing supply cables must not be less than 26 inches in diameter; 24 inches if manhole contains communication cables only and has a fixed ladder that does not obstruct the opening
Rectangular access opening must not be less than 26 inches by 22 inches
Circular opening
Plan View
Square or rectangular opening
Corrosion resistant fixed ladder
Opening should not be directly over cable or equipment
Cross-section
Drain
Where drainage is into sewers, suitable traps or other means should be provided to prevent entrance of sewer gas into manhole.

Figure 2-40: *Specifications of typical manhole*

envelope with at least 3 inches of concrete on all sides. The extremely smooth interior walls of this type of duct facilitates pulling cable through them.

Vitrified clay duct: Vitrified clay tile is sometimes called *hollow brick*. Its main use is in underground systems for low-voltage and communication cables and is especially useful where the duct run must be routed around underground obstacles, because the individual pieces of duct are shorter than other types.

The four-way multiple duct is the type most often encountered. However, vitrified duct is available in sizes up to 16 ducts in one bank. The square ducts are usually 3½ inches in diameter, while round ducts vary from 3½ to 4½ inches.

When vitrified clay ducts are installed, their joints should be staggered to prevent a flame or spark from a defective cable in one duct from damaging cable in an adjacent duct.

Metal conduit: Metal conduit, such as iron, rigid metal conduit, intermediate metal conduit, etc., is relatively more expensive to install than other kinds of underground ductwork. However, it provides better protection than most other types, especially against the hazards caused by future excavation.

Plastic conduit: Plastic conduit is made of polyvinylchloride (PVC), polyethylene (PE), or styrene. Since they are available in lengths up to 30 feet, fewer couplings are needed than with many types of duct systems. PVC conduit is currently very popular for underground electrical systems since it is light in weight, relatively inexpensive, and requires less labor to install.

Monolithic concrete ducts: This type of system is poured at the job site. Multiple duct lines can be formed using tubing cores or spacers. The cores may be removed after the concrete has set. Although relatively expensive, this system has the advantage of creating a very clean duct interior with no residue that can decay. It is also useful when curves or bends in duct systems are necessary.

Cable-in-duct: This is another popular duct type that offers a reduction in labor cost when installed. It is manufactured with cables already installed. Both the duct and the cable it contains is shipped on a reel to facilitate installing the entire system with ease. Once installed, the cables can be withdrawn or replaced in the future if it should become necessary.

Installing Underground Duct

The selection of high-voltage cables and their installation in underground ductwork is not part of the requirements of the *NEC*. However, Appendix B of the *NEC* provides installation criteria "for information purposes only." Furthermore, the National Electrical Safety Code (NESC) covers regulations governing high-voltage underground installations in detail. Section 33 of the NESC covers supply cable including detailed requirements of conductors, insulation, sheaths, jackets, and shielding. Section 34 also covers underground installa-

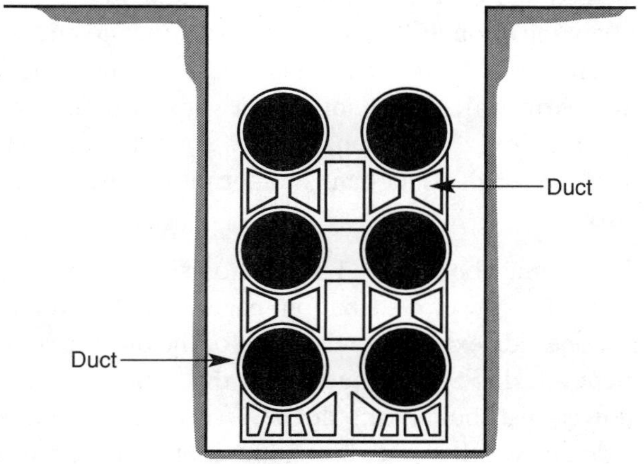
Duct
Duct

Figure 2-41: *Duct bank with spacer separation*

tions. Both of these Sections should be studied by anyone involved in underground duct systems — either design or actual installations.

In practical applications of underground duct systems, present needs and potential growth are both considered. Construction aspects such as trenching and pouring concrete are cost factors, so it is economically sound to provide for future growth as part of an original installation. Often the number of conduits laid is twice the number needed for present usage.

In general, there should be at least 1 to 3 inches of earth or concrete between adjacent conduits containing power cables. This insulation will ensure that the heat radiated by one line will not affect the surrounding ducts. Heat will cause insulation to deteriorate faster and, in general, the hotter the cable, the smaller the amount of load it will carry. Consequently, duct banks should be designed so that heat from conductors is dissipated into either the surrounding concrete envelope or earth.

During the installation of a duct system, spacers should be used to hold the ducts in place while concrete is being poured. Figures 2-41 and 2-42 show how spacers are arranged prior to a concrete pour.

Duct Installation

Excavation is the first order of business; that is, a backhoe or other digging apparatus is employed to dig the trenches and ground openings for manholes. Manholes are then constructed at various intervals throughout the "run." In the case of two-piece, prefabricated manholes, only the bases are installed prior to installing the ductwork.

The bottom of the trenches must be flat and compacted prior to installing the ducts. This is to prevent the trench from settling and putting stress on the duct banks. In most cases, if a concrete envelope is to be used, the trench is first filled with 3 inches of concrete and finished at the appropriate grade. Once hardened, duct lines are then placed in the trench utilizing fiber or plastic spacers at various intervals along the duct run. This process is repeated until the final row of duct is laid and imbedded.

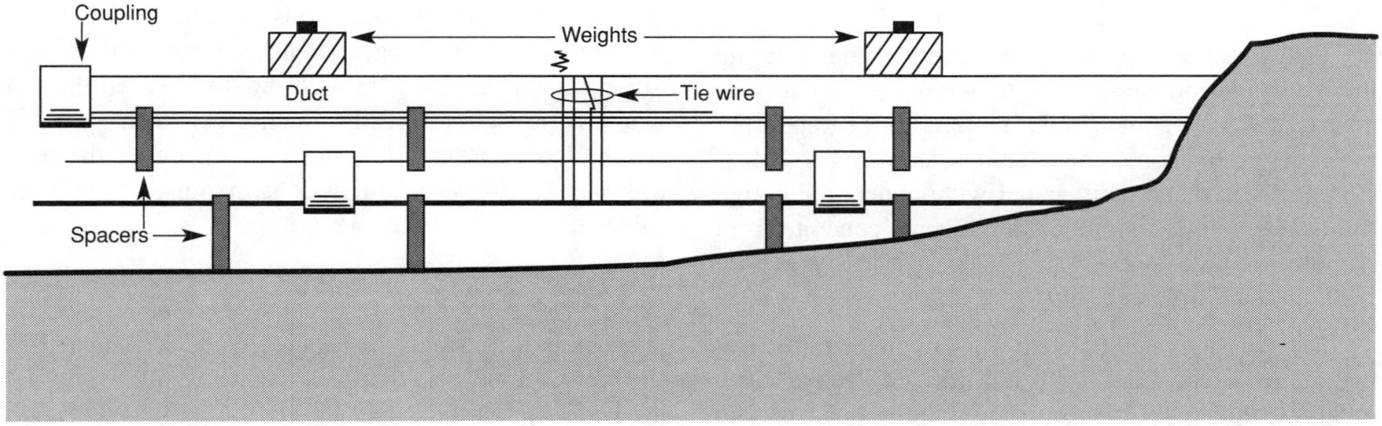
Coupling
Weights
Duct
Tie wire
Spacers

Figure 2-42: *Duct bank, prior to pouring concrete, using spacers, weight, and twine to keep duct in place during the pouring operation*

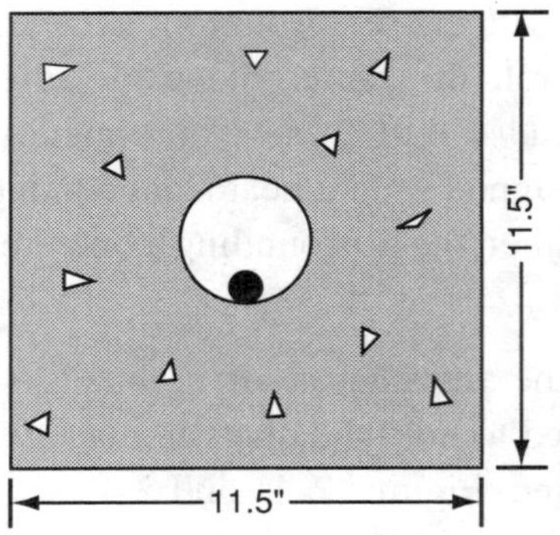
11.5"
11.5"

Figure 2-43: *Electrical duct bank for one electrical duct*

During the pouring process, and before the concrete sets, it is important to make absolutely certain that the ducts line up at the joints and couplings. All duct installations must join in a manner sufficient to prevent solid matter from entering the raceway. Furthermore, the joints must form a sufficiently continuous smooth interior surface between joining duct sections so that supply cable will not be damaged when pulled past the joint.

To ensure alignment, various dowels, mandrels, and scrapers may be used. The mandrel, for example, must be long enough to reach back two joints so that at least three sections of duct will be aligned. A leather or rubber washer is attached to the mandrel which serves to clean out the conduit as the mandrel is pulled through.

A wire brush — slightly larger than the interior duct diameter — or a scraper should be pulled through the ducts after the concrete has been poured. This will eliminate any cement or dirt that has penetrated the duct at the joints.

One method used to prevent concrete from entering the joints during pouring is to wrap each joint with coarse muslin or some similar material prior to pouring the concrete envelope. The muslin is dampened to help it stick to the cut and then coated with cement. Although time-consuming, this procedure prevents concrete from entering into the raceway. It is also best to stagger the joints in a multiple duct run.

Soil conditions will dictate whether concrete encasement is required. If the soil is not firm, concrete encasement is mandatory. Concrete is also required with certain types of duct lines that are not able to withstand the pressure of an earth covering. If the soil is firm, and concrete encasement is still desired (or specified), the trench need only be wide enough for the ducts and concrete encasement. The concrete is then poured between the conduit and the earth wall. If the soil is not very firm and concrete is required, 3 additional inches should be allowed on each side of the duct banks to permit the use of concrete forms.

Ducts may be grouped in any of several different ways, but for power distribution, each duct should have at least one side exposed to the earth or the outside of the concrete envelope. This means that the pattern of ducts for power distribution should be restricted to either a two-conduit width or a two-conduit depth. This permits the heat generated by power transmission to dissipate into the surrounding earth. In other words, ducts for power distribution should not be completely surrounded by other ducts. When this type of situation exists, the inner ducts may be referred to as *dead ducts*. The heat that these ducts radiate is not dissipated as fast as from the ducts surrounding them. While not suited for power cable, these dead ducts may be used for street lighting, control cable, or communication cable. The heat generated by these types of cables is relatively low, so the ducts can be arranged in any convenient configuration.

Figures 2-43 through 2-46 show duct-bank configurations and dimensions given in the *NEC*, and should serve as a guide.

Pulling Cable

There are several preliminary operations prior to pulling cable through a duct bank. *Rodding* involves the use of many short wooden rods or dowels which are joined together on a long, flexible steel rod, stiff enough to be pushed or pulled through the duct. When the rod reaches the far end of the duct, steel "fish" wire is attached to the near end. Then the rod is pulled through, followed by the wire. This wire is then used to pull the cables through the ductwork.

Rigging of some kind must be set up to pull the cable through the duct. In general, a gripping device is attached to the ends of the cable. This device consists of flexible steel mesh so that the harder the pull, the tighter the device grips the cable. Before the pulling operation is started, however, a wire lubrication known as "soap" in the trade is freely applied to the ends of the cable to

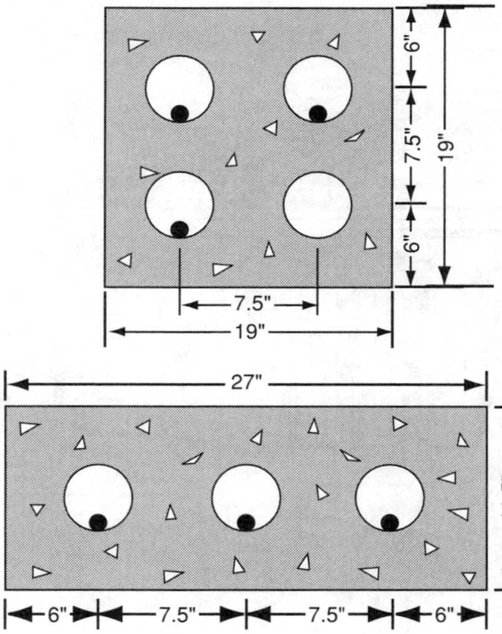
6"
7.5"
6"
19"
7.5"
19"
27"
11.5"
6"
7.5"
7.5"
6"

Figure 2-44: *Two arrangements for three electrical ducts*

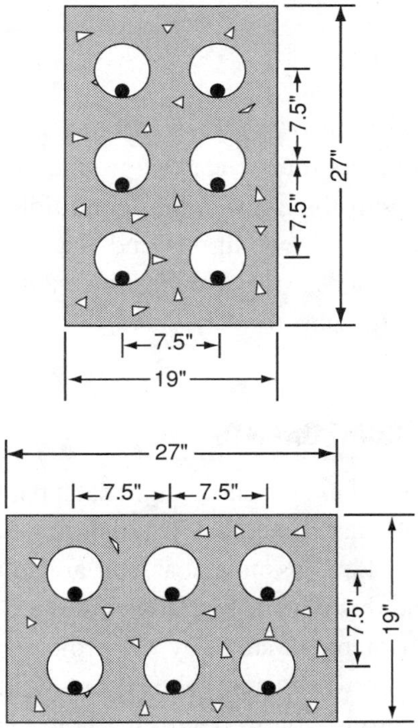
7.5"
7.5"
27"
7.5"
19"
27"
7.5"
7.5"
7.5"
19"

Figure 2-45: *Two arrangements for duct bank containing six electrical ducts*

reduce friction during the pull. A winch or other pulling mechanism is used to move the cable through the duct. More soap or wire lubricant is applied at regular intervals as the cable enters the ducts.

Part 341A of the National Electric Safety Code covers control of bending, pulling, tensions and sidewall pressures during the installation of cable. This section also covers cleaning foreign material from ducts, selection of cable lubricants that will not damage any part of the installation, and restraint of cables in sloping or vertical runs to prevent them from moving downhill. It further specifies that power-supply cables, control cables, and communication cables should not be installed in the same duct unless the same utility maintains or operates them.

SERVICE CLEARANCES

Electricians working mostly on residential and commercial projects will be involved with their service installation from the power company's point of attachment to the building and then to the service equipment, including all wiring and components in between, with the possible exception of the electric meter.

The locations of the service drop, electric meter and the load center should be considered first. It is always wise to consult the local power company to obtain their

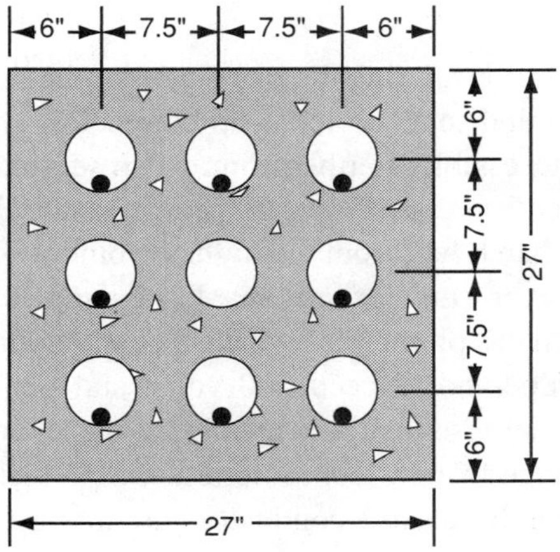
6"
7.5"
7.5"
6"
6"
7.5"
7.5"
6"
27"
27"

Figure 2-46: *Electrical duct bank with nine ducts. The center duct, however, should not be used for power transmission; only control or communication cables*

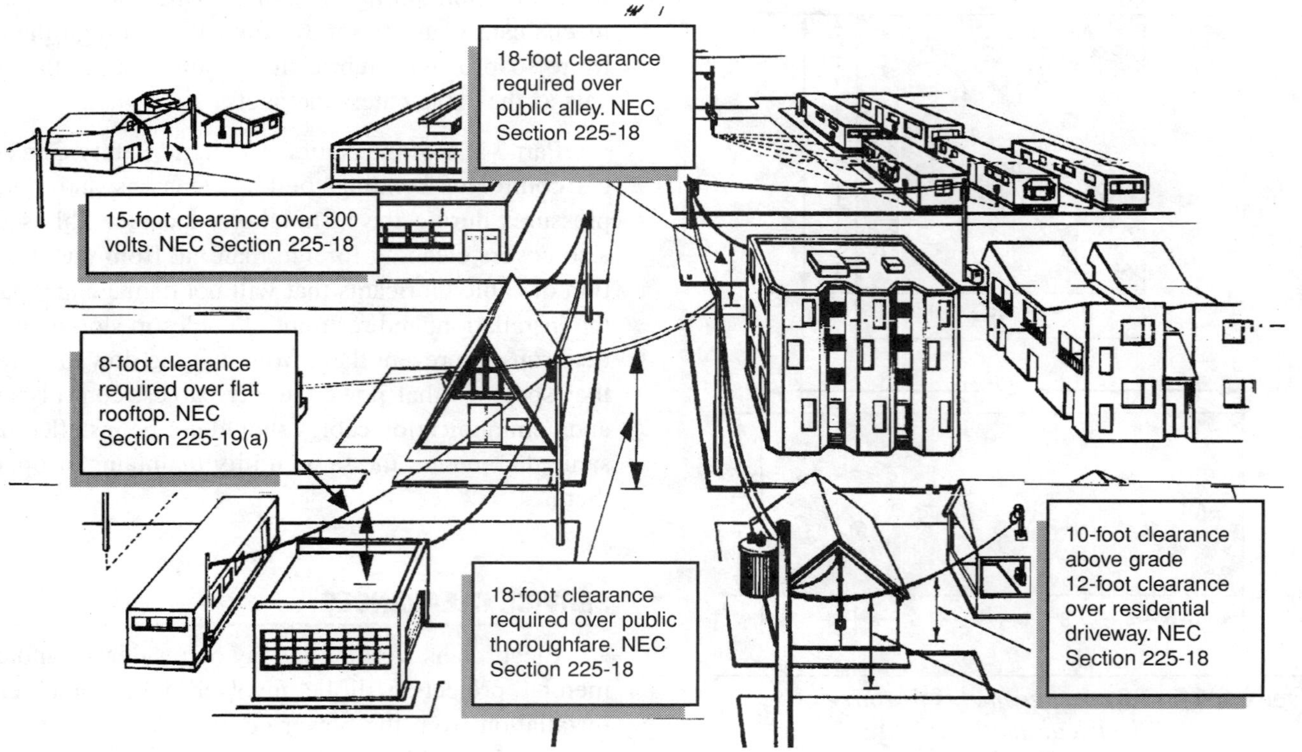
18-foot clearance required over public alley. NEC Section 225-18
15-foot clearance over 300 volts. NEC Section 225-18
8-foot clearance required over flat rooftop. NEC Section 225-19(a)
18-foot clearance required over public thoroughfare. NEC Section 225-18
10-foot clearance above grade
12-foot clearance over residential driveway. NEC Section 225-18

Figure 2-47: *Summary of NEC clearances for service drops*

recommendations; where you want the service drop, and where they want it, may not coincide. A brief meeting with the power company about the location of the service drop can prevent much grief, confusion, and perhaps expense.

While considering the placement of the service drop, keep in mind that service-drop conductors must not be readily accessible. Furthermore, when service-drop conductors pass over rooftops, they must have a clearance of not less than 8 feet from the highest point of roofs. There are, however, some exceptions. For example, in residential or small commercial wiring installations where the voltage does not exceed 300 volts, and the roof has a slope of not less than 4 inches in 12 inches, a reduction in clearance from 8 feet to no less than 3 feet is permitted. When these same conductors do not pass over more than 4 feet of the overhead portion of the roof, and they terminate at a through-the-roof raceway or approved support, this clearance may be further reduced to only 18 inches.

The *NEC* also specifies the distance that service-drop conductors must clear the ground. These distances will vary with the surrounding conditions. Figure 2-47 summarizes these conditions and the required heights; that is, clearances over roofs, driveways, public roads, alleys, and the like.

SERVICE COMPONENTS

To understand the function of each part of an electric service, let's take an actual installation — a commercial retail store. Let's assume that you are in charge of the project and this is your first day on the job. The contractor's superintendent stops by the project site and hands you a set of working drawings and written specifications. It is up to you to determine how the service is to be installed. Furthermore, you will be required to compile a material list and order all necessary items to complete the service installation.

This particular project consists of a rectangular building that is a part of a shopping center complex. The concrete block walls have been erected and the building is under roof. The concrete floor will not be poured until all electrical and plumbing work has been installed. However, the permanent electric service is to be installed immediately to provide temporary power for the workers. The remaining wiring in the building will be installed later.

Consulting The Construction Documents

The first order of business is to consult the working drawings and then perhaps read the appropriate sections in the written specifications. In doing so, the floor plan of the building appears as shown in Figure 2-48. Note that the standard panel symbol is used (a solid rectangle) to indicate the location of Panel "A" — the only power panel used on this project. The panel symbol indicates that the panel is to be surface mounted on the inside rear wall of the building in the storage area. The electric meter, as well as a time clock for controlling night and outside lighting are also shown on this floor plan; the meter installed on the outside rear wall, while the time clock is installed on the inside rear wall, next to panel "A."

Notes and call-out arrows on this floor plan refer to a power-riser diagram and also a panelboard schedule on the same drawing sheet; these appear in Figures 2-49 and 2-50 respectively. This drawing sheet — showing the floor plan, power-riser diagram, and panelboard schedule — provides most of the required information so that the service can be installed to the project specifications. In most cases, electrical workers are not required to design electrical systems; rather, they are required to interpret engineer's designs. Consequently, panelboard schedules will vary with each designer. However, once you have a "feel" for interpreting electrical working drawings, you should have little difficulty in "reading" any schedules that will be encountered.

The written specifications should also be read just to make certain that no conflicts exist, and to further verify the information found on the working drawings. Part of a sample specification appears in Figure 2-51.

From the information obtained from the drawings and specifications, we know that the service for this project is single-phase, 3-wire, 120/240-volt, 200 amperes. The main panel (panel "A") is a surface-mounted Square D Type NQO (or equivalent) with a 200-ampere main circuit breaker. Furthermore, we can

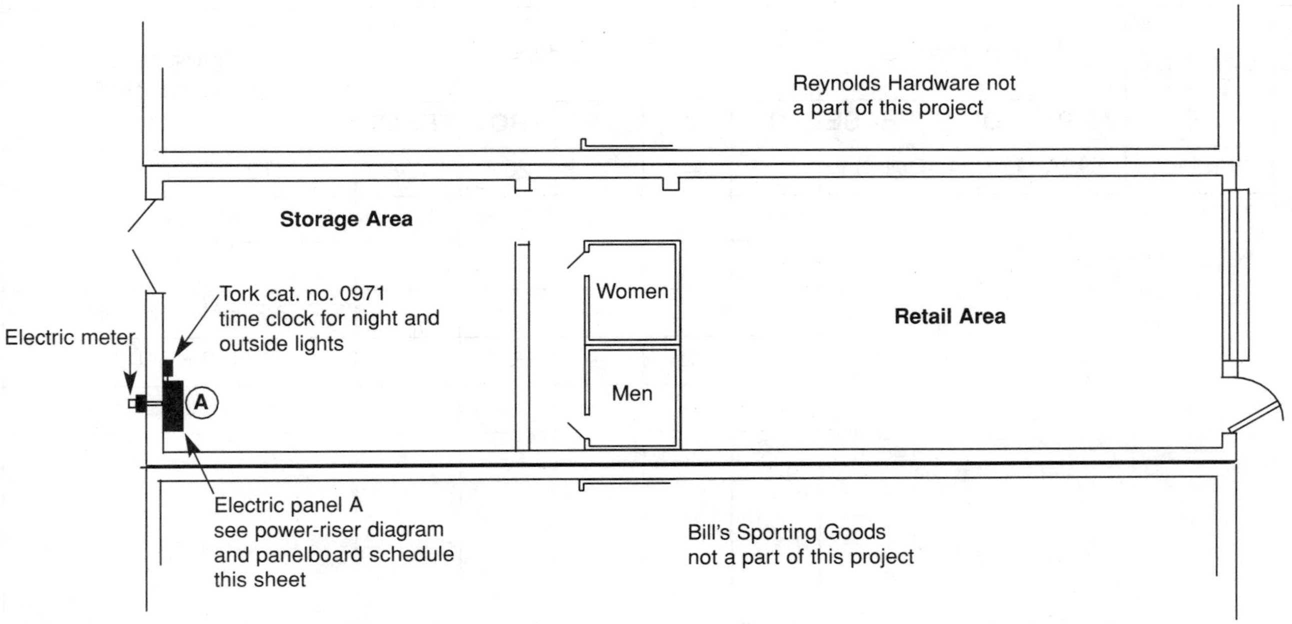
Reynolds Hardware not
a part of this project
Storage Area
Tork cat. no. 0971
time clock for night and
outside lights
Electric meter
A
Women
Men
Retail Area
Electric panel A
see power-riser diagram
and panelboard schedule
this sheet
Bill's Sporting Goods
not a part of this project

Figure 2-48: *Floor plan of a small commercial facility*

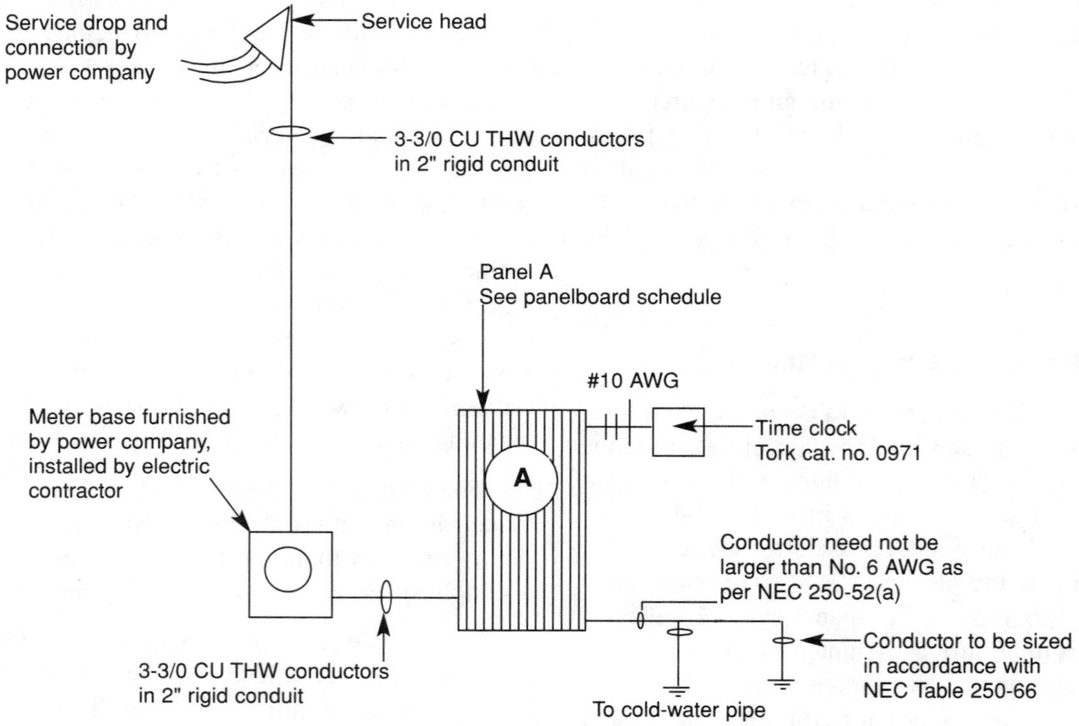
Service drop and
connection by
power company
Service head
3-3/0 CU THW conductors
in 2" rigid conduit
Panel A
See panelboard schedule
#10 AWG
Time clock
Tork cat. no. 0971
Meter base furnished
by power company,
installed by electric
contractor
A
Conductor need not be
larger than No. 6 AWG as
per NEC 250-52(a)
Conductor to be sized
in accordance with
NEC Table 250-66
3-3/0 CU THW conductors
in 2" rigid conduit
To cold-water pipe

Figure 2-49: *Power-riser diagram corresponding to the floor plan in Figure 2-48*

PANELBOARD SCHEDULE

PANEL NO.	CABINET TYPE	PANEL MAINS			BRANCHES					ITEMS FED OR REMARKS
		AMPS	VOLTS	PHASE	1P	2P	3P	PROT.	FRAME	
A	SURFACE	200A	120/240	1Ø, 3-W	12	—	—	20A	70A	LTS., RECEPTS, W.C.
SQUARE "D" TYPE NQO					—	1	—	60A	100A	CONDENSING UNIT
200A MAIN CIRCUIT BREAKER					—	1	—	30A	70A	WATER HEATER
					—	1	—	20A	↓	AIR-HANDLING UNIT
					—	2	—	20A	↓	TOILET HEATERS
					8	—	—	—	↓	PROVISIONS ONLY

Figure 2-50: *Panelboard schedule corresponding to the floor plan in Figure 2-48*

PANELBOARDS — CIRCUIT BREAKER

A. GENERAL

Furnish and install circuit-breaker panelboards as indicated in the panelboard schedule and where shown on the drawings. The panelboard shall be dead front safety type equipped with molded case circuit breakers and shall be the type as listed in the panelboard schedule: Service entrance panelboards shall include a full capacity box bonding strap and approved for service entrance. The acceptable manufacturers of the panelboards are ITE, General Electric, Culter-Hammer, and Square D, provided that they are fully equal to the type listed on the drawings. The panelboard shall be listed by Underwriters' Laboratories and bear the UL label.

B. CIRCUIT BREAKERS

Provide molded case circuit breakers of frame, trip rating and interrupting capacity as shown on the schedule. Also, provide the number of spaces for future circuit breakers as shown in the schedule. The circuit breakers shall be quick-make, quick-break, thermal-magnetic, trip indicating and have common trip on all multipole breakers with internal tie mechanism.

C. WIRING TERMINALS

Terminals for feeder conductors to the panelboard mains and neutral shall be suitable for the type of conductor specified. Terminals for branch circuit wiring, both breaker and neutral, shall be suitable for the type of conductor specified.

D. CABINETS AND FRONTS

The panelboard bus assembly shall be enclosed in a steel cabinet. The size of the wiring gutters and gauge of steel shall be in accordance with NEMA Standards. The box shall be fabricated from galvanized steel or equivalent rust resistant steel. Fronts shall include door and have flush, brushed stainless steel, spring-loaded door pulls. The flush lock shall not protrude beyond the front of the

Figure 2-51: *Part of a sample panelboard specification*

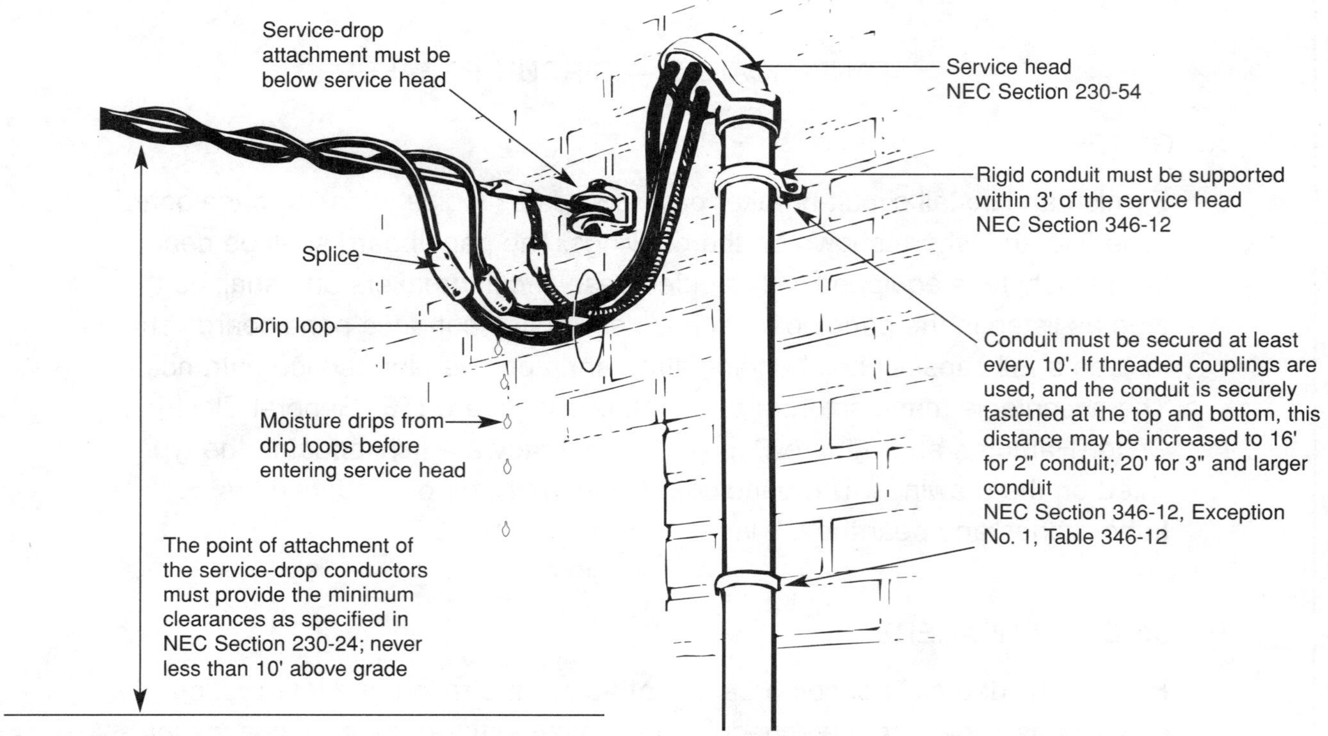
Service-drop
attachment must be
below service head
Splice
Drip loop
Moisture drips from
drip loops before
entering service head
The point of attachment of
the service-drop conductors
must provide the minimum
clearances as specified in
NEC Section 230-24; never
less than 10' above grade
Service head
NEC Section 230-54
Rigid conduit must be supported
within 3' of the service head
NEC Section 346-12
Conduit must be secured at least
every 10'. If threaded couplings are
used, and the conduit is securely
fastened at the top and bottom, this
distance may be increased to 16'
for 2" conduit; 20' for 3" and larger
conduit
NEC Section 346-12, Exception
No. 1, Table 346-12

Figure 2-52: *Service head and related components*

determine the number of spaces required in the panel by totaling the number of circuit breakers listed in the panelboard schedule as follows:

12 1-pole, 20A breakers =	12 spaces
1 2-pole, 60A breaker =	2 spaces
1 2-pole, 30A breaker =	2 spaces
3 2-pole, 20A breakers =	6 spaces
8 Provisions only =	8 spaces
Total =	**30 spaces**

Therefore, a surface-mounted panel — Square D Type NQO with 200 ampere main circuit breaker and provisions for 30 spaces — can be ordered. The required circuit breakers should also be ordered and installed at the same time. This will meet with project specifications.

Service Head

Referring again to Figure 2-49, let's start at the top of the service riser. The first item shown is the service head, sometimes called "weatherhead." Since the service raceway in our example consists of 2-inch rigid conduit, *NEC* Section 230-54 requires this conduit (raceway) to be equipped with a raintight service head at the point of connection to the service-drop conductors.

A service head (Figure 2-52) is a fitting that prevents water from entering the service raceway. This is accomplished by bending the service conductors (contained in the raceway) downward as they exit from the service head so that any water or moisture will drip from the outside conductors before entering the service head. These conductors are also protected by a plastic or fiber strain insulator or bushing — placed at the entrance of the service head — to separate the service conductors as required by *NEC* Section 230-54(e). Two types of service heads are in common use: one type has internal threads that enable the service head to be screwed directly onto the conduit; the other type utilizes a clamp with retaining screws. In this latter type, the service head is placed on top of the service raceways and the clamp tightened with the retaining screws.

Further protection from water and moisture is provided by drip loops as required in *NEC* Section 230-54(f). Service heads are required to be located above the service-drop attachment. Drip loops are then formed where the service-drop conductors are connected to the service conductors and these drip loops must be located below the service head. Figure 2-52 shows how drip loops prevent water from entering the service raceway; that is, water will not flow uphill into the service head, so the water drips from the conductors at the lowest point of the drip loop.

The service-entrance conductors must have a minimum length of 3.5 feet after they leave the service head. This is to ensure a good drip loop and to give adequate length for splicing onto the service drop.

Service-entrance Conductors

The size 3/0 AWG, Type THW conductors shown in the power-riser diagram in Figure 2-49 are service-entrance conductors. These conductors are run from the main disconnect breaker, through the meter, to the service head, and terminate with splices onto the service drop. The conductors must not be spliced at any place between these points except for the following:

- Clamped or bolted connections in metering equipment.
- Where service-entrance conductors are tapped to supply two to six disconnecting means that are grouped at a common location.
- Where service conductors are extended from a service drop to an outside meter location and returned to connect to the service-entrance conductors of an existing installation.
- Where the service-entrance conductors consist of busway, connections are permitted as required to assemble the various section and fittings.

Service conductors are normally installed in two different ways: in a raceway system or in a cable assembly. In our sample, the conductors are installed in 2-inch rigid conduit and extend from the service head down to the threaded weatherproof hub on top of the meter base. The *NEC* permits the conductors to be spliced at this point; that is, connected to the bolted terminals on the meter base. However, no splices are permitted from the service head to the meter base.

Service Equipment

Equipment and components falling under this heading include the main disconnect switch or breaker, circuit breakers, fuses, and other necessary items to meter, control, and cut off the power supply. See Chapter 3 of this book.

Metering Equipment

An electric meter is used by the power company to determine the amount of electrical energy consumed by the customer.

Energy is the product of power (kilowatts) and time (hours). The type of meter connected to most residential and small commercial occupancies provides a reading in kilowatt-hours. For example, if the meter reads a usage of 500 watts for a period of 6 hours, it would register (.5 x 6 =) 3 kilowatt-hours.

There are several different types of metering devices in use. The type used on our sample building is known as the feed-through type. This type of meter is used mostly for services up to 200 amperes, although feed-through meters up to 400 amperes are not uncommon in many locations. Services rated above 400 amperes will almost always use separate current transformers enclosed in a current-transformer cabinet (CT cabinet). Current transformers are discussed in greater detail in Chapter 3.

A typical watt-hour meter consists of a combination of coils, conductors, and gears — all encased in a housing as shown in Figure 2-53. The coils are constructed on the same principle as a split-phase induction motor, in that the stationary current coil and the voltage coil are placed so that they produce a rotating magnetic field. The disc near the center of the meter is exposed to the rotating magnetic field. The torque applied to the disc is proportional to the power in the circuit, and the braking action of the eddy currents in the disc makes the speed of the rotation proportional to the rate at which the

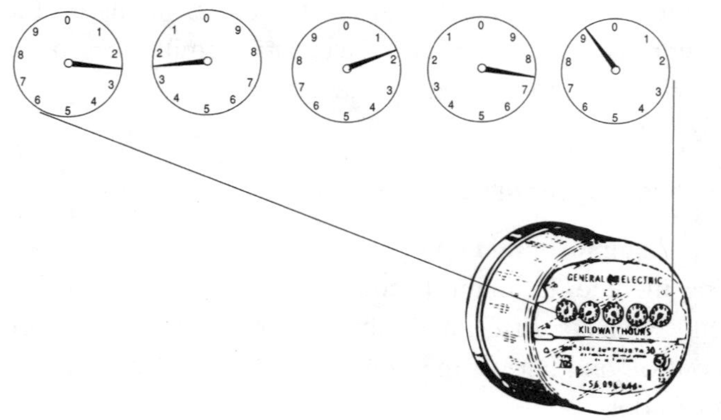

Figure 2-53: *Typical watt-hour meter*

power is consumed. The disc, through a train of gears, moves the pointers on the register dials to record the amount of power used directly in kilowatt hours (kWh).

Most watt-hour meters utilize five dials; again, see Figure 2-53. The dial farthest to the right on the meter counts the kilowatt hours singly. The second dial from the right counts by tens, the third dial by hundreds, the fourth dial from the right by thousands, and the left-hand dial by ten-thousands. The dials may seem a little strange at first, but are actually very simple to read. The number which the dial has passed is the reading. For example, look at the dial on the very left in Figure 2-53. Note that the pointer is about halfway between the number 2 and the number 3. Since it has passed the number 2, but has not yet reached number 3, the dial reading is "2." The same is true of the second dial from the left; that is the pointer is between the number 2 and 3. Consequently, the reading of this dial is also "2." Following this same procedure, the reading in the illustration (Figure 2-53) is 2, 2, 1, 7, 9, or 22,179 kilowatt-hours.

Although knowing how to read an electric meter is interesting, most electricians will be involved only with installing the meter base and making the connections therein. Once these connections are made and inspected, the local power company will install and seal the meter.

Meter bases should be installed securely with anchors sufficient to hold the weight of the meter as well as the raceway system resting upon the meter base. In our example, this base must support the 2-inch conduit, service head, and the copper conductors. Although the conduit will be supported with conduit straps, most of the weight will rest upon the meter base; the straps are used mainly to keep the conduit from moving sideways in this example.

Most single-phase, feed-through meter bases are arranged as shown in Figure 2-54. The ungrounded service conductors from the service drop terminate in the top terminals. These conductors are once again picked up from the bottom terminals. Clips are provided on these terminals to clamp-in the meter itself, allowing current from the ungrounded conductors to pass through the meter for a reading. Since the grounded conductor (neutral) is not metered, one terminal is provided for both the incoming and outgoing conductors.

Two other possible service arrangements for the service in Figure 2-49 are shown in Figure 2-55. All three arrangements actually accomplish the same thing; just in a different manner.

MATERIAL TAKE-OFF

After reviewing the details of construction and the *NEC* requirements so far, most of the necessary information required for a complete material take-off of the service information has been gathered. However, a few more structural details are necessary to obtain exact quantities.

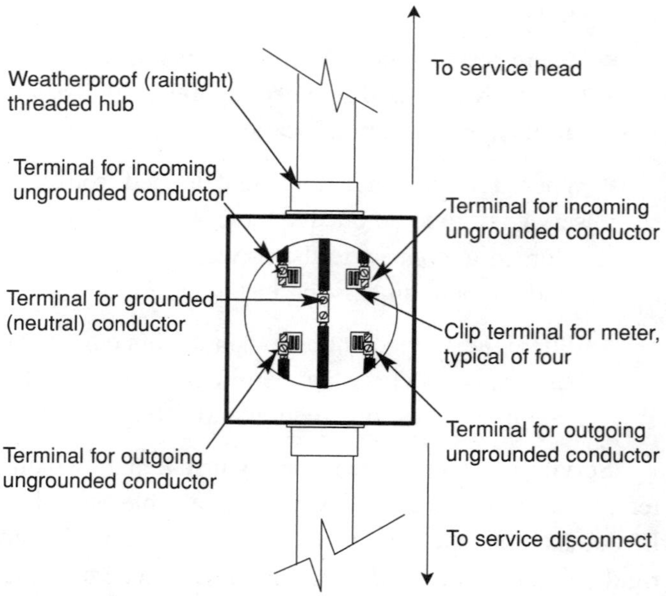
To service head
Weatherproof (raintight) threaded hub
Terminal for incoming ungrounded conductor
Terminal for incoming ungrounded conductor
Terminal for grounded (neutral) conductor
Clip terminal for meter, typical of four
Terminal for outgoing ungrounded conductor
Terminal for outgoing ungrounded conductor
To service disconnect

Figure 2-54: *Arrangement of conductors in a typical single-phase meter base*

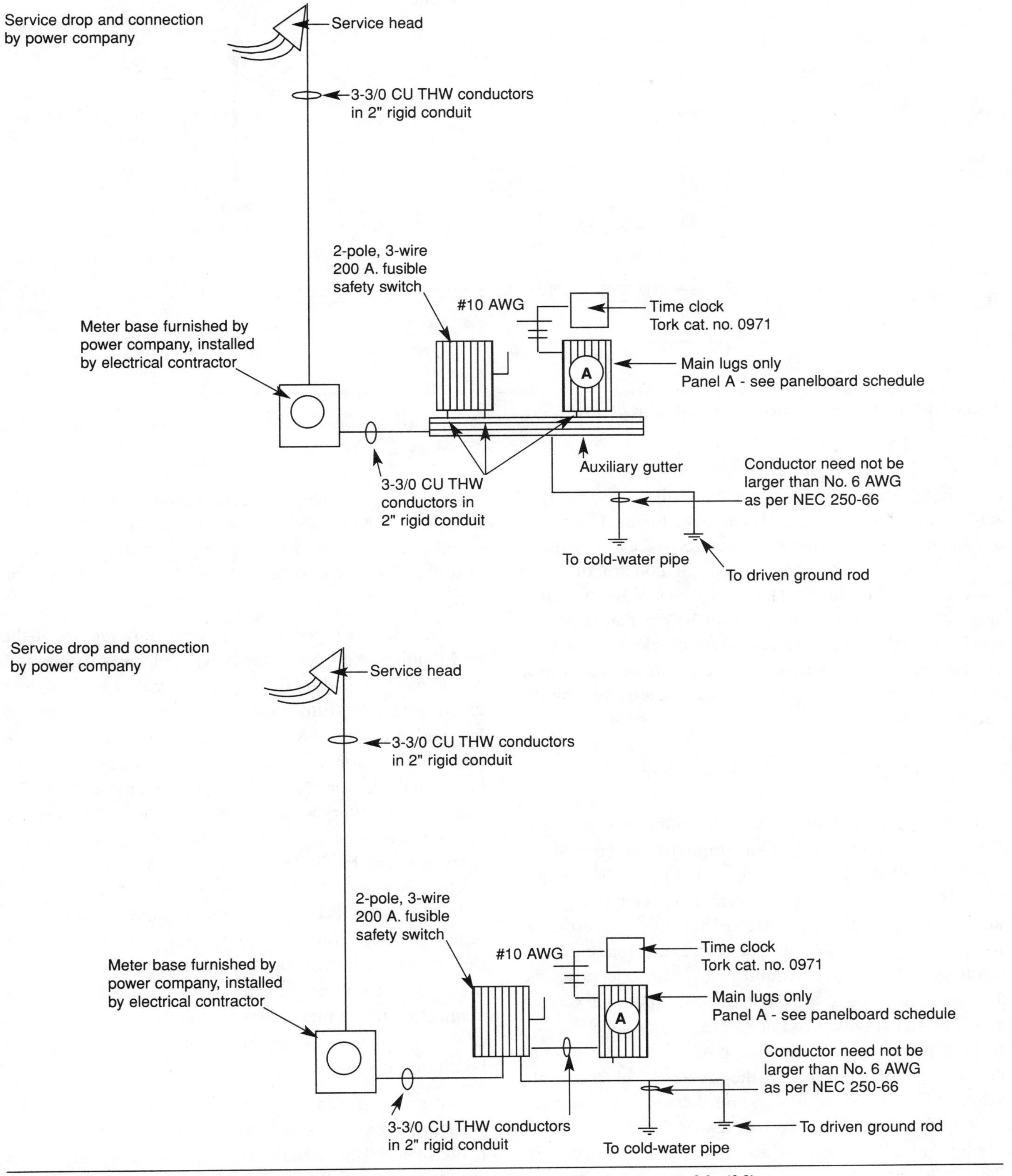

Service drop and connection by power company
Service head
3-3/0 CU THW conductors in 2" rigid conduit
2-pole, 3-wire 200 A. fusible safety switch
#10 AWG
Time clock Tork cat. no. 0971
Meter base furnished by power company, installed by electrical contractor
A
Main lugs only Panel A - see panelboard schedule
Auxiliary gutter
3-3/0 CU THW conductors in 2" rigid conduit
Conductor need not be larger than No. 6 AWG as per NEC 250-66
To cold-water pipe
To driven ground rod
Service drop and connection by power company
Service head
3-3/0 CU THW conductors in 2" rigid conduit
2-pole, 3-wire 200 A. fusible safety switch
#10 AWG
Time clock Tork cat. no. 0971
Meter base furnished by power company, installed by electrical contractor
Main lugs only Panel A - see panelboard schedule
A
Conductor need not be larger than No. 6 AWG as per NEC 250-66
3-3/0 CU THW conductors in 2" rigid conduit
To driven ground rod
To cold-water pipe

Figure 2-55: *Various possible service configurations for the sample commercial building*

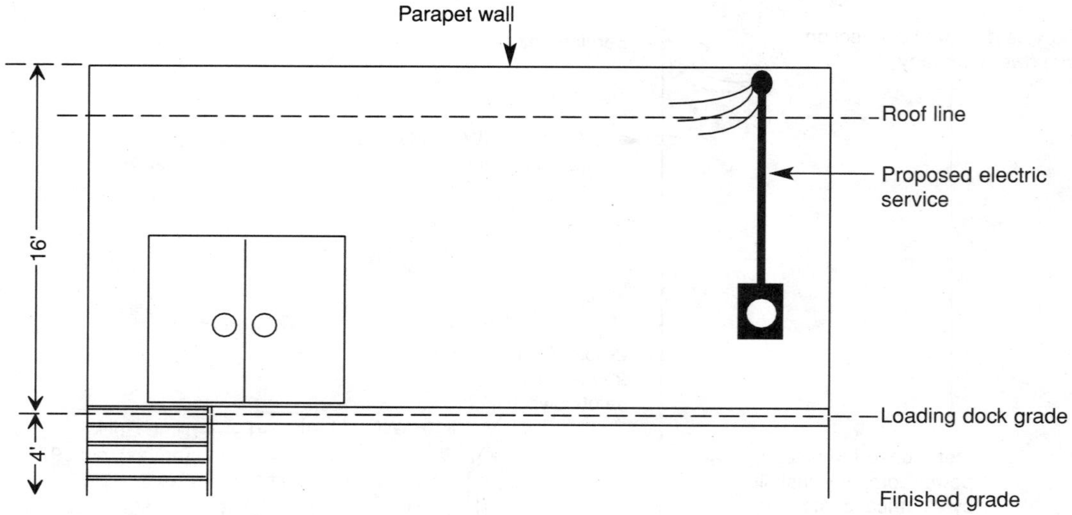
Parapet wall
Roof line
Proposed electric service
16'
4'
Loading dock grade
Finished grade

Figure 2-56: *Rear elevation of the sample building*

The sample building has a flat, built-up gravel roof with a parapet wall around the roof perimeter. The height of the back wall, where the service raceway will be installed, along with other details of construction, are shown on the architect's elevation views of the building. If no plans are available, the wall height can be measured, or since 8-inch high concrete blocks are used for the construction of the wall, one column of blocks may be counted and the height of the wall determined by the following equation:

Number of block x 8 /12 = total height in feet

A rear elevation of the sample building is shown in Figure 2-56. Note that the total height of the rear wall — from ground level — is 20 feet, including the parapet wall. However, a loading platform is 4 feet above grade, and since the meter can be read from this platform, the meter can be installed 4 feet above the loading-dock grade or 8 feet above ground level. Consequently, one 10-foot length of 2-inch conduit will put the service raceway and service head at the very top of the wall — leaving plenty of room for the power company to connect the service drop below the service head to comply with *NEC* Section 230-54(c). Furthermore, the service drop will be more than 10 feet above the deck — complying with *NEC* Section 230-26, and also Section 230-24 — Clearances.

Once the service configuration has been laid out, it is also best to verify the layout with the local power company. This is also a good time to request the meter base that will have to be installed by the electrician on the job.

This is also a good time to start thinking about the tool requirements. Since the service raceway extends 16 feet above the loading dock, workers will need a 24-foot extension ladder. Either a star drill or rotary hammer will be needed to drill the holes for the lead anchors and to penetrate the concrete block to insert the 2-inch nipple between the meter base and the panelboard on the inside wall. Cable cutters will speed up work when cutting the 3/0 service conductors and a bucket of wire-pulling lubricant will facilitate the installation of the 3/0 conductors into the 2-inch conduit. For this short length of pull, no special cable-pulling apparatus should be required. These conductors are merely cut to length, and then two workers can easily feed these conductors from the top of the service raceway down to the meter base for connection to the meter-base terminals. The same is true of the short run between the meter base and the panelboard on the opposite wall.

Much time can be saved on any project if careful planning is exercised before beginning any work. Even if the material is already on the job site, as taken from the electrical estimator's list, the electricians on the job must know how to utilize this material to the best advantage. Traditionally, more time has been wasted on electrical jobs because of material or tool shortages than any other causes. Therefore, before a project is started, make sure that all the necessary materials and tools are at hand. If not, order them immediately. Any time saved on any project — in most cases — means a savings to someone (probably the person who signs your pay check). Saving this person money usually means insuring your own job and welfare . . . and, in the same breath, such actions will show that you are one step closer to becoming a journeyman or master electrician.

A pictorial view of a typical electric service is shown ins Figure 2-57 and a summary of *NEC* installation requirements for electric services is shown in Figure 2-58. For details of the requirements in Figure 2-58, please refer to the *NEC* book.

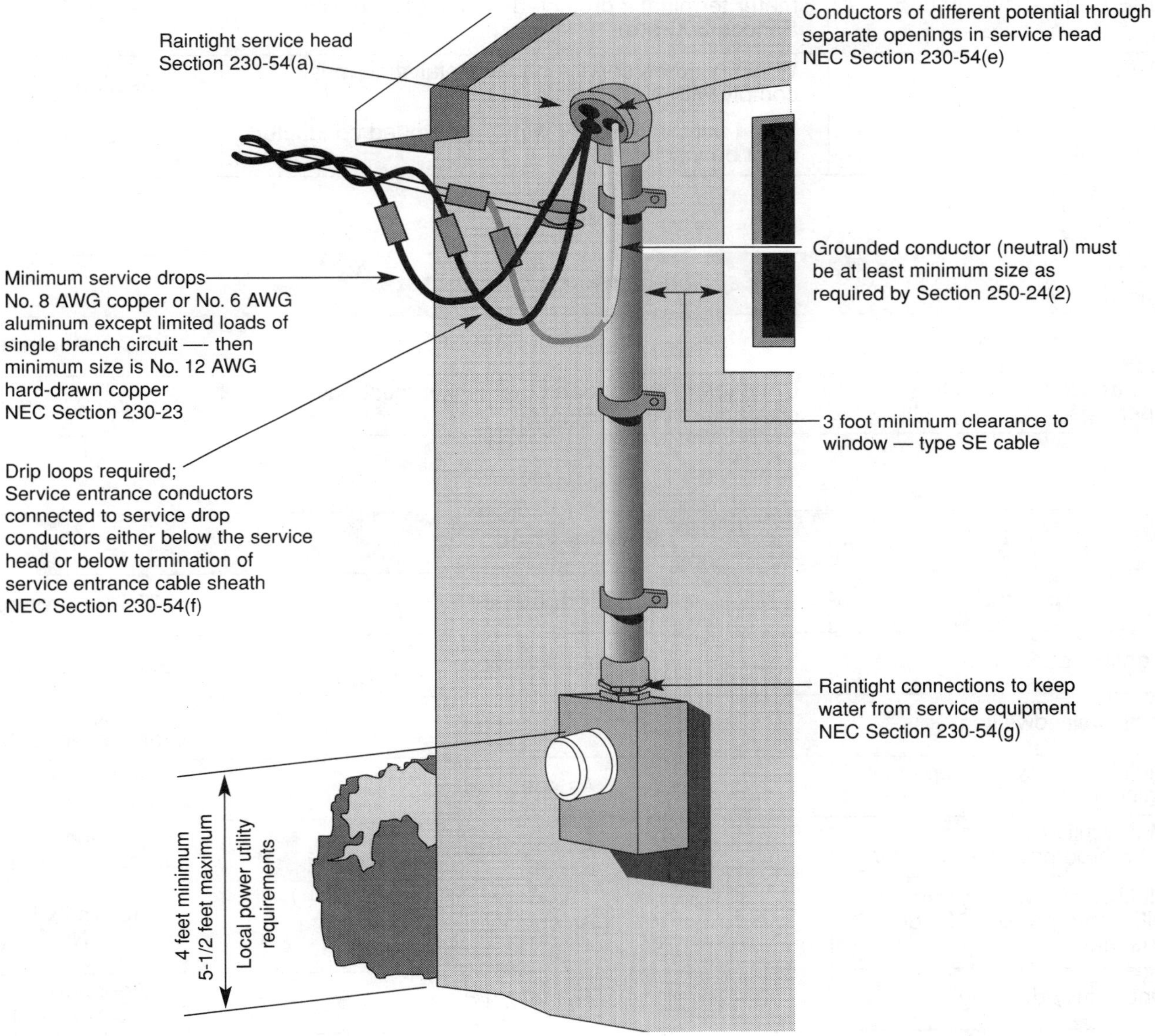
Raintight service head
Section 230-54(a)
Conductors of different potential through separate openings in service head
NEC Section 230-54(e)
Minimum service drops
No. 8 AWG copper or No. 6 AWG aluminum except limited loads of single branch circuit — then minimum size is No. 12 AWG hard-drawn copper
NEC Section 230-23
Grounded conductor (neutral) must be at least minimum size as required by Section 250-24(2)
3 foot minimum clearance to window — type SE cable
Drip loops required;
Service entrance conductors connected to service drop conductors either below the service head or below termination of service entrance cable sheath
NEC Section 230-54(f)
Raintight connections to keep water from service equipment
NEC Section 230-54(g)
4 feet minimum
5-1/2 feet maximum
Local power utility requirements

Figure 2-57: *Typical electric service with NEC installation requirements*

Service-Entrance Cable		
Application	**Requirements**	**NEC Reference**
Bends	Must be made so that protective covering will not be damaged.	Section 338-6
Interior installations	Must also comply with Parts A and B of NEC Articles 300 and 336.	Section 338-4
Uninsulated conductor	If assembly contains two or more conductors, one may be uninsulated.	Section 338-1(c)
USE cable above ground	May emerge above ground from service lateral to meter terminal if protected in accordance with Section 300-5(d).	Section 338-2
Uses	When used as service cable, installation must comply with NEC Article 230.	Section 338-2
	When used for interior wiring, grounded conductor must be insulated.	Section 338-3(a)

Service-Entrance Conductors		
Application	**Requirements**	**NEC Reference**
Clearance from building openings	Conductors without an outer jacket must have clearance of 3 ft. or more.	Section 230-24(c)

Service Loads		
Application	**Requirements**	**NEC Reference**
Lighting loads	As specified.	Section 230-3(a)
Multifamily dwelling	As specified.	Appendix D Example No. D 4(a)
Multifamily dwelling optional calculation	As specified.	Appendix D Example No. D 4(b)
Multifamily dwelling served with three phase 120/280 V	As specified.	Appendix D Example No. D 5(a)
Multifamily dwelling served with three phase 120/280 V optional	As specified.	Appendix D Example No. D 5(b)
One-family dwelling	As indicated.	Appendix D Example No. D 1(a)
One-family dwelling w/air conditioning	As indicated.	Appendix D Example No. D 1(b)

Figure 2-58: *Summary of NEC installation requirements for electric services*

Service Loads (continued)		
Application	**Requirements**	**NEC Reference**
One-family dwelling heating larger than air conditioning	As indicated.	Appendix D Example No. D 2(a)
One-family dwelling air conditioning larger than heating	As indicated.	Appendix D Example No. D 2(b)
One-family dwelling with heat pump	As indicated.	Appendix D Example No. D 2(c)
Range loads, maximum demand for	As indicated.	Appendix D Example No. D 6
Store building	As indicated.	Appendix D Example No. D 3

Service Mast		
Application	**Requirements**	**NEC Reference**
Conductors over roof	Not more than 4 ft. of service-drop conductors pass above the roof overhang.	Section 230-24 *Exception No. 3*
Drip loop	Required as specified.	Section 230-54(f)
Electric meter	As specified.	Section 90-2(b)(5)
Identified for use	All raceway fittings must be identified for use with service mast.	Section 230-28
Point of connection	Not less than 18 in.	Section 230-24(a) *Exception No. 3*
Voltage	Must not exceed 300 V.	Section 230-24 *Exception No. 2*

Figure 2-58: *Summary of NEC installation requirements for electric services (continued)*

Chapter 3
Service Equipment and Grounding

Service equipment usually consists of a circuit breaker or switch and fuses, and their accessories, located near the point of entrance of supply conductors to a building or other structure, or an otherwise defined area. The service equipment is intended to constitute the main control and means of cutoff of the electric service supply.

Service switches or main distribution panelboards are normally installed at a point immediately where the service-entrance conductors enter the building. Branch circuits and feeder panelboards (when required in addition to the main service panelboard) are usually grouped together at one or more centralized locations to keep the length of the branch-circuit conductors at a practical minimum of operating efficiency and to lower the initial installation costs.

Distribution equipment is generally intended to carry and control electrical current, but is not intended to dissipate or utilize energy. Eight basic factors influence the selection of distribution equipment:

1. *Codes and Standards:* Suitability for installation and use, in conformity with the provisions of the *NEC* and all local codes, must be considered. Suitability of equipment may be evidenced by listing or labeling.
2. *Mechanical Protection:* Mechanical strength and durability, including the adequacy of the protection provided must be considered.
3. *Wiring Space:* Wire bending and connection space is provided according to UL standards in all distribution equipment. When unusual wire arrangements or connections are to be made, then extra wire bending space, gutters, and terminal cabinets should be investigated for use.
4. *Electrical Insulation:* All distribution equipment carries labels showing the maximum voltage level that should be applied. The electrical supply voltage should always be equal to, or less than the voltage rating of distribution equipment.
5. *Heat:* Heating effects under normal conditions of use and also under abnormal conditions likely to arise in service must be constantly considered. Ambient heat conditions, as well as wire insulation ratings, along with the heat rise of the equipment must be evaluated during selection.

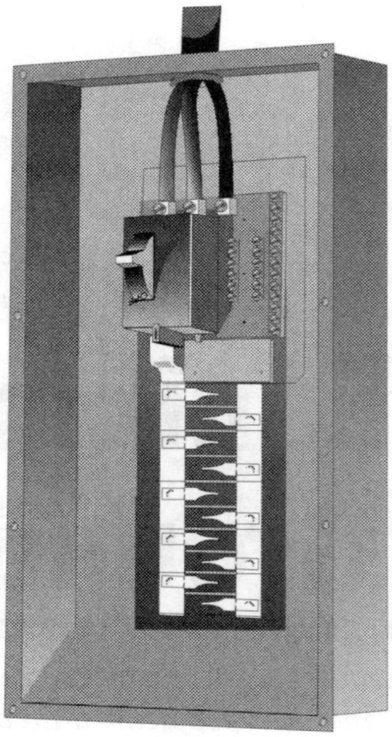

Figure 3-1: *Typical inside view of a panelboard*

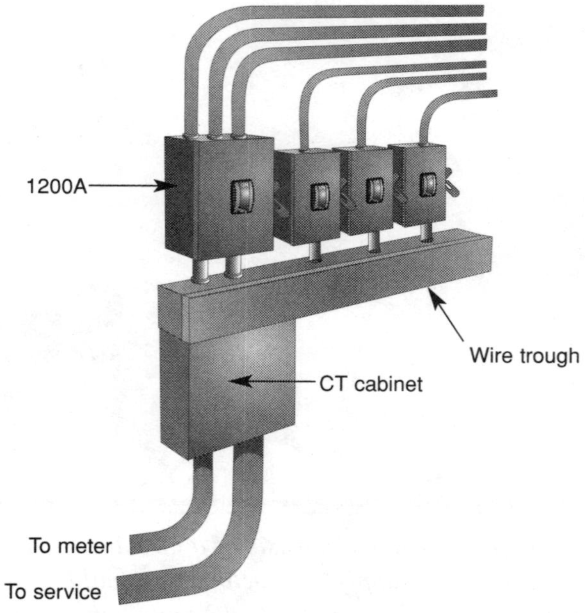
1200A
Wire trough
CT cabinet
To meter
To service

Figure 3-2: *An electric service made up of a wire trough and individual safety switches*

6. *Arcing Effects:* The normal arcing effects of overcurrent protective devices must be considered when the application is in or near combustible materials or vapors. Enclosures are selected to prevent or contain fires created by normal operation of the equipment. Selected locations of equipment must be made when another location may cause a hazardous condition.
7. *Classification:* Classification according to type, size, voltage, current capacity, interrupting capacity and specific use must be considered when selecting distribution equipment. Loads may be continuous or noncontinuous and the demand factor must be determined before distribution equipment can be selected.
8. *Personnel Protection:* Other factors which contribute to the practical safeguarding of persons using or likely to come in contact with the equipment must be considered. The equipment selected for use by only qualified persons may be different from equipment used or applied where unqualified people may come in contact with it.

In electrical wiring installations, overcurrent protective devices, consisting of fuses or circuit breakers, are factory-assembled in a metal cabinet, the entire assembly commonly being called a *panelboard* as shown in Figure 3-1.

Sometimes the main service-disconnecting means will be made up on the job by the workers by assembling individually enclosed fused switches or circuit breakers on a length of metal auxiliary gutter, as shown in Figure 3-2. Note that the various components are connected by means of short conduit nipples in which the insulated conductors are fed. Other services will consist of one large panelboard, often called a main distribution panelboard, which gives a neater appearance.

SAFETY SWITCHES

Enclosed single-throw safety switches are manufactured to meet industrial, commercial, and residential requirements. The two basic types of safety switches are:

- General duty
- Heavy duty

Double-throw switches are also manufactured with enclosures and features similar to the general and heavy-duty, single-throw designs.

The majority of safety switches have visible blades and safety handles. The switch blades are in full view when the enclosure door is open and there is visually no doubt when the switch is OFF. The only exception is Type 7 and 9 enclosures as discussed later; these do not have visible blades. Switch handles on all types of enclosures are an integral part of the box, not the cover, so that the handle is in control of the switch blades under normal conditions.

HEAVY-DUTY SWITCHES

Heavy-duty switches are intended for applications where ease of maintenance, rugged construction, and continued performance are primary concerns. They can be used in atmospheres where general-duty switches would be unsuitable, and are therefore widely used in industrial applications. Heavy-duty switches are rated 30 through 1200 amperes and 240 to 600 volts ac or dc. Switches with horsepower ratings are capable of opening a circuit up to six times the rated current of the switch. When equipped with Class J or Class R fuses for 30 through 600 ampere switches, or Class L fuses in 800 and 1200 ampere switches, many heavy-duty safety switches are UL listed for use on systems with up to 200,000 RMS symmetrical amperes available fault current. This, however, is about the highest short-circuit rating available for any heavy-duty safety switch. Applications include use where the required enclosure is NEMA Type 1, 3R, 4, 4X, 5, 7, 9, 12 or 12K.

Warning! Before changing fuses or performing maintenance on any safety switch, always visibly check the switch blades and jaws to ensure that they are in the OFF position. In addition, all equipment that can possibly be energized *must* be tested with a meter for possible back-feeds or improper wiring before any work is attempted.

Switch Blade And Jaws

Two types of switch contacts are used by the industry in today's safety switches. One is the "butt" contact; the other is a knife-blade and jaw type. On switches with knife-blade construction, the jaws distribute a uniform clamping pressure on both sides of the blade's contact surface. In the event of a high-current fault, the electromagnetic forces which develop tend to squeeze the jaws tightly against the blade. In the butt type contact, only one side of the blade's contact surface is held in tension against the conducting path. Electromagnetic forces due to high current faults tend to force the contacts apart, causing them to burn severely. Consequently, the knife-blade and jaw-type construction is the preferred type for use on all heavy-duty switches. The action of the blades moving in and out of the jaws aids in cleaning the contact surfaces. All current-carrying parts of these switches are plated to reduce heating by keeping oxidation at a minimum. Switch blades and jaws are made of copper for high conductivity. Spring-clamped blade hinges are another feature that help assure good contact surfaces and cool operations. "Visible blades" are utilized to provide visual evidence that the circuit has been opened.

Fuse Clips

Fuse clips are plated to control corrosion and to keep heating to a minimum. All fuse clips on heavy-duty switches have steel reinforcing springs for increased mechanical strength and firmer contact pressure. See Figure 3-3.

Terminal Lugs

Most heavy-duty switches have front removable, screw-type terminal lugs. Switch lugs are suitable for copper or aluminum wire except NEMA Types 4, 4X, 5 stainless and Types 12 and 12K switches which have all copper current-carrying parts and lugs, designated for use with copper wire only. Copper is more suitable for the areas in which these switches are used. Heavy-duty switches are suitable for the wire sizes and number of wires per pole as listed in Figure 3-4.

Insulating Material

As the voltage rating of switches is increased, arc suppression becomes more difficult and the choice of insulation material becomes more critical. Arc suppressors are

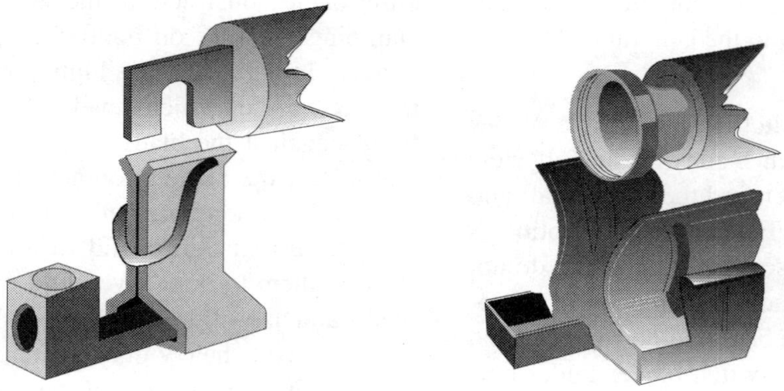

Figure 3-3: *Details of safety-switch fuse clips*

Ampere Rating	Conductors Per Phase	Wire Range and Wire Bending Space Per NEC Table 373-6 (a)	Lug Wire Range
Type 1 and 3R Heavy-Duty Terminal Lug Data			
30	1	#12-6 AWG (Al) or #14-6 AWG (Cu)	#12-2 AWG (Al) or #14-2 AWG (Cu)
60	1	#12-3 AWG (Al) or #14-3 AWG (Cu)	#12-2 AWG (Al) or #14-2 AWG (Cu)
100	1	#12-1 /0 AWG (Al) or #14-1 /0 AWG (Cu)	#12-1 /0 AWG (Al) or #14-1 /0 AWG (Cu)
200	1	#6 AWG-250 kcmil (Al/Cu)	#6 AWG-300 kcmil (Al/Cu)
400	1 or 2	#3/0 AWG-750 kcmil (Al/Cu) or #6 AWG-300 kcmil (Al/Cu)	#3/0 AWG-750 kcmil (Al/Cu) or #6 AWG-300 kcmil (Al/Cu)
600	2	#3/0 AWG-500 kcmil (Al/Cu)	#3/0 AWG-500 kcmil (Al/Cu)
800	3	#3/0 AWG-750 kcmil (Al/Cu)	#3/0 AWG-750 kcmil (Al/Cu)
1200	4	#3/0 AWG-750 kcmil (Al/Cu)	#3/0 AWG-750 kcmil (Al/Cu)
Type 4, 4X, 5 Stainless, and Type 12 and 12K Heavy Duty Terminal Lug Data			
30	1	#14-6 AWG (Cu)	#14-2 AWG (Cu)
60	1	#14-4 AWG (Cu)	#12-2 AWG (Cu)
100	1	#14-1 AWG (Cu)	#14-1 AWG (Cu)
200	1	#6 AWG-250 kcmil (Cu)	#6 AWG-250 kcmil (Cu)
400	1 or 2	#1 /0 AWG-600 kcmil (Cu) or #6 AWG-250 kcmil (Cu)	#1 /0 AWG-600 kcmil (Cu) and #6 AWG-250 kcmil (Cu)
600	2	#4 AWG-350 kcmil (Cu)	#4 AWG-350 kcmil (Cu)

Figure 3-4: *Safety-switch lug specifications*

usually made of insulation material and magnetic suppressor plates when required. All arc suppressor materials must provide proper control and extinguishing of arcs.

Operating Mechanism And Cover Latching

Most heavy-duty safety switches have a spring- driven quick-make, quick-break mechanism. A quick-breaking action is necessary if the switch is to be safely switched OFF under a heavy load.

The spring action, in addition to making the operation quick-make, quick-break, firmly holds the switch blades in the ON or OFF position. The operating handle is an integral part of the switching mechanism and is in direct control of the switch blades under normal conditions.

A one-piece cross bar, connected to all switch blades, should be provided which adds to the overall stability and integrity of the switching assembly by promoting proper alignment and uniform switch blade operation.

Dual cover interlocks are standard on most heavy-duty switches where the NEMA enclosure permits. However, NEMA Types 7 and 9 have bolted covers and obviously cannot contain dual cover interlocks. The purpose of dual interlock is to prevent the enclosure door from being opened when the switch handle is in the ON position and prevents the switch from being turned ON while the door is open. A means of bypassing the interlock is provided to allow the switch to be inspected in the ON position by qualified personnel. However, this practice should be avoided if at all possible. Heavy-duty switches can be padlocked in the OFF position with up to three padlocks.

Enclosures

Heavy-duty switches are available in a variety of enclosures which have been designed to conform to specific industry requirements based upon the intended use. Sheet metal enclosures (that is, NEMA Type 1) are constructed from cold-rolled steel which is usually phosphatized and finished with an electrode deposited enamel paint. The Type 3R rainproof and Type 12 and 12K dusttight enclosures are manufactured from galvannealed sheet steel and painted to provide better weather protection. The Type 4, 4X and 5 enclosures are made of corrosion resistant Type 304 stainless steel and require no painting. Type 7 and 9 enclosures are cast from copper-free aluminum and finished with an enamel paint. Type 1 switches are general purpose and designed for use indoors to protect the enclosed equipment from falling dirt and personnel from live parts. Switches rated through 200 amperes are provided with ample knockouts. Switches rated from 400 through 1200 ampere are provided without knockouts.

The following are the NEMA enclosure Types that will be encountered most often. Always make certain that the proper enclosure is chosen for the application. See Figure 3-5.

Type 3R switches are designated "rainproof" and are designed for use outdoors.

Type 3R enclosures for switches rated through 200 amperes have provisions for interchangeable bolt-on hubs at the top endwall. Type 3R switches rated higher than 200 amperes have blank top endwalls. Knockouts are provided (below live parts only) on enclosures for 200 ampere and smaller Type 3R switches. Type 3R switches are available in ratings through 1200 amperes.

Type 4, 4X, 5 stainless steel switches are designated dusttight, watertight and corrosion resistant and designed for indoor and outdoor use. Common applications include commercial type kitchens, dairies, canneries, and other types of food processing facilities, as well as areas where mildly corrosive liquids are present. All Type 4, 4X, and 5 stainless steel enclosures are provided without knockouts. Use of watertight hubs is required. Available switch ratings are 30 through 600 amperes.

Type 12 and Type 12K switches are designated dusttight (except at knockout locations on Type 12K) and are designed for indoor use. In addition, NEMA Type 12 safety switches are designated as raintight for outdoor use when the supplied drain plug is removed. Common applications include heavy industries where the switch must be protected from such materials as dust, lint, flyings, oil seepage, etc. Type 12K switches have knock-

The Following Chart Offers A Brief Explanation of the NEMA Enclosures Specifications

	Enclosure	Explanation
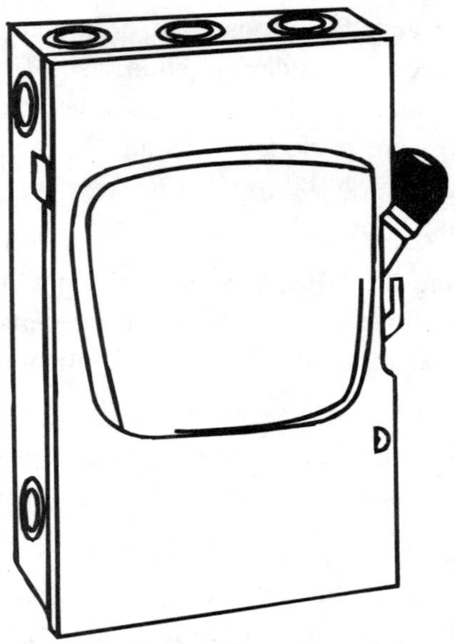 General-duty safety switch	NEMA Type 1 General Purpose	To prevent accidential contact with enclosed apparatus. Suitable for application indoors where not exposed to unusual service conditions
	NEMA Type 3 Weatherproof (Weather Resistant)	Protection against specified weather hazards. Suitable for use outdoors
	NEMA Type 3R Raintight	Protects against entrance of water from rain. Suitable for general outdoor application not requiring sleetproof
	NEMA Type 4 Watertight	Designed to exclude water applied in the form of a hose stream. To protect against a stream of water during cleaning operations
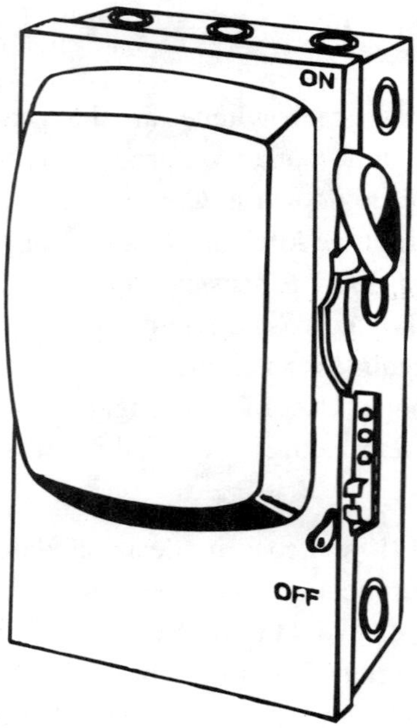 ON OFF Heavy-duty safety switch	NEMA Type 5 Dusttight	Constructed so that dust will not enter the enclosed area. Being replaced in some equipment by NEMA 12 Types
	NEMA Type 7 Hazardous Locations A, B, C, or D Class I -- air break Letter or letters following type number indicates particular groups of hazardous locations as per NEC	Designed to meet application requirements of NEC for Class I hazardous locations (explosive atmospheres). Circuit interruption occurs in air
	NEMA Type 9 Hazardous Locations E, F, or G Class II Letter or letters following type number indicates particular groups of hazardous locations as per NEC	Designed to meet application requirements of NEC for Class II hazardous locations (combustible dust, etc.)
	NEMA Type 12 Industrial Use	For use in those industries where it is desired to exclude dust, fibers and filings, or oil or coolant seepage

Figure 3-5: *NEMA classifications of safety switches*

outs in the bottom and top endwalls only. Available switch ratings are 30 through 600 amperes in Type 12 and 30 through 200 amperes in Type 12K.

NEMA Type 4, 4X, 5, Type 12 and 12K switch enclosures have positive sealing to provide a dusttight and raintight (watertight with stainless steel) seal. Enclosure doors are supplied with oil resistant gaskets. Switches rated 30 through 200 amperes incorporate spring loaded, quick-release latches; 400 and 600 ampere switches feature single-stroke sealing by operation of a cover mounted handle, while 30, 60, and 100 ampere switches in these enclosures are provided with factory installed fuse pullers.

Interlocked Receptacles

Heavy-duty, 60 ampere Type 1 and Type 12 switches within an interlocked receptacle are also available. This receptacle provides a means for connecting and disconnecting loads directly to the switch. A non-defeating interlock prevents the insertion or removal of the receptacle plug while the switch is in the ON position. It also prevents operation of the switch if an incorrect plug is used. Such arrangements are frequently used in industrial establishments for connecting 440-volt electric welders.

Accessories

Accessories available for field installation include Class R fuse kits, fuse pullers, insulated neutrals with grounding provisions, equipment grounding kits, watertight hubs for use with Type 4, 4X, 5 stainless or Type 12 switches, and interchangeable bolt-on hubs for Type 3R switches.

Electrical interlock consists of auxiliary contacts for use where control or monitoring circuits need to be switched in conjunction with the safety switch operation. Kits can be either factory or field installed, and they contain either one normally open and one normally closed contact, or two normally open and two normally closed contacts. The electrical interlock is actuated by a pivot arm which operates directly from the switch mechanism. The electrical interlock is designed so that its contacts disengage before the blades of the safety switch open and engage after the safety switch blades close.

GENERAL-DUTY SWITCHES

General-duty switches for residential and light commercial applications are used where operation and handling are moderate and where the available fault current is 10,000 RMS symmetrical amperes or less. Some general-duty safety switches, however, exceed this specification in that they are UL listed for application on systems having up to 100,000 RMS symmetrical amperes of available fault current when Class R fuses and Class R fuse kits are used. Class T fusible switches are also available in 400, 600 and 800 ampere ratings. These switches accept 300 VAC Class T fuses only. Some examples of general-duty switch applications include residential, farm, and small business service- entrances, and light-duty branch circuit disconnects.

General-duty switches are rated up to 600 amperes at 240 volts ac in general-purpose (Type 1) and rainproof (Type 3R) enclosures. Some general-duty switches are horsepower rated and capable of opening a circuit up to six times the rated current of the switch; others are not. Always check the switch's specifications before using under a horsepower-rated condition.

Switch Blades And Jaws

All current carrying parts of general-duty switches are plated to minimize oxidation and reduce heating. Switch jaws and blades are made of copper for high conductivity. Where required, a steel reinforcing spring increases the mechanical strength of the jaws and contact pressure between the blade and jaw. Good pressure contact maintains the blade-to-jaw resistance at a minimum, which in turn, promotes cool operation. All general-duty switch blades feature visible blade construction. With the door open, there is visually no doubt when the switch is OFF.

Fuse Clips

Fuse clips are normally plated to control corrosion and keep heating to a minimum. Where required, steel reinforcing springs are provided to increase the mechanical strength of the fuse clip. The result is a firmer, cooler connection to the fuses as well as superior fuse retention.

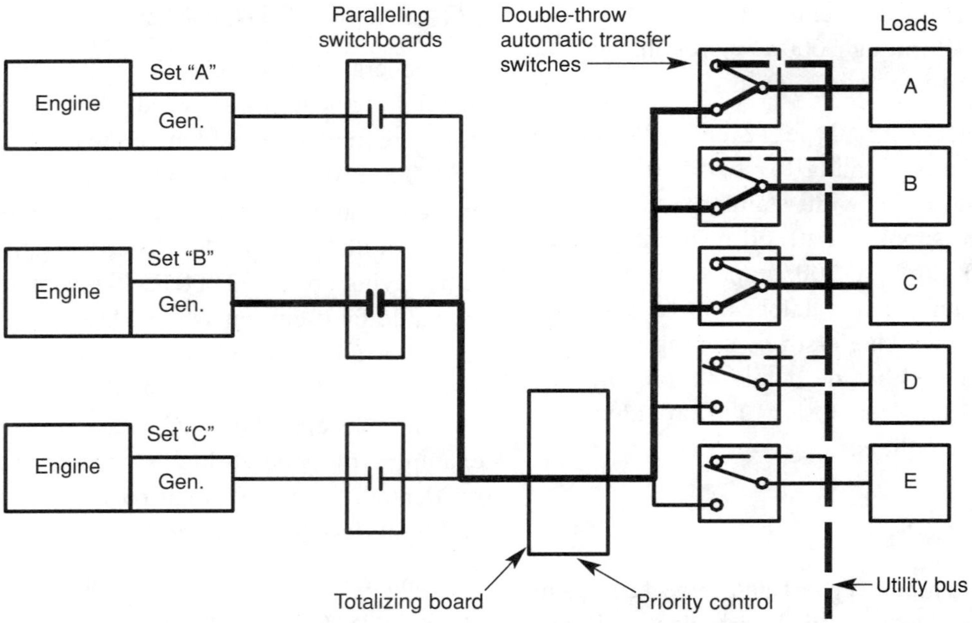
Paralleling switchboards
Double-throw automatic transfer switches
Loads
Engine
Set "A"
Gen.
A
B
Engine
Set "B"
Gen.
C
D
Engine
Set "C"
Gen.
E
Totalizing board
Priority control
Utility bus

Figure 3-6: *Practical application of a double-throw safety switch*

Terminal Lugs

Most general-duty safety switches are furnished with mechanical set-screw lugs which are suitable for aluminum or copper conductors.

Insulating Material

Switch and fuse bases are made of a strong, noncombustible, moisture-resistant material which provides the required phase-to-phase and phase-to-ground insulation for applications on 240VAC systems.

Operating Mechanism And Cover Latching

Although not required by either the UL or NEMA standards, some general-duty switches have spring-driven quick-make, quick-break operating mechanisms. Operating handles are an integral part of the operating mechanism and are not mounted on the enclosure cover. The handle provides indication of the status of the switch. When the handle is up, the switch is ON. When the handle is down, the switch is OFF. A padlocking bracket is provided which allows the switch handle to be locked in the OFF position. Another bracket is provided which allows the enclosure to be padlocked closed.

Enclosures

General-duty safety switches are available in either Type 1 for general purpose, indoor applications, or Type 3R for rainproof, outdoor applications. See earlier information in this chapter.

DOUBLE-THROW SAFETY SWITCHES

Double-throw switches are used as manual transfer switches and are not intended for use as motor circuit switches; thus, horsepower ratings are generally unavailable.

Double-throw switches are available as either fused or nonfusible devices and two general types of switch operation are available:

- Quick-make, quick-break
- Slow-make, slow-break

Figure 3-6 shows a practical application of a double-throw safety switch used as a transfer switch in conjunction with a stand-by emergency generator system.

NEC SAFETY-SWITCH REQUIREMENTS

Safety switches, in both fusible and nonfusible types, are used as a disconnecting means for services, feeders, and branch circuits. Installation requirements involving safety switches are found in several places throughout the *NEC*, but mainly in the following articles and sections:

- *NEC* Article 373
- *NEC* Article 380
- *NEC* Article 430-H
- *NEC* Article 440-B
- *NEC* Section 450-8(c)

When used as a service disconnecting means, the major installation requirements are listed in Figure 3-7.

PANELBOARDS

Panelboards consist of assemblies of overcurrent protective devices, with or without disconnecting devices, placed in a metal cabinet. The cabinet includes a cover or trim with one or two doors to allow access to the overcurrent and disconnecting devices and, in some types, access to the wiring space in the panelboard.

There is some confusion concerning the definition of "load center" and "panelboard." It's almost like the statement, "All Cognacs are brandy, but not all brandies are

	Safety Switches for Service Disconnects	
Application	**Requirements**	**NEC Reference**
Connection to terminals	Service conductors must be connected to the disconnecting means by pressure connectors, clamps, or other approved means. Soldered connections are forbidden. Each building must have an individual disconnecting means and each disconnecting means must be suitable for use as service equipment. A second service drop is permitted where the existing service capacity is over 2000 amperes	Section 230-81 Section 230-2 Exception 4
General	A means must be provided to disconnect all conductors serving the premises from service-entrance conductors. Each disconnecting means must be identified. Disconnecting means must be installed at a readily accessible location near point of entrance of service-entrance conductors In multiple-occupancy buildings, access to the disconnecting means for each occupant must be provided.	Section 230-70
More than one occupancy in same building	Buildings with more than one occupancy are permitted to have one set of service-entrance conductors run to each occupancy or to a group of occupancies.	Section 230-40 Exception No. 1
Number of disconnects	Service disconnecting means can consist of more than six switches for each service	Section 230-71
Type of disconnects	A manually- or power-operated safety switch meets NEC requirements	Section 230-76
Working space	Requirements for any electrical equipment apply; that is, the minimum headroom of working spaces about service equipment must be 6.5 feet or more. The dimensions of working space in the direction of access to live parts operating at 600 volts, nominal, or less to ground and likely to require examination, adjustment, servicing, or maintenance while energized must not be less than indicated in NEC Table 110-16(a).	Section 110-16

Figure 3-7: *NEC installation requirements governing switches used for service disconnects*

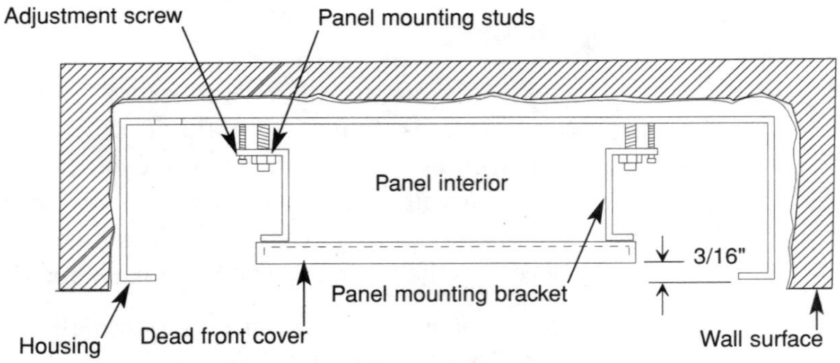
Adjustment screw
Panel mounting studs
Panel interior
3/16"
Panel mounting bracket
Housing
Dead front cover
Wall surface

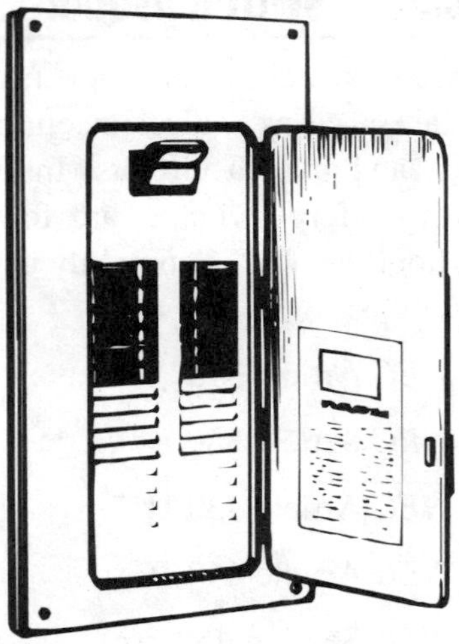

Figure 3-8: *Cross-sectional plan and pictorial views of a flush-mounted panelboard*

Cognac." Typically, load centers are fuse or circuit-breaker cabinets used on residential or small commercial projects. They are preassembled units with the interior busses installed at the factory. Upon installation, the required number of plug-in circuit breakers or fuse holders are installed, the circuit conductors terminated, and the front cover installed.

Many electrical workers classify "panelboard" as an enclosure for overcurrent protective devices used on larger commercial and industrial installations. Furthermore, the "can" or housing usually consists of "raw" unpainted galvanized metal. Frequently, the circuit breakers are factory installed using bolt-in circuit breakers.

A person would probably be correct in calling all load centers a *panelboard*, but not all panelboards are *load centers*!

Panelboards fall into two mounting classifications:

- Flush mounting, wherein the trim extends beyond the outside edges of the cabinet to provide a neat finish with the wall surface. See Figure 3-8.
- Surface mounting, wherein the edge of the trim is flush with the edge of the cabinet. See Figure 3-9.

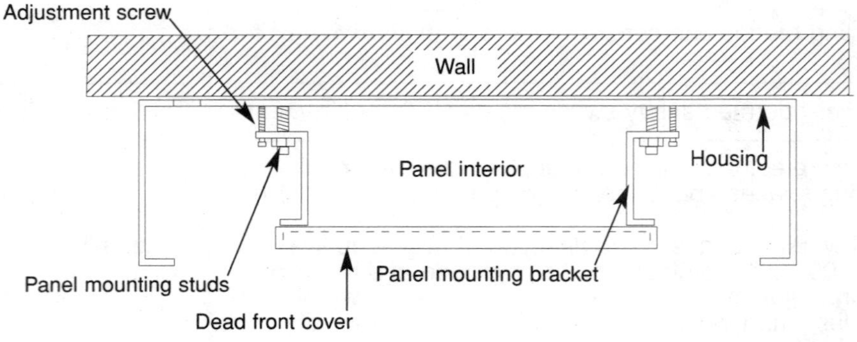
Adjustment screw
Wall
Panel interior
Housing
Panel mounting studs
Panel mounting bracket
Dead front cover

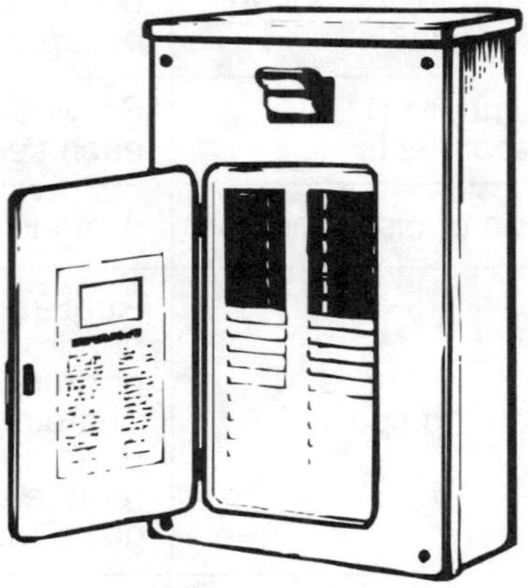

Figure 3-9: *Cross-sectional plan and pictorial views of a surface-mounted panelboard*

Furthermore, panelboards fall into two general classifications with regard to overcurrent devices:

- Circuit breaker
- Fused

Panel Installation

Prior to installing a panel, the selected location must be examined to verify that proper clearances exist and that the environment is proper for the panel installation.

In general, all panelboards must have a rating not less than the minimum feeder capacity required for the load computed in accordance with *NEC* Article 220. Panelboards must be durably marked by the manufacturer with the voltage and the current rating and the number of phases for which they are designed. Furthermore, the manufacturer's name or trademark must appear on the panelboard and be visible after installation, without disturbing the interior parts or wiring. All panelboard circuits and circuit modifications must be legibly identified as to purpose or use on a circuit directory located on the face or inside of the panel doors. See *NEC* Section 384-13.

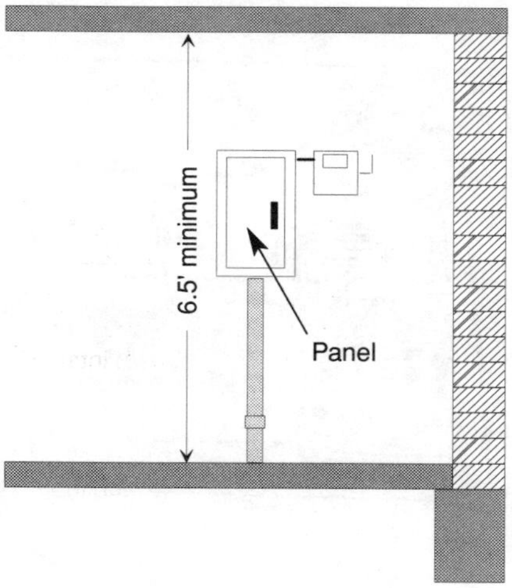
6.5' minimum
Panel

Figure 3-10: *NEC headroom requirements for service equipment*

The working height about panelboards must be as shown in Figure 3-10. Other space requirements appear in Figures 3-11 and 3-12.

Once a proper location has been determined, the panel is removed from its packing boxes, assembled, and installed. When removing the panel from its packing,

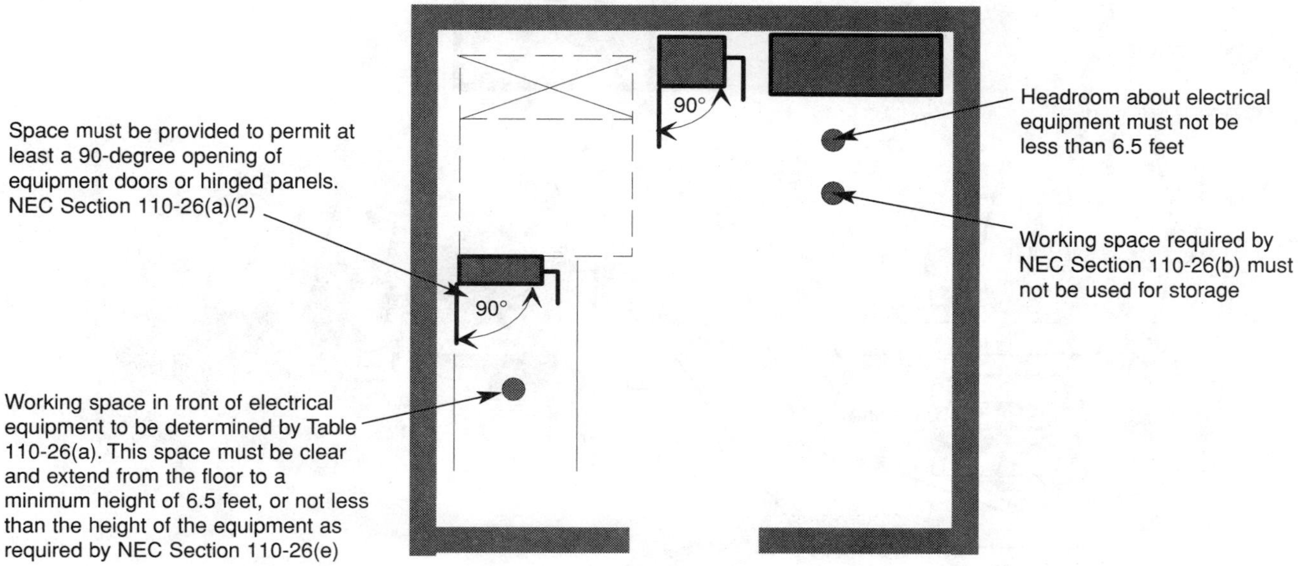
Space must be provided to permit at least a 90-degree opening of equipment doors or hinged panels. NEC Section 110-26(a)(2)
90°
90°
Headroom about electrical equipment must not be less than 6.5 feet
Working space required by NEC Section 110-26(b) must not be used for storage
Working space in front of electrical equipment to be determined by Table 110-26(a). This space must be clear and extend from the floor to a minimum height of 6.5 feet, or not less than the height of the equipment as required by NEC Section 110-26(e)

Figure 3-11: *NEC mounting requirements for service-entrance equipment*

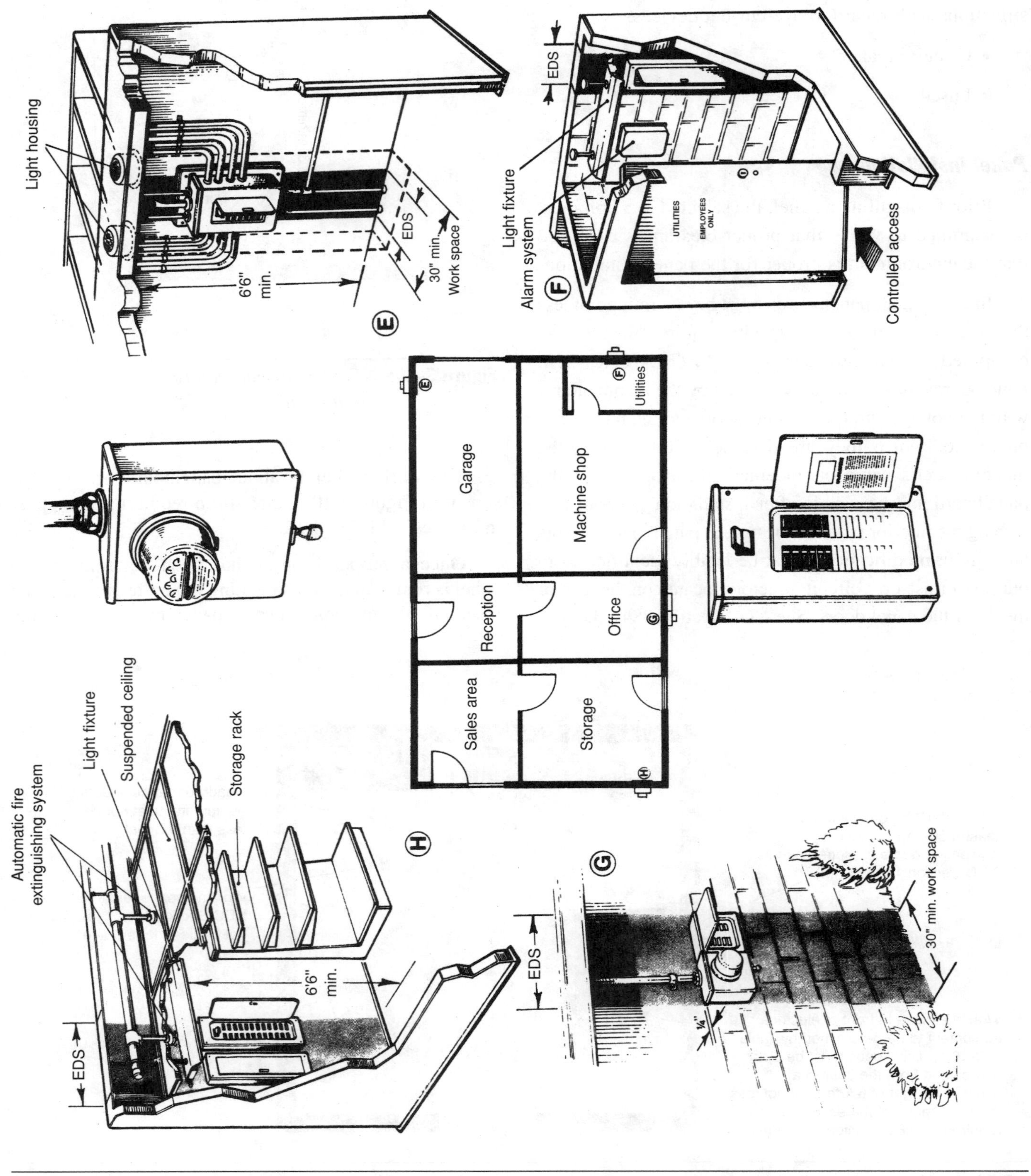
Light housing
6'6" min.
EDS
30" min. Work space
E
Light fixture
Alarm system
F
EDS
UTILITIES
EMPLOYEES ONLY
Controlled access
Garage
Machine shop
Utilities
Reception
Office
Sales area
Storage
Automatic fire extinguishing system
Light fixture
Suspended ceiling
Storage rack
H
EDS
6'6" min.
G
EDS
¼"
30" min. work space

Figure 3-12: *Summary of NEC installation requirements for service equipment*

verify that all necessary components have been delivered and make sure that any stray packing material has been removed from the panel. Check to make sure that the right panel is to be installed. A checklist might include the following items:

- Is the panel to be top fed or bottom fed? This information should be obtained from the drawings or from the project supervisor.
- Check to verify that the voltage rating of the panel is as specified on the drawings.
- Check to verify that the ampacity of the panel is as specified on the drawings.
- Check to verify that the phase and number of conductors is as specified on the drawings.
- Verify that the panel was not damaged during shipping.

Installing Flush-Mounted Enclosures

Flush-mounted enclosures installed in noncombustible material must be mounted so that the front edge of the enclosure is not set back further than ¼inch from the finished surface. If installed in other than noncombustible walls, the panel edge must be flush with the finished wall. *NEC* Section 373-3.

Installing Surface-Mounted Enclosures

Surface-mounted enclosures must be securely fastened in place. If the wall structure offers little structural support, as in the case of ¼-inch wood paneling or ½-inch gypsum board, the enclosure must be located so that it may be attached to framing members inside the wall covering. In some cases, a framing structure will have to be built to support the panel.

Installing The Panel Interior

Prior to installing the panel interior, check to verify that the enclosure is securely fastened in place and is free of all foreign material. Obtain and study the specifications and instructions that are included with the panel. If no instructions are available, the following is a general installation procedure that may be used.

- Mount the interior to the enclosure using the four mounting studs installed on the enclosure back.
- Adjust the depth of the interior with the adjustment screws. The dead front cover should be no further than 3/16 inch from the wall surface for a flush panel, or the same distance from the enclosure face for a surface-mounted panel.

Knockouts

A series of concentric or eccentric circular partial openings are usually cut in the top, bottom, and sides of both load center and panelboard housings; some may also be cut in the back. These openings are cut in such a manner that they may be removed by tapping (knocking) them out — usually with a screwdriver blade and hammer.

The direction from which the knockouts can be removed alternates from inside the enclosure to outside the enclosure. That is, ½-inch knockouts are knocked outward from inside the enclosure, ¾-inch knockouts are knocked inward from outside the enclosure, 1-inch knockouts are knocked outward from inside the enclosure, and so forth.

In most cases, raceways connected to panelboards using the concentric knockout openings have poor equipment grounding connections. Consequently, *NEC* Section 250-94(4) requires bonding jumpers to be used around concentric or eccentric knockouts that are punched or otherwise formed so as to impair the electrical connection to ground. This is accomplished by using a grounding locknut or bushing and then connecting a bonding jumper to either another grounding locknut or bushing, or else to the equipment grounding terminal inside the panelboard.

Panel Connections

Electrical connections in a panelboard fall under two categories:

- Line connections, which include termination and routing of the service and feeder conductors.

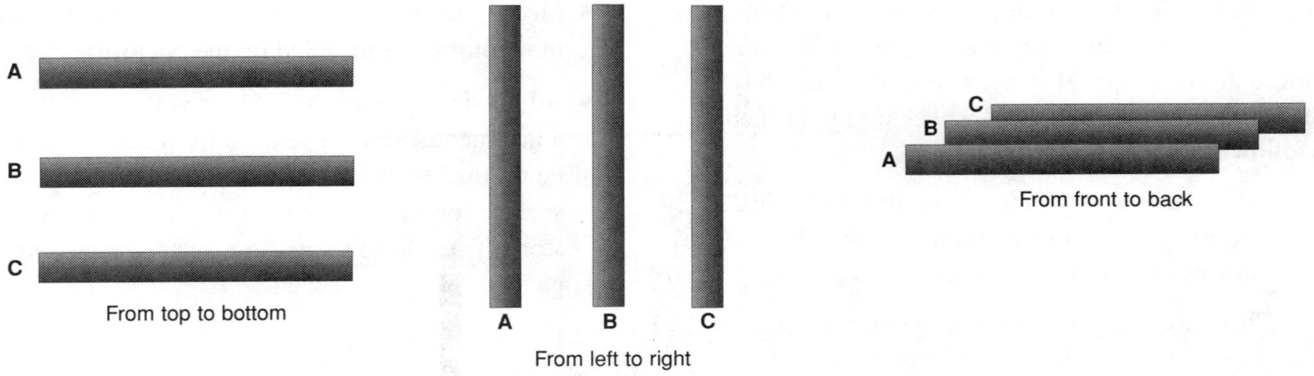
A
B
C
From top to bottom
A B C
From left to right
C
B
A
From front to back

Figure 3-13: *NEC approved phase arrangements*

- Load connections, which include termination and routing of the branch-circuit and feeder conductors.

When installing the line connections, verify that the lugs are stamped "CU/AL" or a label is inside the panel which states that the connection of aluminum conductors is permitted prior to terminating aluminum conductors.

When installing the load connections, again verify that the lugs are suitable for both copper and aluminum if aluminum conductors are used. Due to the difficulty in keeping aluminum conductors tight in their termination lugs, copper is usually specified for most industrial installations.

NEC Section 384-3(f) requires the phase arrangement on 3-phase buses to be A, B, C from front to back, top to bottom, or left to right, as viewed from the front of the panel. The B phase must be the phase with the highest voltage to ground on a 3-phase, 4-wire delta-connected system. See Figure 3-13.

CURRENT TRANSFORMERS

Meters used by power companies to record the amount of current used by customers usually respond to a current which varies from zero to five amperes. To respond to the actual current of the service, each meter is provided with current transformers. If the peak demand of the service is 100 amperes, a 100:5 current transformer is used. If the peak current demand is expected to be 200 amperes, a 200:5 current transformer is used.

Services above 400 amperes usually utilize a group of current transformers — one for each ungrounded conductor in the service. There are two basic types of current transformers: the busbar type and the "doughnut" type. These latter current transformers encircle the ungrounded conductors in the system to read the current flow, much the same as a clamp-on ammeter. They are sometimes called "doughnuts" due to their appearance. The busbar type current transformer has each transformer connected in series with a busbar, and does not encircle the conductor.

Current transformers are normally housed in an enclosure called a "CT cabinet." The letters "CT," of course, stand for "current transformer." Figure 3-14 shows a typical CT cabinet with current transformers and their related wiring. In some rare cases, the current transformers may be mounted exposed on overhead conductors, but this is more the exception than the rule.

Power companies have different requirements for sizes of CT cabinets but the dimensions shown in Figure 3-15 are typical for several service sizes - including both single-and three-phase services.

Power companies also have different specifications for the location and wiring of CTs and CT cabinets, depending on the locale. The following are the require-

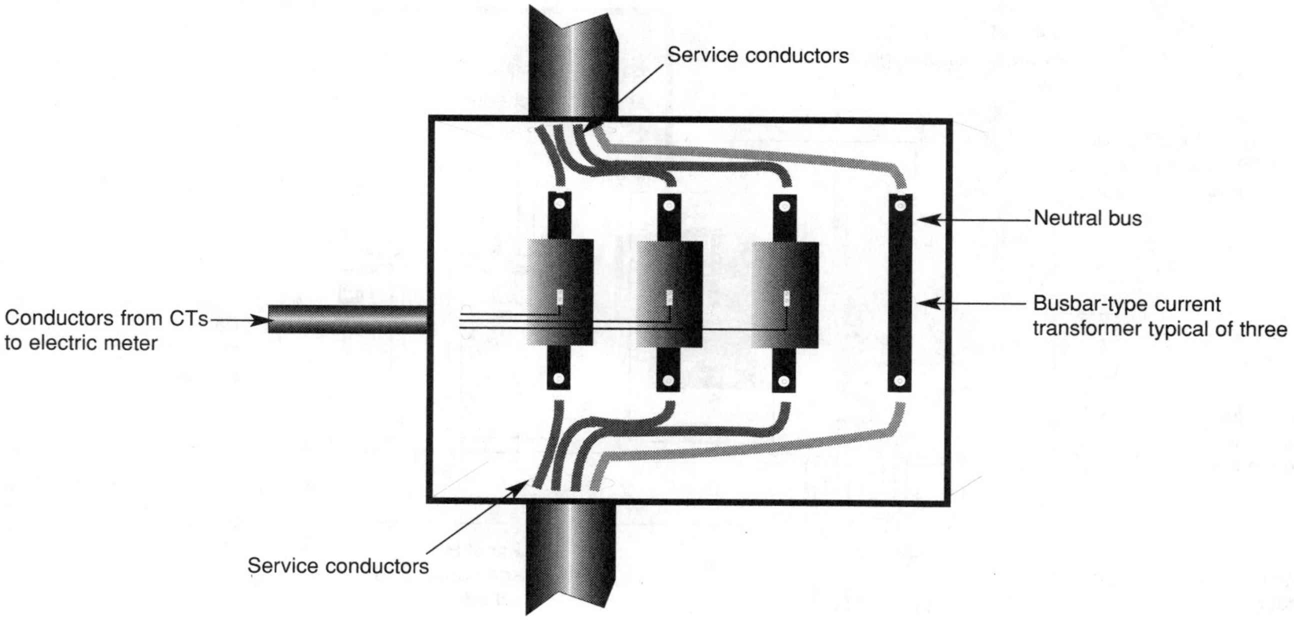
Service conductors
Neutral bus
Conductors from CTs to electric meter
Busbar-type current transformer typical of three
Service conductors

Figure 3-14: *Typical CT cabinet arrangement*

ments of one power company. However, always check locally to verify their requirements.

1. The meter base and meter may be located on either side or top of the current transformer cabinet, or it may be located at a distance away, if approved by power company and the conduit containing the instrument wiring is run exposed.

2. In no instance shall more than one set of conductors terminate in the instrument transformer cabinet. Sub-feeders and branch circuits are to terminate at the customer's distribution panel. The instrument transformer cabinet shall not be used as a junction box.

3. When service-entrance conduits enter or leave through the back of the current-transformer cabinet, the size of the CT cabinet must be increased to provide additional working space.

4. For services at higher voltages, additional space must be provided in the transformer cabinet for mounting potential transformers. Consult the local power company for dimensions.

Phase	Service Characteristics	Cabinet Size in Inches
Single	120/240 volts, 3-wire	10 x 24 x 32
Three	120/240 volts, 4-wire	10 x 36 x 42
Three	120/208 volts, 4-wire	10 x 36 x 42
Three	480 volts, 3-wire	10 x 36 x 42
Three	480/277 volts, 4-wire	10 x 36 x 48

Figure 3-15: *Typical CT cabinet sizes*

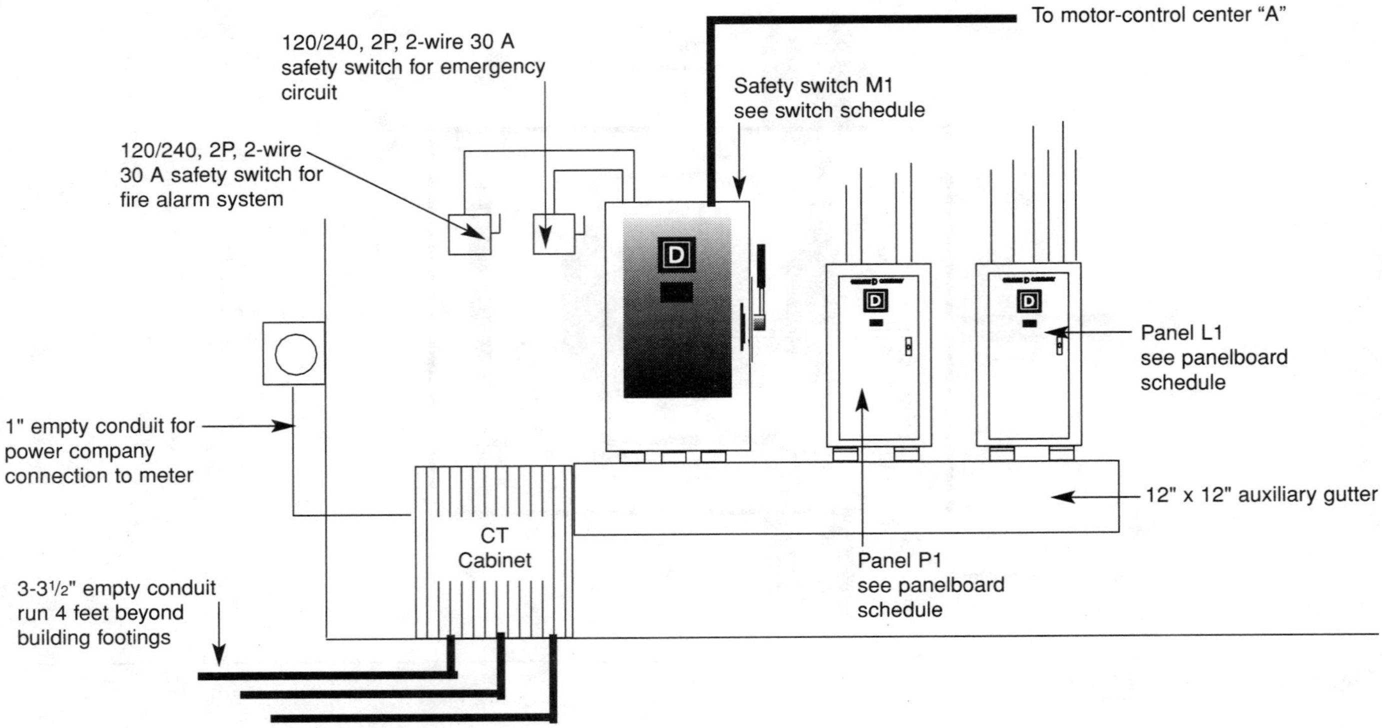
To motor-control center "A"
120/240, 2P, 2-wire 30 A safety switch for emergency circuit
Safety switch M1 see switch schedule
120/240, 2P, 2-wire 30 A safety switch for fire alarm system
Panel L1 see panelboard schedule
1" empty conduit for power company connection to meter
12" x 12" auxiliary gutter
CT Cabinet
3-3½" empty conduit run 4 feet beyond building footings
Panel P1 see panelboard schedule

Figure 3-16: *Power-riser diagram for a three-phase, 4-wire, 1200-ampere service*

5. If kilovar metering is required, increase the width of meter mounting from 18 to 36 inches.
6. If recording demand instruments are required, increase the height of meter mounting from 36 to 48 inches.

GUTTERS

The auxiliary gutter shown in Figure 3-16 is used to route the service conductors and also to provide an enclosure for tapping these conductors for the safety switch and two 200-ampere panels. In this type of arrangement, appropriate connectors are normally used to make the taps. However, many electrical contractors have found that a bussed gutter saves labor and provides for a neater installation.

A bussed gutter is an assembly of bus bars in an enclosure. The enclosure may be rated for outdoor (weatherproof) or indoor installations. Busbars installed in the gutter may be made of aluminum or copper and must have an ampacity rating for the application; that is, if the service conductors are rated for 1200 amperes (in our sample) the busbars must be rated for at least the same ampacity. Furthermore, it must be UL listed.

From an installation or a maintenance/modification viewpoint, bussed gutters are one of the favorite types of wiring methods for use with multiswitch services. An advantage of a bussed gutter is the ease of installation and modification. Adding disconnect switches or changing switches is relatively easy. No connectors have to be untaped and reconnected as with systems using wire connectors on the conductors for taps.

In our example — using service conductors with an ampacity of 1200 amperes — the rating of the busbars in bussed gutters must also be rated at 1200 amperes, if bussed gutters were to be used. See Figure 3-17. Furthermore, the bussed gutter must have AIC rating sufficient for the available fault current and must have sufficient wire-bending space per *NEC* Section 300-34.

Bus Bracing

One characteristic of ground-faults is an induced torque in conductors carrying the fault. Because of this torque, the busbars in a bussed gutter must be attached to the enclosure in such a manner as to prevent their being dislodged and/or making contact with the gutter frame during the fault. When busbars are attached in such a manner as to withstand the torque created by the available amount of ground-fault current, they are said to be "braced" for that amount of current. For instance, busbars may be braced for 20,000 amperes, 30,000 amperes or whatever level of ground-fault current required up to 200,000 amperes. The bussed gutter must be labeled by the manufacturer for the amount of fault current the busses are braced to withstand.

GROUNDING

The grounding system is a major part of the electrical system. Its purpose is to protect life and equipment against the various electrical faults that can occur. It is sometimes possible for higher-than-normal voltages to appear at certain points in an electrical system or in the electrical equipment connected to the system. Proper grounding ensures that the high electrical charges that cause these high voltages are channeled to earth or ground before damaging equipment or causing danger to human life. Therefore, circuits are grounded to limit the voltage on the circuit, improve overall operation of the electrical system and the continuity of service. Grounding also provides the following:

- Rapid operation of fuses, circuit breakers, and circuit interrupters.
- Minimizes the magnitude and duration of step and touch potentials in substations.
- Decreases the duration and magnitude of lightning current effects on the electrical system and transfers fault current into the earth.
- Voltage surges higher than that for which the circuit is designed are transferred into the earth.
- Transfers fault current into the earth.
- Helps prevent low-voltage problems when a grounded (neutral) conductor opens up.
- Minimizes heat damage to the electrical system and related equipment.
- Increases the load capacity of the primary line of electrical equipment such as motor starters.
- Substantially lowers the neutral-to-earth voltages.

When we refer to *ground*, we are talking about ground potential or earth ground. If a conductor is connected to the earth or to some conducting body that

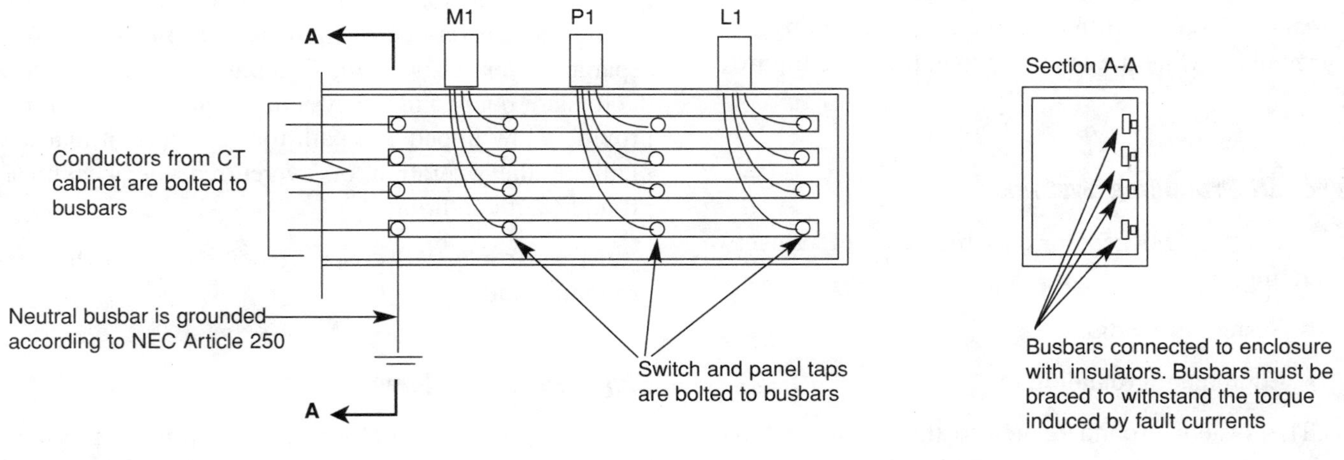
A
M1
P1
L1
Section A-A
Conductors from CT cabinet are bolted to busbars
Neutral busbar is grounded according to NEC Article 250
A
Switch and panel taps are bolted to busbars
Busbars connected to enclosure with insulators. Busbars must be braced to withstand the torque induced by fault currrents

Figure 3-17: *Three-phase bussed gutter*

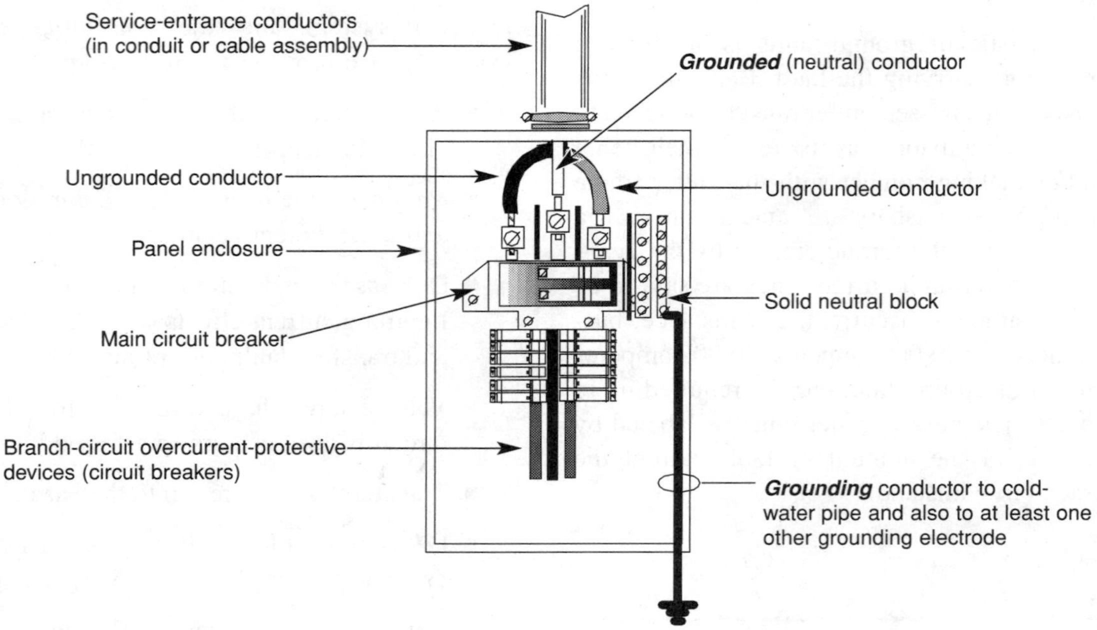
Service-entrance conductors
(in conduit or cable assembly)
Grounded (neutral) conductor
Ungrounded conductor
Ungrounded conductor
Panel enclosure
Solid neutral block
Main circuit breaker
Branch-circuit overcurrent-protective devices (circuit breakers)
Grounding conductor to cold-water pipe and also to at least one other grounding electrode

Figure 3-18: *Panelboard showing grounded and grounding conductors connected to their proper terminals*

serves in place of the earth, such as a driven ground rod (electrode) or cold-water pipe, the conductor is said to be *grounded*. The neutral conductor in a three- or four-wire service, for example, is intentionally grounded and therefore becomes a *grounded conductor*. However, a wire used to connect this neutral conductor to a grounding electrode or electrodes is referred to a *grounding conductor*. Note the difference in the two meanings; one is grounded, while the other is grounding. See Figure 3-18.

Types Of Grounding Systems

There are two general classifications of protective grounding:

- System grounding
- Equipment grounding

The system ground relates to the service-entrance equipment and its interrelated and bonded components. That is, system and circuit conductors are grounded to limit voltages due to lighting, line surges, or unintentional contact with higher voltage lines, and to stabilize the voltage to ground during normal operation.

Equipment grounding conductors are used to connect the noncurrent-carrying metal parts of equipment, conduit, outlet boxes, and other enclosures to the system grounded conductor, the grounding electrode conductor, or both, at the service equipment or at the source of a separately derived system. Equipment grounding conductors are bonded to the system grounded conductor to provide a low impedance path for fault current that will facilitate the operation of overcurrent devices under ground-fault conditions.

NEC Article 250 covers general requirements for grounding and bonding.

Single-Phase Systems

To better understand a complete grounding system, let's take a look at a conventional residential or small commercial system beginning at the power company's

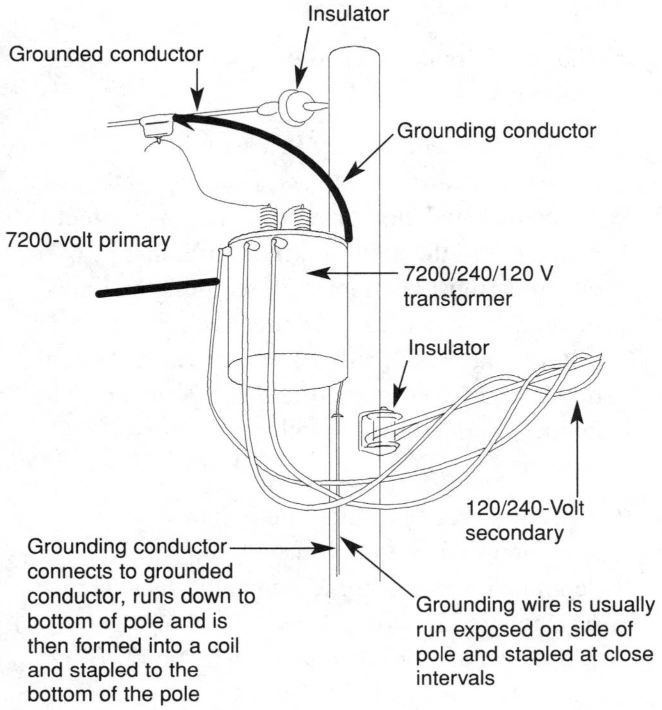
Insulator
Grounded conductor
Grounding conductor
7200-volt primary
7200/240/120 V transformer
Insulator
120/240-Volt secondary
Grounding conductor connects to grounded conductor, runs down to bottom of pole and is then formed into a coil and stapled to the bottom of the pole
Grounding wire is usually run exposed on side of pole and stapled at close intervals

Figure 3-19: *Pole-mounted transformer that reduces transmission voltage to usable house current*

high-voltage lines and transformer as shown in Figure 3-19. The pole-mounted transformer is fed with a 2-wire 7200-volt system which is transformed and stepped down to a 3-wire, 120/240-volt, single-phase electric service suitable for residential use. A wiring diagram of the transformer connections is shown in Figure 3-20. Note that the voltage between phase A and phase B is 240 volts. However, by connecting a third wire (neutral) on the secondary winding of the transformer - between the other two - the 240 volts are split in half, giving 120 volts between either phase A or phase B and the neutral conductor. Consequently, 240 volts are available for household appliances such as ranges, hot-water heaters, clothes dryers, and the like, while 120 volts are available for lights, small appliances, TVs, and similar electrical appliances.

Referring again to the diagram in Figure 3-20, conductors A and B are ungrounded conductors, while the neutral is a grounded conductor. If only 240-volt loads were connected, the neutral (grounded conductor) would carry no current. However, since 120-volt loads are present, the neutral will carry the unbalanced load and becomes a current-carrying conductor. For example, if phase A carries 60 amperes and phase B carries 50 amperes, the neutral conductor would carry only 10

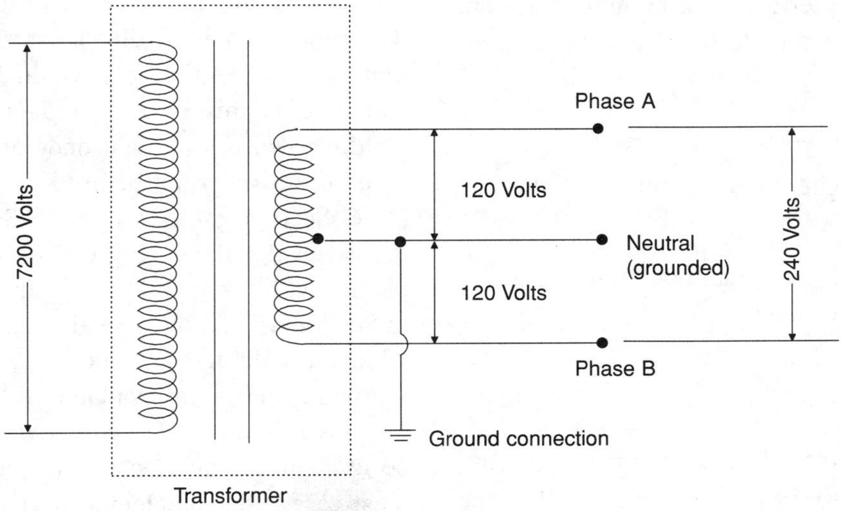
7200 Volts
Phase A
120 Volts
Neutral
(grounded)
120 Volts
240 Volts
Phase B
Ground connection
Transformer

Figure 3-20: *Wiring diagram of a 7200-volt - 120/240-volt, single-phase transformer connection*

amperes (60 − 50 = 10). This is why the *NEC* allows the neutral conductor in an electric service to be smaller than the ungrounded conductors.

The typical pole-mounted service-entrance is normally routed by messenger cable from a point on the pole to a point on the building being served, terminating in a meter housing. Another service conductor is installed between the meter housing and the main service switch or panelboard. This is the point where most systems are grounded - the neutral bus in the main panelboard.

Caution! Exercise extreme caution when lifting a ground. Never grab a disconnected ground wire with one hand and the grounding electrode with the other. Your body will act as a conductor for any fault current; the results could be fatal.

OSHA And NEC Requirements

The grounding equipment requirements established by Underwriters' Laboratories, Inc., has served as the basis for approval for grounding of the *NEC*. The *NEC*, in turn, provides the grounding premises of the Occupational Safety and Health Act (OSHA).

All electrical systems must be grounded in a manner prescribed by the *NEC* to protect personnel and valuable equipment. To be totally effective, a grounding system must limit the voltage on the electrical system and protect it from:

- Exposure to lightning.
- Voltage surges higher than that for which the circuit is designed.
- An increase in the maximum potential to ground due to abnormal voltages.

Grounding Methods

Methods of grounding an electric service are covered in *NEC* Section 250, Part C. In general, all of the following (if available) and any made electrodes must be bonded together to form the grounding electrode system:

- An underground water pipe in direct contact with the earth for no less than 10 feet that is within 5 feet of the point of entrance.
- The metal frame of a building where effectively grounded.
- An electrode encased by at least 2 inches of concrete, located within and near the bottom of a concrete foundation or footing that is in direct contact with the earth. Furthermore, this electrode must be at least 20 feet long and must be made of electrically conductive coated steel reinforcing bars or rods of not less than ½-inch diameter, or consisting of at least 20 feet of bare copper conductor not smaller than No. 4 AWG wire size.
- A ground ring encircling the building or structure, in direct contact with the earth at a depth below grade not less than 2½ feet. This ring must consist of at least 20 feet of bare copper conductor not smaller than No. 2 AWG wire size.

In some structures only the water pipe will be available, and this water pipe must be supplemented by an additional electrode as specified in *NEC* Sections 250-50 or 250-52. With these facts in mind, let's take a look at a typical electric service, and the available grounding electrodes.

The building in Figure 3-21 has a metal underground water pipe that is in direct contact with the earth for more than 10 feet, and is within 5 feet of the point of entrance, so this is one valid grounding source. The building also has a metal underground gas-piping system, but this may not be used as a grounding electrode [*NEC* Section 250-52(a)]. *NEC* Section 250-50(a) further states that the underground water pipe must be supplemented by an additional electrode of a type specified in Section 250-50 or in Section 250-52. Since a grounded metal building frame, concrete-encased electrode, or a ground ring is not available in this application, *NEC* Section 250-52 — *Made and Other Electrodes* — must be used in determining the supplemental electrode. In most cases, this supplemental electrode will consist of either a driven rod or a pipe electrode, the specifications for which follow:

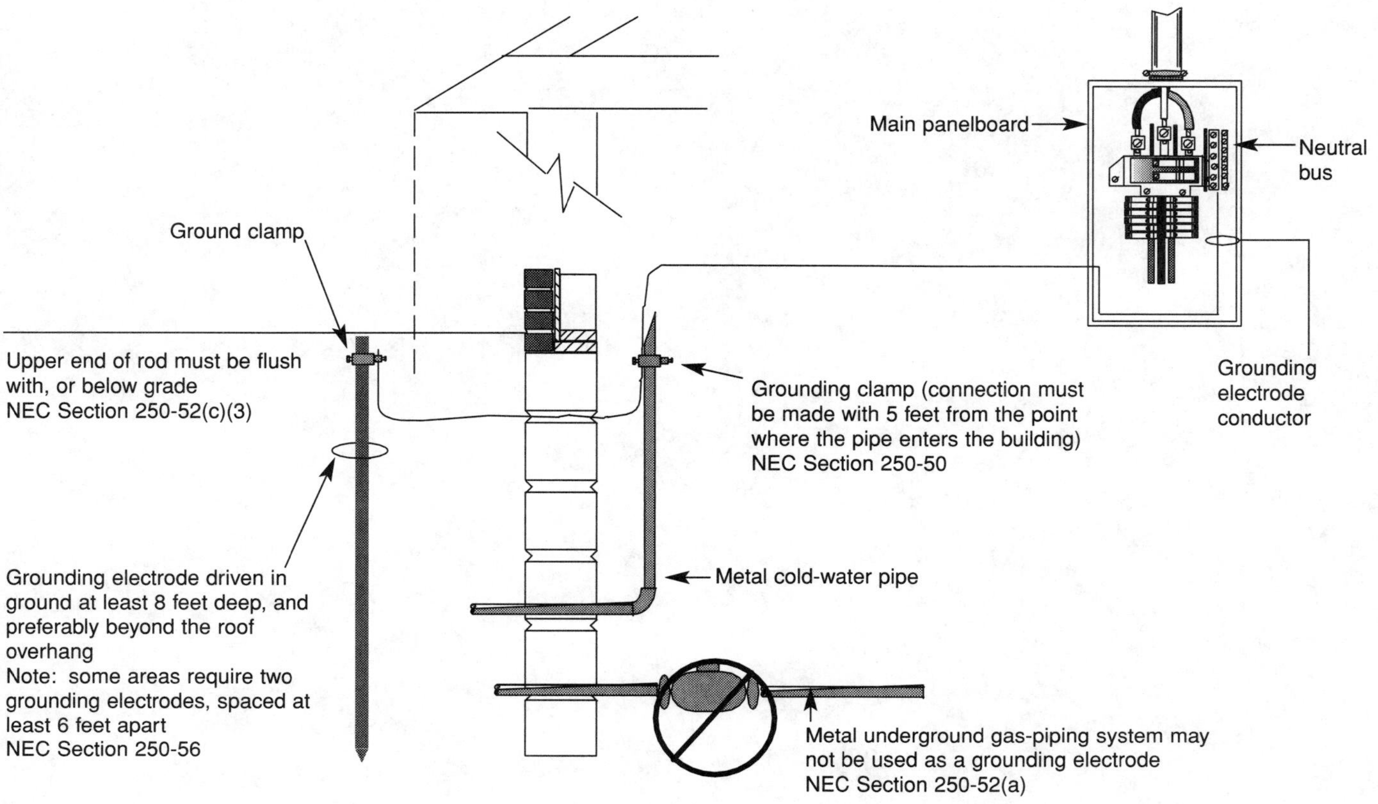
Main panelboard
Neutral bus
Ground clamp
Upper end of rod must be flush with, or below grade NEC Section 250-52(c)(3)
Grounding clamp (connection must be made with 5 feet from the point where the pipe enters the building) NEC Section 250-50
Grounding electrode conductor
Grounding electrode driven in ground at least 8 feet deep, and preferably beyond the roof overhang
Note: some areas require two grounding electrodes, spaced at least 6 feet apart
NEC Section 250-56
Metal cold-water pipe
Metal underground gas-piping system may not be used as a grounding electrode
NEC Section 250-52(a)

Figure 3-21: *Summary of NEC grounding requirements for residential and commercial construction*

- Provide a low-impedance path to ground for personnel and equipment protection and effective circuit relaying.
- Withstand and dissipate repeated fault and surge circuits.
- Provide corrosion resistance to various soil chemistries to ensure continuous performance for the life of the equipment being protected.
- Provide rugged mechanical properties for easy driving with minimum effort and rod damage.

An alternate method to the pipe or rod method is a plate electrode. Each plate electrode must expose not less than 2 square feet of surface to the surrounding earth. Plates made of iron or steel must be at least ¼ inch thick, while plates of nonferrous metal like copper need only be .06 inch thick.

Either type of electrode must have a resistance to ground of 25 ohms or less. If not, they must be augmented by an additional electrode spaced not less than 6 feet from each other. In fact, many locations require two electrodes regardless of the resistance to ground. This, of course, is not an *NEC* requirement, but is required by some power companies and local ordinances in some cities and counties. Always check with the local inspection authority for such rules that surpass the requirements of the *NEC*.

Chapter 4
Overcurrent Protection

To protect electrical conductors and equipment against abnormal operating conditions and their consequences, protective devices are used in the circuits.

All electrical circuits, and their related components, are subject to destructive overcurrents. Harsh environments, general deterioration, accidental damage or damage from natural causes, excessive expansion or overloading of the electrical system are factors that contribute to the occurrence of such overcurrents. Reliable protective devices prevent or minimize costly damage to transformers, conductors, motors, equipment, and the many other components and loads that make up the complete electrical system. Therefore, reliable circuit protection is essential to avoid the severe monetary losses which can result from power blackouts and prolonged downtime of facilities. Two types of automatic overload devices normally used in electrical circuits to prevent fires or the destruction of the circuit and its associated equipment are fuses and circuit breakers.

PLUG FUSES

Plug fuses have a screw-shell base and are sometimes used in small commercial buildings for circuits that supply lighting and 120-volt power outlets. Plug fuses are supplied with standard screw bases (Edison-base) or Type S bases. An Edison-base fuse consists of a strip of fusible (capable of being melted) metal in a small porcelain or glass case, with the fuse strip, or link, visible through a "window" in the top of the fuse. The screw base corresponds to the base of a standard medium-base incandescent lamp. Edison-base fuses are permitted only as replacements in existing installations; all new work must use the S-base fuses (*NEC* Section 240-51(b).

- *"Plug fuses of the Edison-base type shall be used only for replacements in existing installations where there is no evidence of overfusing or tampering."*

The chief disadvantage of the Edison-base plug fuse is that it is made in several ratings from 0 to 30 A, all with the same size base — permitting unsafe replacement of one rating by a higher rating. Type S fuses were developed to reduce the possibility of overfusing a circuit (inserting a fuse with a rating greater than that required by the circuit). There are 15 classifications of Type S fuses based on current rating: 0-30 amperes. Each Type S fuse has a base of a different size and a matching adapter. Once an adapter is screwed into a standard Edison-base fuseholder, it locks into place and is not readily removed without destroying the fuseholder. As a result, only a Type S fuse with a size the same as that of the adapter may be inserted. Two types of plug fuses are shown in Figure 4-1; a Type S adapter is also shown.

Plug fuses also are made in time-delay types that permit a longer period of overload flow before operation, such as on motor inrush current and other higher-than-normal rated currents. They are available in ratings up to 30 amperes, both in Edison base and Type S. Their principle use is in motor circuits, where the starting

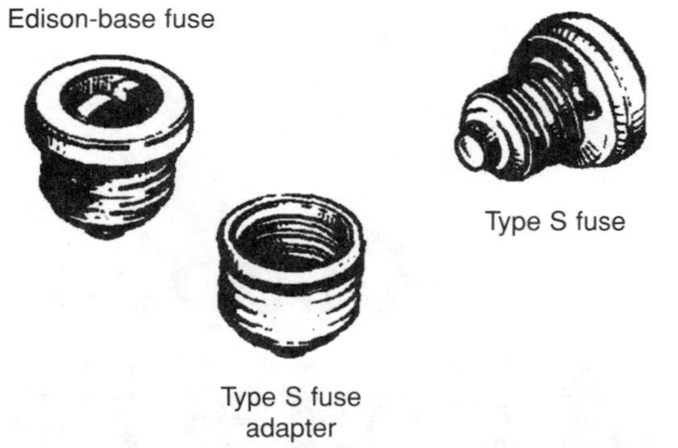
Edison-base fuse
Type S fuse
Type S fuse adapter

Figure 4-1: *Types of common plug fuses*

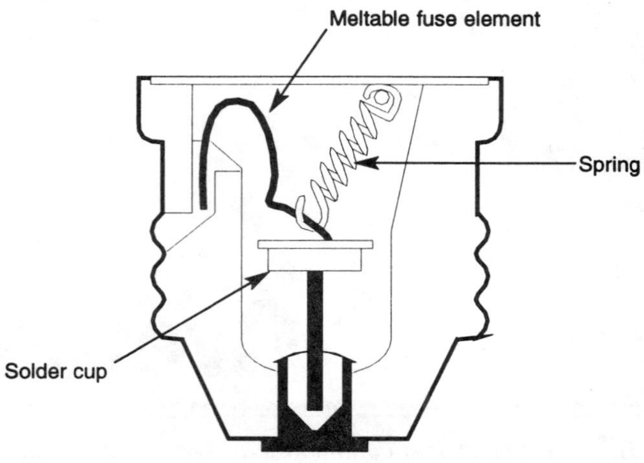
Meltable fuse element
Spring
Solder cup

Figure 4-2: *Time-delay plug fuse*

inrush current to the motor is much higher than the running, or continuous, current. The time-delay fuse will not open on the inrush of high-starting current. If, however, the high current persists, the fuse will open the circuit just as if a short circuit or heavy overload current had developed. All Type S fuses are time-delay fuses. Figure 4-2 shows the internal construction of a time-delay fuse.

Plug fuses are permitted to be used in circuits of no more than 125 volts between phases, but they may be used where the voltage between any ungrounded conductor and ground is not more than 150 volts. The screw-shell of the fuseholder for plug fuses must be connected to the load side circuit conductor; the base contact is connected to the line side or conductor supply. A disconnecting means (switch) is not required on the supply side of a plug fuse.

The plug fuse is a *nonrenewable* fuse; that is, once it has opened the circuit because of a fault or overload, it cannot be used again or renewed. It must be replaced by a new fuse of the same rating and characteristics for safe and effective restoration of circuit operation.

The "window" of a plug fuse can tell much about the condition of the fuse, and also the probable cause if the fuse blows. For example, Figure 4-3 shows window views of a plug fuse in three conditions. If the fuse is good, then the element or link will be whole and clearly visible, as shown in A. If the current through the fuse only slightly exceeded the current rating over a period of time, then a melt-through would occur as shown in B. However, if very high currents occurred — as in a short-circuit — the fuse will melt more violently and will spray the zinc or alloy over the inside viewing window, as shown in C. The window will also be smudged with black.

CARTRIDGE FUSES

In most industrial and commercial applications, cartridge fuses are used because they have a wider range of types, sizes, and ratings than do plug fuses. Many cartridge fuses are also provided with a means to renew the fuse by unscrewing the end caps and replacing the links. In large industrial installations where thousands of cartridge fuses are in use to protect motors and machinery, this replacement feature can be a significant savings over a period of a year. There are several different types of

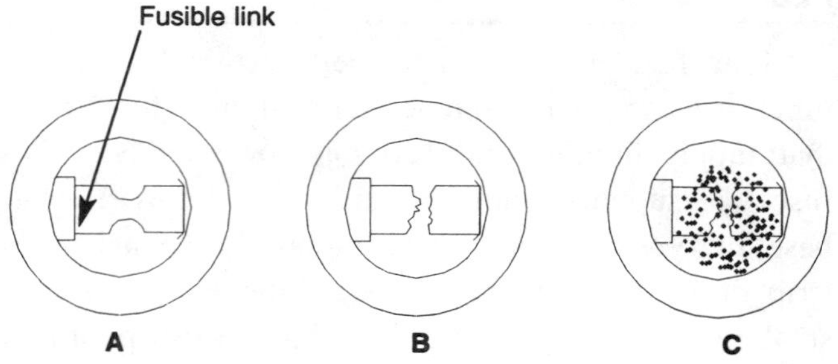
Fusible link
A
B
C

Figure 4-3: *Window view of plug fuse showing various conditions of the element*

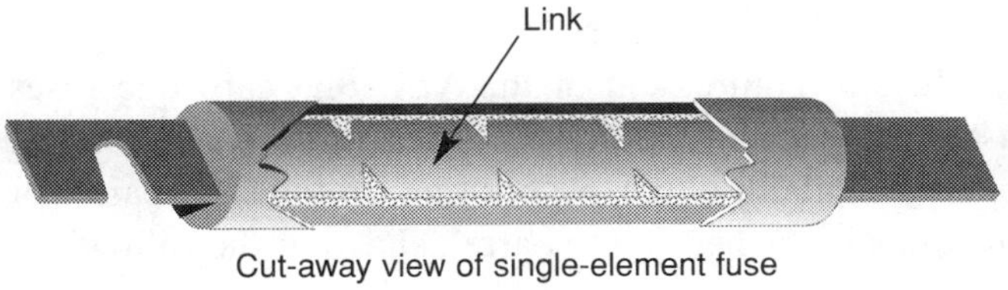
Link
Cut-away view of single-element fuse

Figure 4-4: *Single-element cartridge fuse*

cartridge fuses, and although all operate in a similar fashion, all have slightly different characteristics. Each is described in the paragraphs to follow.

Single-Element Cartridge Fuses: The basic component of a fuse is the link. Depending upon the ampere rating of the fuse, the single-element fuse may have one or more links. They are electrically connected to the end blades (or ferrules) and enclosed in a tube or cartridge surrounded by an arc-quenching filler material.

Under normal operation, when the fuse is operating at or near its ampere rating, it simply functions as a conductor. However, if an overload current occurs and persists for more than a short interval of time, the temperature of the link eventually reaches a level that causes a restricted segment of the link to melt; as a result, a gap is formed and an electric arc established. See Figure 4-4. However, as the arc causes the link metal to burn back, the gap becomes progressively larger. Electrical resistance of the arc eventually reaches such a high level that the arc cannot be sustained and is extinguished; the fuse will have then completely cut off all current flow in the circuit. Suppression or quenching of the arc is accelerated by the filler material.

Single-element fuses have a very high speed of response to overcurrents. They provide excellent short-circuit component protection. However, temporary harmless overloads or surge currents may cause nuisance openings unless these fuses are oversized. They are best used, therefore, in circuits not subject to heavy transient surge currents and the temporary overload of circuits with inductive loads such as motors, transformers, and solenoids. Because single-element fuses have a high speed-of-response to short-circuit currents, they are particularly suited for the protection of circuit breakers with low interrupting ratings.

Dual-Element Cartridge Fuses: Unlike single-element fuses, the dual-element fuse (Figure 4-5) can be applied in circuits subject to temporary motor overload and surge currents to provide both high performance short-circuit and overload protection. Oversizing in order to prevent nuisance openings is not necessary. The dual-element fuse contains two distinctly separate types of elements. Electrically, the two elements are series connected. The fuse links similar to those used in the single-element fuse perform the short-circuit protection function; the overload element provides protection against low-level overcurrents or overloads and will hold an overload that is five times greater than the ampere rating of the fuse for a minimum time of ten seconds.

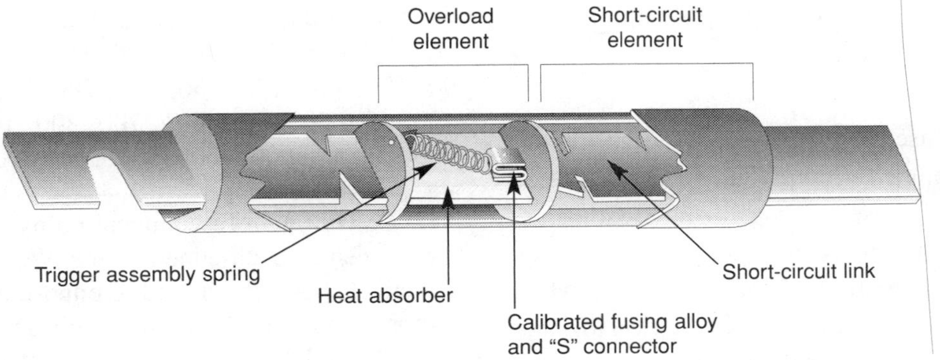
Overload element
Short-circuit element
Trigger assembly spring
Heat absorber
Calibrated fusing alloy and "S" connector
Short-circuit link

Figure 4-5: *Dual-element cartridge fuse*

As shown in Figure 4-5, the overload section consists of a copper heat absorber and a spring-operated trigger assembly. The heat-absorber strip is permanently connected to the short-circuit link and to the short-circuit link on the opposite end of the fuse by the S-shaped connector of the trigger assembly. The connector electronically joins the one short-circuit link to the heat absorber in the overload section of the fuse. These elements are joined by a "calibrated" fusing alloy. An overload current causes heating of the short-circuit link connected to the trigger assembly. Transfer of heat from the short-circuit link to the heat absorbing strip in the midsection of the fuse begins to raise the temperature of the heat absorber. If the overload is sustained, the temperature of the heat absorber eventually reaches a level that permits the trigger spring to "fracture" the calibrated fusing alloy and pull the connector free. The short-circuit link is electrically disconnected from the heat absorber, the conducting path through the fuse is opened, and overload current is interrupted. A critical aspect of the fusing alloy is that it retains its original characteristic after repeated temporary overloads without degradation. The main purposes of dual-element fuses are as follows:

- Provide motor overload, ground-fault and short-circuit protection.
- Permit the use of smaller and less costly switches.
- Give a higher degree of short-circuit protection (greater current limitation) in circuits in which surge currents or temporary overloads occur.
- Simplify and improve blackout prevention (selective coordination).

Dual-element fuses may also be used in circuits other than motor branch circuits and feeders, such as lighting circuits and those feeding mixed lighting and power loads. The low-resistance construction of the fuses offers cooler operation of the equipment, which permits higher loading of fuses in switches and panel enclosures without heat damage and without nuisance openings from accumulated ambient heat.

Fuse Markings

It is a requirement of the *NEC* that cartridge fuses used for branch-circuit or feeder protection must be plainly marked, either by printing on the fuse barrel or by a label attached to the barrel, showing the following:

- Ampere rating
- Voltage rating
- Interrupting rating (if other than 10,000A)
- "Current limiting," where applicable
- The name or trademark of the manufacturer

CONDUCTOR PROTECTION

All conductors are to be protected against overcurrents in accordance with their ampacities as set forth in *NEC* Section 240-3. They must also be protected against short-circuit current damage as required by *NEC* Sections 240-1 and 110-10.

Ampere ratings of overcurrent-protective devices must not be greater than the ampacity of the conductor. There is, however, an exception. *NEC* Section 240-3(b) states that if the conductor rating does not correspond to a standard size overcurrent-protective device, the next larger size overcurrent-protective device may be used provided its rating does not exceed 800 amperes and the conductor is not part of a multioutlet branch circuit supplying receptacles for cord-and-plug connected portable loads. When the ampacity of busway or cablebus does not correspond to a standard overcurrent-protective device, the next larger standard rating may be used even though this rating may be greater than 800 amperes (*NEC* Sections 364-10 and 365-5).

Standard fuse sizes stipulated in *NEC* Section 240-6 are: 1, 3, 6, 10, 15, 20, 25, 30, 35, 40, 45, 50, 60, 70, 80, 90, 100, 110, 125, 150, 175, 200, 225, 250, 300, 350, 400, 450, 500, 600, 601, 700, 800, 1000, 1200, 1600, 2000, 2500, 3000, 4000, 5000, and 6000 amperes.

NOTE: The small fuse ampere ratings of 1, 3, 6, and 10 have recently been added to the *NEC* to provide more effective short-circuit and ground-fault protection for motor circuits in accordance with Sections 430-40 and 430-52, and UL requirements for protecting the overload relays in controllers for very small motors.

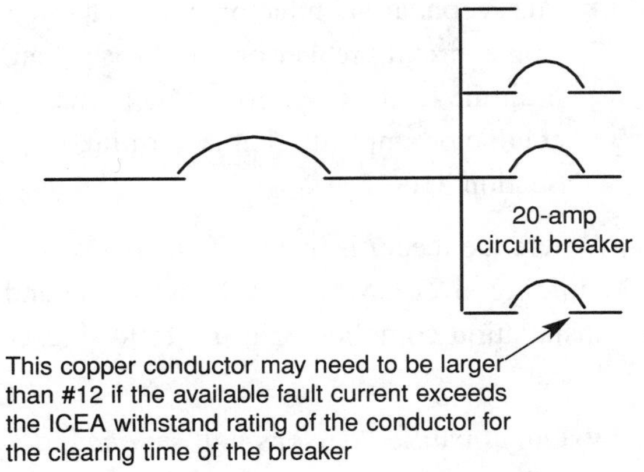
20-amp
circuit breaker
This copper conductor may need to be larger than #12 if the available fault current exceeds the ICEA withstand rating of the conductor for the clearing time of the breaker

Figure 4-6: *Conductor protection with noncurrent-limiting devices*

Protection of conductors under short-circuit conditions is accomplished by obtaining the maximum short-circuit current available at the supply end of the conductor, the short-circuit withstand rating of the conductor, and the short-circuit let-through characteristics of the overcurrent device.

When a noncurrent-limiting device is used for short-circuit protection, the conductor's short-circuit withstand rating must be properly selected based on the overcurrent protective device's ability to protect. See Figure 4-6.

It is necessary to check the energy let-through of the overcurrent device under short-circuit conditions and select a wire size of sufficient short-circuit withstand ability.

In contrast, the use of a current-limiting fuse permits a fuse to be selected which limits short-circuit current to a level less than that of the conductor's short-circuit withstand rating — doing away with the need of oversized ampacity conductors. See Figure 4-7.

In many applications, it is desirable to use the convenience of a circuit breaker for a disconnecting means and general overcurrent protection, supplemented by current-limiting fuses at strategic points in the circuits.

Flexible cord, including tinsel cord and extension cords, must be protected against overcurrent in accordance with their ampacities.

Location Of Fuses In Circuits

In general, fuses must be installed at points where the conductors receive their supply; that is, at the beginning or line side of a branch circuit or feeder. Exceptions to this rule are given in *NEC* Section 240-21, (a) through (g).

(a) Branch circuit tap conductors meeting the requirements specified in *NEC* Section 210-19 shall be permitted to have overcurrent protection located as specified in that Section.

(b) Feeder taps. Conductors shall be permitted to be tapped, without overcurrent protection at the tap, to a feeder as specified in (1) through (5).

(1) Where the length of the tap conductors does not exceed 10 feet and the tap conductors comply with the following:

- The ampacity of the tap conductor is not less than the combined computed loads on the circuit supplied by the tap conductors, and not less than the rating of the device supplied by the tap conductors or not less than the rating of the overcurrent-protective device at the termination of the tap conductors;

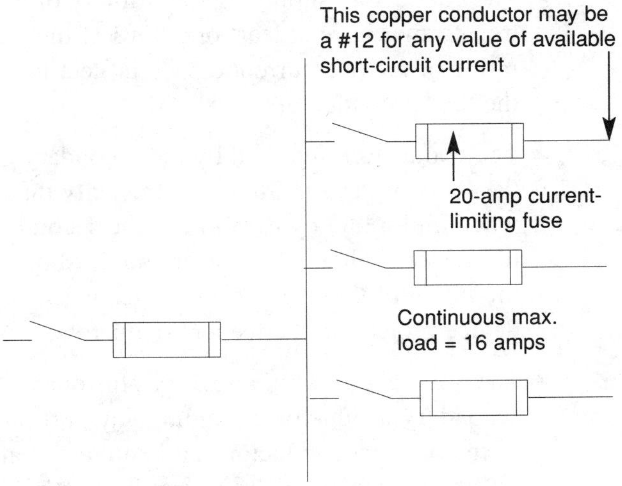
This copper conductor may be a #12 for any value of available short-circuit current
20-amp current-limiting fuse
Continuous max. load = 16 amps

Figure 4-7: *Circuits protected by current-limiting devices*

- The tap conductors do not extend beyond the switchboard, panelboard or control device which it supplies;
- Except at the point of connection to the feeder, the tap conductor is enclosed in raceway;
- For field installed taps, the rating of the overcurrent device on the line side of the tap conductors shall not exceed 10 times the ampacity of the tap conductor.

(2) Where the length of the tap connection does not exceed 25 feet and the tap conductors comply with all of the following:

- The ampacity of the tap conductor is not less than one-third of the rating of the overcurrent device protecting the feeder conductors;
- The tap conductors terminate in a single circuit breaker or a single set of fuses that will limit the load to the ampacity of the tap conductors;
- The tap conductors are suitably protected from physical damage.

(3) Where the tap conductors supply a transformer (primary plus secondary not over 25 feet long) and comply with the following conditions:

- The conductors supplying the primary of a transformer have at least one-third of the rating of the overcurrent device protecting the feeder conductors;
- The conductors supplied by the secondary of the transformer shall have an ampacity that, when multiplied by the ratio of the second-to-primary voltage, is at least one-third of the rating of the overcurrent protective device protecting the feeder conductors;
- The total length of one primary plus one secondary conductor, excluding any portion of the primary conductor that is protected at its ampacity, is not over 25 feet in length;
- The primary and secondary conductors are suitably protected from physical damage;
- The secondary conductors terminate in a single circuit breaker or set of fuses that will limit the load current to not more than the conductor ampacity that is permitted by Section 310-15.

(4) Where the feeder is in a high bay manufacturing building over 35 feet high at walls, and the installation complies with the following conditions:

- Only qualified persons will service the system;
- The tap conductors are not over 25 feet long horizontally and not over 100 feet total length;
- The ampacity of the tap conductors is not less than one-third of the rating of the overcurrent device protecting the feeder conductors;
- The tap conductors are continuous from end-to-end, contain no splices, and are suitably protected from physical damage;
- The tap conductors are at least No. 6 AWG copper or No. 4 AWG aluminum;
- The tap conductors do not penetrate walls, floors, or ceilings, and are tapped no less than 30 feet from the floor.

(c)(3) Transformer secondary conductors of separately derived systems do not require fuses at the transformer terminals when all of the following conditions are met:

- Must be an industrial location;
- Secondary conductors must be less than 25 feet long;
- Secondary conductor ampacity must be at least equal to secondary full-load current of transformer and sum of terminating, grouped, overcurrent devices;
- Secondary conductors must be protected from physical damage.

Warning! Smaller conductors tapped to larger conductors can be a serious hazard. If not protected against short-circuit conditions, these unprotected conductors can vaporize or incur severe insulation damage.

Lighting/Appliance Loads

The branch-circuit rating must be classified in accordance with the rating of the overcurrent protective device. Classifications for those branch circuits, other than individual loads, must be: 15, 20, 30, 40, and 50 amperes as specified in *NEC* Section 210-3.

Branch-circuit conductors must have an ampacity of the rating of the branch circuit and not less than the load to be served (*NEC* Section 210-19). The minimum size branch-circuit conductor that can be used is No. 14 (*NEC* Section 210-19). However, there are some exceptions as specified in *NEC* Section 210-19.

Branch-circuit conductors and equipment must be protected by a fuse whose ampere rating conforms to *NEC* Section 210-20. Basically, the branch-circuit conductor and fuse must be sized for the actual noncontinuous load and 125 percent for all continuous loads. The fuse size must not be greater than the conductor ampacity. Branch circuits rated 15 through 50 amperes with two or more outlets (other than receptacle circuits) must be fused at their rating and the branch-circuit conductor sized according to *NEC* Table 210-24.

Feeder Circuits With No Motor Load

The feeder fuse ampere rating and feeder conductor ampacity must be at least 100 percent of the noncontinuous load plus 125 percent of the continuous load as calculated per *NEC* Article 220. The feeder conductor must be protected by a fuse not greater than the conductor ampacity. Motor loads shall be computed in accordance with Article 430.

Service Equipment

Each ungrounded service-entrance conductor must have a fuse in series with a rating not higher than the ampacity of the conductor. These service fuses shall be part of the service disconnecting means or be located immediately adjacent thereto (*NEC* Section 230-91).

Service disconnecting means can consist of one to six switches or circuit breakers for each service or for each set of service-entrance conductors permitted in *NEC* Section 230-2. When more than one switch is used, the switches must be grouped together (*NEC* Section 230-71).

Transformer Secondary Conductors

Field installations indicate nearly 50 percent of transformers installed do not have secondary protection. The *NEC* recommends that secondary conductors be protected from damage by the proper overcurrent-protective device. For example, the primary overcurrent device protecting a three-wire transformer cannot offer protection of the secondary conductors. Also see *NEC* exception in Section 240-3 for two-wire primary and secondary circuits.

Motor Circuit Protection

Motors and motor circuits have unique operating characteristics and circuit components. Therefore, these circuits must be dealt with differently from other types of loads. Generally, two levels of overcurrent protection are required for motor branch circuits:

- Overload protection — Motor running overload protection is intended to protect the system components and motor from damaging overload currents.
- Short-circuit protection (includes ground-fault protection) — Short-circuit protection is intended to protect the motor circuit components such as the conductors, switches, controllers, overload relays, motor, etc. against short-circuit currents or grounds. This level of protection is commonly referred to as motor branch-circuit protection applications. Dual-element fuses are designed to provide this protection provided they are sized correctly.

There are a variety of ways to protect a motor circuit — depending upon the user's objective. The ampere rating of a fuse selected for motor protection depends on whether the fuse is of the dual-element time-delay type or the nontime-delay type. If circuit breakers are used, the type and time-delay rating must also be considered.

In general, nontime-delay fuses can be sized at 300 percent of the motor full-load current for ordinary motors so that the normal motor starting current does not affect the fuse. Dual-element, time-delay fuses are able to withstand normal motor-starting current and can be sized closer to the actual motor rating than can nontime-delay fuses.

CIRCUIT BREAKERS

Basically, a circuit breaker is a device for closing and interrupting a circuit between separable contacts under both normal and abnormal conditions. This is done manually (normal condition) by use of its "handle" by switching to the ON or OFF positions. However, the circuit breaker also is designed to open a circuit automatically on a predetermined overload or ground-fault current without damage to itself or its associated equipment. As long as a circuit breaker is applied within its rating, it will automatically interrupt any "fault" and therefore must be classified as an inherently safe overcurrent protective device.

The operating characteristics of a circuit breaker are shown in Figure 4-8. Note that the handle on a circuit breaker resembles an ordinary toggle switch. On an overload, the circuit breaker opens itself or *trips*. In a tripped position, the handle jumps to the middle position (Figure 4-8). To reset, turn the handle to the OFF position and then turn it as far as it will go beyond this position; finally, turn it to the ON position.

A standard molded-case circuit breaker usually contains:

- A set of contacts
- A magnetic trip element
- A thermal trip element
- Line and load terminals
- Bussing used to connect these individual parts
- An enclosing housing of insulating material

The circuit breaker handle manually opens and closes the contacts and resets the automatic trip units after an interruption. Some circuit breakers also contain a manually operated "push-to-trip" testing mechanism.

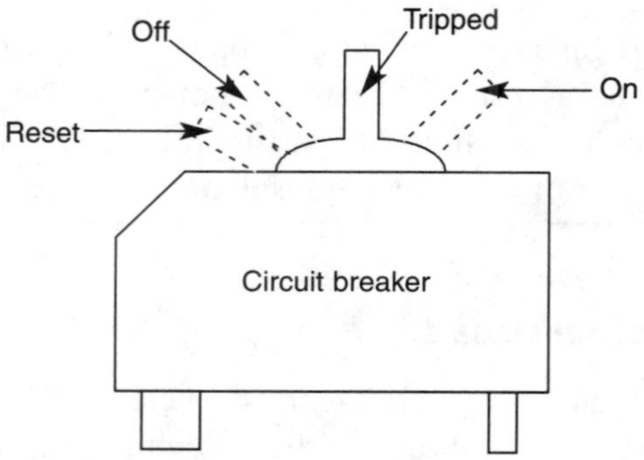
Off
Tripped
On
Reset
Circuit breaker

Figure 4-8: *Operating characteristics of a circuit breaker*

Circuit breakers are grouped for identification according to given current ranges. Each group is classified by the largest ampere rating of its range. These groups are:

- 15-100 amperes
- 125-225 amperes
- 250-400 amperes
- 500-1,000 amperes
- 1,200-2,000 amperes

Therefore, they are classified as 100, 225, 400, 1,000 and 2,000-ampere frames. These numbers are commonly referred to as "frame classification" or "frame sizes" and are terms applied to groups of molded case circuit breakers which are physically interchangeable with each other.

CIRCUIT BREAKER RATINGS

The established voltage rating of a circuit breaker is based on its clearances or space, both through air and over surfaces, between components of the electrical circuit and between the electrical components and ground. Circuit breaker voltage ratings indicate the maximum electrical system voltage on which they can be applied.

A circuit breaker can be rated for either alternating current (ac) or direct current (dc) system applications or for both. Single-pole circuit breakers, rated at 120/240 volts ac or 125/250 volts dc can be used singly and in pairs on three-wire circuits having a neutral connected to the mid-point of the load. Single-pole circuit breakers rated at 120/240 volts ac or 125/250 volts dc also can be used in pairs on a two-wire circuit connected to the ungrounded conductors of a three-wire system. Two-pole or three-pole circuit breakers rated at 120/240 volts ac or 125/250 volts dc can be used only on a three-wire, direct current, or single-phase, alternating current system having a grounded neutral. Circuit breaker voltage ratings must be equal to or greater than voltage of the electrical system on which they are used.

Circuit breakers have two types of current ratings. The first — and the one that is used most often — is the *continuous current rating*. The second is the *ground-fault current-interrupting capacity*.

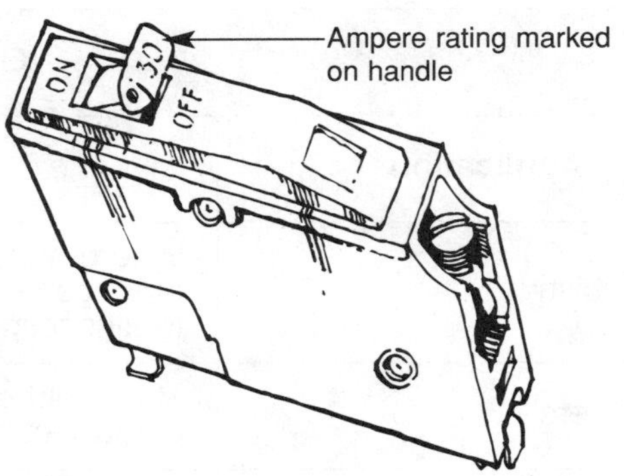
Ampere rating marked on handle
ON
30
OFF

Figure 4-9: *The current rating of a circuit breaker is located on the handle*

Current Rating

The rated continuous current of a device is the maximum current in amperes which it will carry continuously without exceeding the specified limits of observable temperature rise. Continuous current ratings of circuit breakers are established based on standard UL ampere ratings. These are 15, 20, 25, 30, 35, 40, 45, 50, 60, 70, 80, 90, 100, 110, 125, 150, 175, 200, 225, 250, 300, 350, 400, 450, 500, 600, 700, 800, 1000, 1200, 1600, 2000, 2500, 4000, 5000, and 6000 amperes. The ampere rating of a circuit breaker is located on the handle of the device and the numerical value alone is shown. See Figure 4-9.

General application requires that the circuit breaker current rating must be equal to or less than the load circuit conductor current-carrying capacity (ampacity). Consequently, in most cases, the conductor size dictates the circuit-breaker ampacity and rating.

Most overcurrent protective devices are labeled with the following current ratings:

- Normal current rating
- Intrerrupting rating

Summary

Reliable overcurrent-protective devices prevent or minimize costly damage to transformers, conductors, motors, and the other many components and electrical loads that make up the complete electrical distribution system. Consequently, reliable circuit protection is essential to avoid the severe monetary losses which can result from power blackouts and prolonged downtime of various types of facilities. Knowing these facts, the NFPA — via the *NEC* — has set forth various minimum requirements dealing with overcurrent-protective devices, and how they should be installed in various types of electrical systems.

Knowing how to select the type of overcurrent devices for specific applications is one of the basic requirements of every electrician. It is one of the best ways to ensure a safe, fault-free electrical installation that will give years of maintenance-free service.

Always remember that the conductor size and type must match the load, and then overcurrent protection must be provided for both; that is, the conductor and the load. These two items dictate the size of overcurrent protection. A summary of *NEC* requirements for overcurrent protective devices is shown in Figure 4-10.

Overcurrent Protection		
Application	**Requirements**	**NEC Reference**
Accessibility	Where more than one building is served under one management, overcurrent devices must be accessible to all occupants	Section 230-92
Arc welders	Must not be rated or set more than 200% of the rated primary current of the welder	Section 630-12
Bonding	Must be provided where necessary to assure electrical continuity and have the capacity to conduct any fault current safely	Section 250-90
Circuit impedance and other characteristics	The overcurrent device must be selected and coordinated as to permit them to clear faults without causing extension damage to any electrical components	Section 110-10
Conductors	Must be protected at their source of supply	Section 240-21
	Additional protection is not required when a smaller conductor (tap) is protected by the overcurrent device protecting the larger conductor	Section 240-21(b)(1-5)
Grounding	The path to ground must have capacity to conduct any fault current safely	Section 250-136
Installation	Listed or labeled equipment must be installed as per instructions in the listing or labeling	Section 110-3(b)
Interrupting rating	Must be sufficient for the system voltage and current	Section 110-9
Location in circuit	Device must be in series with each ungrounded conductor	Section 230-90(a) Exception No. 3
Location in building, electric service	Overcurrent device must be part of the service disconnecting means	Section 230-91
Number of devices, electric service	Up to six circuit breakers or sets of fuses may be used	Section 230-90(a)
Short-circuit current	Service equipment must be suitable for the short-circuit current at its terminals	Section 230-65
Protection required	All circuits and devices must be protected, except as otherwise specified	Section 230-94
Protection required, electric service	Each ungrounded conductor must have overcurrent protection	Section 230-90(a)
Resistance welders	Must not be rated or set more than 300% of the rated primary current of the welder	Section 630-32(b)
Scope (*FPN*)	Overcurrent protection for conductors and equipment must be provided	Section 240-1

Figure 4-10: *Summary of NEC requirements for overcurrent-protective devices*

Chapter 5
Branch Circuits and Feeders

The point at which electrical equipment is connected to the wiring system is commonly called an outlet. Conductors between the service equipment and the final overcurrent device, as in a subpanel, are called feeders. The conductors extending from the last overcurrent device to an outlet or outlets are called branch circuits.

A completely roughed-in electrical system for any type of building includes the following:

- All outlet boxes properly secured to the building structure — that is, outlet boxes for receptacles, wall switches, lighting, fixtures, junction boxes, pull boxes, and the like.
- All concealed cable wiring feeding the outlet boxes should be in place and properly secured, and all splices made in the outlet boxes. Wiring that will be partially exposed and partially concealed should also be installed during the rough-in wiring stage of the electrical installation. For raceway systems, only the outlet boxes and conduit is installed during the rough-in stage; conductors are pulled in the raceway when the system is complete.
- Flush-mounted panelboards, electric fans, electric heaters, and other flush-mounted equipment should be mounted and all wires connected to the housing of the equipment. The wires do not necessarily need to be connected to the equipment terminals as long as they are accessible for connection later on. For example, where a flush-mounted panelboard is used, only the housing will be installed and all cables and conduits connected to this housing. The loose wires are then left inside the empty housing until the wall covering is applied and finished. Once finished, the panel interiors (circuit breakers, fuse blocks, cable terminals, etc.) are installed, and the wiring is connected to their respective terminals. Unless the panel is energized before the final inspection, the panel cover is also left off until the final inspection.
- Service conductors from flush-mounted panels should be installed as well as service conductors running through a concealed area. If a conduit raceway system is used, only the raceway or conduit needs to be installed during the rough-in inspection; the conductors may be pulled in later.

BOXES

On every job, a great number of boxes is required for outlets, switches, pull and junction boxes. All of these must be sized, installed and supported to meet current *NEC* requirements. Since the *NEC* limits the number of conductors allowed in each outlet or switch box — according to its size — electricians must install boxes large enough to accommodate the number of conductors that must be spliced in the box or fed through. Therefore, a knowledge of the various types of boxes and the volume of each is essential.

A box or fitting must be installed at:

- Each conductor splice point
- Each outlet, switch point, or junction point
- Each pull point for the connection of conduit or other raceways

Furthermore, boxes or other fittings are required when a change is made from conduit to open wiring or cable. Electrical workers also install pull boxes in raceway systems to facilitate the pulling of conductors.

In each case — raceways, outlet boxes, pull and junction boxes — the *NEC* specifies specific maximum fill requirements; that is, the area of conductors in relation to the area of the box, fitting, or raceway system.

Sizing Outlet Boxes

In general, the maximum number of conductors permitted in standard outlet boxes is listed in Table 370-16(a) of the *NEC*. These figures apply where no fittings or devices such as fixture studs, cable clamps, switches, or receptacles are contained in the box and where no grounding conductors are part of the wiring within the box. Obviously, in all modern residential wiring systems there will be one or more of these items contained in the outlet box. Therefore, where one or more of the above mentioned items are present, the number of conductors must be one less than shown in the tables. For example, a deduction of two conductors must be made for each strap containing a device such as a switch or duplex receptacle; a further deduction of one conductor must be made for one or more grounded conductors entering the box. A 3-inch x 2-inch x 2¾-inch box for example, is listed in the table as containing a maximum number of six No. 12 wires. If the box contains cable clamps and a duplex receptacle, three wires will have to be deducted from the total of six — providing for only three No. 12 wires. If a ground wire is used, only two No. 12 wires may be used.

Figure 5-1 illustrates one possible wiring configuration for outlet boxes and the maximum number of conductors permitted in them as governed by Section 370-16 of the *NEC*. This example shows two single-gang switch boxes joined or "ganged" together to hold a single-pole toggle switch and a duplex receptacle. This type of arrangement is likely to be found above kitchen countertops where the duplex receptacle is provided for small appliances and the single-pole switch could be used to control a garbage disposal. This arrangement is also useful above a workbench — the receptacle for small power tools and the switch to control lighting over the bench.

Since Table 370-16(a) gives the capacity of one 3 x 2 x 2½-inch device box as 12.5 cubic inches, the total capacity of both boxes in Figure 5-1 is 25 cubic inches. These two boxes have a capacity to allow 10 No. 12 AWG conductors, or 12 No. 14 AWG conductors, less the deductions as listed below.

- Two conductors must be deducted for each strap-mounted device. Since there is one duplex receptacle (X) and one single-pole toggle switch (Y), four conductors must be deducted from the total number stated in the above paragraph.
- Since the combined boxes contain one or more cable clamps (Z), another conductor must be deducted. Note that only one deduction is made for similar clamps, regardless of the number. However, any unused clamps may be removed to facilitate the electrical worker's job; that is, allowing for more work space.
- The equipment grounding conductors, regardless of the number, count as one conductor only.

To comply with the *NEC*, and considering the combined deduction of six conductors, only four No. 12 AWG conductors (six No. 14 AWG conductors) may be installed in the outlet-box configuration in Figure 5-1.

Figure 5-1 shows three nonmetallic-sheathed (NM or Romex) cables, designated 12/2 with ground, entering the ganged outlet boxes. This is a total of six current-carrying conductors and three ground wires, for a total of nine. Is this arrangement in violation of the *NEC*? Yes, because the total number of conductors exceeds the *NEC* limits. However, if No. 14 AWG conductors were installed rather than No. 12, the configuration will comply with the 1999 *NEC*. Another alternative is to go to 3 x 2 x 3½-inch device boxes which would then have a total of 36 cubic inches for the two boxes.

Also note the jumper wire in Figure 5-1; this is numbered "8" in the drawing. Conductors that both originate and end in the same outlet box are exempt from being counted against the allowable capacity of an outlet box. This jumper wire (8) taps off one terminal of the duplex

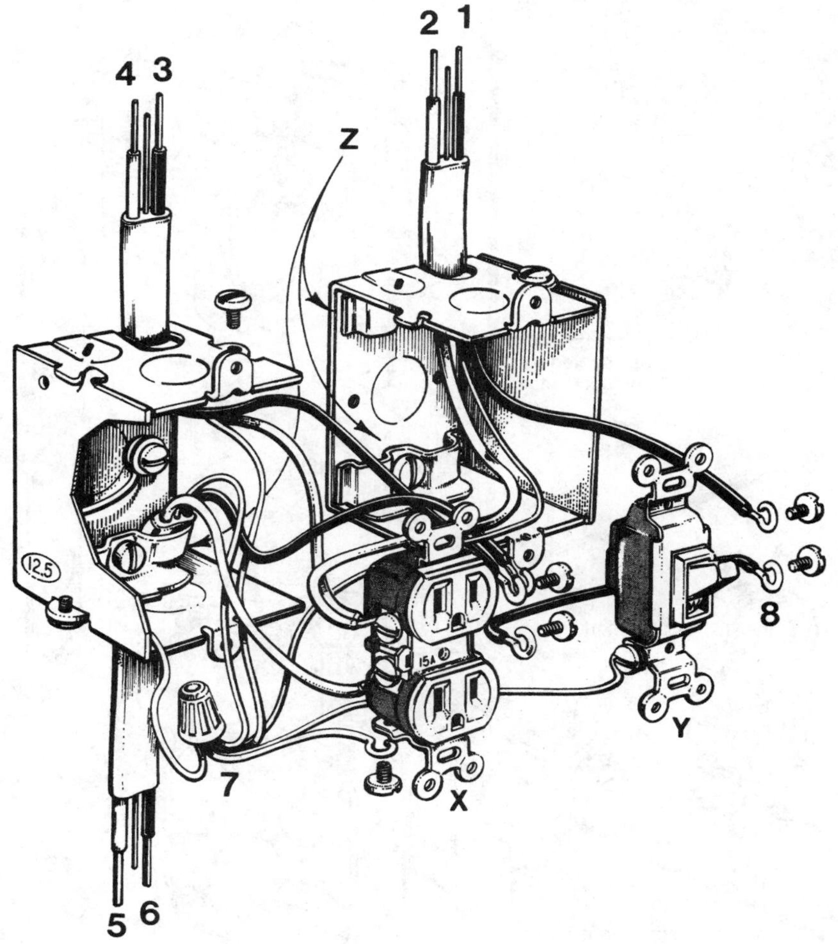
2 1
4 3
Z
12.5
15A
8
Y
7
X
5 6

Figure 5-1: *One possible box-sizing configuration for outlet boxes*

receptacle to furnish a "hot wire" to the single-pole toggle switch. Therefore, this wire originates and terminates in the same set of ganged boxes and is not counted against the total number of conductors. By the same token, the three grounding conductors extending from the wire nut to the individual grounding screws on the devices originate and terminate in the same set of boxes. These conductors are also exempt from being counted with the total.

A pictorial definition of stipulated conditions as they apply to Section 370-16 of the *NEC* is shown in Figure 5-2. Figure 5-2A illustrates an assortment of raised covers and outlet box extensions. These components, when combined with the appropriate outlet boxes, serve to increase the usable work space. Each type is marked with their cubic-inch capacity which may be added to the figures in *NEC* Table 370-16(a) to calculate the increased number of conductors allowed.

Figure 5-2B shows components that may be used in outlet boxes without affecting the total number of conductors. Such items include grounding clips and screws, wire nuts and cable connectors when the latter is inserted through knockout holes in the outlet box and secured with lockouts. Prewired fixture wires are not counted against the total number of allowable conductors in an outlet box where there are not over four wires smaller than No. 14 AWG. Conductors originating and ending in the same box need not be counted.

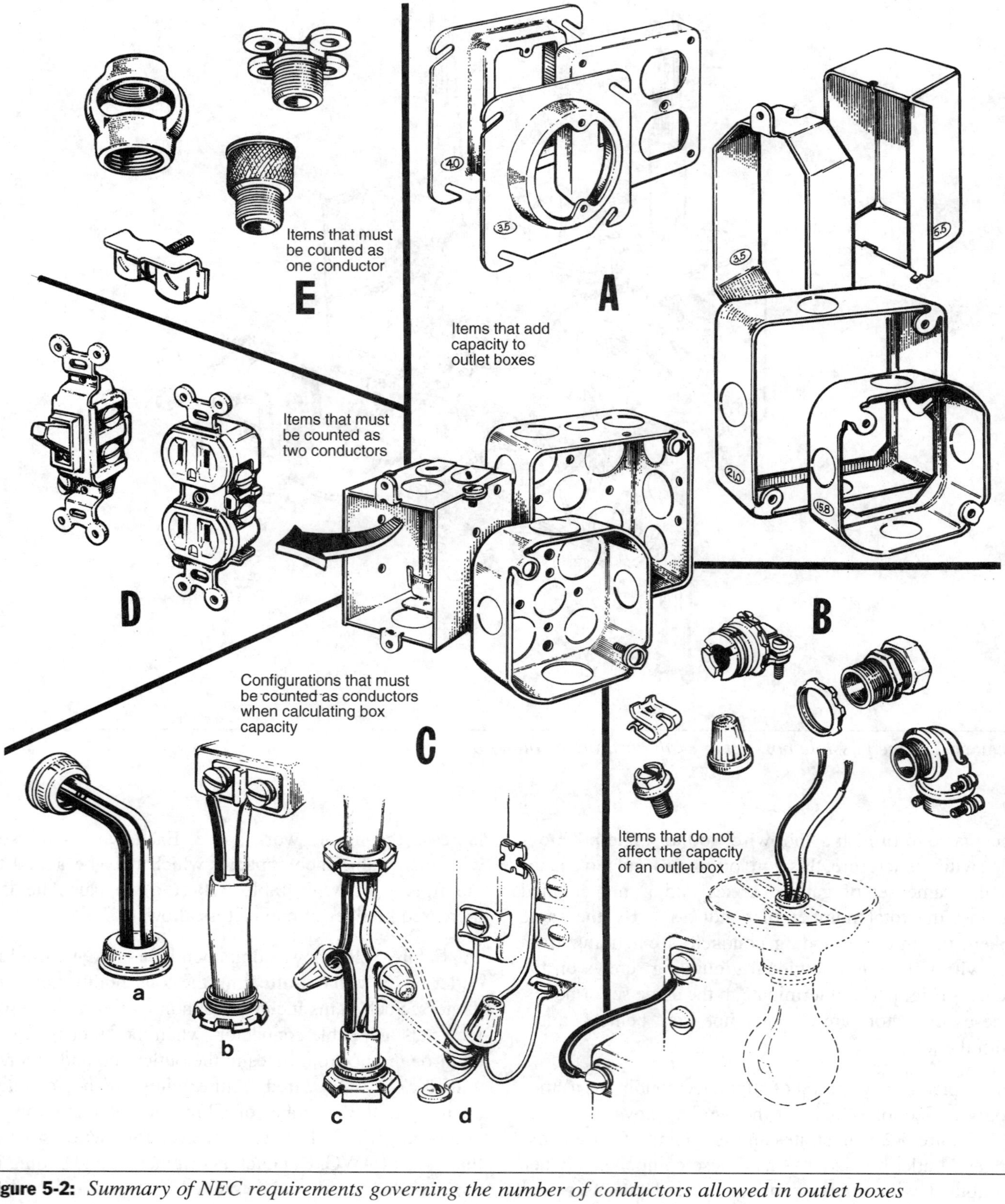
Items that must
be counted as
one conductor
E
Items that add
capacity to
outlet boxes
A
Items that must
be counted as
two conductors
D
B
Configurations that must
be counted as conductors
when calculating box
capacity
C
Items that do not
affect the capacity
of an outlet box
a
b
c
d

Figure 5-2: *Summary of NEC requirements governing the number of conductors allowed in outlet boxes*

Figure 5-2C shows typical wiring configurations that must be counted as conductors when calculating the total capacity of outlet boxes. Further details of these configurations are as follows:

- a: Each wire passing through a box without a splice or tap is counted as one conductor. Therefore, a cable containing two wires (shown in the drawing) that passes in and out of an outlet box without a splice or tap is counted as two conductors.
- b: Wires that enter and terminate in the same box are charged as individual conductors and in this case, the total charge would be two conductors.
- c: Each wire that enters a box and is either spliced or connected to a terminal, and then exits again, is counted as two conductors. In the case of two 2-wire cables, the total conductors charged will be four.
- d: When one or more grounding wires enter the box and are joined, a deduction of only one conductor is required. Where an additional set of equipment grounding conductors is present in a box, as permitted by Section 250-146(d), an additional volume allowance shall be made based on the largest equipment grounding conductor in that additional set.

Two conductors must be deducted for each strap-mounted device, like duplex receptacles and wall switches as shown in Figure 5-2D.

Figure 5-2E shows further components that require deduction adjustments from those specified in Table 370-16(a). Such items include fixture studs, hickeys, and fixture-stud extensions and one conductor must also be deducted from the total for each type of fitting used. A deduction of one conductor is made when one or more internally-mounted cable clamps are used.

To better understand how outlet boxes are sized, let's take three No. 12 AWG conductors installed in ½-inch EMT and terminating into a metallic outlet box containing one duplex receptacle. What size of outlet box will meet *NEC* requirements?

The first step is to count the total number of conductors and equivalents that will be used in the box — following the requirements specified in *NEC* Section 370-16.

Step 1. Calculate the total number of conductors and equivalents.

One receptacle = 2 conductors

Three No. 12 conductors = 3 conductors

Total No. 12 conductors = 5

Step 2. Determine amount of space required for each conductor.

NEC Table 370-16(b) gives the box volume required for each conductor:

No. 12 AWG = 2.25 cubic inches

Step 3. Calculate the outlet-box space required by multiplying the number of cubic inches required for each conductor by the number of conductors found in No. 1 above.

5 x 2.25 = 11.25 cubic inches

Once you have determined the required box capacity, again refer to *NEC* Table 370-16(a) and note that a 3 x 2 x 2½-inch box comes closest to our requirements. This box size is rated for 12.5 cubic inches.

Where six No. 12 conductors enter the box, two additional No. 12 conductors must be added to our previous count for a total of (6 + 2 =) 8 conductors.

8 x 2.25 = 18.0 cubic inches

Again, refer to *NEC* Table 370-16(a) and note that a 3 x 2 x 3-inch device box, with a rated capacity of 18.0 cubic inches, is the closest device box that meets *NEC* requirements. Of course, any box with a larger capacity is permitted.

CONDUIT BODIES, PULL BOXES AND JUNCTION BOXES

The *NEC* specifically states that at each splice point, or pull point for the connection of conduit or other raceways, a box or fitting must be installed. The *NEC* specifically considers conduit bodies, pull boxes, and junction boxes and specifies the installation requirements as listed in the table in Figure 5-3.

CONDUIT BODIES, PULL AND JUNCTION BOXES		
Application	**Requirements**	**NEC Reference**
Accessible	Must provide access to wiring inside of box without removing any part of the building.	Section 370-29
Cellular floor raceways	Must be leveled to the floor grade and sealed to prevent free entry of water or concrete.	Section 354-13
Conduit bodies, size of	Those enclosing No. 6 or smaller conductors must have a cross-sectional area not less than twice that of the raceway system to which the conduit bodies are attached.	Section 370-16(c)
Conduit bodies, maximum number of conductors	Must not exceed the allowable fill for the attached conduit.	Section 370-16(c)
Junction and pull boxes	Must comply with (a) through (d) of this NEC Section.	Section 370-28
Junction and pull boxes over 6 ft	Conductors must be cabled or racked up.	Section 370-28(b)
Over 600 V	Special requirements apply to boxes used on systems of over 600 V.	Article 370 Part E
Size, straight pulls	Not less than 8 times the trade diameter of the largest raceway.	Section 370-28(a)(1)
Size, angle or U pulls	Not less than 6 times the trade diameter of the largest raceway.	Section 370-28(a)(2)

Figure 5-3: *Summary of NEC installation requirements for conduit bodies, pull and junction boxes*

Conduit bodies provide access to the wiring through removable covers. Typical examples are Types T, C, X, L, and LB. Conduit bodies enclosing No. 6 or smaller conductors must have an area twice that of the largest conduit to which they are attached, but the number of conductors within the body must not exceed that allowed in the conduit. If a conduit body has entry for three or more conduits such as Type T or X, splices may be made within the conduit body. Splices may not be made in conduit bodies having one or two entries unless the volume is sufficient to qualify the conduit body as a junction box or device box.

When conduit bodies or boxes are used as junction boxes or as pull boxes, a minimum size box is required to allow conductors to be installed without undue bending. The calculated dimensions of the box depend on the type of conduit arrangement and on the size of the conduits involved.

Sizing Pull And Junction Boxes

Figure 5-4 shows a junction box with several conduits entering it. Since 4-inch conduit is the largest size in the group, the minimum length required for the box can be determined by the following calculation:

Trade size of conduit x 8 (as per *NEC*) = minimum length of box

4" x 8 = 32"

Therefore, this particular pull box must be at least 32 inches in length. The width of the box, however, need be only of sufficient size to enable locknuts and bushings to be installed on all the conduits or connectors entering the enclosure.

Junction or pull boxes in which the conductors are pulled at an angle, as shown in Figure 5-5 on the next page, must have a distance of not less than six times the

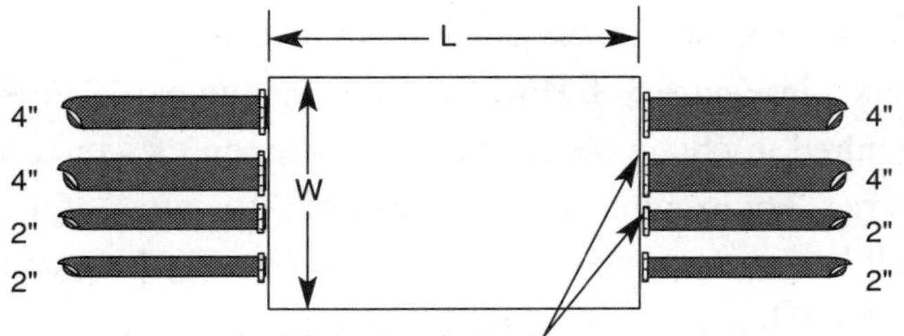
L
W
4"
4"
2"
2"
4"
4"
2"
2"
Allow enough space between conduits to install locknuts and bushings

Figure 5-4: *Pull box used on straight pulls*

trade diameter of the largest conduit. The distance must be increased for additional conduit entries by the amount of the sum of the diameter of all other conduits entering the box on the same side, that is, the wall of the box. The distance between raceway entries enclosing the same conductors must not be less than six times the trade diameter of the largest conduit.

Since the 4-inch conduit is the largest of the lot in this case:

$$L_1 = 6 \times 4 + (3 + 2) = 29"$$

Since the same number and sizes of conduit is located on the adjacent wall of the box, L_2 is calculated in the same way; therefore, L_2 = 29 inches.

The distance (D) = 6 x 4 or 24 inches and this is the minimum distance permitted between conduit entries enclosing the same conductor.

The depth of the box need only be of sufficient size to permit locknuts and bushings to be properly installed. In this case, a 6-inch deep box would suffice.

If the conductors are smaller than No. 4, the length restriction does not apply.

Figure 5-6 shows another straight-pull box. What is the minimum length if the box has one 3-inch conduit and two 2-inch conduits entering and leaving the box? Again, refer to *NEC* Section 370-28(a)(1) and find that the minimum length is 8 times the largest conduit size which in this case is:

8 x 3 inches = 24 inches

Let's review the installation requirements for pull or junction boxes with angular or U pulls. Two conditions must be met in order to determine the length and width of the required box.

The minimum distance to the opposite side of the box from any conduit entry must be at least six times the trade diameter of the largest raceway.

The sum of the diameters of the raceways on the same wall must be added to this figure.

Figure 5-7 shows the minimum length of a box with two 3-inch conduits, two 2-inch conduits, and two 1½-inch conduits in a right-angle pull. The minimum length based on this configuration is:

6 x 3 inches	=	18 inches
1 x 3 inches	=	3 inches
2 x 2 inches	=	4 inches
2 x 1½ inches	=	3 inches
		28 inches

Since the number and size of conduits on the two sides of the box are equal, the box is square and has a minimum size dimension of 28 inches. However, the distance between conduit entries must now be checked to ensure that *all NEC* requirements are met; that is, the spacing (D) between conduits enclosing the same con-

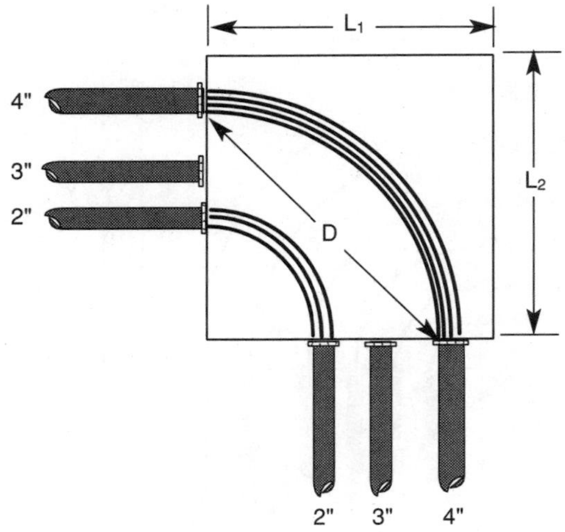
L_1
L_2
4"
3"
2"
D
2"
3"
4"

Figure 5-5: *Pull box with conduit runs entering and leaving at right angles*

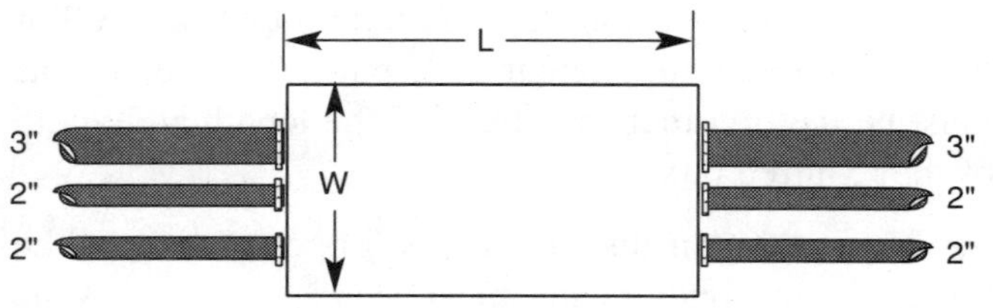

L
3"
2"
2"
W
3"
2"
2"

Figure 5-6: *Typical straight-pull box*

ductor must not be less than six times the conduit diameter. Again refer to Figure 5-7 and note that the 1½-inch conduits are the closest to the left-hand corner of the box. Therefore, the distance (D) between conduit entries must be:

$$6 \times 1\tfrac{1}{2} \text{ inches} = 9 \text{ inches}$$

The next group is the two 2-inch conduits which is calculated in a similar fashion; that is:

$$6 \times 2 = 12 \text{ inches}$$

The remaining raceways in this example are the two 3-inch conduits and the minimum distance between the 3-inch conduit entries must be:

$$6 \times 3 = 18 \text{ inches}$$

A summary of the conduit-entry distances is presented in Figure 5-8. However, some additional math is required to obtain the spacing (w) between the conduit entries. For example, the distance from the corner of the pull box to the center of the conduits (w) may be found by the following equation:

$$\text{Spacing} = \frac{\text{Diagonal distance (D)}}{\sqrt{2}}$$

Consequently, the spacing (w) for the 1½-inch conduit may be determined using the following equation:

$$\frac{9}{\sqrt{2}} = \frac{9}{1.414} = 6.4 \text{ inches}$$

Therefore, the spacing (w) is 6.4 inches. This distance is measured from the left lower corner of the box in each direction — both vertically and horizontally — to obtain the center of the first set of 1½-inch conduits. This distance must be added to the spacing of the other conduits including locknuts or bushings.

A rule-of-thumb is to allow ½ inch clearance between locknuts.

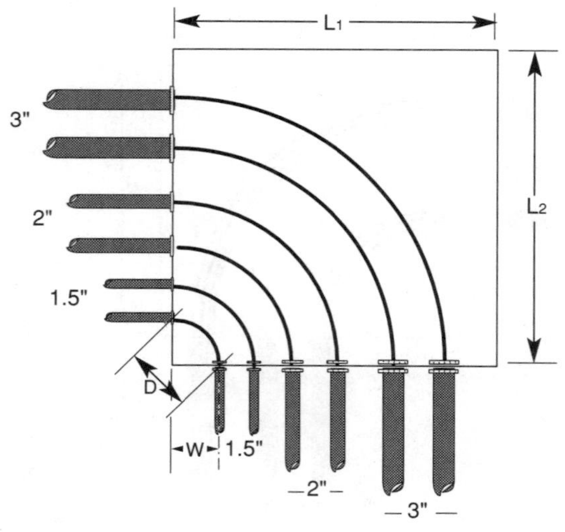

L1
L2
3"
2"
1.5"
D
W
1.5"
2"
3"

Figure 5-7: *Minimum size pull box for angle conduit entries*

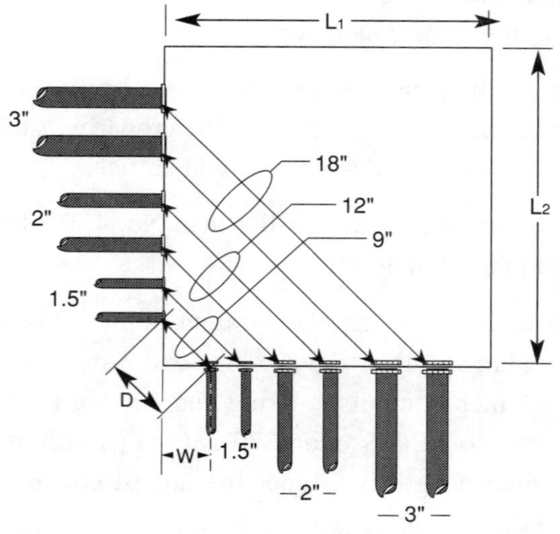

L1
L2
3"
2"
1.5"
18"
12"
9"
D
W
1.5"
2"
3"

Figure 5-8: *Required distances between conduit entries*

Using all information calculated thus far, and using Figure 5-8 as reference, the required measurements of the pull box may be further calculated as follows:

Step 1. Calculate space (w):

D = 6 x 1½ inches = 9 inches

Step 2. Divide this number (9 inches) by the square root of 2 (1.414) and make the following calculation:

$$w = \frac{9 \text{ inches}}{1.414} = 6.4 \text{ inches}$$

Step 3. Measure from the left, lower corner of the pull box over 6.4 inches to obtain the center of the knockout for the first 1½-inch conduit. Measure up (from the lower, left corner) to obtain the center of the knockout for this same cable run on the left side of the pull box.

Step 4. Since there are two 3-inch (inside diameter) conduits, each with an outside diameter of approximately 4.25 inches, the space for these two conduits can be found by the following equation:

2 x 4.25 = 8.5 inches

Step 5. The space required for the two 2-inch (inside diameter) conduits, each with an outside diameter of approximately 3.12 inches, may be determined in a similar manner; that is:

2 x 3.12 = 6.24 inches

Step 6. The space required for the two 1.5-inch (inside diameter) conduits, each with an outside diameter of approximately 2.62 inches, may be determined using the same equation:

2 x 2.62 = 5.24 inches

Step 7. To find the required space for locknuts and bushings, multiply 0.5 inch by the total number of conduit entries on one side of the box. Since there are a total of 6 conduit entries, use the following equation:

6 x .5 = 3.0 inches

Step 8. Add all figures obtained in Steps 2 through 7 together to obtain the total required length of the pull box.

Clear space (w)	=	6.4 inches
1.5-inch conduits	=	5.24 inches
2-inch conduits	=	6.24 inches
3-inch conduits	=	8.5 inches
Space between locknuts	=	3.0 inches
Total length of box	=	29.38 inches

Since the same number and size of conduits enter on the bottom side of the pull box and leave, at a right angle, on the left side of the pull box, the box will be square. Furthermore, although a box exactly 29.38 inches will suffice for this application, the next larger standard size is 30 inches; this should be the size pull box selected. Even if a "custom" pull box is made in a sheet-metal shop, the workers will still probably make it an even 30 inches unless specifically ordered otherwise.

Cabinets And Cutout Boxes

NEC Article 373 deals with the installation requirements for cabinets, cutout boxes, and meter sockets. In general, where cables are used, each cable must be secured to the cabinet or cutout box by an approved method. Furthermore, the cabinets or cutout boxes must have sufficient space to accommodate all conductors installed in them without crowding.

NEC Table 373-6(a), which is reproduced in Figure 5-9, gives the minimum wire-bending space at terminals along with the width of wiring gutter in inches. Figure 5-10 gives a summary of *NEC* requirements for the installation of cabinets and cutout boxes.

AWG or Circular-Mil Size of Wire	Wires per Terminal				
	1	2	3	4	5
14-10	Not Specified	—	—	—	—
8-6	1½	—	—	—	—
4-3	2	—	—	—	—
2	2½	—	—	—	—
1	3	—	—	—	—
1/0-2/0	3½	5	7	—	—
3/0-4/0	4	6	8	—	—
250 kcmil	4½	6	8	10	—
300-350 kcmil	5	8	10	12	—
400-500 kcmil	6	8	10	12	14
600-700 kcmil	8	10	12	14	16
750-900 kcmil	8	12	14	16	18
1000-1250 kcmil	10	—	—	—	—
1500-2000 kcmil	12	—	—	—	—

Figure 5-9: *Minimum wire-bending space in inches*

Other basic *NEC* requirements for cabinets and cutout boxes are as follows:

- Table 373-6(a) must apply where the conductor does not enter or leave the enclosure through the wall opposite its terminal.
- *Exception No. 1 states:* A conductor must be permitted to enter or leave an enclosure through the wall opposite its terminal provided the conductor enters or leaves the enclosure where the gutter joins an adjacent gutter that has a width that conforms to Table 373-6(b) for that conductor.
- *Exception No. 2 states:* A conductor not larger than 350 kcmil must be permitted to enter or leave an enclosure containing only a meter socket(s) through the wall opposite its terminal, provided the terminal is a lay-in type where either: (a) The terminal is directly facing the enclosure wall and offset is not greater than 50 percent of the bending space specified in Table 373-6(a); or (b) The terminal is directed toward the opening in the enclosure and is within a 45-degree angle of directly facing the enclosure wall.
- Table 373-6(b) must apply where the conductor enters or leaves the enclosure through the wall opposite its terminal.

NEC Article 374 covers the installation requirements for auxiliary gutters, which are permitted to supplement wiring spaces at meter centers, distribution centers, and similar points of wiring systems and may enclose conductors or busbars but must not be used to enclose switches, overcurrent devices, appliances, or other similar equipment.

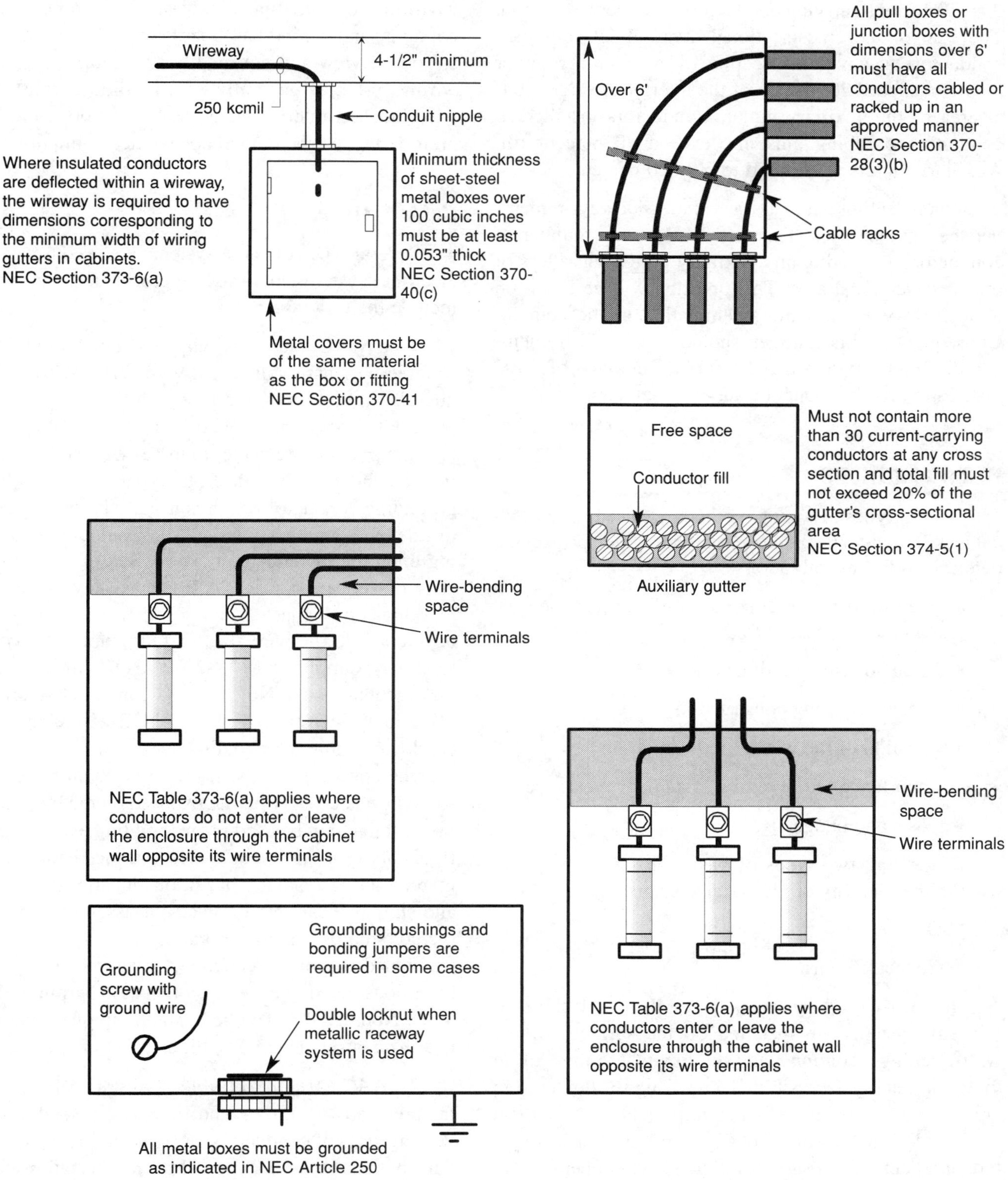
Wireway
4-1/2" minimum
250 kcmil
Conduit nipple
Where insulated conductors are deflected within a wireway, the wireway is required to have dimensions corresponding to the minimum width of wiring gutters in cabinets. NEC Section 373-6(a)
Minimum thickness of sheet-steel metal boxes over 100 cubic inches must be at least 0.053" thick NEC Section 370-40(c)
Metal covers must be of the same material as the box or fitting NEC Section 370-41
Over 6'
All pull boxes or junction boxes with dimensions over 6' must have all conductors cabled or racked up in an approved manner NEC Section 370-28(3)(b)
Cable racks
Free space
Conductor fill
Auxiliary gutter
Must not contain more than 30 current-carrying conductors at any cross section and total fill must not exceed 20% of the gutter's cross-sectional area NEC Section 374-5(1)
Wire-bending space
Wire terminals
NEC Table 373-6(a) applies where conductors do not enter or leave the enclosure through the cabinet wall opposite its wire terminals
Wire-bending space
Wire terminals
NEC Table 373-6(a) applies where conductors enter or leave the enclosure through the cabinet wall opposite its wire terminals
Grounding bushings and bonding jumpers are required in some cases
Grounding screw with ground wire
Double locknut when metallic raceway system is used
All metal boxes must be grounded as indicated in NEC Article 250

Figure 5-10: *Summary of NEC installation requirements for cabinets, cutout boxes, and auxiliary gutters*

In general, auxiliary gutters must not contain more than 30 current-carrying conductors at any cross section. The sum of the cross-sectional areas of all contained conductors at any cross section of an auxiliary gutter must not exceed 20 percent of the interior cross-sectional area of the auxiliary gutter. Conductors installed in conduits and tubing must not exceed 40 percent fill. Auxiliary gutters are limited to only 20 percent.

When dealing with auxiliary gutters, always remember the number "30." This is the maximum number of conductors allowed in any auxiliary gutter regardless of the cross-sectional area. This question will be found on almost every electrician's examination in the country. Consequently, this number should always be remembered. Refer to Figure 5-10 for a summary of *NEC* installation requirements for auxiliary gutters.

WIRING METHODS

Several types of wiring methods are used for electrical installations. The methods used on a given project are determined by several factors:

- The installation requirements set forth in the *National Electrical Code* (*NEC*)
- Local codes and ordinances
- Type of building construction
- Location of the wiring in the building
- Importance of the wiring system's appearance
- Costs and budget

In general, two types of basic wiring methods are used in the majority of electrical systems:

- Open wiring
- Concealed wiring

In open-wiring systems, the outlets and cable or raceway systems are installed on the surfaces of the walls, ceilings, columns, and the like where they are in view and readily accessible. Such wiring is often used in areas where appearance is not important and where it may be desirable to make changes in the electrical system at a later date. You will frequently find open-wiring systems in mechanical rooms and in interior parking areas of commercial buildings.

Concealed wiring systems have all cable and raceway runs concealed inside of walls, partitions, ceilings, columns, and behind baseboards or molding where they are out of view and not readily accessible. This type of wiring system is generally used in all new construction with finished interior walls, ceilings, floors and is the preferred type where good appearance is important.

Cable Systems

Several types of cable systems are used to construct electrical systems in various types of occupancies, and include the following:

Type NM Cable: This cable is manufactured in two- or three-wire assemblies, and with varying sizes of conductors. In both two- and three-wire cables, conductors are color-coded: one conductor is black while the other is white in two-wire cable; in three-wire cable, the additional conductor is red. Both types will also have a grounding conductor which is usually bare, but is sometimes covered with a green plastic insulation — depending upon the manufacturer. The jacket or covering consists of rubber, plastic, or fiber. Most will also have markings on this jacket giving the manufacturer's name or trademark, the wire size, and the number of conductors. For example, "NM 12-2 W/GRD" indicates that the jacket contains two No. 12 AWG conductors along with a grounding wire; "NM 12-3 W/GRD" indicates three conductors plus a grounding wire. This type of cable may be concealed in the framework of buildings, or in some instances, may be run exposed on the building surfaces. It may not be used in any building exceeding three floors above grade; as a service-entrance cable; in commercial garages having hazardous locations; in theaters and similar locations; in places of assembly; in motion picture studios; in storage battery rooms; in hoistways; embedded in poured concrete, or aggregate; or in any hazardous location except as otherwise permitted by the *NEC*. Nonmetallic-sheathed cable is frequently referred to as *Romex* on the job. See Figure 5-11 on the next page.

Type AC (Armored) Cable: Type AC cable is manufactured in two-, three-, and four-wire assemblies, with varying sizes of conductors, and is used in locations similar to those where Type NM cable is allowed. The metallic spiral covering on AC cable offers a greater degree of mechanical protection than with NM cable,

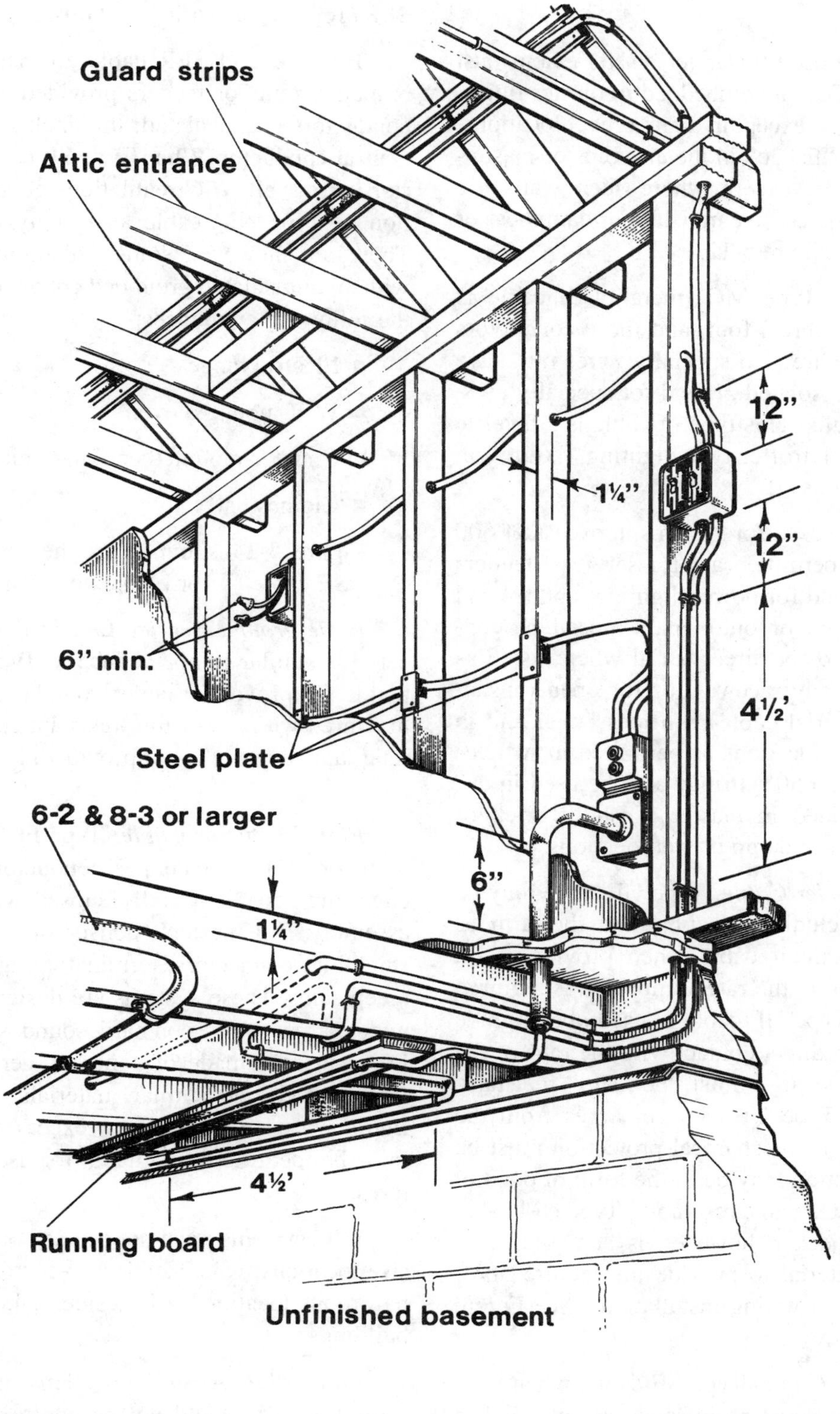
Guard strips
Attic entrance
12"
1¼"
12"
6" min.
4½'
Steel plate
6-2 & 8-3 or larger
6"
1¼"
4½'
Running board
Unfinished basement

Figure 5-11: *Summary of NEC installation requirements for nonmetallic-sheathed cable*

and the metal jacket also provides a continuous grounding bond without the need for additional grounding conductors.

AC cable may be used for under-plaster extensions, as provided in the *NEC*, and embedded in plaster finish, brick, or other masonry, except in damp or wet locations. It may also be run or "fished" in the air voids of masonry block or tile walls, except where such walls are exposed or subject to excessive moisture or dampness or are below grade. See Figure 5-12.

Type MC Cable: Type MC (metal-clad) cable is manufactured in two, three, four, and more conductors with a ground. There are also specialty wires that have both a ground and an isolated ground for specialty uses, such as hospital circuits. This type of cable is a favorite for connecting 2 x 4 troffer-type lighting fixtures in commercial installations.

MC cable may be used for systems in excess of 600 volts nominal. Uses permitted are for services, feeders and branch circuits, and for power, lighting, control and signal circuits. Indoors or outdoors, exposed or concealed, it may be used for direct burial where listed as such, in cable trays, any raceway, or for open runs. It may be used as an aerial cable on a messenger and in hazardous (classified) locations as permitted in Articles 501, 502, 503, 504 and 505. It may also be used in dry locations and embedded in plaster finish or brick or other masonry, except in damp or wet locations.

Underground Feeder Cable: Type UF cable may be used underground, including direct burial in the earth, as a feeder or branch-circuit cable when provided with overcurrent protection at the rated ampacity as required by the *NEC*. When Type UF cable is used above grade where it will come in direct contact with the rays of the sun, its outer covering must be sun resistant. Furthermore, where Type UF cable emerges from the ground, some means of mechanical protection must be provided. This protection may be in the form of conduit or guard strips. Type UF cable resembles Type NM cable in appearance. The jacket, however, is constructed of weather resistant material to provide the required protection for direct-burial wiring installations. See Figure 5-13.

Service-Entrance Cable: Type SE cable, when used for electrical services, must be installed as specified in *NEC* Article 230. This cable is available with the grounded conductor bare for outside service conductors, and also with an insulated grounded conductor (Type SER) for interior wiring systems.

Type SE and SER cable are permitted for use on branch circuits or feeders provided all current-carrying conductors are insulated; this includes the grounded or neutral conductor. When Type SE cable is used for interior wiring, all *NEC* regulations governing the installation of Type NM cable also apply to Type SE cable. There are, however, some exceptions. Type SE cable with an uninsulated grounded conductor may be used on the following appliances:

- Electric range
- Wall-mounted oven
- Counter-mounted cooking unit
- Clothes dryer

Figure 5-14 summarizes the installation rules for Type SE cable — for both exterior and interior wiring.

Underground Service-Entrance Cable: Type USE cable is similar in appearance to Type SE cable except that it is approved for underground use and must be manufactured with a moisture-resistant covering. If a flame-retardant covering is not provided, it is not approved for indoor use.

Flat Conductor Cable: Type FCC cable consists of three or more flat copper conductors placed edge-to-edge and separated and enclosed within an insulating assembly. FCC systems consist of cable and associated shielding, connectors, terminators, adapters, boxes and receptacles. These systems are designed for installation under carpet squares on hard, sound, smooth, continuous floor surfaces made of concrete, ceramic, composition floor, wood, and similar materials. If used on heated floors with temperatures in excess of 86°F, the cable must be identified as suitable for use at these temperatures.

FCC systems must not be used outdoors or in corrosive locations; where subject to corrosive vapors; in any hazardous location; or in residential, school, or hospital buildings.

Flat-Cable Assemblies: This is Type FC cable assembly and should not be confused with Type FCC cable; there is a big difference. A Type FC wiring system

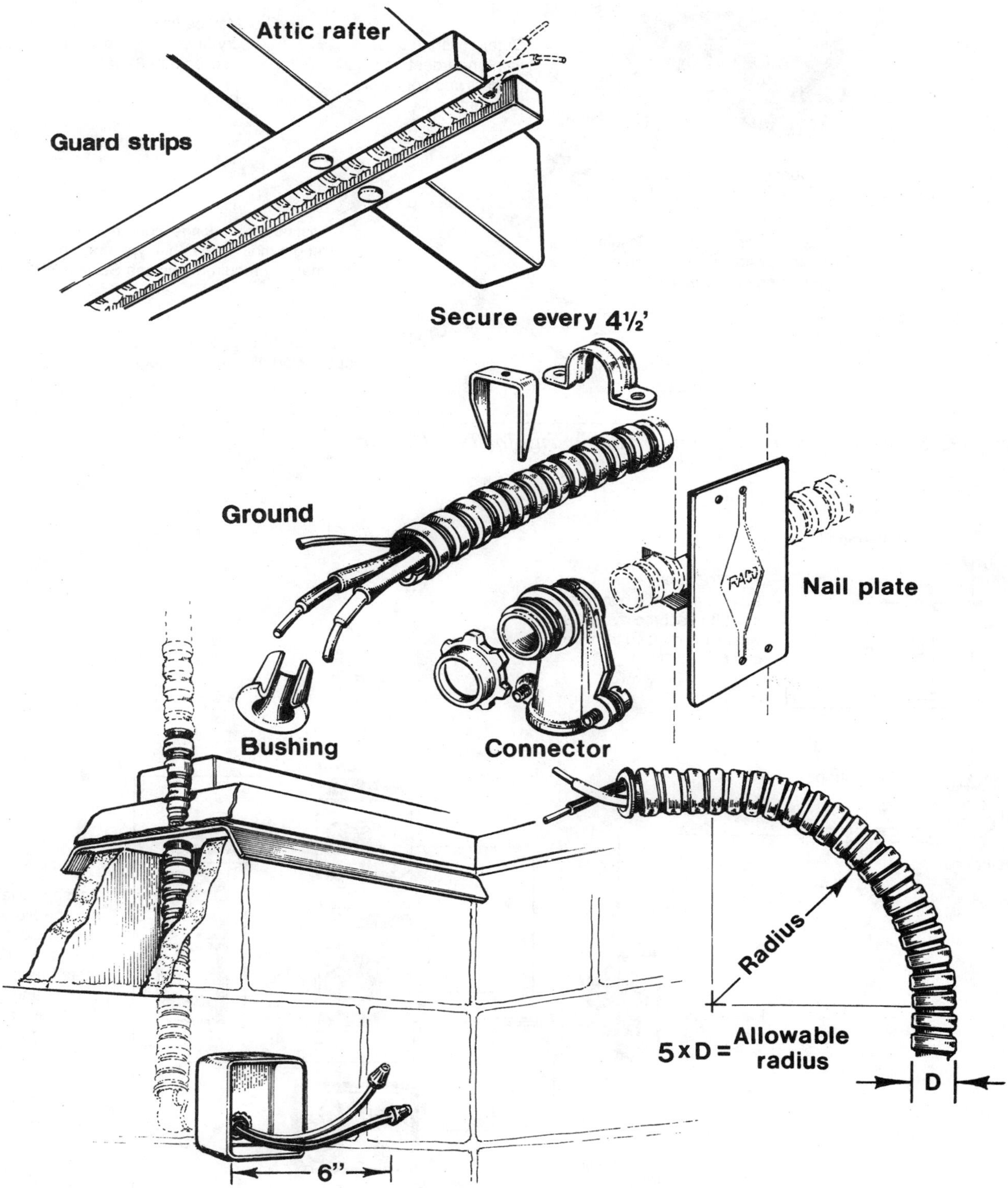
Attic rafter
Guard strips
Secure every 4½'
Ground
Nail plate
Bushing
Connector
Radius
5 x D = Allowable radius
D
6"

Figure 5-12: *Summary of NEC installation requirements for Type AC (MC) cable*

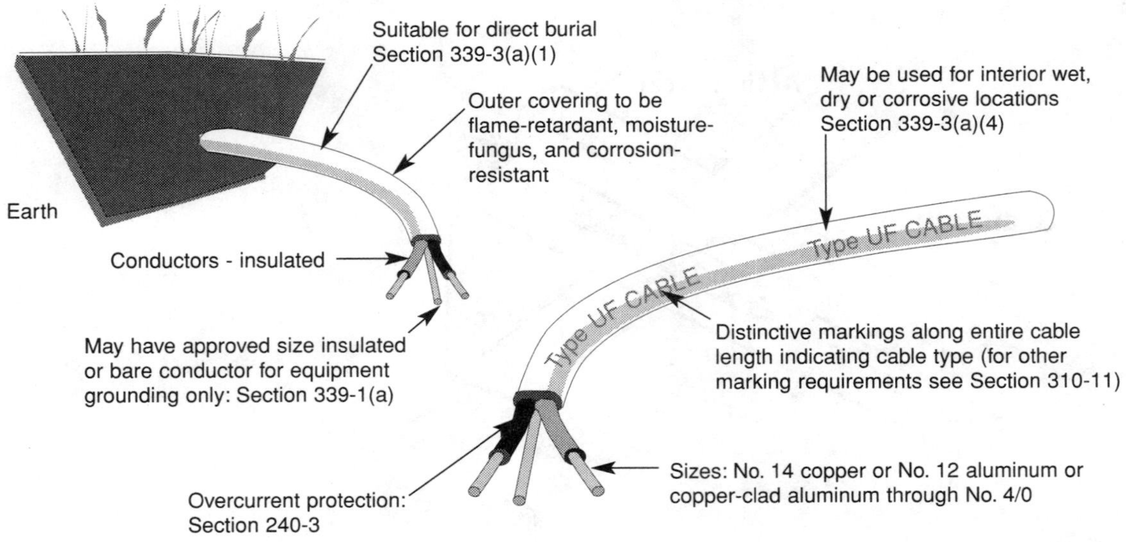
Suitable for direct burial
Section 339-3(a)(1)
Outer covering to be flame-retardant, moisture-fungus, and corrosion-resistant
May be used for interior wet, dry or corrosive locations
Section 339-3(a)(4)
Earth
Conductors - insulated
Type UF CABLE
Type UF CABLE
May have approved size insulated or bare conductor for equipment grounding only: Section 339-1(a)
Distinctive markings along entire cable length indicating cable type (for other marking requirements see Section 310-11)
Sizes: No. 14 copper or No. 12 aluminum or copper-clad aluminum through No. 4/0
Overcurrent protection:
Section 240-3

Figure 5-13: *Summary of NEC installation requirements for Type UF cable*

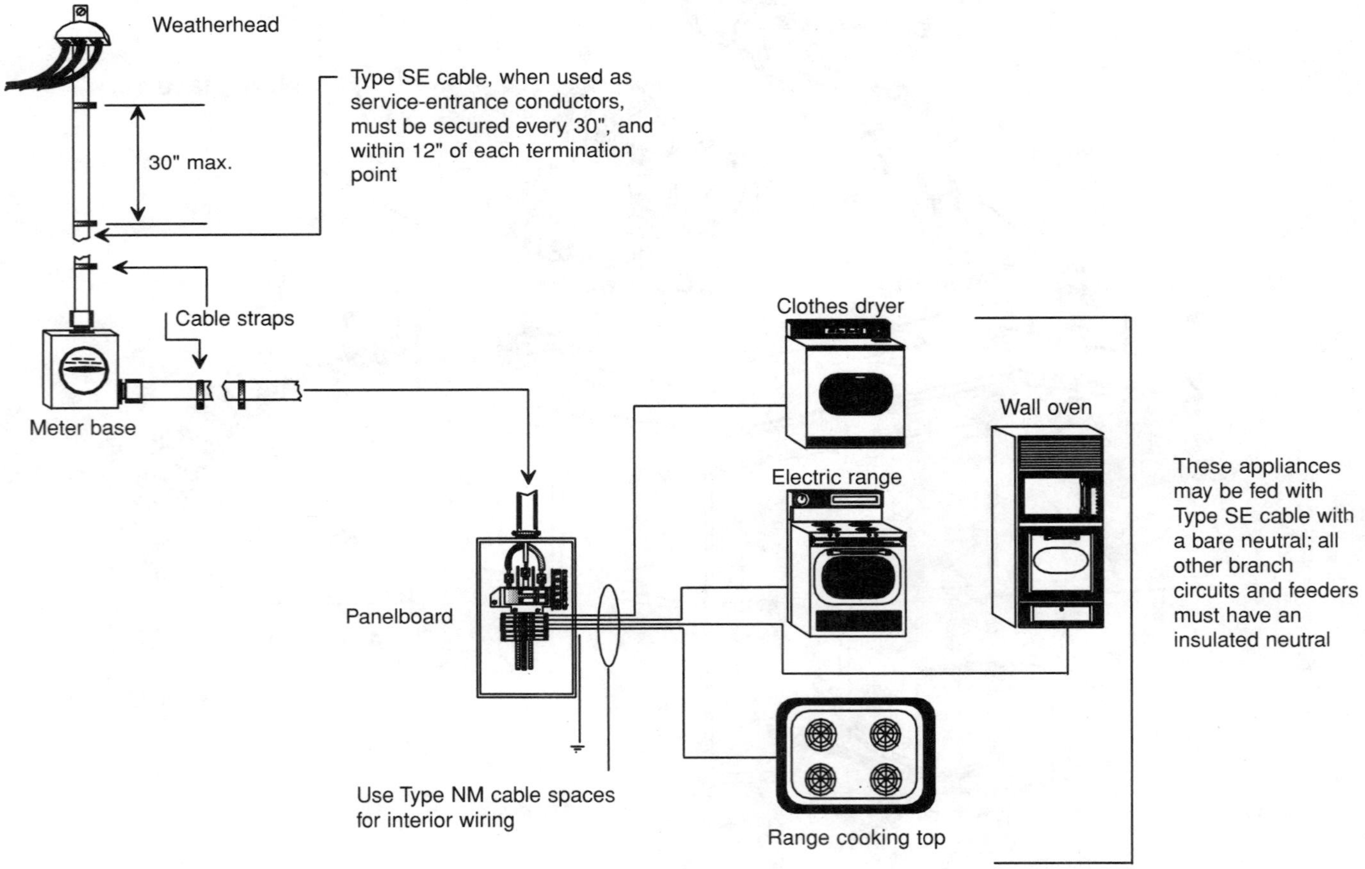
Weatherhead
Type SE cable, when used as service-entrance conductors, must be secured every 30", and within 12" of each termination point
30" max.
Cable straps
Meter base
Clothes dryer
Wall oven
Electric range
These appliances may be fed with Type SE cable with a bare neutral; all other branch circuits and feeders must have an insulated neutral
Panelboard
Use Type NM cable spaces for interior wiring
Range cooking top

Figure 5-14: *Summary of NEC installation requirements for Type SE cable*

is an assembly of parallel, special-stranded copper conductors formed integrally with an insulating material web specifically designed for field installation in surface metal raceway. The assembly is made up of three- or four-conductor cable, cable supports, splicers, circuit taps, fixture hangers, insulating end caps and other fittings. Guidelines for the use of this system are given in *NEC* Article 363. In general, the assembly is installed in an approved U-channel surface-metal raceway with one side open. Tap devices can be inserted anywhere along the channel. Connections from the tap devices to the flat-cable assembly are made by pin-type contacts when the tap devices are secured in place. The pin-type contacts penetrate the insulation of the cable assembly and contact the multistranded conductors in a matched phase sequence. These taps can then be connected to either lighting fixtures or power outlets.

Flat-cable assemblies must be installed for exposed work only and must not be installed in locations where they will be subjected to severe physical damage.

Mineral-Insulated Metal-Sheathed Cable: Type MI cable is a factory assembly of one or more conductors insulated with a highly compressed refractory mineral insulation and enclosed in a liquid-tight and gas-tight continuous copper sheath. It may be used for electric services, feeders, and branch circuits in dry, wet, or continuously moist locations. Furthermore, it may be used indoors or outdoors, embedded in plaster, concrete, fill, or other masonry, whether above or below grade. This type of cable may also be used in hazardous locations, where exposed to oil or gasoline, where exposed to corrosive conditions not deteriorating to the cable's sheath, and in underground runs where suitably protected against physical damage and corrosive conditions. In other words, MI cable may be used in practically any electrical installation.

Power and Control Tray Cable: Type TC power and control tray cable is a factory assembly of two or more insulated conductors, with or without associated bare or covered grounding conductors, under a nonmetallic sheath, approved for installation in cable trays, in raceways, or where supported by a messenger wire. The use of this cable is limited to commercial and industrial applications where the conditions of maintenance and supervision assure that only qualified persons will service the installation.

Raceway Systems

A raceway wiring system consists of an electrical wiring system in which one or more individual conductors are pulled into a conduit or similar housing, usually after the raceway system has been completely installed. The basic raceways are rigid steel conduit, electrical metallic tubing (EMT), PVC (polyvinyl chloride) plastic, and electrical nonmetallic tubing (ENT). Other raceways include surface metal moldings and flexible metallic conduit.

These raceways are available in standardized sizes and serve primarily to provide mechanical protection for the wires run inside and, in the case of metallic raceways, to provide a continuously grounded system. Metallic raceways, properly installed, provide the greatest degree of mechanical and grounding protection and provide maximum protection against fire hazards for the electrical system. However, they are more expensive to install.

Most electricians prefer to use a hacksaw with a blade having 18 teeth per inch for cutting rigid conduit and 32 teeth per inch for cutting the smaller sizes of conduit. For cutting larger sizes of conduit (1½ inches and above), a special conduit cutter should be used to save time. While quicker to use, the conduit cutter almost always leaves a hump inside the conduit and the burr is somewhat larger than made by a standard hacksaw. If a power band saw is available on the job, it is preferred for cutting the larger sizes of conduit. Abrasive cutters are also popular for the larger sizes of conduit.

Conduit cuts should be made square and the inside edge of the cut must be reamed to remove any burr or sharp edge that might damage wire insulation when the conductors are pulled inside the conduit. After reaming, most experienced electricians feel the inside of the cut with their finger to be sure that no burrs or sharp edges are present.

Lengths of conduit to be cut should be accurately measured for the size needed and an additional ½ inch should be allowed on the smaller sizes of conduit for terminations; the larger sizes of conduit will require approximately ¾ inch for locknuts, bushings, and the like at terminations.

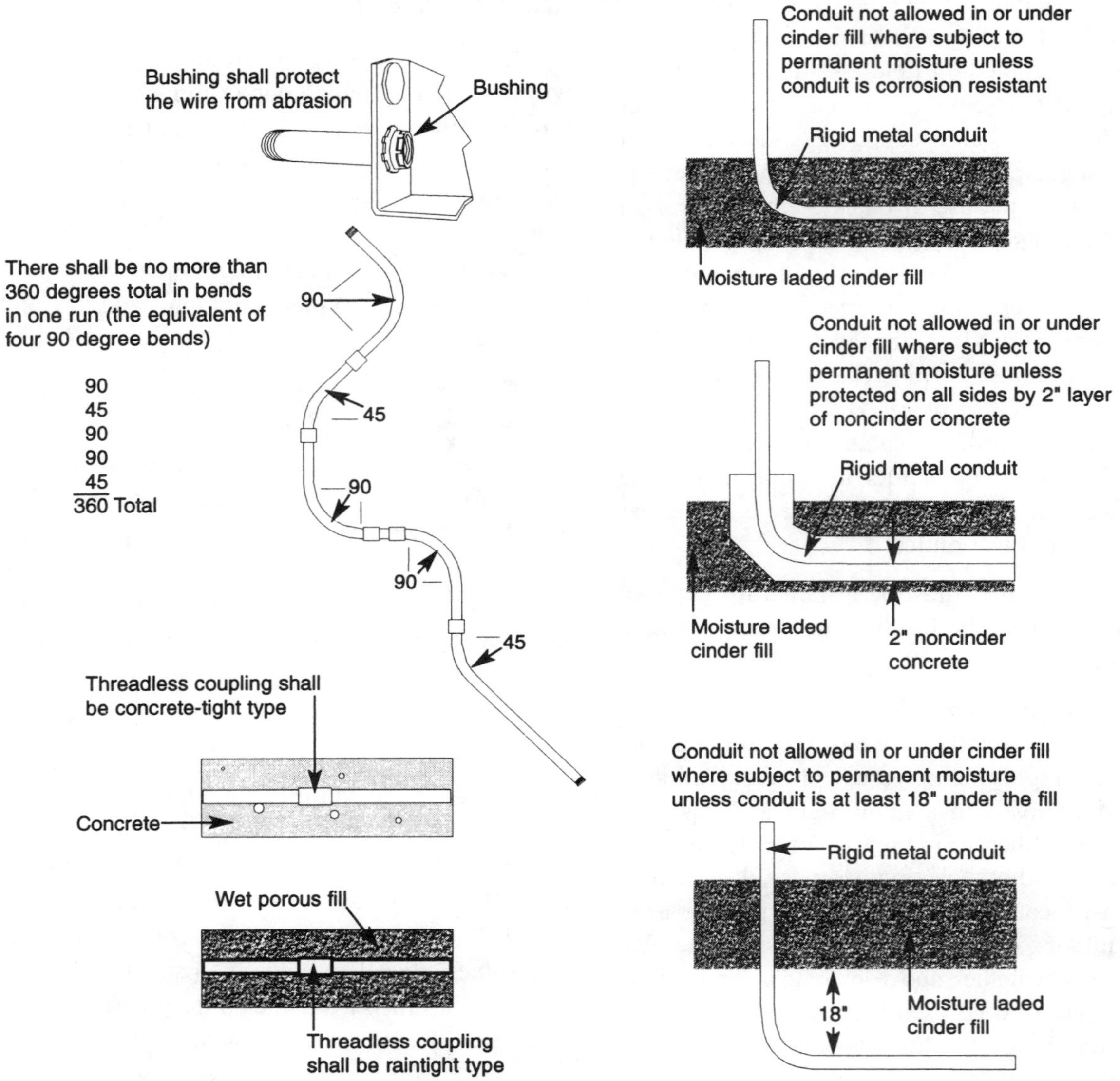
Bushing shall protect the wire from abrasion
Bushing
There shall be no more than 360 degrees total in bends in one run (the equivalent of four 90 degree bends)
90
45
90
90
45
360 Total
90
45
90
90
45
Threadless coupling shall be concrete-tight type
Concrete
Wet porous fill
Threadless coupling shall be raintight type
Conduit not allowed in or under cinder fill where subject to permanent moisture unless conduit is corrosion resistant
Rigid metal conduit
Moisture laded cinder fill
Conduit not allowed in or under cinder fill where subject to permanent moisture unless protected on all sides by 2" layer of noncinder concrete
Rigid metal conduit
Moisture laded cinder fill
2" noncinder concrete
Conduit not allowed in or under cinder fill where subject to permanent moisture unless conduit is at least 18" under the fill
Rigid metal conduit
18"
Moisture laded cinder fill

Figure 5-15: *Summary of NEC installation requirements for rigid metal conduit*

A good lubricant (cutting oil) is then used liberally during the thread-cutting process. If sufficient lubricant is used, cuts may be made cleaner and sharper, and the cutting dies will last much longer.

Full threads must be cut to allow the conduit ends to come close together in the coupling or to firmly seat in the shoulders of threaded hubs of conduit bodies. To obtain a full thread, run the die up on the conduit until the conduit barely comes through the die. This will give a good thread length adequate for all purposes. Anything longer will not fit into the coupling and will later corrode because threading removes the zinc or other protective coating from the conduit.

Clean, sharply cut threads also make a better continuous ground and save much trouble once the system is in operation.

Some *NEC* regulations governing the installation of rigid metal conduit are shown in Figure 5-15.

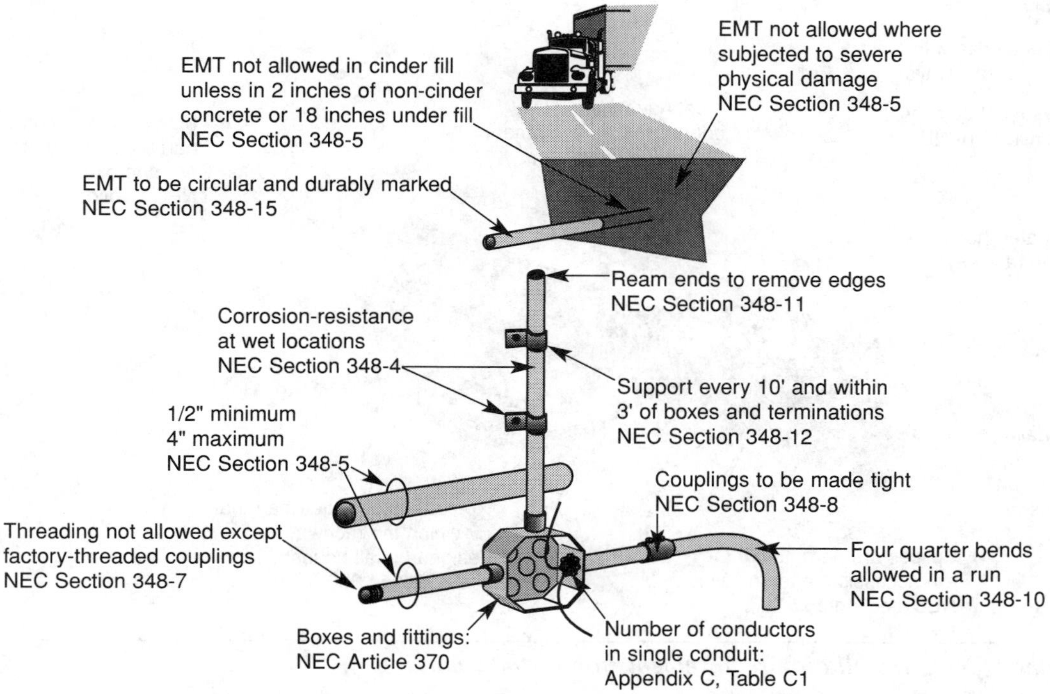
EMT not allowed in cinder fill
unless in 2 inches of non-cinder
concrete or 18 inches under fill
NEC Section 348-5
EMT not allowed where
subjected to severe
physical damage
NEC Section 348-5
EMT to be circular and durably marked
NEC Section 348-15
Ream ends to remove edges
NEC Section 348-11
Corrosion-resistance
at wet locations
NEC Section 348-4
Support every 10' and within
3' of boxes and terminations
NEC Section 348-12
1/2" minimum
4" maximum
NEC Section 348-5
Couplings to be made tight
NEC Section 348-8
Threading not allowed except
factory-threaded couplings
NEC Section 348-7
Four quarter bends
allowed in a run
NEC Section 348-10
Boxes and fittings:
NEC Article 370
Number of conductors
in single conduit:
Appendix C, Table C1

Figure 5-16: *Summary of NEC installation requirements for electrical metallic tubing*

Electrical Metallic Tubing

Electrical metallic tubing (EMT) may be used for both exposed and concealed work, *except* where it will be subject to severe damage during use; or in cinder concrete or cinder fill where subject to permanent moisture unless it is protected by 2 inches of noncinder fill or is installed a minimum of 18 inches under the fill.

Threadless couplings and connectors are used for EMT installation and these should be installed so that the tubing will be made up tight. Both set-screw and compression types are commonly in use. Where buried in masonry or installed in wet locations, couplings and connectors, as well as supports, bolts, straps, and screws, should be of a type approved for the conditions.

Bends in the tubing should be made with a tubing bender so that no injury will occur and so the internal diameter of the tubing will not be effectively reduced. The bends between outlets or termination points should contain no more than the equivalent of four quarter-bends (360° total), including those bends located immediately at the outlet or fitting (offsets).

All cuts in EMT are made with a hacksaw, power hacksaw, tubing cutter, or other approved device. Once out, the tubing ends should be reamed with a screwdriver handle or pipe reamer to remove all burrs and sharp edges that might damage conductor insulation. See Figure 5-16.

Flexible Metal Conduit

Flexible metal conduit generally is manufactured in two types, a standard metal-clad type and a liquid-tight type. The former type cannot be used in wet locations unless the conductors pulled in are of a type specially approved for such conditions. Neither type may be used where they will be subjected to physical damage or where any combination of ambient and/or conductor temperature will produce an operating temperature in excess of that for which the material is approved. Other uses are fully described in Articles 350 and 351 of the *NEC*.

When this type of conduit is installed, it should be secured by an approved means at intervals not exceeding 4½ feet and within 12 inches of every outlet box, fitting,

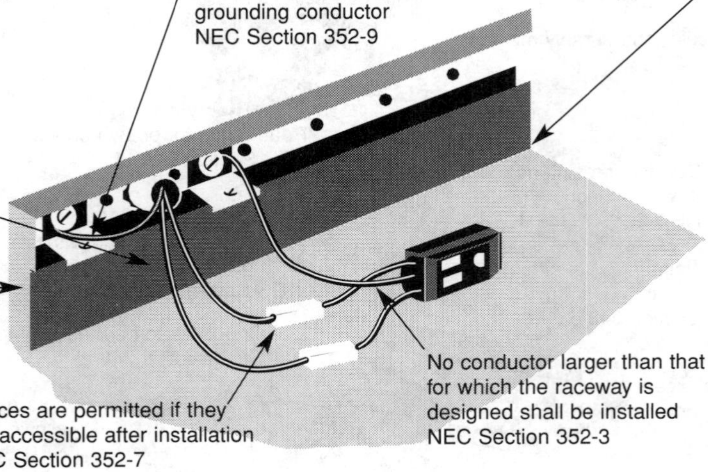
Surface Metal Raceway
Must be of such construction as will distinguish them from other raceways
Surface metal raceways must comply with the applicable provisions of NEC Article 300
The number of conductors installed must be no greater than the number for which the raceway is designed NEC Section 352-4
Must have means for connecting an equipment grounding conductor NEC Section 352-9
May extend through dry walls, partitions, and floors if lengths are unbroken NEC Section 352-5
Where used both for signaling and for lighting and power, the different systems must be run in separate compartments of the raceway system
Splices are permitted if they are accessible after installation NEC Section 352-7
No conductor larger than that for which the raceway is designed shall be installed NEC Section 352-3

Figure 5-17: *Summary of NEC installation requirements for surface molding*

or other termination points. In some cases, however, exceptions exist. For example, when flexible metal conduit must be finished in walls, ceilings, and the like, securing the conduit at these intervals would not be practical. Also, where more flexibility is required, lengths of not more than 3 feet may be utilized at termination points.

Flexible metal conduit may be used as a grounding means where both the conduit and the fittings are approved for the purpose. In lengths of more than 6 feet, it is best to install an extra grounding conductor within the conduit for added insurance.

Liquidtight flexible metal conduit is used in damp or wet locations and is covered in *NEC* Article 351. Please see the *NEC* book for further details.

Surface Metal Molding

When it is impractical to install the wiring in concealed areas, surface metal molding (Figure 5-17) is a good compromise. Even though it is visible, proper painting to match the color of the ceiling and walls makes it very inconspicuous. Surface metal molding is made from sheet metal strips drawn into shape and comes in various shapes and sizes with factory fittings to meet nearly every application found in finished areas of commercial buildings. A complete list of fittings can be obtained at your local electrical equipment supplier.

The running of straight lines of surface molding is simple. A length of molding with the coupling is slipped in the end, out enough so that the screw hole is exposed, and then the coupling is screwed to the surface to which the molding is to be attached. Then another length of molding is slipped on the coupling.

Factory fittings are used for corners and turns or the molding may be bent (to a certain extent) with a special bender. Matching outlet boxes for surface mounting are also available, and bushings are necessary at such boxes to prevent the sharp edges of the molding from injuring the insulation on the wire.

Clips are used to fasten the molding in place. The clip is secured by a screw and then the molding is slipped into the clip, wherever extra support of the molding is needed, and fastened by screws. When parallel runs of molding are installed, they may be secured in place by means of a multiple strap. The joints in runs of molding are covered by slipping a connection cover over the joints. Such runs of molding should be grounded the same as any other metal raceway, and this is done by use of grounding clips. The current-carrying wires are normally pulled in after the molding is in place.

The installation of surface metal molding requires no special tools unless bending the molding is necessary. The molding is fastened in place with screws, toggle bolts, and the like, depending on the materials to which it is fastened. All molding should be run straight and parallel with the room or building lines, that is, baseboards, trims, and other room moldings. The decor of the room should be considered first and the molding made as inconspicuous as possible.

It is often desirable to install surface molding not used for wires in order to complete a pattern set by other surface molding containing current-carrying wires, or to continue a run to make it appear to be part of the room's decoration.

Wireways

Wireways are sheet-metal troughs with hinged or removable covers for housing and protecting wires and cables, and in which conductors are held in place after the wireway has been installed as a complete system. They may be used only for exposed work, and shouldn't be installed where they will be subject to severe physical damage, corrosive vapor, or in any hazardous location, as per Section 501-4(b) for Class I, Division 2 locations, and Section 502-4(b)(1) for Class II, Division 2 locations.

The wireway structure must be designed to safely handle the sizes of conductors used in the system. Furthermore, the system should not contain more than 30 current-carrying conductors at any cross section. The sum of the cross-sectional areas of all contained conductors at any cross section of a wireway shall not exceed 20 percent of the interior cross-sectioned area of the wireway.

Splices and taps, made and insulated by approved methods, may be located within the wireway provided they are accessible. The conductors, including splices and taps, shall not fill the wireway to more than 75 percent of its area at that point.

Wireways must be securely supported at intervals not exceeding 5 feet, unless specially approved for supports at greater intervals, but in no case shall the distance between supports exceed 10 feet.

Busways

There are several types of busways or duct systems for electrical transmission and feeder purposes. Lighting duct, trolley duct, and distribution bus duct are just a few. All are designed for a specific purpose, and the electrician or electrical designer should become familiar with all types before an installation is laid out.

Lighting duct, for example, permits the installation of an unlimited amount of footage from a single working platform. As each section and the lighting fixtures are secured in place, the complete assembly is then simply transported to the area of installation and installed in one piece.

Trolley duct is widely used for industrial applications, and where the installation requires a continuous polarization to prevent accidental reversal, a polarizing bar is used. This system provides polarization for all trolley, permitting standard and detachable trolleys to be used on the same run.

Plug-in bus duct is also widely used for industrial applications, and the system consists of interconnected prefabricated sections of bus duct so formed that the complete assembly will be rigid in construction and neat and symmetrical in appearance.

Cable Trays

Cable trays are used to support electrical conductors used mainly in industrial applications, but are sometimes used for communication and data processing conductors in large commercial establishments. The trays themselves are usually made up into a system of assembled, interconnected sections and associated fittings, all of which are made of metal or other noncombustible material. The finished system forms into a rigid structural run to contain and support single, multiconductor, or other wiring cables. Several styles of cable trays are available, including ladder, trough, channel, solid-bottom trays, and similar structures. See Chapter 7 in this book.

Underfloor Raceway

Underfloor raceway (Figure 5-18) consists of ducts laid below the surface of the floor and interconnected by means of fittings and outlet or junction boxes. Both metallic and nonmetallic ducts are used. Obviously, this system must be installed prior to the floor being finished.

IDENTIFYING CONDUCTORS

The *NEC* specifies certain methods of identifying conductors used in wiring systems of all types. For example, the high leg of a 120/240-volt grounded three-phase delta system must be marked with an orange color for identification; a grounded conductor must be identified either by the color of its insulation, by markings at the terminals, or by other suitable means. Unless allowed by *NEC* exceptions, a grounded conductor must have a white or natural gray finish. When this is not practical for conductors larger than No. 6 AWG, marking the terminals with white color is an acceptable method of identifying the conductors.

Color Coding

Conductors contained in cables are color-coded so that identification may be easily made at each access point. The following lists the color-coding for cables up through five-wire cable:

- Two-conductor cable: one white wire, one black wire, and a grounding conductor (usually bare)
- Three-conductor cable: one white, one black, one red, and a grounding conductor
- Four-conductor cable: fourth wire blue
- Five-conductor cable: fifth wire yellow
- The grounding conductor may be either green or green with yellow stripes

Although some control-wiring and communication cables contain 60, 80, or more pairs of conductors — using a combination of colors — the ones listed are the most common and will be encountered the most on electrical installations.

When conductors are installed in raceway systems, any color insulation is permitted for the ungrounded phase conductors except the following:

White or gray	Reserved for use as the grounded circuit conductor
Green	Reserved for use as a grounding conductor only

Changing Colors

Should it become necessary to change the actual color of a conductor to meet *NEC* requirements or to facilitate maintenance on circuits and equipment, the conductors may be reidentified with colored tape or paint.

For example, assume that a two-wire cable containing a black and white conductor is used to feed a 240-volt, two-wire single-phase motor. Since the white-colored conductor is supposed to be reserved for the grounded conductor, and none is required in this circuit, the white conductor may be marked with a piece of red tape at each end of the circuit. By doing that, everyone will know that this wire is not a grounded conductor.

LOCATING OUTLETS

The *NEC* specifies specific locations for most outlets in various occupancies, especially in dwellings. For conventional residential-outlet layouts, most experienced electricians will make notes of the general wiring arrangement and then select outlet boxes to comply with the number of conductors required in each. For the more complex layouts, and if working drawings are not available, a sketch should be made.

The outlet boxes are then secured to the framing of the house. Where wood studs are used, the outlet boxes are usually secured with nails driven through the nail holes provided in the box and then directly into the wood studs. Some electrical workers prefer boxes with mounting brackets. For use on concrete blocks — as in a residential basement — masonry fasteners are used. Once all outlet boxes are secured, the final layout should appear similar to the one in Figure 5-19.

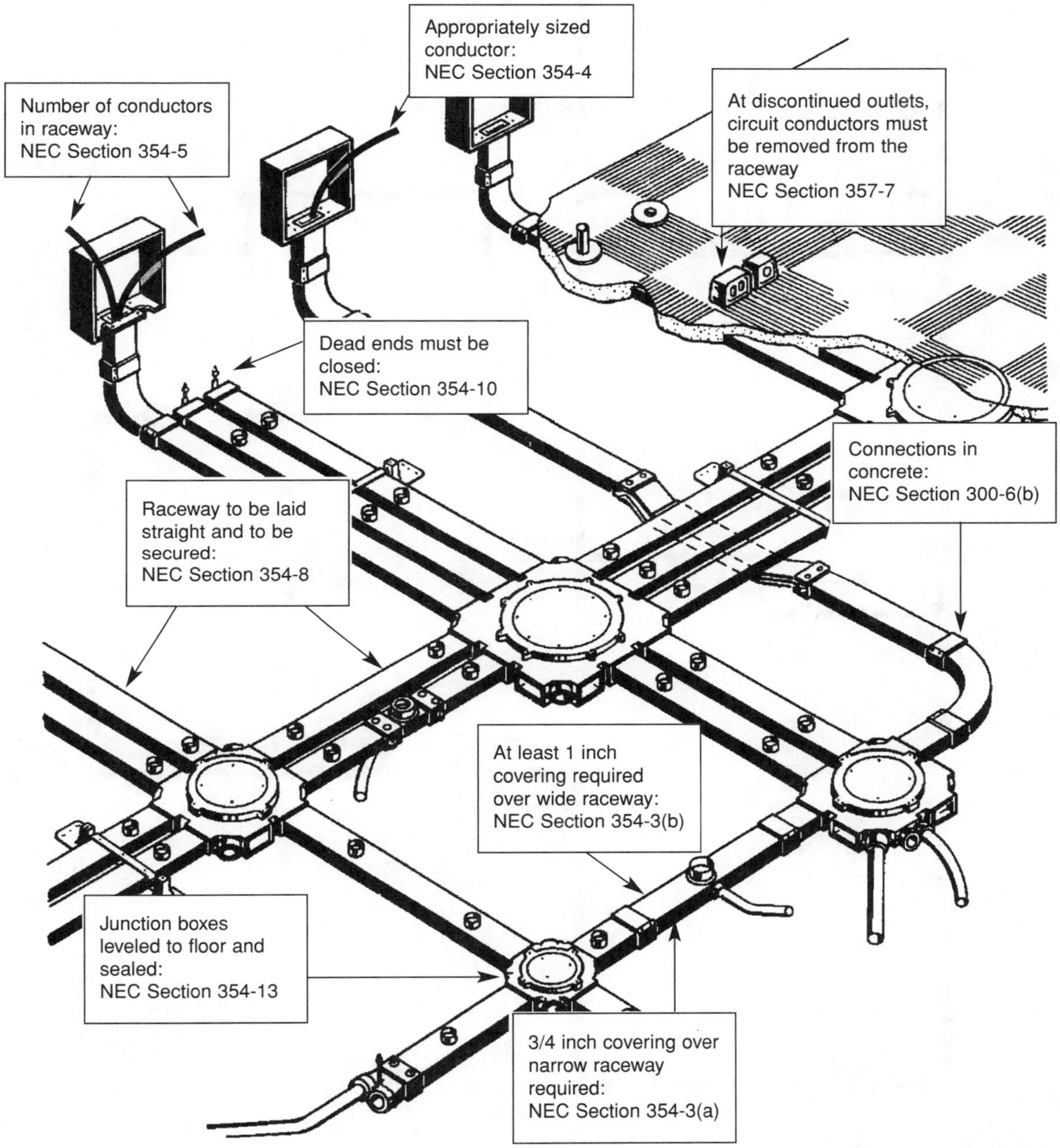
Appropriately sized conductor: NEC Section 354-4
Number of conductors in raceway: NEC Section 354-5
At discontinued outlets, circuit conductors must be removed from the raceway NEC Section 357-7
Dead ends must be closed: NEC Section 354-10
Connections in concrete: NEC Section 300-6(b)
Raceway to be laid straight and to be secured: NEC Section 354-8
At least 1 inch covering required over wide raceway: NEC Section 354-3(b)
Junction boxes leveled to floor and sealed: NEC Section 354-13
3/4 inch covering over narrow raceway required: NEC Section 354-3(a)

Figure 5-18: *Summary of NEC installation requirements for underfloor raceway*

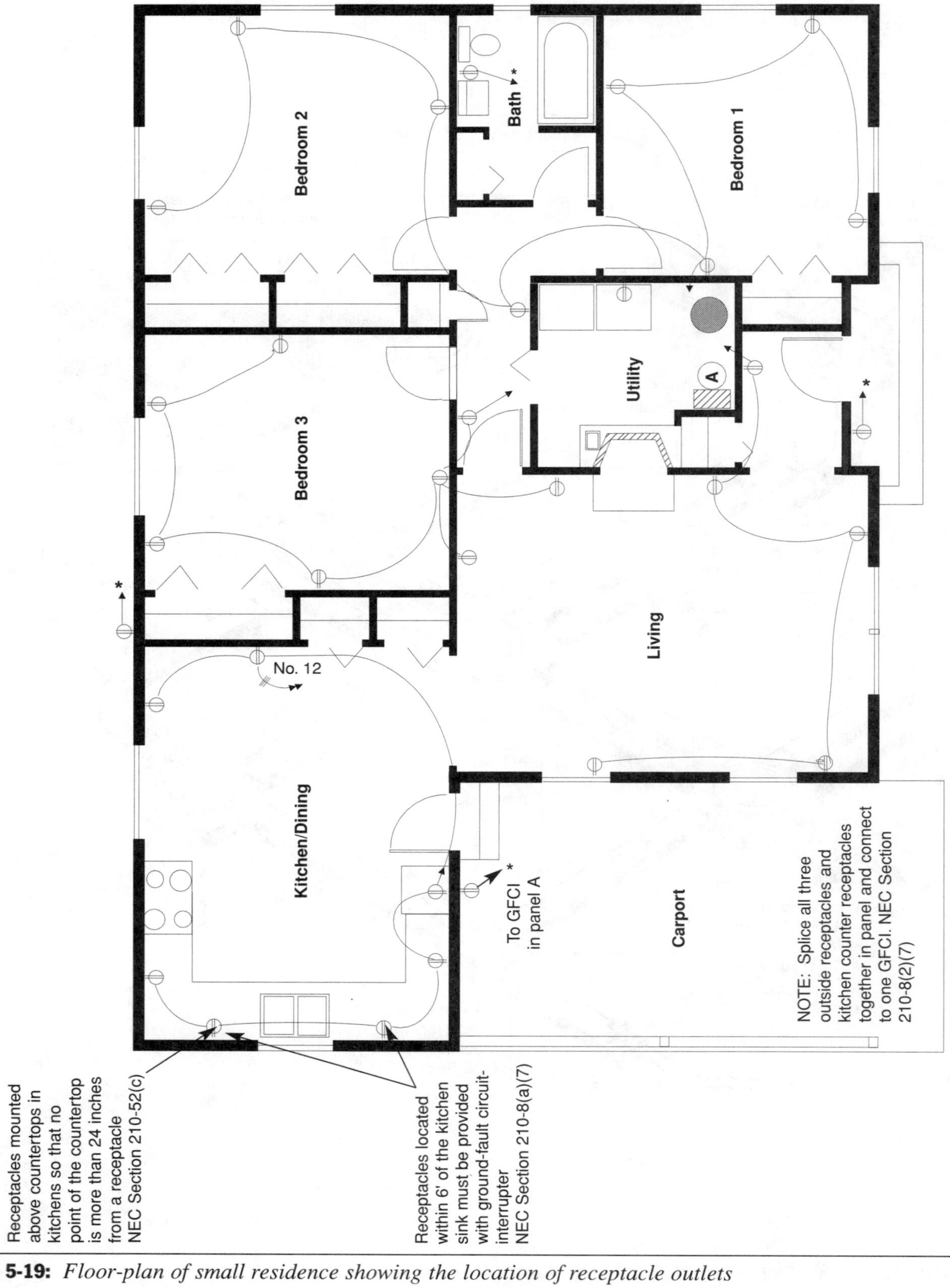
Bedroom 2
Bath
Bedroom 1
Utility
A
Bedroom 3
Living
No. 12
Kitchen/Dining
To GFCI in panel A
Carport
NOTE: Splice all three outside receptacles and kitchen counter receptacles together in panel and connect to one GFCI. NEC Section 210-8(2)(7)
Receptacles mounted above countertops in kitchens so that no point of the countertop is more than 24 inches from a receptacle NEC Section 210-52(c)
Receptacles located within 6' of the kitchen sink must be provided with ground-fault circuit-interrupter NEC Section 210-8(a)(7)

Figure 5-19: *Floor-plan of small residence showing the location of receptacle outlets*

Before outlet boxes and branch-circuit wiring are installed, the panelboard must be sized and selected (see Chapters 2 and 3). The residence shown in Figure 5-19 has the panelboard located in the utility room, designated by the letter "A" on the drawings. A small residential building such as this will usually have only one panelboard to service the entire electrical system. In larger buildings, there may be dozens of subpanels throughout the building. When more than one panelboard is used, the next panelboard will be designated "B" and then "C" and so forth.

NEC Section 210-52(a), General Provisions, specifically states the minimum requirement for the location of receptacles in dwelling units.

> *In every kitchen, family room, dining room, . . . or similar room or area of dwelling units, receptacle outlets shall be installed in accordance with the general provisions specified in (1) through (3).*
>
> *Receptacles shall be installed so that no point along the floor line in any wall space is more than 6 feet, measured horizontally, from an outlet in that space.*

The *NEC* defines "wall space" as a wall unbroken along the floor line by doorways, fireplaces, and similar openings. Each wall space 2 or more feet wide must be treated individually and separately from other wall spaces within the room.

The purpose of the above *NEC* requirement is to minimize the use of cords across doorways, fireplaces, and similar openings.

With this *NEC* requirement in mind, outlets for our sample residence will be laid out. In laying out these receptacle outlets, the floor line of the wall is measured (also around corners) but not across doorways, fireplaces, passageways, or other spaces where a flexible cord extended across the space would be unsuitable. In general, duplex receptacle outlets must be no more than 12 feet apart. When spaced in this manner, a 6-foot extension cord will reach a receptacle from any point along the wall line — complying with the latest edition of the *NEC*.

Figure 5-19 shows all of the duplex receptacles laid out in our sample residence. Note that at no point along the wall line is any receptacle more than 12 feet apart or more than 6 feet from any door or room opening. Where practical, no more than eight receptacles are connected to one circuit. However *NEC* Section 220-3(b)(9) specifies a demand factor of 180 volt-amperes per receptacle or group of receptacles. Since a 15-ampere branch-circuit is rated at 1800 volt-amperes (15 amps x 120 volts), 10 general-purpose duplex receptacles may be connected to one circuit.

The utility room has two receptacles: one for the washer on a separate circuit in order to comply with *NEC* Section 210-52(f), and another for ironing.

One duplex receptacle, connected to the living-room circuit, is located in the vestibule for cleaning purposes, such as feeding a portable vacuum cleaner or similar appliance.

The living-room outlets are split-wired — the lower half of each duplex receptacle is "hot" (energized) all the time, while the upper half can be switched on or off. The reason for this is that a great deal of the illumination for this area will be provided by portable table lamps and the split-wired receptacles provide a means of control for them. Note that these split-wired receptacles can be controlled from several locations — at each entry to the living room.

To comply with *NEC* Section 210-52(b)(3), the kitchen receptacles are laid out according to the following specifications:

> *In addition to the number of branch circuits determined previously, two or more 20-ampere small appliance branch circuits must be provided to serve all receptacle outlets, including refrigeration equipment, in the kitchen, pantry, breakfast room, dining room, or similar area of the house. Such circuits, whether two or more are used, must have no other outlets connected to them.*

To further meet *NEC* requirements, one duplex receptacle is installed in each bathroom on a ground-fault circuit-interrupter that supplies only these bathroom receptacles. All outside receptacles and basement receptacles need to be on a separate GFCI.

240-Volt Circuits

The electric range, clothes dryer, and water heater in our sample residence all operate at 240 volts ac. Each will be fed with a separate circuit and connected to a 2-pole circuit breaker of the appropriate rating in the panelboard. To determine the conductor size and overcurrent protection for the range, proceed as follows:

- Find the nameplate rating of the electric range. This has previously been determined to be 12 kVA.
- Refer to *NEC* Table 220-19. Since Column A of this table applies to ranges rated not over 12 kVA, this will be the column to use in this example.
- Under the "Number of Appliances" column, locate the appropriate number of appliances (1 in this case), and find the maximum demand given for it in Column A. Column A states that the circuit should be sized for 8 kVA (not the nameplate rating of 12 kVA).
- Calculate the required conductor ampacity as follows:

$$\frac{8000 \text{ va}}{240 \text{ V}} = 33.33 \text{ amperes}$$

The minimum branch circuit, however, must be rated at 40 amperes since common residential circuit breakers are rated in steps of 15-, 20-, 30-, 40-, etc. amperes. A 30-ampere circuit breaker is too small, so a 2-P, 40-ampere circuit breaker is selected. The conductors must have a current-carrying capacity equal to or greater than the overcurrent protection. Therefore, No. 8 AWG conductors will be used.

If a cooktop and wall oven were used instead of the electric range, the circuit would be sized similarly. The *NEC* specifies that a branch circuit for a counter-mounted cooking unit and not more than two wall-mounted ovens, all supplied from a single branch circuit and located in the same room, is computed by adding the nameplate ratings of the individual appliances, and then treating this total as equivalent to one range. Therefore, two appliances of 6 kVA each may be treated as a single range with a 12 kVA nameplate rating.

The connection may be made directly to the range junction box, but more often a 50-ampere range receptacle is mounted at the range location whereas a range cord-and-plug set is used to make the connection. This facilitates moving the appliance later for maintenance or cleaning.

The branch circuit for the water heater in the residence under consideration must be sized for its full capacity because there is no diversity or demand factor for this appliance. Since the nameplate rating on the water heater in the residence under consideration indicates two heating elements of 4500 watts each, the first inclination would be to size the circuit for a total load of 9000 watts (volt-amperes). However, only one of the two elements operates at a time. Each element is controlled by a separate thermostat. The lower element becomes energized when the thermostat calls for heat, and at the same time the thermostat opens a contact to prevent the upper element from operating. When the lower-element thermostat is satisfied, the lower contact opens and (at the same time) the thermostat closes the contact for the upper element to become energized to maintain the water temperature.

With this information in hand, the circuit for the water heater may be sized by the equation:

$$\frac{4500 \text{ va}}{240 \text{ V}} = 18.75 \text{ A} \times 1.25 = 23.44 \text{ amperes}$$

Since a water heater will more than likely fall under the "continuous load" category, the conductor and overcurrent protection should be rated at 125 percent. Therefore, No. 10 AWG wire should be used and protected with a 2-P, 30-ampere circuit breaker. A direct connection is made to water heaters at the integral junction box on top of the heater.

The *NEC* specifies that electric clothes dryers must be rated at 5 kVA or the nameplate rating — whichever is greater. In our case, the dryer is rated at 5.5 kVA and the conductor current-carrying capacity is calculated as follows:

$$\frac{5500 \text{ va}}{240 \text{ V}} = 22.92 \text{ amperes}$$

A 3-wire, 30-ampere circuit will be provided (No. 10 AWG wire) protected by a 2-P, 30-ampere circuit breaker. The dryer may be connected directly, but a 30-ampere dryer receptacle is normally provided for the same reasons as mentioned for the electric range.

Large appliance outlets — rated at 240 volts — are frequently shown on the electrical floor plan, using lines and symbols to indicate the outlets and circuits. In other cases, no drawings are provided at all.

GROUND-FAULT CIRCUIT-INTERRUPTERS

Although the placement of the various outlets has not been done at this point, the experienced electrician knows that circuits providing power to certain areas of the home require ground-fault circuit-interrupters (GFCIs) to be installed for additional protection of people using these circuits. Such areas include:

- All outside receptacles
- Receptacles used in bathrooms
- Receptacles located in residential garages
- Receptacles located in unfinished basements
- Receptacles located in crawl spaces
- Receptacles installed within 6 feet of a kitchen or bar sink

Since there is no basement or crawl space in our sample residence, these two areas do not apply to this project. However, there is a bathroom and kitchen. Furthermore, outside receptacles will be provided. Therefore, at least one small-appliance circuit will be provided with GFCI protection, along with one circuit supplying outdoor receptacles. The bathroom receptacle must be connected to its own ground-fault circuit-interrupter, or to a separate circuit with a GFCI receptacle. This brings the total number to at least three circuits that will require GFCI protection.

GFCI circuit breakers require one space in the load center or panelboard — the same as a 1-pole circuit breaker.

Chapter 6
Utilization Equipment

The NEC defines "utilization equipment" as equipment that uses electric energy for mechanical, chemical, heating, lighting, or other useful purposes. Examples are electric motors, large and small appliances, electric heaters, air-conditioning equipment, and the like.

Each piece of utilization equipment, or groups of equipment, must have a disconnecting means so that the equipment is disconnected from all ungrounded conductors when servicing is necessary. The type of disconnecting means required and its location will vary with the type of equipment. The *NEC* also specifies requirements for grounding noncurrent-carrying metal parts of utilization equipment.

The following list covers most types of utilization equipment recognized by the *NEC*, and also gives the *NEC* Articles for reference.

The *NEC* installation requirements covered in each Article apply to typical occupancies. Any special conditions, such as when used in hazardous locations, require referral to other *NEC* Sections and Articles.

UTILIZATION EQUIPMENT

Application	NEC Reference
Air-conditioning and refrigeration equipment	Article 440
Appliances	Article 422
Capacitors	Article 460
Circuits motor	Article 430
Computers electronic	Article 645

UTILIZATION EQUIPMENT (cont.)

Application	NEC Reference
Controllers, motor	Article 430
Converters, phase	Article 455
Cranes and hoists	Article 610
Data-processing equipment	Article 645
Dielectric heating equipment	Article 665
Dumbwaiters	Article 620
Electric deicing and snow-melting equipment fixed	Article 426
Electric sign and outline lighting	Article 600
Electric space heating, fixed	Article 424
Electric heating, fixed	Article 427
Electric vehicle charging system equipment	Article 625
Electric welders	Article 630
Electrolytic cells	Article 668
Electronic computers	Article 645
Electroplating	Article 669

UTILIZATION EQUIPMENT (cont.)

Application	NEC Reference
Elevators	Article 620
Escalators	Article 620
Fire pumps*	Article 695
Fixed electric heating	Article 427
Fixed electric space heating	Article 424
Fixed outdoor electric deicing and snow-melting equipment	Article 426
Fountains	Article 680 Part E
Generators	Article 445
Hoists	Article 610
Hot tubs	Article 680 Part D
Hydromassage bathtubs	Article 680 Part G
Induction and dielectric heating equipment	Article 665
Industrial machinery	Article 670
Information technology equipment	Article 645
Information technology equipment in general areas	Article 645
Irrigation machines	Article 675
Integrated electrical systems	Article 685
Lighting fixtures	Article 410
Machine tools	Article 670
Manufactured wiring systems	Article 604
Motors	Article 430
Motor circuits	Article 430
Motor controllers	Article 430
Moving walks	Article 620
Office furnishings	Article 605
Organs, pipe	Article 650

UTILIZATION EQUIPMENT (cont.)

Application	NEC Reference
Outdoor electric deicing and snow-melting equipment	Article 426
Outline lighting	Article 600
Phase converters	Article 455
Pipe organs	Article 650
Reactors	Article 470
Refrigeration equipment	Article 440
Resistors	Article 470
Signs and outline lighting, electric	Article 600
Solar photovoltaic systems	Article 690
Sound-recording equipment	Article 640
Spas	Article 680 Part D
Stairway chair lifts	Article 620
Storable pools	Article 680 Part C
Storage batteries	Article 480
Swimming pools	Article 680
Transformers	Article 450
Therapeutic pools and tubs	Article 680 Part F
Wading pools	Article 680 Part C
Welders electric	Article 630
Wheelchair lifts	Article 620
Wiring systems, manufactured	Article 604
X-ray equipment	Article 660

AIR-CONDITIONING AND REFRIGERATION EQUIPMENT

NEC Article 440 covers electric motor-driven air-conditioning and refrigeration equipment, including the complete wiring system. Furthermore, this Article covers circuits supplying hermetic refrigerant motor compressors.

It must be noted, however, that the installation requirements specified in *NEC* Article 440 are in addition to, or are amendments to, the rules given in the basic *NEC* Article dealing with electric motors; that is, *NEC* Article 430.

Where refrigeration compressors are driven by conventional motors (not the hermetic type), the motors and controls are subject to *NEC* Article 430.

Household refrigerators and freezers, drinking-water coolers and beverage dispensers are considered by the *NEC* to be appliances, and their application must comply with *NEC* Article 422, but must also satisfy *NEC* Article 440 if such appliances contain sealed motor-compressors.

Hermetic refrigerant motor-compressors, controllers, and equipment must also comply with the applicable provisions for the following:

- Capacitors - *NEC* Section 460-9
- Special occupancies - *NEC* Articles 511, 513 through 517 (Part D), and 530.
- Hazardous (Classified) locations - *NEC* Articles 500 through 504.
- Resistors and reactors - *NEC* Article 470

NEC installation requirements for air-conditioning and refrigeration equipment are summarized in Figure 6-1. Also refer to Figure 6-2 and 6-3 on the pages to follow, and to the *NEC* book for further details and exact specifications.

Air Conditioning and Refrigeration Equipment		
Application	**Requirements**	**NEC Reference**
Ampacity and rating	Conductors must be sized according to NEC Tables 310-16 through 310-19 or calculated in accordance with NEC Section 310-15.	Section 440-6
Application and selection of controllers	Each motor-compressor must be protected against overload and failure to start by one of the means specified in NEC Section 440-52(a) (1) through (4).	Section 440-52
Combination load	Conductors must be sufficiently sized for the other loads plus the required ampacity for the compressor as required in NEC Section 440-33.	Section 440-34
Compressor branch-circuit conductors	Branch-circuit conductors supplying a single compressor must have an ampacity not less than 125% of either the motor-compressor rated-load current or the branch-circuit selection current, whichever is greater.	Section 440-32
Controller rating	Must have both a continuous-duty full-load current rating, and a locked-rotor current rating, not less than the nameplate rated-load current.	Section 440-41

Figure 6-1: *Summary of NEC installation requirements for air-conditioning and refrigeration equipment*

Air Conditioning and Refrigeration Equipment (cont.)		
Application	**Requirements**	**NEC Reference**
Cord-connected equipment	An attachment plug and receptacle is permitted to serve as the disconnecting means.	Section 440-13
Highest rated motor	The largest motor is considered to be the motor with the highest rated-load current.	Section 440-7
Location	Disconnecting means must be located within sight of equipment. The disconnecting means may be mounted on or within the HVAC equipment.	Section 440-14
Marking on hermetic compressors	Must be provided with a nameplate containing manufacturer's name; trademark or symbol; identifying designation; phase, voltage; frequency; rated-load current; locked-rotor current, and the words "thermally protected."	Section 440-4
Marking on controller	Must be provided with maker's name, trademark or symbol; identifying designation; voltage; phase; full-load and locked-rotor current (or hp rating).	Section 440-5
Multimotor-load equipment	Conductors must be sized to carry the circuit ampacity marked on the equipment as specified in NEC Section 440-4(b)	Section 440-35
Overload relays	Overload relays and other devices for motor overload protection, that are not capable of opening short circuits must be protected by a suitable fuse or inverse time circuit breaker.	Section 440-53
Single machine	The entire HVAC system is considered to be one machine, regardless of the number of motors involved.	Section 440-8
Rating and interrupting capacity	Must be selected on the basis of the nameplate rated-load current or branch-circuit selection current, whichever is greater.	Section 440-12
Short-circuit and ground-fault protection	Amendments to NEC Article 240 are provided here for circuits supplying hermetically-sealed compressors.	Section 440-21
Rating of short-circuit and ground-fault protective device	Rating must not exceed 175% of the compressor rated-load current; if necessary for starting device may be increased to a maximum of 225%.	Section 440-22(a)

Figure 6-1: *Summary of NEC installation requirements for air-conditioning and refrigeration equipment (cont.)*

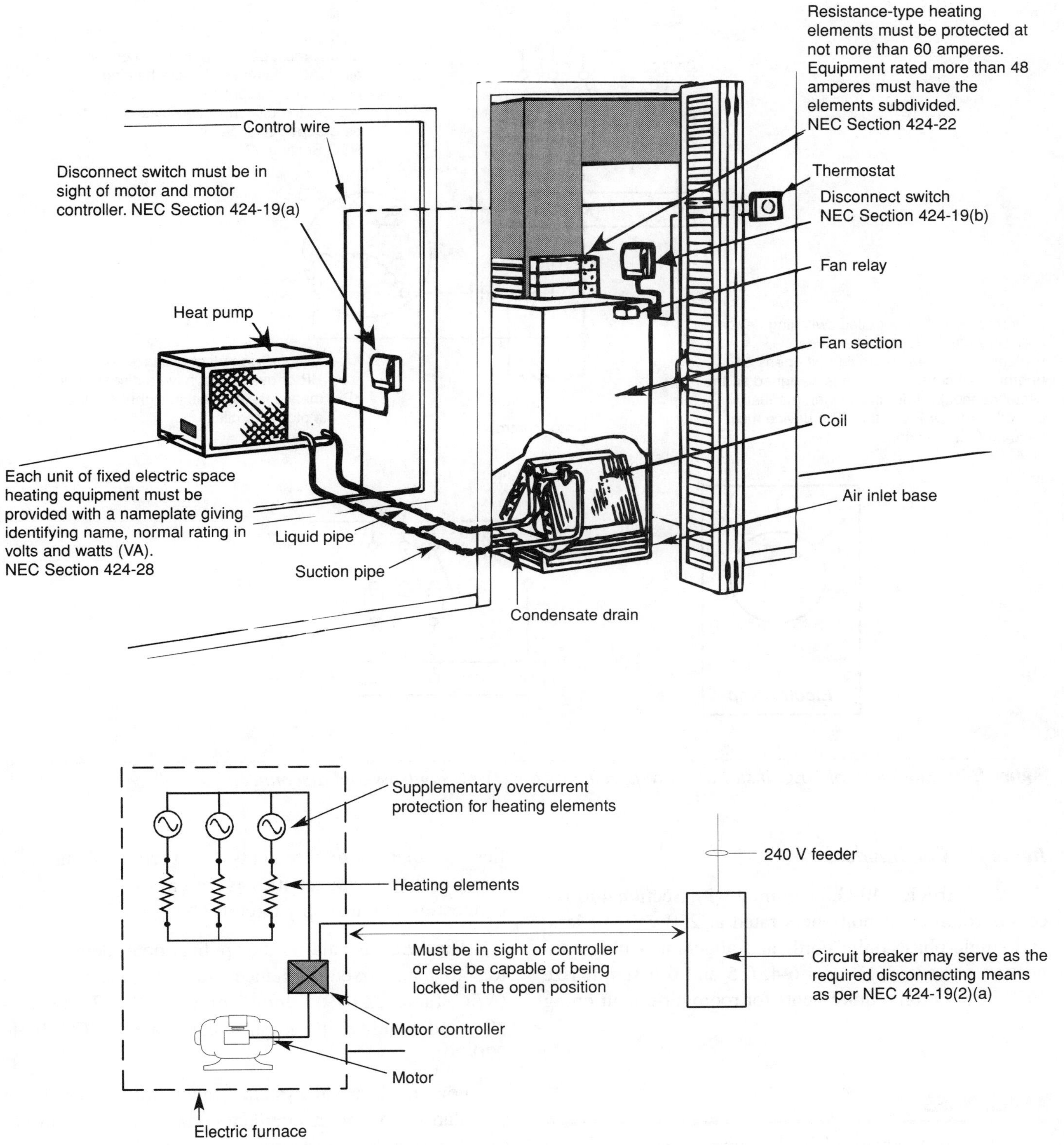
Resistance-type heating elements must be protected at not more than 60 amperes. Equipment rated more than 48 amperes must have the elements subdivided. NEC Section 424-22
Control wire
Disconnect switch must be in sight of motor and motor controller. NEC Section 424-19(a)
Thermostat
Disconnect switch NEC Section 424-19(b)
Fan relay
Heat pump
Fan section
Coil
Each unit of fixed electric space heating equipment must be provided with a nameplate giving identifying name, normal rating in volts and watts (VA). NEC Section 424-28
Air inlet base
Liquid pipe
Suction pipe
Condensate drain
Supplementary overcurrent protection for heating elements
240 V feeder
Heating elements
Must be in sight of controller or else be capable of being locked in the open position
Circuit breaker may serve as the required disconnecting means as per NEC 424-19(2)(a)
Motor controller
Motor
Electric furnace

Figure 6-2: *Summary of NEC installation requirements for HVAC equipment*

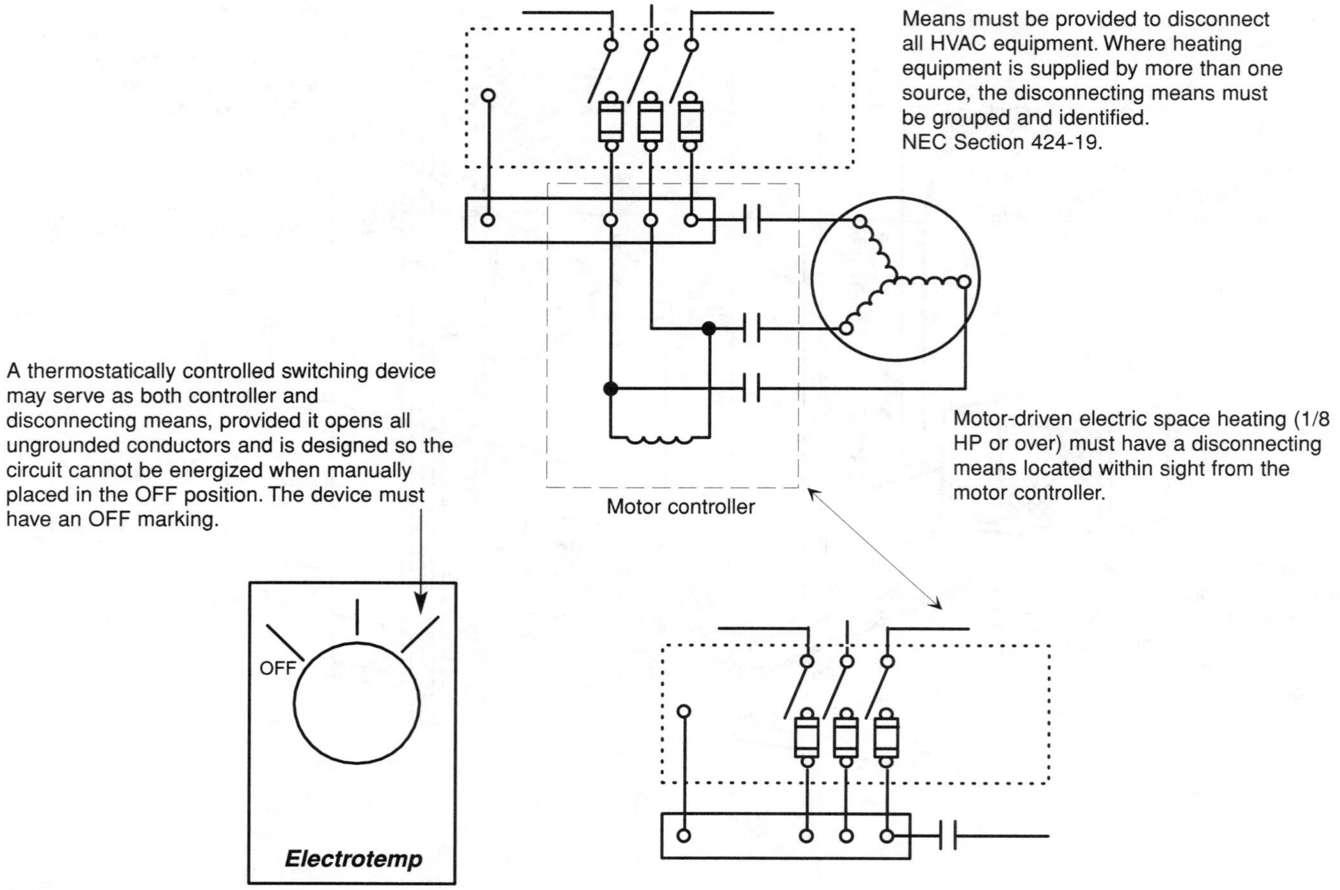
Means must be provided to disconnect all HVAC equipment. Where heating equipment is supplied by more than one source, the disconnecting means must be grouped and identified.
NEC Section 424-19.
A thermostatically controlled switching device may serve as both controller and disconnecting means, provided it opens all ungrounded conductors and is designed so the circuit cannot be energized when manually placed in the OFF position. The device must have an OFF marking.
Motor controller
Motor-driven electric space heating (1/8 HP or over) must have a disconnecting means located within sight from the motor controller.
OFF
Electrotemp

Figure 6-3: *Summary of NEC installation requirements for HVAC controls and disconnects*

Room Air Conditioners

NEC Article 440-G, beginning with Section 440-60, covers room air conditioners rated at 250 volts or less, and single-phase only; cord- and attachment plug-connection permitted. Figures 6-4, 6-5 and 6-6 summarize *NEC* installation requirements for room air-conditioning units.

APPLIANCES

Appliances are used in dwellings, restaurants, and other occupancies to perform specific functions such as cooking, cooling, washing, and drying. The *NEC* specifies installation requirements for such appliances in terms of their characteristics as well as by the method of connection to the supply circuits.

The *NEC* also allows the application of demand factors to many fixed appliances such as electric ranges (*NEC* Table 220-19), clothes dryers *(NEC* Table 220-18), and nondwelling kitchen equipment (*NEC* Table 220-20).

For an electrical system branch- or feeder-circuit calculation to obtain conductor sizes and overcurrent protection, a demand factor is a ratio of the amount of connected load (in kVA or amperes) that will be operating at the same time to the total amount of connected

Room Air Conditioning		
Application	**Requirements**	**NEC Reference**
Connection to a 15-, 20-, or 30-ampere branch circuit.	Must not exceed 80% of the branch-circuit rating.	Section 440-62(b
Connection to branch circuit also supplying lighting or other appliances.	Must not exceed 50% of the branch-circuit rating.	Section 440-62(c)
Disconnecting means	An attachment plug and receptacle may be used.	Section 440-63
Grounding	Grounded receptacle may be used.	Section 440-61 Section 250-110 Section 250-112 Section 250-114

Figure 6-4: *Summary of NEC installation requirements for room air conditioners*

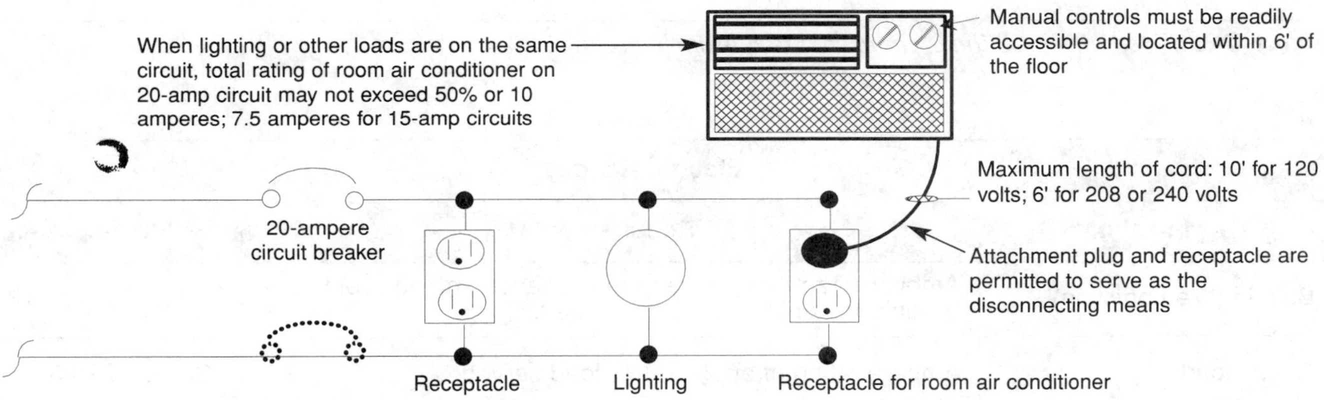
When lighting or other loads are on the same circuit, total rating of room air conditioner on 20-amp circuit may not exceed 50% or 10 amperes; 7.5 amperes for 15-amp circuits
Manual controls must be readily accessible and located within 6' of the floor
Maximum length of cord: 10' for 120 volts; 6' for 208 or 240 volts
20-ampere circuit breaker
Attachment plug and receptacle are permitted to serve as the disconnecting means
Receptacle
Lighting
Receptacle for room air conditioner

Figure 6-5: *Branch circuits for room air conditioners*

load on the circuit. An 80% demand factor, for instance, indicates that only 80% of the connected load on a circuit will ever be operating at the same time. Conductor capacity can be based on that amount of load in certain situations as stipulated in the *NEC*, but in most cases, the conductor size should be capable of carrying the full load of each individual appliance.

Cooking Appliances

Electric Range: A cooking appliance, either free-standing or built-in, usually equipped with top heating elements, an oven, and controls. Electric ranges may be connected to its branch circuit by a cord-and-plug connection or it may be permanently connected. The means of connection, however, does not identify the appliance as either being fastened in place or portable, although most portable appliances utilize a cord and plug for their connection to a supply circuit.

Figure 6-7 summarizes the *NEC* installation requirements for electric ranges.

Wall-Mounted Oven: A flush- or surface-mounted oven installed in or on a wall or other surface. The unit is designed for cooking and contains one or more electric heating elements, internal wiring, and built-in or

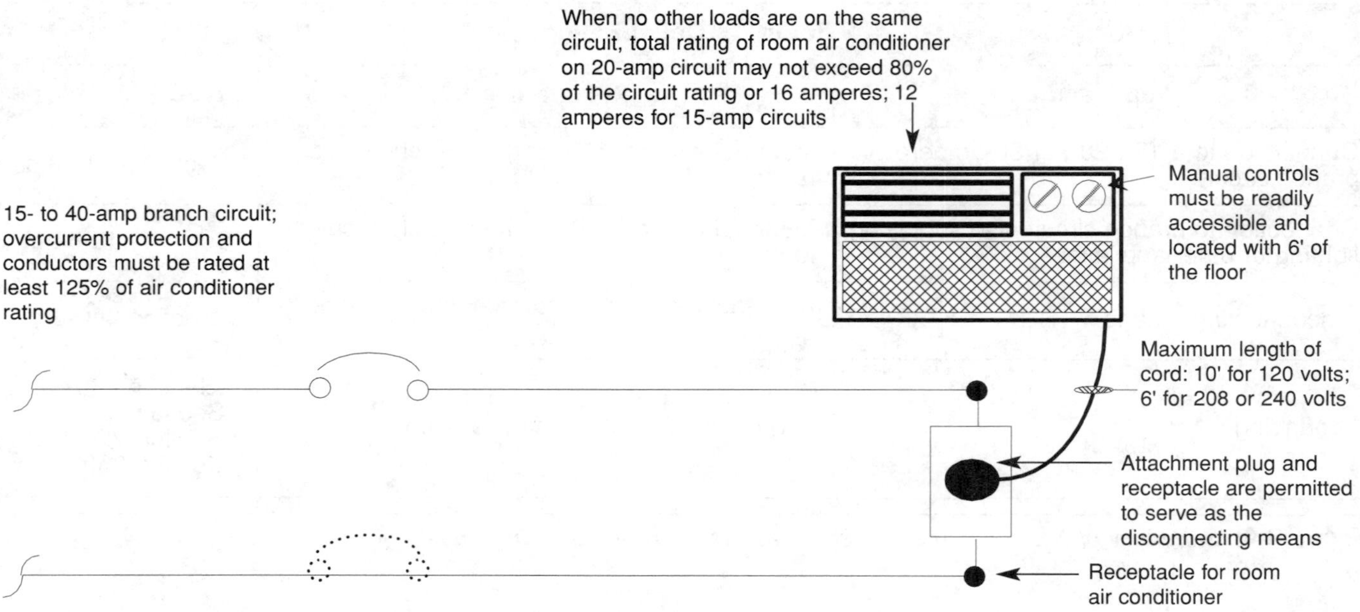
When no other loads are on the same circuit, total rating of room air conditioner on 20-amp circuit may not exceed 80% of the circuit rating or 16 amperes; 12 amperes for 15-amp circuits
15- to 40-amp branch circuit; overcurrent protection and conductors must be rated at least 125% of air conditioner rating
Manual controls must be readily accessible and located with 6' of the floor
Maximum length of cord: 10' for 120 volts; 6' for 208 or 240 volts
Attachment plug and receptacle are permitted to serve as the disconnecting means
Receptacle for room air conditioner

Figure 6-6: *Fixed air conditioner connected to a branch circuit*

Electric Ranges		
Application	**Requirements**	**NEC Reference**
Built-in (fixed position)	Ample protection between the appliance and adjacent combustible materials must be provided.	Section 422-17
Circuit load	Ampere rating of appliance or load served.	Section 220-3(b)(1)
Conductor size minimum	Not less than the maximum load to be served.	Section 210-19(a)
Demand factors	As specified in table.	Table 220-19
Disconnect	Attachment plug and receptacle may be used as the disconnecting means.	Section 422-32(a)
Disconnecting means	Must be provided to disconnect each appliance.	Section 422-30
Feeders	Must have sufficient ampacity to supply the load served.	Section 220-10
Grounding	Frame must be grounded as specified in NEC Sections 250-140 or Sections 250-134 and 250-138.	Section 250-60
Overcurrent protection	May be used as the disconnecting means.	Section 422-31(b)

Figure 6-7: *Summary of NEC installation requirements for electric ranges*

Wall-Mounted Oven		
Application	**Requirements**	**NEC Reference**
Ampacity of conductors	Not less than the load to be served unless otherwise specified in this NEC Section.	Section 210-19(c)
Branch circuits and feeders	Type SE cable permitted as branch circuit to supply range or oven.	Section 338-3
Calculations	Total nameplate rating used in mobile homes for wall-mounted oven.	Section 550-13
Definition	As indicated.	Article 100-A
Demand loads	Computed as per NEC Table 220-19.	Section 220-19
Dwelling, optional calculations for	As indicated in this NEC Section.	Section 220-30

Figure 6-8: *Summary of NEC installation requirements for wall-mounted ovens*

separately mountable controls. A wall-mounted oven is separated from the counter-mounted cooktop as opposed to an electric range which has the oven and cooking burners installed in one housing or unit.

Figure 6-8 summarizes the *NEC* installation requirements for wall-mounted ovens.

Portable Appliances: As the name implies, portable appliances consist of movable devices such as coffee makers, mixers, toasters, and the like. Most are cord- and plug-connected for easy removal from the electric circuit so they may be moved from one location to another, or stored in a cabinet or drawer.

Figure 6-9 summarizes the *NEC* installation requirements for portable appliances.

CAPACITORS

Capacitors are devices used to improve the power factor of an electrical installation or an individual piece of electrically-operated equipment. The efficiency lowers the cost of power. See Chapter 11 of this book.

COMPUTER/DATA PROCESSING

Rooms or areas specifically designed for the operation of electronic computer/data processing equipment fall under *NEC* Article 645. This Article covers equipment, power-supply wiring, equipment interconnecting wiring, and grounding of electronic computer/data processing equipment and systems, including terminal units, in an electronic computer/data processing room. See Chapter 7 of this book.

DIELECTRIC HEATING EQUIPMENT

The heating of an insulating material by ac induced internal losses; normally frequencies above 10 mHz are used. See induction heating equipment in Chapter 7 of this book.

ELECTRIC SIGN AND OUTLINE LIGHTING

Fixed, stationary, or portable self-contained, electrically illuminated utilization equipment with words or symbols designed to convey information or attract attention. See Chapter 7 of this book.

Portable Appliances		
Application	**Requirements**	**NEC Reference**
Color of grounding conductor	May be bare, but if insulated, it must have a continuous outer finish that is either green or green with one or more yellow stripes.	Section 250-119
Grounding of	All exposed metal noncurrent-carrying parts must be grounded.	Section 250-110
Hand lamps	Must be wired with flexible cord, recognized by NEC Section 400-4, and provided with an attachment plug of the polarized or grounding type.	Section 410-42(b)
Immersion heaters	Must be effectively insulated from electrical contact with the substance in which they are immersed.	Section 422-9

Figure 6-9: *Summary of NEC installation requirements for portable appliances*

ELECTRIC DEICING AND SNOW-MELTING EQUIPMENT

Equipment designed for melting unwanted ice and snow, using the heat caused by electrical resistance. See Figure 6-10 for a summary of *NEC* installation requirements.

ELECTRIC WELDER

An electrical utilization equipment that produces an arc between an electrode and the base metal. The electric arc generates an intense heat of 6000 to 10,000 degrees Fahrenheit. The arc melts the base metal and the electrode, forming a weld. There are two types of welding currents: ac (alternating current) and dc (direct current). See Chapter 7 of this book.

ELECTROLYTIC CELLS

Electrolytic cells are receptacles or vessels in which electrochemical reactions are caused by applying electrical energy for the purpose of refining or producing usable materials such as aluminum, cadmium, chlorine, copper, fluorine, hydrogen peroxide, magnesium, sodium, sodium chlorate, and zinc. A summary of *NEC* installation requirements appears in Figure 6-11.

ELEVATORS, ESCALATORS, MOVING WALKS AND DUMBWAITERS

NEC Article 620 covers the installation of electric equipment and wiring used in connection with elevators, dumbwaiters, moving walks, wheelchair lifts, and stairway chair lifts. Figure 6-12 summarizes *NEC* installation requirements for these types of utilization equipment. See the *NEC* book for additional details and exact specifications.

FIRE PUMPS

A fire pump is a water pump designed specifically for pumping water to sprinkler systems and building fire hoses. Fire pumps are covered in *NEC* Article 695. In general, this article covers the installation of:

- Electrical power sources and interconnecting circuits for fire pump devices
- Switching and control equipment dedicated to fire pump devices

Figure 6-13 on page 161 summarizes *NEC* installation requirements for fire pumps. Please refer to the *NEC* book for additional details.

Deicing and Snow-Melting Equipment		
Application	**Requirements**	**NEC Reference**
Branch-circuit sizing	Must not be less than 125% of the total load.	Section 426-4
Cord- and plug-connected deicing equipment	Must be listed.	Section 426-54
Disconnecting means	A means to disconnect all ungrounded conductors must be provided.	Section 426-50(a)
	Where readily accessible, the branch-circuit switch or circuit breaker may act as a disconnecting means.	Section 426-50(a)
	Attachment plug may be used as disconnecting means under certain conditions.	Section 426-50(b)
Embedded equipment	Must comply with (a) through (e) of this NEC Section.	Section 426-20
Equipment protection	GFCI protection must be provided.	Section 426-28
Exposed equipment	Must comply with (a) through (d) of this NEC Section.	Section 426-21
Identification	Appropriate caution signs must be posted.	Section 426-13
Overcurrent protection of fixed equipment	Located where supplied by branch circuit.	Section 426-52
Special permission	Required for all installation methods not indicated in the NEC.	Section 426-14
Thermal protection	Must be guarded, isolated, or thermally insulated.	Section 426-12

Figure 6-10: *Summary of NEC installation requirements for deicing and snow-melting equipment*

Electrolytic Cells		
Application	**Requirements**	**NEC Reference**
Cell-line conductor connections	Joined by bolted, welded, clamped, or compression connectors.	Section 668-12(c)
Cell-line conductors	Must be either bare, covered, or insulated and of copper, aluminum, copper-clad aluminum, steel, or other suitable material.	Section 668-12(a)
Cell-line conductor size	Must be large enough to maintain a safe working temperature under maximum load conditions.	Section 668-12(b)

Figure 6-11: *Summary of NEC installation requirements for electrolytic cells*

Elevators, Escalators, Moving Walks, Dumbwaiters and Wheelchair Lifts		
Application	**Requirements**	**NEC Reference**
Conductor insulation	Must have insulation rated at the maximum voltage rating.	Section 620-11(d)
	Must have flame-retardant insulation.	Section 620-11(c)
Conductor insulation type	Must be either Type MTW, TF, TFF, TFN, TFFN, THHN, THW, THWN, TW, or XHHW.	Section 620-11(c)
Conductors, minimum size for control equipment and signaling circuits	Traveling cables for lighting circuits: No. 14 or equivalent No. 20 or larger paralleled conductors.	Section 620-12(a)(1)
	Operating control and signaling circuits: No. 20.	Section 620-12(a)(2)
	Must be of the types of elevator cables listed in NEC Table 400-4 or other approved types.	Section 620-11(b)
Live electrical parts	Must be enclosed to protect against accidental contact.	Section 620-4
Voltage limitations for control and signaling circuits	Must not exceed 300 volts.	Section 620-3
Voltage limitations for driving-machine motors, machine brakes, and motor-generator sets	Must not exceed 600 volts except for motor-generator sets.	Section 620-3(a)

Figure 6-12: *Summary of NEC Article 620*

FIXED ELECTRIC SPACE HEATING

The use of electric heating in all types of occupancies has risen tremendously over the past couple of decades, and electric heating — including the use of heat pumps — shows all signs of remaining a popular fuel type.

Two general types of fixed electric heating equipment are recognized by the *NEC*:

- Fixed electric heating equipment for pipelines and vessels
- Fixed electric space heating equipment

Most electricians will deal with space heating; that is, installing electric heating systems to heat the interiors of buildings and such installations are covered in *NEC* Article 424. This Article, however, does not apply to process heating nor room air conditioning. Equipment used in hazardous locations must comply with Articles 500 through 517, and equipment incorporating a hermetic refrigerant motor-compressor must also comply with *NEC* Article 440.

Electric Baseboard Heaters

All requirements of the *NEC* apply for the installation of electric baseboard heaters, especially *NEC* Article 424 — Fixed Electric Space Heating Equipment. In general, electric baseboard heaters must not be used where they will be exposed to severe physical damage unless they are adequately protected from such possible damage. Heaters and related equipment installed in damp or wet locations must be approved for such locations and must be constructed and installed so that water cannot enter or accumulate in or on wired sections, electrical components, or duct work.

Baseboard heaters must be installed to provide the required spacing between the equipment and adjacent combustible material, and each unit must be adequately grounded in accordance with *NEC* Article 250.

Figure 6-14 summarizes the *NEC* regulations governing the installation of electric baseboard heaters, while Figure 6-15 on the next page shows a residential floor-plan layout for electric heat.

Electric Space-Heating Cables

Radiant ceiling heat is acknowledged to be one of the greatest advances in structural heating since the Franklin stove, or so say the manufacturers, and thousands of homeowners all over the country who have chosen this type of heat for their homes.

The enormous heating surface precludes the necessity of raising air temperatures to a high degree. Rather, gentle warmth flows downward (or upward in the case of cable embedded in concrete floors) from the surfaces, heating the entire room or area evenly, and usually leaving no cold spots or drafts.

There is no maintenance with a radiant heating system as there are no moving parts, nothing to get clogged up, nothing to clean, oil, or grease, and nothing to wear out.

The installation of this system is within reach of even the smallest electrical contractor. The most difficult part of the entire project is the layout of the system; that is, how far apart to string the cable on the ceiling or in a concrete slab.

An ideal application of electric radiant heating cable would be during the renovation of an area within an existing residence, where the old ceiling plaster is beginning to crack and this ceiling will be recovered with dry-

Fire Pumps		
Application	**Requirements**	**NEC Reference**
Connection at service	One of the disconnects (from two to six) as specified in this NEC Section may be located remote from the others if it serves a fire pump and is used only for that purpose.	Section 230-72(a) Exception
	May be connected to the supply side of the electric service.	Section 230-82 Exception No. 4
	May be connected to the service supply side if a separate means of overcurrent protection is provided.	Section 230-94 Exception No. 4
Control wiring	External control circuits must be arranged so that failure of any external circuit will not prevent the operation of the pumps.	Section 695-14
Emergency power supply	As specified.	Article 700
Equipment location	Must be located as specified in (a) through (f) of this NEC Section.	Section 695-7
Motor overcurrent protection	Not required where it might introduce additional or increased hazards.	Section 430-31

Figure 6-13: *Summary of NEC installation requirements for fire pumps*

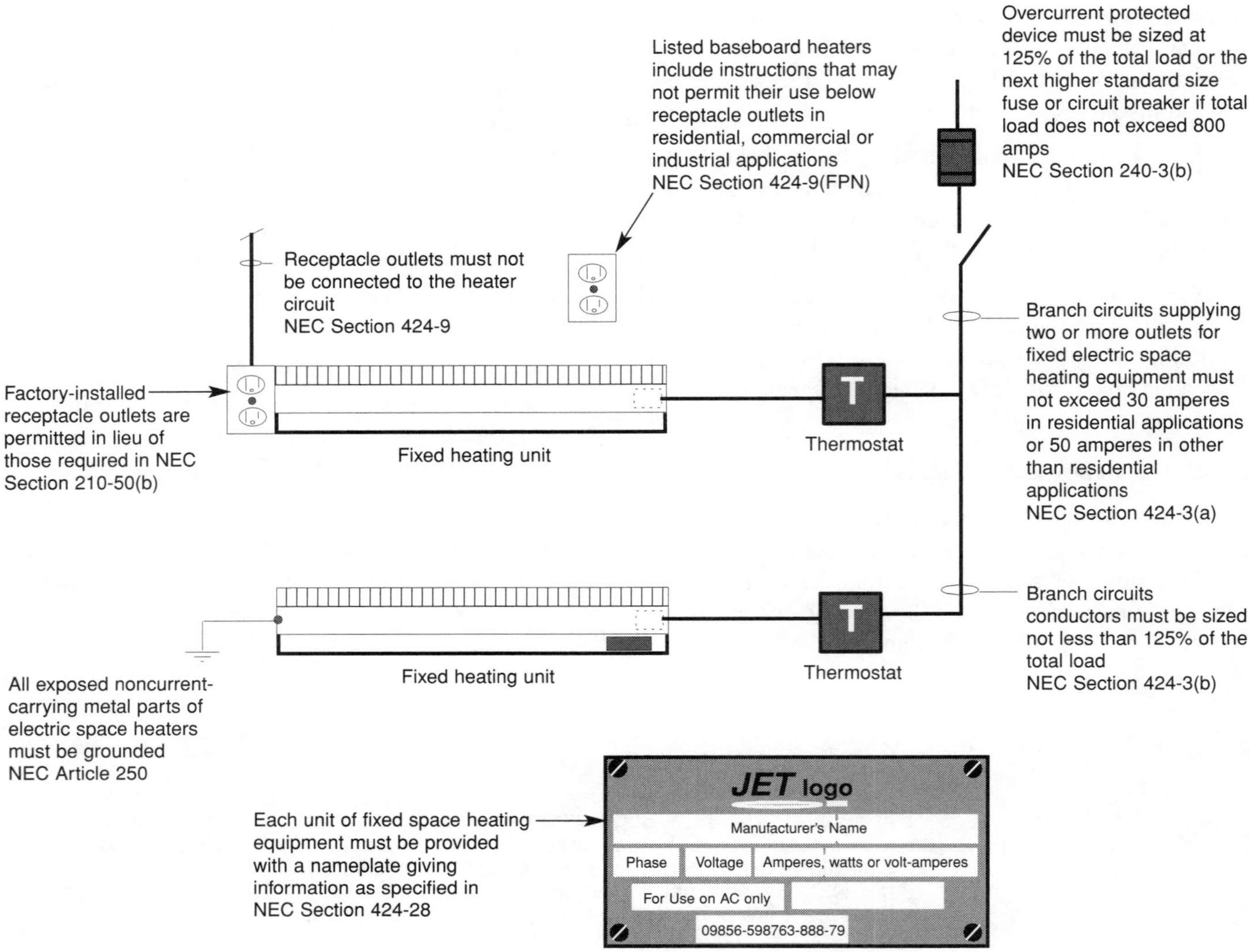
Listed baseboard heaters include instructions that may not permit their use below receptacle outlets in residential, commercial or industrial applications NEC Section 424-9(FPN)
Overcurrent protected device must be sized at 125% of the total load or the next higher standard size fuse or circuit breaker if total load does not exceed 800 amps NEC Section 240-3(b)
Receptacle outlets must not be connected to the heater circuit NEC Section 424-9
Branch circuits supplying two or more outlets for fixed electric space heating equipment must not exceed 30 amperes in residential applications or 50 amperes in other than residential applications NEC Section 424-3(a)
Factory-installed receptacle outlets are permitted in lieu of those required in NEC Section 210-50(b)
Fixed heating unit
T
Thermostat
Branch circuits conductors must be sized not less than 125% of the total load NEC Section 424-3(b)
Fixed heating unit
T
Thermostat
All exposed noncurrent-carrying metal parts of electric space heaters must be grounded NEC Article 250
JET logo
Each unit of fixed space heating equipment must be provided with a nameplate giving information as specified in NEC Section 424-28
Manufacturer's Name
Phase
Voltage
Amperes, watts or volt-amperes
For Use on AC only
09856-598763-888-79

Figure 6-14: *Summary of NEC installation requirements for electric baseboard heaters*

wall or other type of plaster board. Or, perhaps the basement floor needs repair and 3 inches of additional concrete will be poured over the existing floor. These are ideal locations to install radiant heating cable.

Installation In Plaster Ceilings

To determine the spacing of the cable on a given ceiling, deduct one foot from the room length and one foot from the room width and multiply this new length by the new width, which will give the usable ceiling area in square feet. Multiply the square feet of the ceiling by 12 to get the ceiling area in inches before dividing by the length of the heating cable. The result will be the number of inches apart to space the cable.

For example, assume that a room 14 x 12 feet has a calculated heat loss of 2,000 watts. Therefore, (14 ft. – 1 ft.) (12 ft. - 1 ft.) x 12 in. = 13 x 11 x 12 = 1,716. We then look at manufacturers' tables and see that a 2,000-watt heating cable is 728 feet in length. Dividing the usable area (1,716 sq. in.) by this length (728), we find that the cable should be spaced 2.4 inches apart.

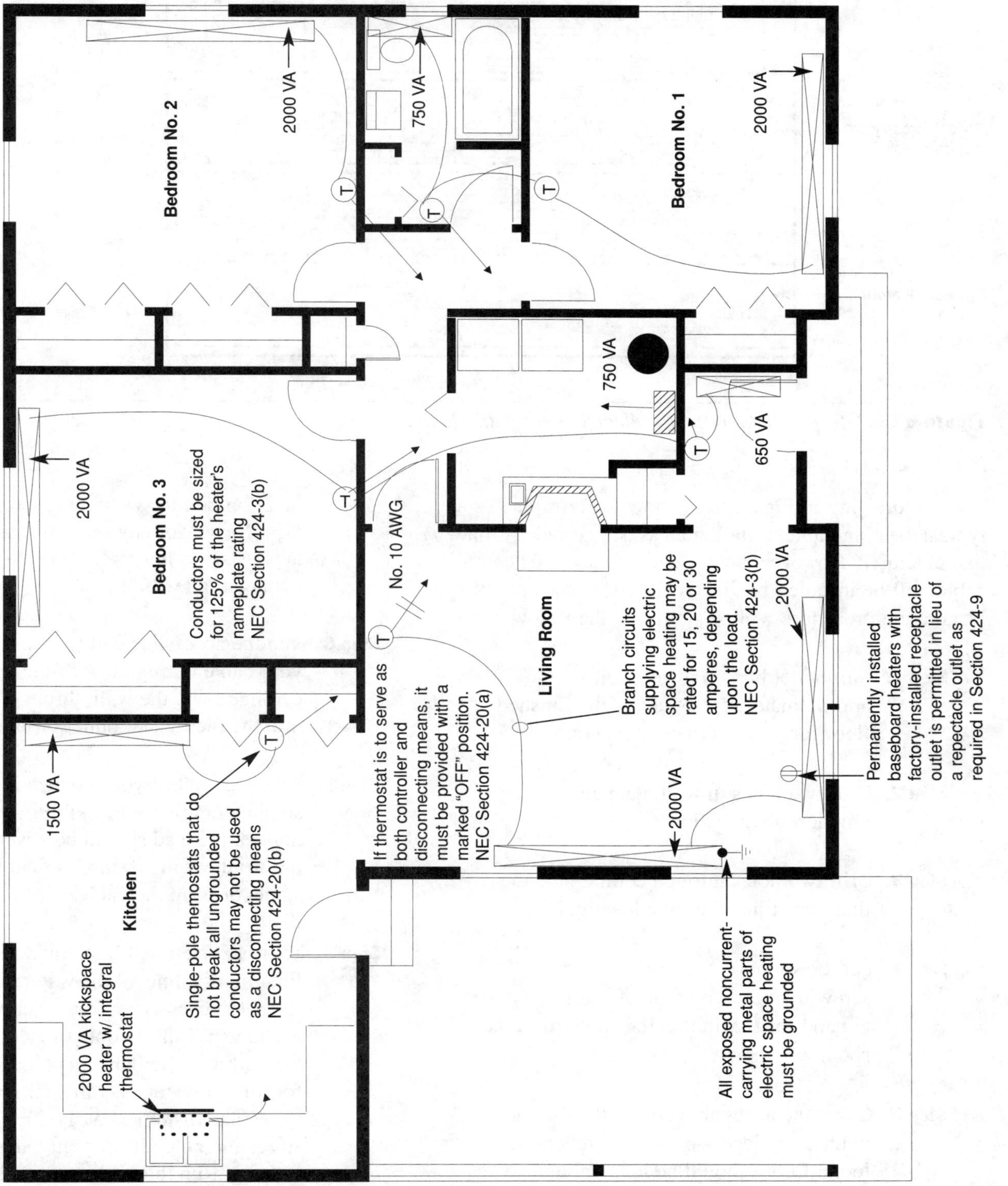
Bedroom No. 2
2000 VA
750 VA
Bedroom No. 1
2000 VA
750 VA
650 VA
Bedroom No. 3
2000 VA
Conductors must be sized for 125% of the heater's nameplate rating NEC Section 424-3(b)
No. 10 AWG
Living Room
Branch circuits supplying electric space heating may be rated for 15, 20 or 30 amperes, depending upon the load. NEC Section 424-3(b)
2000 VA
Permanently installed baseboard heaters with factory-installed receptacle outlet is permitted in lieu of a repectacle outlet as required in Section 424-9
1500 VA
If thermostat is to serve as both controller and disconnecting means, it must be provided with a marked "OFF" position. NEC Section 424-20(a)
2000 VA
Kitchen
Single-pole thermostats that do not break all ungrounded conductors may not be used as a disconnecting means NEC Section 424-20(b)
2000 VA kickspace heater w/ integral thermostat
All exposed noncurrent-carrying metal parts of electric space heating must be grounded

Figure 6-15: *Floor-plan layout for a residential electric-heating application*

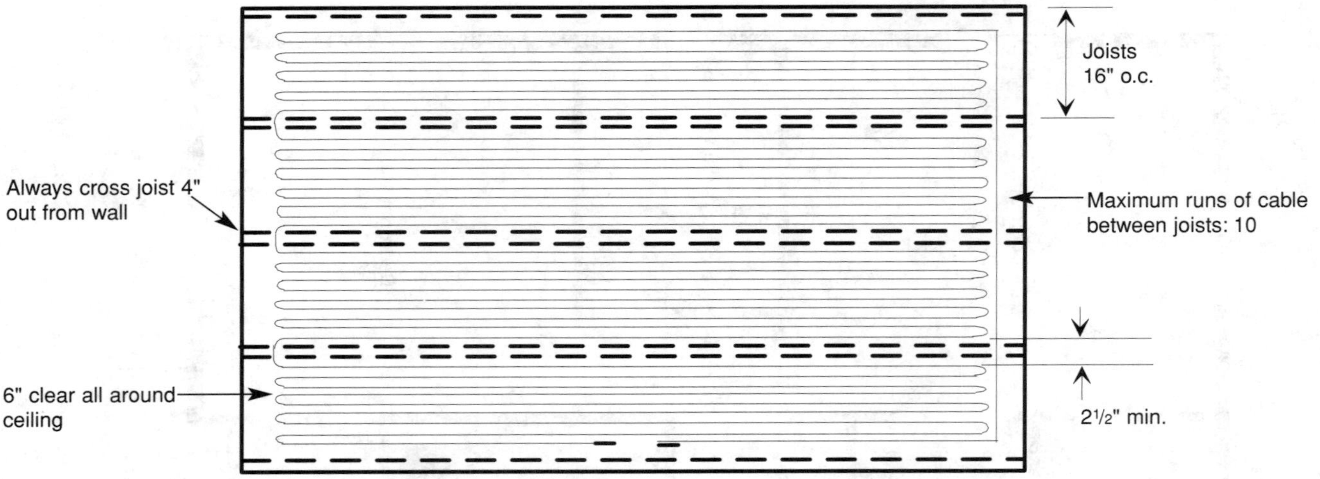
Joists
16" o.c.
Always cross joist 4"
out from wall
Maximum runs of cable
between joists: 10
6" clear all around
ceiling
2 1/2" min.

Figure 6-16: *Recommendations for electric heat-cable layout*

The drawing in Figure 6-16 shows a floor plan of a typical heating-cable installation as suggested by one manufacturer. However, nearly every brand of heating cable will be installed in exactly the same way, and the installation procedure will be similar to the following:

Step 1. Nail outlet box on inside wall approximately 5 feet above the finished floor for your thermostat location.

Step 2. Drill two holes in wall plate above this junction box location.

Step 3. Drill two holes through ceiling lath above thermostat junction box location.

Step 4. Put the spool of heat cable on a nail, screwdriver, or any type of shaft you have at hand for unwinding the cable from the spool.

Step 5. Cover the accessible end of the 8-foot nonheating lead wire with nonmetallic loom. Loom should be long enough so at least 2 inches will go on ceiling surface and reach to the thermostat outlet box, but leave 6 inches of lead wire inside of this junction box for viewing the identification tags. Never, for any reason, remove those tags. Also do not cut or shorten the nonheating leads.

Step 6. Run the accessible end (loom and all) of your cable through one of the holes in the ceiling, down the wall, through the plate, into the thermostat outlet or junction box.

Step 7. Pull slack out of your nonheating leads and staple securely to the ceiling. Any excess nonheating lead should be covered with plaster the same as the heat cable. Do not staple or bend the cable.

Step 8. Your next step will be to mark the ceiling. First mark a line all the way around the room 6 inches away from each wall; this accounts for the 1 foot you deducted from the width and length. A chalk line is best for this. Now take a yardstick, often given to regular customers at the local hardware store, and notch it for your calculated spacing. Run the cable along the ceiling 6 inches out from the wall to an outside wall, where you attach it in parallel spacing. In this example the spacing is 2.3 inches.

Always keep the cable at least 2 inches from metal corner lath, or other metal reinforcing.

Step 9. When you are down to the return lead wire, you are back to the starting wall. Cover this lead with the same length of loom as you did the starting lead. Then staple this return lead securely to the ceiling, run through the other hole in the ceiling, down the wall and into the thermostat outlet box.

Step 10. Connect the thermostat, which should be fed by a circuit of proper wire size according to the current, in amperes, drawn by the heating cable. Divide the wattage by the rated voltage. Your answer will be the load in amperes. Then use the following table to size your wire.

Amperes	Type TW Wire
15 or lower	10 AWG
15 – 20	8 AWG
20 – 30	6 AWG
30 – 50	4 AWG

Always make certain that the heating cable is connected to the proper voltage. A 120-volt cable connected to a 240-volt circuit will melt the cable, while a 240-volt heating cable connected to 120 volts will produce only 25 percent of the rated wattage of the cable. See Figure 6-17 for a summary of *NEC* requirements governing the installation of electric space-heating cable.

Installing Concrete Cable

The heat loss is calculated in the same manner as for any other area. For best results, the heating cable should never be spaced less than 2½ inches apart when installing in concrete floor except around outside walls, which may be spaced on 1½-inch minimum centers for the first 2 feet out from the wall. The concrete thickness above the cable should be from ½ to 1 inch.

The spacing of the cable is found exactly as previously shown for the spacing of the cable on the ceiling, and then the installation procedure is as follows:

Step 1. Secure junction box on an inside wall approximately 5 feet from the finished floor to house the thermostat.

Step 2. Install a piece of rigid conduit from the switch or thermostat junction box to house the nonheating leads, between the concrete slab and the switch or thermostat outlet box.

Step 3. Approximately 6 inches of the lower end of the conduit should be embedded in the concrete and a smooth porcelain bushing should be on this end of the conduit to protect the nonheating leads where they leave the conduit.

Step 4. Place the spool of heating cable on a nail, screwdriver, or other shaft you have at hand for unwinding the cable from the spool.

Step 5. Run the accessible end of the 8-feet nonheating lead through the conduit to the thermostat junction box, leaving 6 inches extending out of the box. Never remove the identification tags or shorten the nonheating leads. They should be embedded in the concrete the same as the heating portion of the cable.

Step 6. Run the cable along the floor 6 inches out from the wall to the outside or exposed wall, fastening the cable to the floor either with staples or masking tape.

Step 7. The cable usually is spaced 1½ inches apart for the first 2 feet around the exposed wall and never less than 2½ inches apart for the remaining area.

Step 8. Run the return nonheating lead wire through the conduit to the thermostat junction box in the same manner as the starting nonheating lead was run in Step 5.

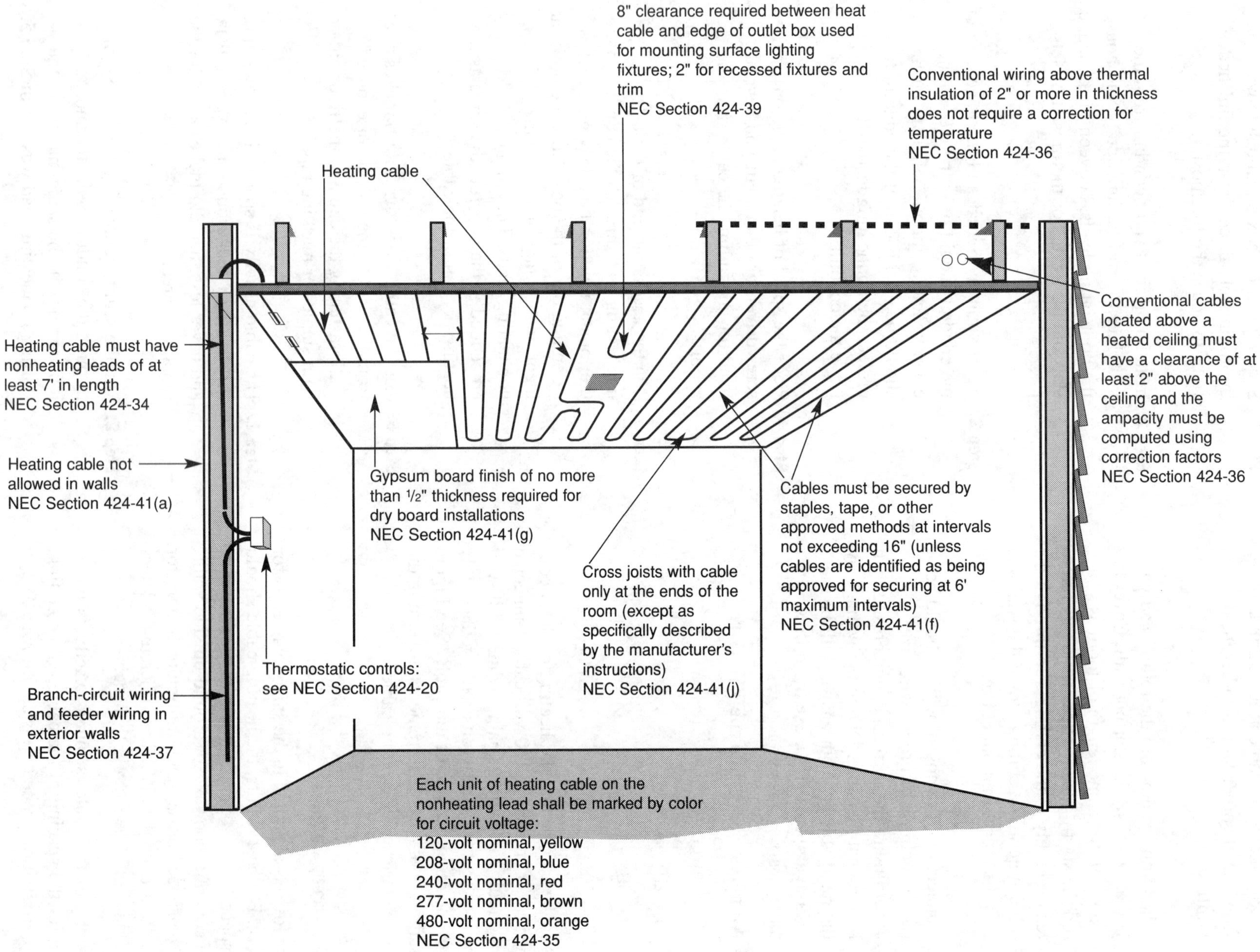
8" clearance required between heat cable and edge of outlet box used for mounting surface lighting fixtures; 2" for recessed fixtures and trim
NEC Section 424-39
Conventional wiring above thermal insulation of 2" or more in thickness does not require a correction for temperature
NEC Section 424-36
Heating cable
Conventional cables located above a heated ceiling must have a clearance of at least 2" above the ceiling and the ampacity must be computed using correction factors
NEC Section 424-36
Heating cable must have nonheating leads of at least 7' in length
NEC Section 424-34
Heating cable not allowed in walls
NEC Section 424-41(a)
Gypsum board finish of no more than 1/2" thickness required for dry board installations
NEC Section 424-41(g)
Cables must be secured by staples, tape, or other approved methods at intervals not exceeding 16" (unless cables are identified as being approved for securing at 6' maximum intervals)
NEC Section 424-41(f)
Cross joists with cable only at the ends of the room (except as specifically described by the manufacturer's instructions)
NEC Section 424-41(j)
Thermostatic controls: see NEC Section 424-20
Branch-circuit wiring and feeder wiring in exterior walls
NEC Section 424-37
Each unit of heating cable on the nonheating lead shall be marked by color for circuit voltage:
120-volt nominal, yellow
208-volt nominal, blue
240-volt nominal, red
277-volt nominal, brown
480-volt nominal, orange
NEC Section 424-35

Figure 6-17: *Summary of NEC installation requirements for installing electric space-heating cable*

In general, the concrete slab should be prepared by applying a vapor barrier over 4 to 6 inches of gravel. Then pour 4 inches of vermiculite or other insulating concrete over the gravel after the outside edges of the slab have been insulated, in accordance with good building practice. The heating cable is then installed as described previously.

At this time, workers should inspect and test the cable before the final layer of concrete is installed, because once the concrete is poured, it's an expensive matter to repair. First, visually inspect the cable for any possible damage to the insulation during the application. Then, with a suitable ohmmeter, check for continuity and capacity of the cable. Concealed breaks may be found by leaving an ohmmeter connected to the cable, and then brushing the cable lightly with the bristles of a broom. Any erratic movement of the meter dial will indicate a fault.

During the pouring of the final coat of concrete, it is recommended that the ohmmeter be left connected to the heating cable leads to detect any possible damage to the cable during the pouring of the finish layer of ordinary concrete (do not use insulating concrete). If an ohmmeter is not available at the time, a 100-watt lamp may be connected in series with the cable to immediately detect any damage during the installation of the concrete. The lamp will glow as long as the circuit is complete and no damage occurs. However, if a break does occur, the lamp will go out and the break can be repaired before the concrete hardens.

Repairs to a broken cable are made by stripping the ends of the broken cable and rejoining the ends with a No. 14 AWG pressure-type connector provided and approved for this purpose. The splice must then be insulated with thermoplastic tape to a thickness equal to the insulation of the cable. Use any thermoplastic tape listed by the Underwriters' Laboratories as suitable for a temperature of 176°F.

Once the finish layer of concrete sets, then asphalt tile, linoleum tile, or linoleum can be laid on the concrete in the normal manner.

Besides the heating of interior spaces, electric-heating cable also has many other uses. It can be embedded in concrete or asphalt surfaces for the removal of ice and snow; provide freeze protection for water pipes exposed to cold weather; used for roof and gutter deicing; and it can heat soil in a hotbed or window box — keeping the temperature at a constant 70°F. The cost of heat cable is relatively inexpensive and the installation goes fast.

FORCED-AIR SYSTEMS

Electric forced-air heating systems are usually of three types:

- Heat pumps
- Electric furnaces
- Air conditioner with duct heaters

In the majority of installations utilizing central, forced-air systems, the system is usually designed by mechanical engineers and installed by mechanical contractors. Branch circuits, feeders, motor starters, and some control wiring is frequently installed by electrical workers. Consequently, every electrician should have a basic working knowledge of forced-air systems, along with applicable *NEC* installation requirements.

Duct Heaters

Duct heater is a term applied to any heater mounted in the air stream of a forced-air system where the air-moving (fan-coil) unit is not provided as an integral part of the equipment. Duct heaters are used in electric furnaces, combination electric heating/cooling systems, and most of the time in heat pumps to offer auxiliary heat when the pumps themselves cannot supply the demand.

In general, heaters installed in an air duct must be identified as suitable for the installation, and some means must be provided to assure uniform and adequate airflow over the face of the heater in accordance with the manufacturer's instructions. This latter requirement is normally accomplished by airflow controls and other components involving turning vanes, pressure plates, or other devices on the inlet side of the duct heater to assure an even distribution of air over the face of the heater.

Duct heaters installed closer than 4 feet to a heat pump or air conditioner must have both the duct heater and heat pump or air conditioner identified as suitable for such installation and must be so marked.

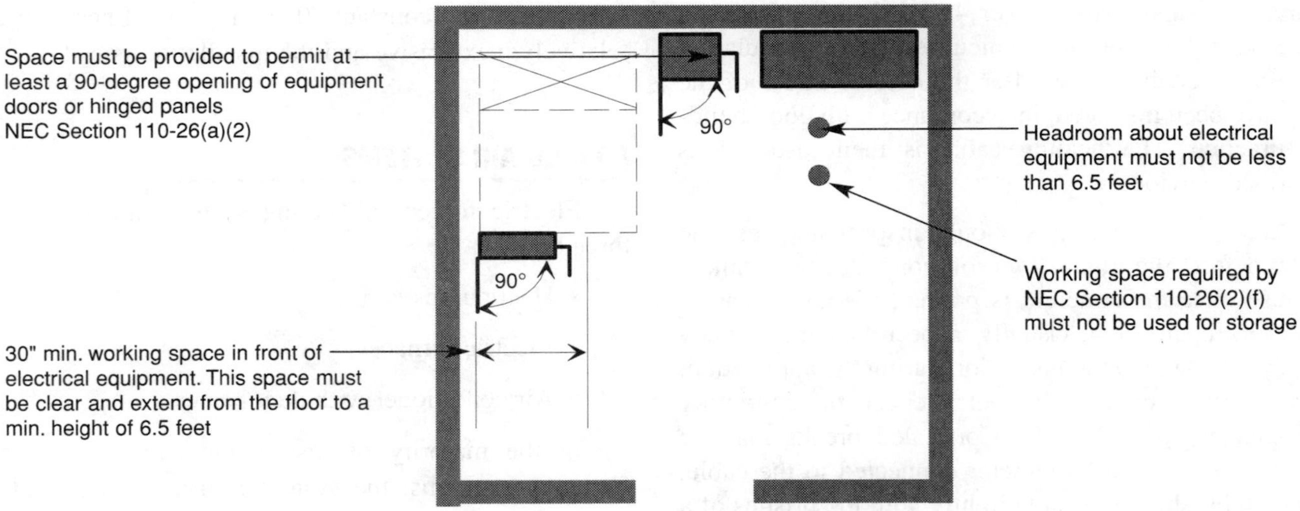
Space must be provided to permit at least a 90-degree opening of equipment doors or hinged panels NEC Section 110-26(a)(2)
90°
Headroom about electrical equipment must not be less than 6.5 feet
Working space required by NEC Section 110-26(2)(f) must not be used for storage
90°
30" min. working space in front of electrical equipment. This space must be clear and extend from the floor to a min. height of 6.5 feet

Figure 6-18: *Summary of NEC Section 110-26*

Duct heaters intended for use with elevated inlet air temperature must be identified as suitable for use at the elevated temperatures. Furthermore, duct heaters used with air conditioners or other air-cooling equipment that may result in condensation of moisture must be identified as suitable for use with air conditioners.

The *NEC* requires that all duct heaters are installed according to the manufacturer's instructions. Furthermore, duct heaters must be located with respect to building construction and other equipment so as to permit access to the heater. Sufficient clearance must be maintained to permit replacement of controls and heating elements and for adjusting and cleaning of controls and other parts requiring such maintenance. See Figure 6-18.

Control requirements — including disconnecting means — are specified in *NEC* Sections 424-57 through 424-66. In general, a fan-circuit interlock is one of the requirements. Such a control ensures that the fan circuit is energized when any heater circuit is energized. It would be a waste of energy, and perhaps also be a hazard, if the duct heaters became energized and no air flowed over or through them. However, the *NEC* permits a slight time- and temperature-delay before the fan may be energized. This prevents the system from blowing cold air into the conditioned space. In other words, such a control gives the duct heaters time to "warm-up" before air is induced in the system.

Furthermore, each duct heater must be provided with an automatic-reset limit control to deenergize the duct-heater circuit or circuits in case overheating or other faults occur. In addition, an integral independent supplementary control or controller shall be provided in each duct heater that will disconnect a sufficient number of conductors to interrupt current flow. This device must be manually resettable or replaceable.

The disconnect for duct heaters — the same as for other types of HVAC equipment — must be accessible and within sight of the controller. All control equipment must also be accessible.

ELECTRIC LIGHTING

NEC Article 410 covers lighting fixtures, lampholders, pendants, receptacles, incandescent filament lamps, arc lamps, electric-discharge lamps, and the related wiring to each. While the rules in this *NEC* Article provide for a safe installation, guidelines are not given as to good, efficient lighting designs.

In general, lighting layouts for any type of building should be designed to provide the highest visual comfort and performance that is consistent with the type of area to be lighted and the budget provided. However, since individual tastes and opinions vary, there can be many solutions to any given lighting situation. Some of these solutions can be commonplace, while others will show imagination and resourcefulness.

General-purpose branch circuits are provided in all occupancies to supply lighting outlets for illumination and receptacle outlets for small appliances and office equipment. When lighting circuits are separate from circuits that supply receptacles the *NEC* provides rules for the design of each type of branch circuit. This is usually the case in commercial and industrial occupancies.

Applicable Rules: The lighting load to be used in the branch-circuit calculations for determining the required number of circuits must be the larger of the values obtained by using one of the following:

- The actual load
- A minimum load in volt-amperes or watts per square foot as specified in the *NEC*

Table 220-3(a) of the *NEC* specifies the minimum unit load in volt-amperes (watts) per square foot of floor area based on outside dimensions for the occupancies listed. If the actual lighting load is known and if it exceeds the minimum determined by the watts per square foot basis, the actual load must be used because the *NEC* specifies that branch-circuit conductors shall have an ampacity not less than the maximum load to be served. A store building, for example, with 4000 square feet of floor space (outside dimensions of 40 feet by 100 feet) would have a minimum lighting load of 12 kilovolt-amperes (12,000 watts) based on the 3 watts per square foot unit load specified by the *NEC*. If the actual connected load happened to be one hundred 150-watt lamps, or 15 kilowatts, and the *NEC* calculation requires only 12 kilowatts of lighting, the actual load that must be used in service-entrance, feeders and branch circuit calculations is 15 kilowatts. Therefore, since 15,000 watts is the larger value, this figure would be used to calculate the required number of branch circuits.

Branch Circuits for Lighting: The *NEC* permits only 15- or 20-ampere branch circuits to supply lighting units with standard lampholders. Branch circuits of greater than 20-ampere rating are permitted to supply fixed lighting units with heavy-duty lampholders in other than dwelling occupancies. In other words, branch circuits of greater than 20-ampere rating are not permitted to supply lighting units in dwellings.

In certain design and installation applications, the designer or worker must determine the number of branch circuits that are necessary to supply a given load. The number of branch circuits as determined by the load is:

$$\text{number of circuits} = \frac{\text{total load in volt-amperes}}{\text{capacity of each circuit in volt-amperes}}$$

A 15-ampere, 120-volt circuit has a capacity of 15 amperes x 120 volts = 1800 watts. If the circuit is rated at 20 amperes, the capacity is 20 amperes x 120 volts = 2400 watts. By comparison, a 480/277-volt, three-phase circuit rated at 20 amperes has a capacity of 1.73 x 480 volts x 20 amperes = 16,627 watts, which is a considerable increase over the 120-volt single-phase circuits. This is the way most large office buildings utilize 277-volt lighting whenever practical.

To determine the number of 120-volt, 20-ampere branch circuits to supply a 60,000 volt-ampere lighting load, when each 20-ampere circuit has a capacity of 2400 watts, use the following equation:

$$\frac{\text{total load in volt-amperes}}{2400} = \text{number of 20-amp circuits required}$$

Thus, the number of circuits is

$$\frac{60{,}000\text{ va}}{2400} = 25\text{ circuits}$$

If the number of lamps per circuit is known to be 400, 150-watt lamps, two methods may be used to determine the result. When the watts per lamp is known and when the capacity of the circuit has been determined, the number of lamps per circuit is:

$$\frac{\text{capacity of each circuit in volt-amperes (2400 va)}}{\text{watts per lamp (150 W)}} = 16\text{ lamps per circuit}$$

With each circuit supplying 16 lamps, the total number of circuits required would be:

$$\frac{\text{400 lamps}}{\text{16 lamps per circuit}} = \text{25 circuits}$$

This problem may be checked by noting that each circuit ampacity of 20 amperes must not be exceeded. The current drawn by each lamp at 120 volts is:

$$I = \frac{150\text{ W}}{120\text{ V}} = 1.25\text{ A}$$

The 20-ampere circuit, then, can supply:

$$\frac{\text{20 A}}{\text{1.25 A per lamp}} = \text{16 lamps}$$

or 16 lamps as before. The confidence in the result is high since the answer has been determined by several different ways.

The illustrations and tables in Figures 6-19 through 6-21 summarize *NEC* installation requirements for lighting fixtures, their related components, and circuits — including conductor sizes and overcurrent protection.

It is also recommended that electrical workers dealing with new types of lighting fixtures, obtain installation instructions and brochures from the lighting-fixture manufacturers. Manufacturers of electrical lamps also have brochures available that contain helpful design and installation data.

Lighting		
Application	**Requirements**	**NEC Reference**
Appliances	Branch circuits feeding both lighting and appliances must limit the appliance load to 50% of the circuit rating.	Section 210-23(a)
Bushings for neon lighting	Must be listed for the purpose.	Section 600-32(f)
Calculation of load	As per (a) through (c) of this NEC Section.	Section 220-3
Clothes closets	As per (a) through (d) of this NEC Section.	Section 410-8
Cove	Must have adequate space for proper installation and maintenence.	Section 410-9
Dressing rooms	All lamps must be controlled by wall switches.	Section 520-73
Elevator lighting circuits	Must comply with the requirements of NEC Article 410.	Section 620-3(b)
Emergency, battery	Only emergency lighting may be connected to emergency lighting circuits with a battery power supply.	Section 700-15
Emergency systems	Must be installed as per NEC Section 700-12.	Section 700-17
Exit	See Life Safety Code, NFPA 101-1994,Section 5-10.	Section 517-32(b)
Festoon	Conductors must not be smaller than No. 12	Section 225-6(b)

Figure 6-19: *Summary of NEC installation requirements for occupancy lighting*

Lighting (cont.)		
Application	**Requirements**	**NEC Reference**
Inductive lighting	The computed load must be based on the total ampere rating of all unit components and not on the total watts of the lamps.	Section 210-22(b)
Panelboard rating	Must have a rating not less than the minimum feeder capacity required for the load.	Section 384-13
Show window	Externally wired fixtures must not be used.	Section 410-7
Show window, calculation of	Not less than 200 va per linear ft must be allowed.	Section 220-12
Signs	As specified in this NEC Article.	Article 600
Spray booths	Must comply with (1) through (5) of this NEC Section.	Section 516-3(c)
Theaters, border lights	Must comply with (a) and (b) of this NEC Section.	Section 520-44(a)
Theaters, side lights	Must comply with (a) and (b) of this NEC Section.	Section 520-44(a)
Theaters, scenery lights	Must be wired internally with stems carried through to the back of the scenery where a bushing must be placed on the end of the stem.	Section 520-63(a)

Figure 6-19: *Summary of NEC installation requirements for occupancy lighting (cont.)*

Lighting Assembly		
Application	**Requirements**	**NEC Reference**
Cord type	Must be of the hard-service type and must be marked "water resistant."	Section 680-56(b)
GFCI	Must be protected by GFCI when installed in wall of hot tub, spa, or storage pool.	Section 680-56(a)
Sealing	Cord conductor termination within equipment must be arranged to prevent the entry of water.	Section 680-56(c)
Terminations	Connections with flexible cords must be permanent except as permitted in this NEC Section.	Section 680-56(d)

Figure 6-20: *Summary of NEC installation requirements for lighting assembly*

Lighting Fixtures		
Application	**Requirements**	**NEC Reference**
Construction of	As specified.	Section 410-34 through Section 410-46
Cord-connected	As specified in (a) through (c) of this NEC Section.	Section 410-30
Fixtures as raceways	Must not be used as a raceway for circuit conductors unless fixtures are listed for the purpose or as specified in Exception No. 2 through No. 3 of this NEC Section.	Section 410-31
Grounding	As specified.	Section 410-17 through Section 410-21
Location of	As specified.	Section 410-4 through Section 410-9
Provisions for	As specified.	Section 410-10 through Section 410-14
Support	As specified.	Section 410-15 through Section 410-16
Wiring	As specified.	Section 410-22 through Section 410-31

Figure 6-21: *Summary of NEC installation requirements for lighting fixtures*

Chapter 7 Miscellaneous Electrical Systems

The NEC specifies installation requirements for certain special electrical equipment which supplement or modify the general rules. Some of the items falling under this category include: electric signs and outline lighting, manufactured wiring systems, office furnishings, cranes and hoists, elevators, dumbwaiters, escalators, moving walks, electric welders, induction and dielectric heating equipment, industrial machines, and similar applications.

ELECTRIC SIGNS AND OUTLINE LIGHTING

Electric signs and outline lighting are covered in *NEC* Article 600, and many changes have been made to this article in the 1999 *NEC*. Such equipment is used for decorative and advertising purposes and is usually self-contained and attached to, rather than a part of, the building or wiring system. The *NEC* defines an electric sign as illuminated utilization equipment designed to convey information or attract attention. Outline lighting is designed to outline or call attention to certain features of a building or other structure. Various types of lamps are used in both cases: incandescent, electric-discharge, neon tubing, etc. Figure 7-1 on the next page summarizes important installation requirements for electric signs and outline lighting.

In general, lighting and the related electrical wiring for outdoor signs, neon tubing, and outline lighting must be weatherproof so that exposure to the elements will not adversely affect their operation. External conductors, terminals and devices must be enclosed in metal or other noncombustible material. Controllers known as flashers are often used with electric signs to obtain a blinking effect for further decoration or to gain attention. Any cutouts, flashers, disconnects, and other control devices must be enclosed in weatherproof metal boxes with accessible doors.

Signs and outline lighting installations must be clear of open conductors and be elevated at least 14 feet above areas accessible to vehicles, except when the installation is protected from damage at a lower height as per *NEC* Section 600-9.

A branch circuit supplying lamps, ballasts, and transformers, or combinations of these, is limited in its rating to 20 amperes. If the circuit supplies only electric-discharge lighting transformers, however, it may have a rating of up to 30 amperes. While the rating of a branch circuit is limited, no restrictions are placed on the number of branch circuits that may be used to supply an outdoor sign. Each sign and outline lighting system must be controlled by an externally operable switch or circuit breaker that will open all ungrounded conductors regardless of the number of branch circuits used, and each disconnect must be within sight of the sign or outline lighting that it controls. If it is not within sight, it must be capable of being locked in the open position [*NEC* Section 600-6(a)]. Such an arrangement, however, is not required for an exit directional sign located within a building.

Switches, flashers, and other devices controlling the primary circuits of transformers may be subject to damage from arcing at the contacts, especially if neon tubes are utilized. These control devices, therefore,

ELECTRIC SIGNS AND OUTLINE LIGHTING		
Application	**Requirements**	***NEC* Reference**
Branch circuits, commercial buildings	At least one outlet at each entrance to each tenant space for sign lighting.	Section 600-5
Clearances	Bottom of enclosure must be at least 14 ft above areas accessible to vehicles unless protected or guarded.	Section 600-9(a)
Conductors for neon secondary circuits	Must be installed on insulators or enclosed in conduit.	Section 600-32
Controls	Switches, flashers, and other devices controlling transformers must be rated for purpose or have an ampere rating of twice that of the transformer.	Section 600-6(b)
Disconnecting means	A means must be provided to disconnect all ungrounded conductors and must be within sight of the sign. If it is not within sight, it must be capable of being locked in the open positon.	Section 600-6(a)
Enclosures	Must be constructed of metal or other listed noncombustible material.	Section 600-8(b)
Enclosures, outdoor	Must be weatherproof and have ample drain holes of 1/4 to 1/2 in diameter.	Section 600-9(d)
Grounding	Signs and boxes must be grounded unless insulated from ground.	Section 600-7
Portable signs	Must have supply cords as specified in *NEC* Table 400-4.	Section 600-10(c)(1)
Transformer	Maximum secondary voltage: 15,000	Section 600-23(c)
Wiring methods, 1000 V or less	Any wiring method suitable for the condition that is listed in *NEC* Chapter 3 may be used.	Section 600-31(a)
Wiring methods, over 1000 V	Open wiring on insulators, metallic conduit, nonmetallic rigid conduit, EMT, or Type MC cable.	Section 600-32

Figure 7-1: *Summary of NEC installation requirements for electric signs and outline lighting*

must be of a type approved for the purpose or have at least twice the ampere rating of the transformer. The exception to this would be in the case of ac general-use snap switches which may control ac circuits with an inductive load as great as the rating of the switch as per *NEC* Section 380-14(a)(1).

The wiring methods, types of lampholders, and type of transformers used in the construction of a sign are defined in detail by the *NEC*, according to the range of operating voltage of the lamps used. In general, signs operating at 600 volts or less use either incandescent lamps or electric-discharge lamps. In most cases, signs operating at over 600 volts employ electric-discharge lamps only and the *NEC* requirements are more extensive because of the increased fire and shock hazard associated with high voltages (up to 15,000 volts).

ELECTRIC VEHICLE CHARGING SYSTEMS

Article 625 of the *NEC* has been extensively updated in the 1999 edition. It deals with the electrical conductors and the equipment external to an electric vehicle that connect the vehicle to a supply source, and the installation of equipment and devices related to electric vehicle charging. This section reflects the ongoing growth of this new area of electric work and will continue to change dramatically with each revision of the code as new demands and problems become apparent. This edition of the *NEC* includes several new definitions in Section 625-2, as well as major revisions to wiring methods (Part B), equipment construction (Part C), control and protection (Part D), and vehicle supply equipment locations (Part E).

Since there are currently no industry-wide standards for electric vehicles (at the manufacturing level), it is very important that you check each new edition of the *NEC* for any changes.

ELECTRIC WELDERS

NEC Article 630 covers electric arc welding, resistance welding apparatus, and other similar welding equipment that is connected to an electric supply system. The following *NEC* Sections apply to the various types of electric welders:

- Transformer and rectifier arc welders, *NEC* Article 630-B
- Motor-generator arc welders, *NEC* Article 630-C
- Resistance welders, *NEC* Article 630-D
- Welding cable, *NEC* Article 630-E

Energizing Electrically Powered Welding Machines

Electrically powered welding machines usually receive electric power by means of a cord-and-plug arrangement. The electrical requirements (primary current) will be on the equipment's nameplate or specification tag displayed prominently on the machine. Most machines will require single-phase, 240-volt current or three-phase, 480-volt current. Machines requiring single-phase, 240-volt power will have a three-prong plug; those requiring three-phase, 480-volt power will have a four-prong plug. See Figure 7-2.

AC Transformer And DC Arc Welders

Figure 7-3 on the next page shows a typical electrical system with ac transformer and dc rectifier arc welders connected. In general, the supply conductors for these welders must not be less than the current values determined by multiplying the rated primary current in amperes given on the welding machine's nameplate by the appropriate factor for the machine's duty cycle as listed in *NEC* Section 630-11(a).

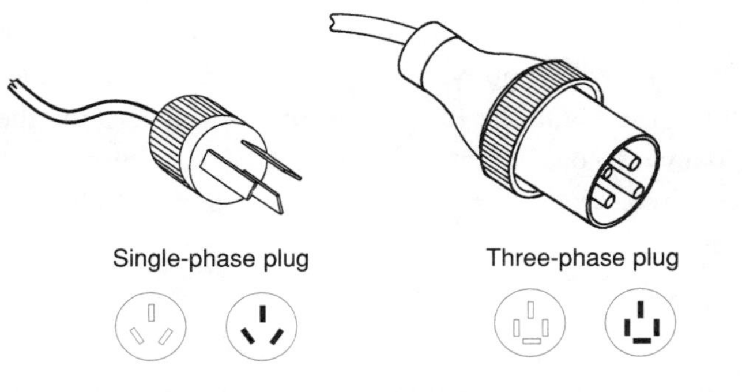
Single-phase plug
Three-phase plug

Figure 7-2: *Plugs and receptacle configurations designed for electric welding machines*

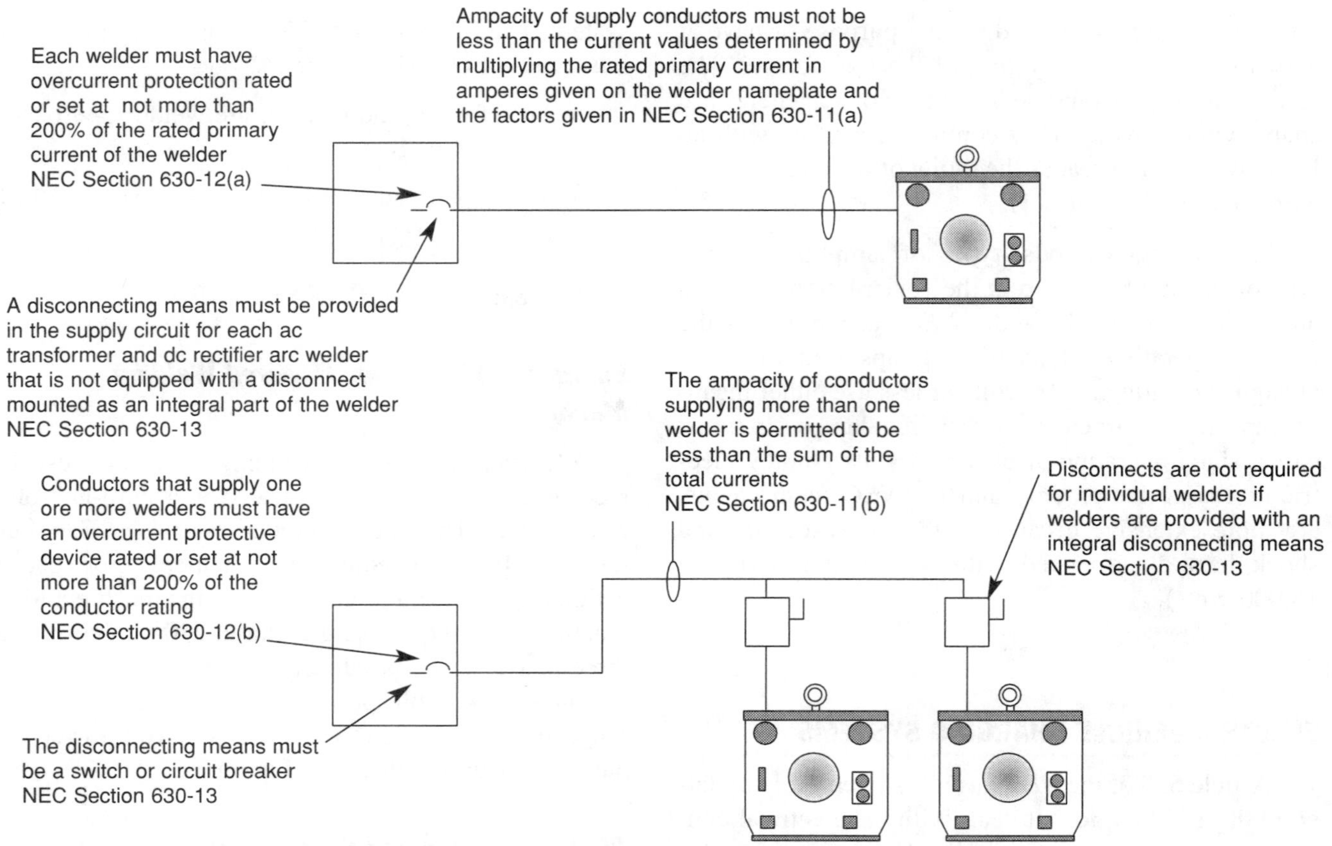
Ampacity of supply conductors must not be less than the current values determined by multiplying the rated primary current in amperes given on the welder nameplate and the factors given in NEC Section 630-11(a)
Each welder must have overcurrent protection rated or set at not more than 200% of the rated primary current of the welder NEC Section 630-12(a)
A disconnecting means must be provided in the supply circuit for each ac transformer and dc rectifier arc welder that is not equipped with a disconnect mounted as an integral part of the welder NEC Section 630-13
The ampacity of conductors supplying more than one welder is permitted to be less than the sum of the total currents NEC Section 630-11(b)
Disconnects are not required for individual welders if welders are provided with an integral disconnecting means NEC Section 630-13
Conductors that supply one ore more welders must have an overcurrent protective device rated or set at not more than 200% of the conductor rating NEC Section 630-12(b)
The disconnecting means must be a switch or circuit breaker NEC Section 630-13

Figure 7-3: *Summary of NEC installation requirements for ac transformer and dc rectifier arc welders*

For example, if a 240-volt, single-phase transformer welder has a full-load primary ampere rating of 50 amperes, and the welder is rated at 60 percent duty cycle, what size of conductors must be used for primary or supply circuit?

Referring to the table in *NEC* Section 630-11, we see that the *multiplier* for a transformer type welding machine with a 60 percent duty cycle is .78. This factor is used with the full-load amperes to determine the "demand load" for the conductors. Therefore, the required size of conductors may be found as follows:

50 A x .78 = 39 amperes

Let's assume that Type THHN insulated conductors will be used for the supply circuit. Refer to *NEC* Table 310-16, and look under the column heading "90° C" which is the column where Type THHN conductors may be found. Scan down this column until either 39 amperes is found or the next higher ampere listing. In doing so, we see the number "40" which seems exactly what we are looking for. Scanning in this row (to the left) shows that the wire size indicated is No. 10 AWG. However, there is an asterisk (*) next to this number, which means that we have to refer to a footnote at the bottom of the page. The footnote tells us to see Section 240-3. Looking back at that section, we find that it says under (d) that:

unless specifically permitted in (e) through (g), the overcurrent protection shall not exceed . . . 30 amperes for No. 10 copper . . .

Since we will need a conductor rating higher than 30 amperes to meet our 39-ampere requirement, we must go to a No. 8 AWG THHN conductor, which is listed in the table as having a 55 ampere current-carrying capacity. Since the conductor fill does not exceed three conductors for this single-phase circuit, no derating of the conductors is necessary in this case.

The next concern for the circuit in question is the overcurrent protection. *NEC* Section 630-12 requires that the overcurrent protection may not exceed 200 percent of the rated primary current of the welder and also not more than 200 percent of the current rating of the conductors — whichever is smaller. *NEC* Table 310-16 rates No. 8 THHN conductors at 55 amperes. This number (55) multiplied by 2.0 (200%) = 110 amperes. However, since the machine is rated at 50 amperes, the largest overcurrent device may not exceed 50 x 2.0 (200%) = 100 amperes. A fuse or circuit breaker with a lesser rating may be used, but not larger than 100 amperes.

NEC Section 630-13 requires a disconnect in the supply circuit for each ac transformer and dc rectifier arc welder that is not equipped with a disconnect mounted as an integral part of the welder. The disconnecting means must be a switch or circuit breaker, and its rating must not be less than that necessary to accommodate the overcurrent protection as discussed previously.

A nameplate is also required on all electric welding machines. This nameplate must provide the following information: name of manufacturer, frequency, number of phases, primary voltage, rated primary current, maximum open-circuit voltage, rated secondary current, and basis of rating, such as duty cycle or time rating.

The *NEC* also allows a diversity in sizing branch circuits or feeders for a group of welders. The specifications for ac transformer and dc rectifier arc welders may be found in *NEC* Section 630-11(b). In general, the ampacity of conductors that supplies a group of welders is permitted to be less than the sum of the currents, as determined in *NEC* Section 630-11(a). The conductor's rating must be determined, in each case, according to the welder loading — based on the use to be made of each welder. Load value used for each welder must take into account both the magnitude and the duration of the load while the welding machine is in use. *NEC* Section 630-11(b) FPN recommends the following:

- 100 percent of the current of the two largest welders
- 85 percent for the third largest welder
- 70 percent for the fourth largest welder
- 60 percent for all remaining welders

Resistance Welders

The ampacity of the supply conductors for resistance welders necessary to limit the voltage drop to a value permissible for the satisfactory performance of the welder must be considered when sizing branch-circuit and feeder conductors.

The rated ampacity for conductors for individual welders must comply with the following:

- The ampacity of the supply conductors for a welder that may be operated at different times at different values of primary current or duty cycle, shall not be less than 70 percent of the rated primary current for seam and automatically fed welders, and 50 percent of the rated primary current for manually operated nonautomatic welders.
- The ampacity of the supply conductors for a welder wired for a specific operation for which the actual primary current and duty cycle are known and remain unchanged, shall not be less than the product of the actual primary current and the multiplier given for the duty cycle at which the welder will be operated.

The ampacity of conductors that supply two or more welders must not be less than the sum of the value obtained in accordance with *NEC* Section 630-31(a) for the largest welder supplied and 60 percent of the values obtained for all the other welders supplied.

The *NEC* offers a *fine-print note* (FPN) under *NEC* Section 630-31(b) that explains the terms used for calculating circuit ratings for electric welders:

- The rated primary current is the rated kVA multiplied by 1000 and divided by the rated primary voltage, using values given on the nameplate.
- The actual primary current is the current drawn from the supply circuit during each welder operation at the particular heat tap and control setting used.

- The duty cycle is the percentage of the time during which the welder is loaded. For instance, a spot welder supplied by a 60-hertz system (216,000 cycles per hour) making four hundred 15-cycle welds per hour would have a duty cycle of 2.8 percent (400 multiplied by 15, dividing by 216,000, multiplied by 100). A seam welder operating 2 cycles "on" and 2 cycles "off" would have a duty cycle of 50 percent.

Overcurrent Protection For Welders and Conductors

Overcurrent protection for resistance welders must comply with *NEC* Section 630-32. Where the nearest standard rating of the overcurrent device used is under the value specified in this section, or where the rating or setting specified results in unnecessary opening devices, the next higher standard rating or setting is permitted.

Each welder must have an overcurrent device rated or set at not more than 300 percent of the rated primary current of the welder.

Conductors that supply one or more welders must be protected by an overcurrent device rated or set at not more than 300 percent of the conductor rating.

A switch or circuit breaker must be provided by which each resistance welder and its control equipment can be disconnected from the supply circuit. The ampere rating of this disconnecting means must not be less than the supply conductor ampacity determined in accordance with *NEC* Section 630-31. The supply circuit switch is permitted as the welder disconnecting means where the circuit supplies only one welder.

A nameplate must also be provided for each resistance welder giving the following information: name of manufacturer; frequency; primary voltage rated kVA at 50 percent duty cycle; maximum and minimum open-circuit secondary voltage; short-circuit secondary current at maximum secondary voltage and specified throat and gap setting.

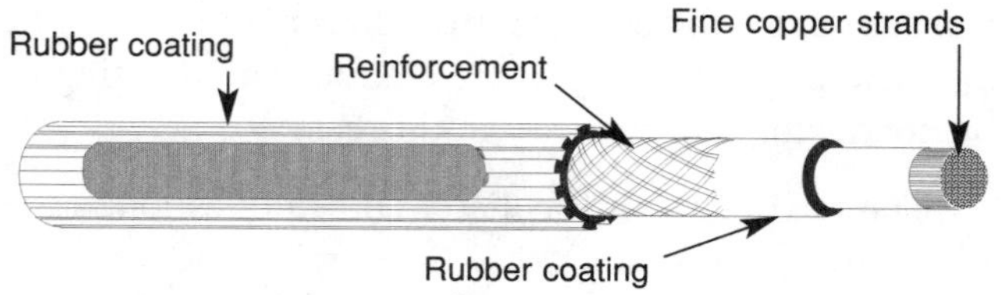
Rubber coating
Reinforcement
Fine copper strands
Rubber coating

Figure 7-4: *Typical welding cable*

Installing Welding Cables

Welding cables (Figure 7-4) are normally installed temporarily for the job at hand and then moved to another location for additional work. In some industrial installations, welding cables are installed in cable tray for use by workers throughout the plant. The *NEC* gives specific requirements for installing welding cable in cable tray. These requirements are summarized below and in Figure 7-5.

Insulation of conductors intended for use in the secondary circuit of electric welders shall be flame-retardant. Cables are permitted to be installed in a dedicated cable tray as follows:

- The cable tray shall provide support at not greater than 6-inch intervals.
- The installation shall comply with *NEC* Section 300-21.
- A permanent sign shall be attached to the cable tray at intervals not greater than 20 feet. The sign shall read "Cable tray for welding cables only."

Summary Of Electric Welders

AC transformer and dc rectifier arc welders must be protected by an overcurrent protective device rated at not more than 200 percent of the rated primary current. The device protecting the supply conductors can serve as the welder protection, if the fuse is rated at not more than 200 percent of the welder rated primary current [*NEC* Section 630-12(a)]. Conductors supplying one or more welders must be protected by an overcurrent protective device that is no more than 200 percent of the conductor rating as specified in *NEC* Section 630-12(b).

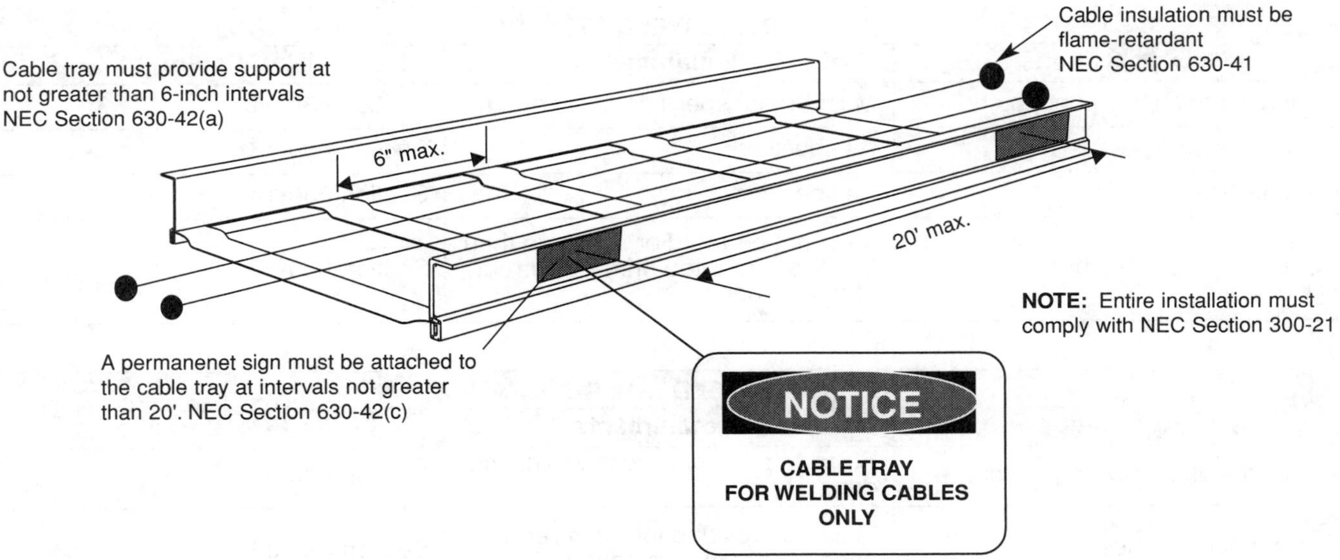
Cable insulation must be flame-retardant NEC Section 630-41
Cable tray must provide support at not greater than 6-inch intervals NEC Section 630-42(a)
6" max.
20' max.
NOTE: Entire installation must comply with NEC Section 300-21
A permanenet sign must be attached to the cable tray at intervals not greater than 20'. NEC Section 630-42(c)
NOTICE
CABLE TRAY
FOR WELDING CABLES
ONLY

Figure 7-5: *NEC requirements for installing welding cable in trays*

Resistance welders must be protected by an overcurrent protective device rated at not more than 300 percent of the rated primary current of the welder. The overcurrent device protecting the supply conductors can serve as the welder protection if the device is rated at not more than 300 percent of the welder rated primary current as specified in *NEC* Section 630-32(a). Conductors supplying one or more welders must be protected by an overcurrent protective device at not more than 300 percent of the conductor rating [*NEC* Section 630-32(b)]. See Figure 7-6.

AUDIO SIGNAL PROCESSING, AMPLIFICATION, AND REPRODUCTION EQUIPMENT

This is an all-new article updating the old Article 640 — Sounding Recording and Similar Equipment. It includes all of the most commonly-used equipment, new definitions, and a complete breakdown of uses permitted and not permitted by location. It brings this field of electrical wiring into the scope of *NEC* enforcement, with new articles covering:

- Access to electrical equipment behind panels designed to allow access (Section 640-5)
- Mechanical executions of work (Section 640-6)
- Grounding (Section 640-7)
- Wiring methods (Section 640-9)
- Audio systems near bodies of water (Section 640-10)

Another important new inclusion in this article is Part C, Portable and Temporary Audio Systems Installations, covering the use of flexible cords and cables (Section 640-42), wiring of equipment racks (Section 640-43), environmental protection of equipment (Section 640-44), and equipment access (Section 640-46).

INDUSTRIAL MACHINERY

Control wiring and feeder connection terminals on nonportable, electrically-driven machines are usually installed at the factory. In most cases, due to the areas in which the equipment is used, the wiring method is restricted to rigid conduit except for short lengths of flexible conduit where necessary for final connection to the equipment. Continuously moving parts of the machine are interconnected with approved type, extra flexible, nonmetallic covered cable. The size of the conductors, type of mounting of control equipment, overcurrent protection, and grounding are covered in Article 670 of the *NEC*.

ELECTRIC WELDERS, ARC		
Application	**Requirements**	***NEC* Reference**
Capacity of supply conductors	Must be as specified in (a) and (b).	Section 630-11
Disconnecting means	As specified.	Section 630-13
Nameplate	As specified.	Section 630-14
Overcurrent protection	Must not be rated or set more than 200% of the rated primary current of the welder.	Section 630-12

ELECTRIC WELDERS, RESISTANCE		
Application	**Requirements**	***NEC* Reference**
Capacity of supply conductors	Sized to prevent excessive voltage drop.	Section 630-31
Disconnecting means	Must be provided for each resistance welder and its controls.	Section 630-33
Nameplate	Must be provided giving electrical characteristics as specified.	Section 630-34
Overcurrent protection	Must not be rated or set more than 300% of the rated primary current of the welder.	Section 630-32

Figure 7-6: *Summary of NEC installation requirements for electric welders*

The electric supply for industrial machines may be from conventional branch circuits or feeders or in the form of bus ducts or wireways. See Figure 7-7. These two latter methods provide a very flexible type of installation allowing the moving of machines from one part of the plant or shop to another. Their reconnection to another part of the bus duct system is almost instantaneous, eliminating changes in the raceway wiring. Figure 7-8 summarizes the *NEC* installation requirements for industrial machinery. Please refer to the *NEC* book for details.

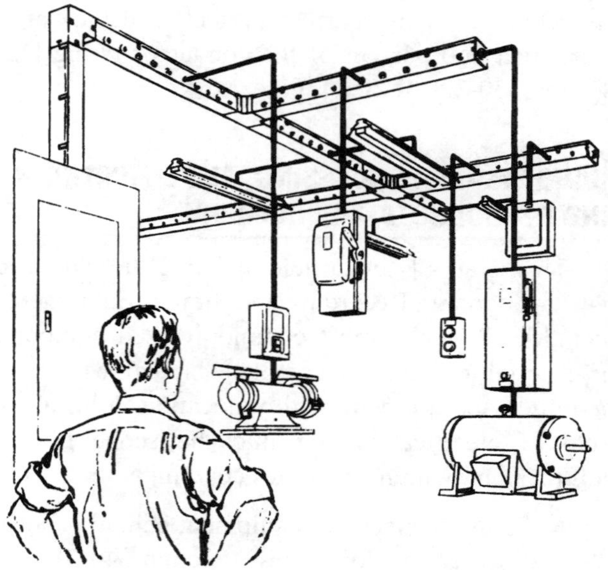

Figure 7-7: *Typical application of wireway to feed industrial machinery*

CABLE TRAY

NEC Section 318-2 defines *cable tray system* as a unit or assembly of units or sections, and associated fittings, forming a rigid structural system used to securely fasten or support cables and raceways. *NEC* Article 318 — containing *NEC* Sections 318-1 through 318-13

INDUSTRIAL MACHINERY		
Application	**Requirements**	***NEC* Reference**
Clearances for qualified persons	Must not be less than 2½ ft in the direction of access to live parts.	Section 670-5
Disconnect	Each must be provided with a disconnecting means.	Section 670-4(b)
Markings, where overcurrent protection is provided as per *NEC* Section 670-4(b)	Must be marked "overcurrent protection provided at machine supply terminals."	Section 670-3(b)
Nameplate data	Must be attached to machine where plainly visible.	Section 670-3(a)
Overcurrent protection	As specified in this *NEC* Section.	Section 670-4(b)

Figure 7-8: *Summary of NEC installation requirements for industrial machinery*

— covers cable-tray installations, along with the types of conductors to be used in such systems. Whenever a question arises concerning cable-tray installations, this is the *NEC* Article to use.

Cable trays are the usual means of supporting cable systems in industrial applications. The trays themselves are usually made up into a system of assembled, interconnected sections and associated fittings, all of which are made of metal or noncombustible units. The finished system forms into a continuous rigid assembly for supporting and carrying single, multiconductor, or other electrical cables and raceways from their origin to their point of termination, frequently over considerable distances. Several styles of cable tray are available, including ladder, trough, channel, solid-bottom trays, and similar structures.

Cable tray is fabricated from both aluminum and steel. Some manufacturers provide an aluminum cable tray that is coated with PVC for installation in caustic environments. Relatively new all-nonmetallic trays are also available; this type of tray is ideally suited for use in corrosive areas and in areas requiring voltage isolation. Furthermore, cable tray is available in three basic forms:

- Ladder
- Trough
- Solid bottom

Ladder tray, as the name implies, consists of two parallel channels connected by rungs, similar in appearance to a conventional straight or extension ladder. Trough types consist of two parallel channels (side rails) having a corrugated, ventilated bottom. The solid-bottom cable tray is similar to the trough except that this type of trough has a corrugated, solid bottom. All of these types are shown in Figure 7-9 on the next page. Ladder, trough, and solid-bottom trays are completely interchangeable; that is, all three types can be used in the same run when needed.

Cable tray is manufactured in 12- and 24-foot lengths. Common widths include 6, 9, 12, 24, 30 and 36 inches. All sizes are provided in 3-, 4-, 5-, or 6-inch depths.

Cable-tray sections are interconnected with various types of fittings. Fittings are also used to provide a means of changing the direction or dimension of the cable-tray system. Some of the more common fittings include:

- Horizontal and vertical tees
- Horizontal and vertical bends
- Horizontal crosses
- Reducers
- Barrier strips
- Covers
- Splice plates
- Box connectors

Again, all of these fittings are shown in Figure 7-9.

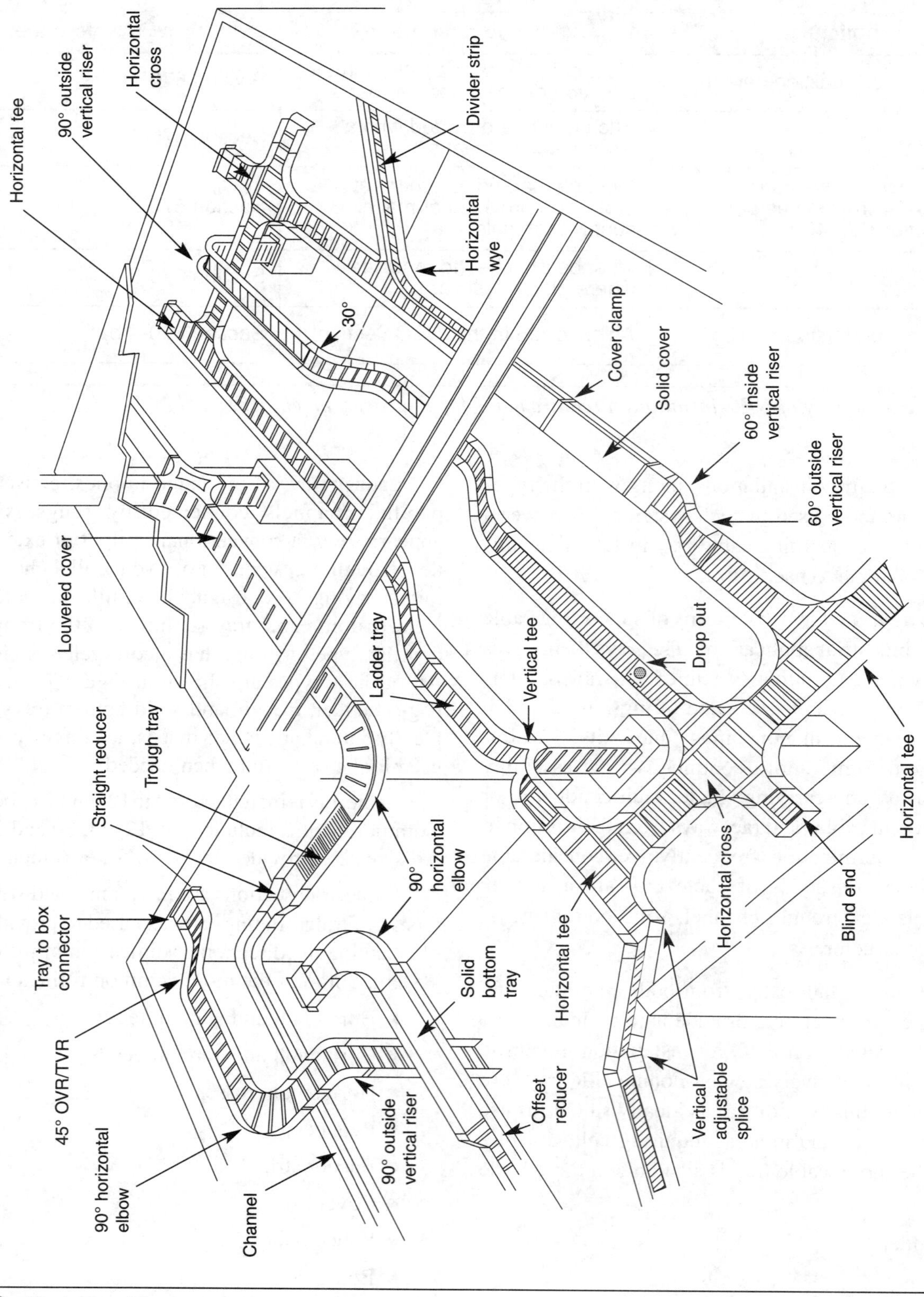
Horizontal tee
90° outside vertical riser
Horizontal cross
Divider strip
Horizontal wye
30°
Cover clamp
Solid cover
60° inside vertical riser
60° outside vertical riser
Louvered cover
Ladder tray
Vertical tee
Drop out
Horizontal tee
Straight reducer
Trough tray
90° horizontal elbow
Horizontal cross
Blind end
Tray to box connector
Solid bottom tray
Horizontal tee
45° OVR/TVR
Offset reducer
Vertical adjustable splice
90° horizontal elbow
Channel
90° outside vertical riser

Figure 7-9: *Typical cable-tray system*

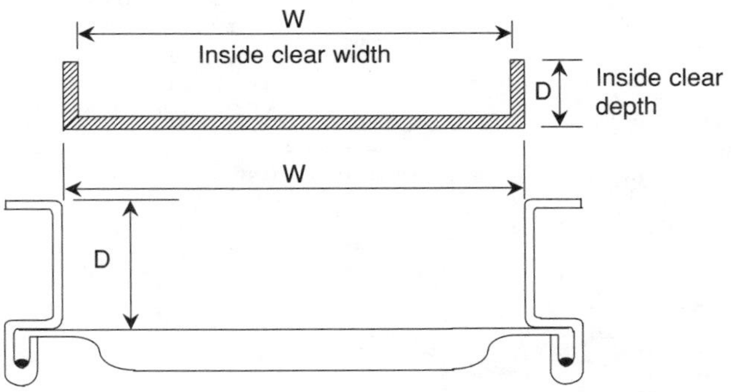
W
Inside clear width
D
Inside clear depth
W
D

Figure 7-10: *Cross-sectional view of cable tray comparing usable to overall dimensions*

The area of a cable tray cross-section which is usable for cables is defined by width (W) x depth (D), as shown in Figure 7-10. The overall dimensions of a cable tray, however, are greater than W and D because of the side flanges and seams. Therefore, overall dimensions vary according to the tray design. Cables rest upon the bottom of the tray and are held within the tray area by two longitudinal side rails as shown in Figure 7-11.

A *channel* is used to carry one or more cables from the main tray system to the vicinity of the cable termination. Conduit is then used to finish the run from the channel to the actual termination.

Certain *NEC* regulations and NEMA (National Electrical Manufacturers Association) standards should be followed when designing or installing cable tray. Consequently, practically all projects of any great size will have detailed drawings and specifications for the workers to follow. Shop drawings may also be provided.

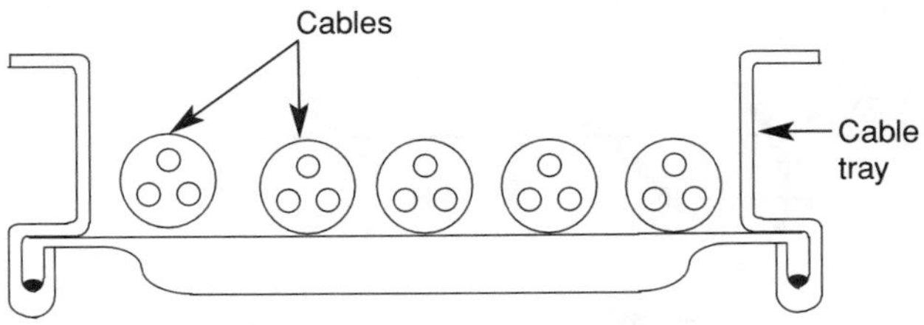
Cables
Cable tray

Figure 7-11: *Cables rest upon the bottom of the tray and are held in place by the longitudinal side rails*

NEC Requirements

NEC Article 318 deals with cable-tray systems along with the related wire, cable, and raceway installations therein. Article 340 covers *NEC* regulations governing power and control cable for use in cable tray. Everyone involved in industrial electrical systems should be thoroughly familiar with these Articles (and their related Sections) of the *NEC*.

Although the *NEC* allows the use of cable-tray installations in other than industrial establishments, many of the allowable practices are based on certain conditions for use in industrial projects only. For example, *NEC* Section 318-3(b) states:

> *In industrial establishments only, where conditions of maintenance and supervision ensure that only qualified persons will service the installed cable tray system, any of the cables in (1) [single-conductor Type TC cable] and (2) [multiconductor Type TC cable] shall be permitted to be installed in ladder, ventilated trough, or ventilated channel cable trays.*

And then in Section 318-3(c), the *NEC* allows the use of metal cable tray as the equipment grounding conductor provided, ". . . where continuous maintenance and supervision ensure that qualified persons will service the installed cable tray system . . ."

Consequently, it is extremely important that electricians interpret the intent of the *NEC* in regard to where cable tray may be used in commercial buildings, many of which have no continuous maintenance program to ensure that the system will be serviced only by qualified personnel. On the other hand, most industrial establishments have electrical maintenance crews working 24 hours per day — ensuring that only qualified personnel will service the system.

Figure 7-12 summarizes *NEC* installation requirements for cable tray installations. For in-depth coverage, however, you will want to refer to the *NEC* book. Cable-tray manufacturers also have some excellent reference material available that details the installation of their products. Such material is usually available from local electrical material distributors. If not, contact the manufacturer directly.

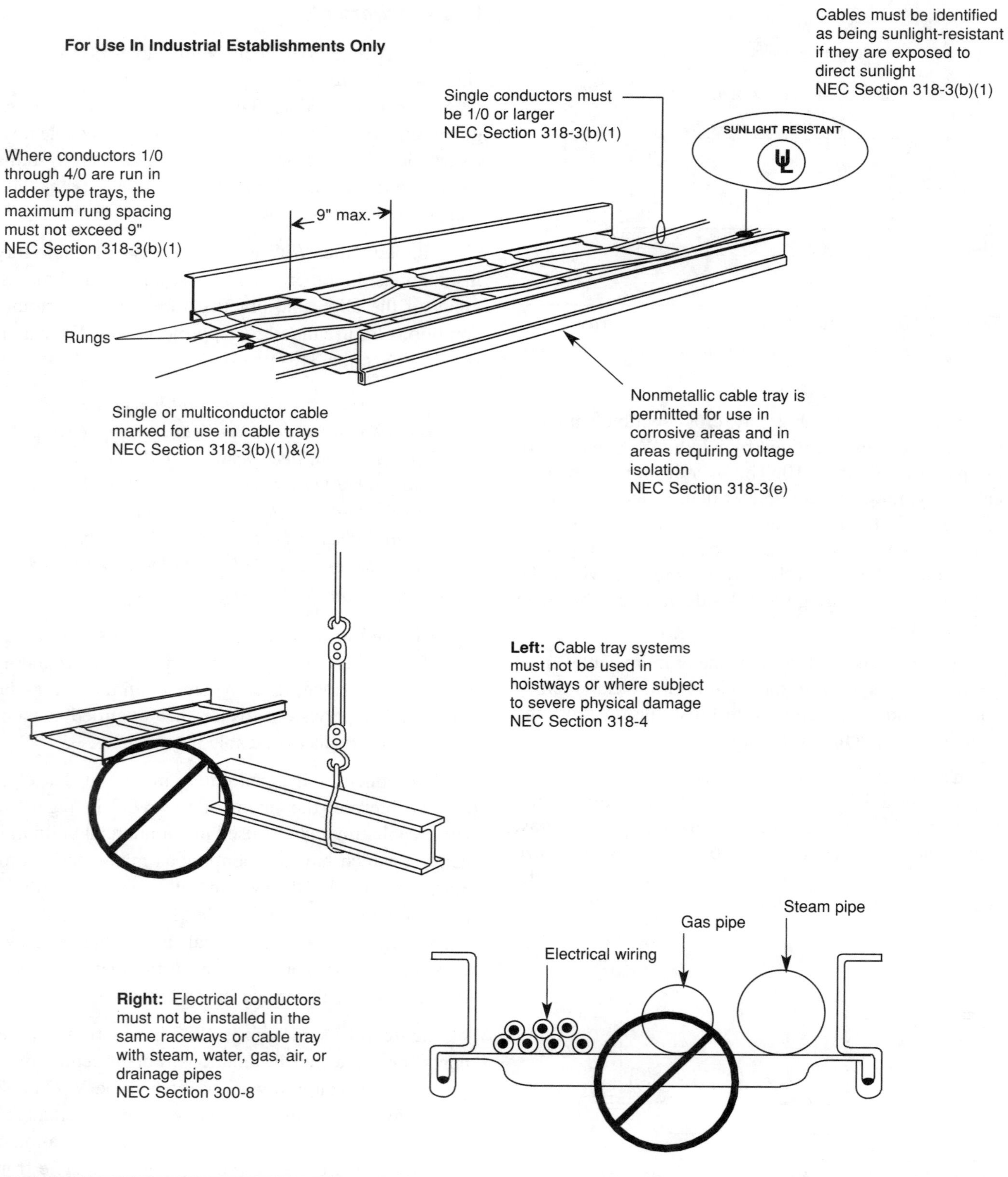
For Use In Industrial Establishments Only
Cables must be identified as being sunlight-resistant if they are exposed to direct sunlight
NEC Section 318-3(b)(1)
Single conductors must be 1/0 or larger
NEC Section 318-3(b)(1)
SUNLIGHT RESISTANT
Where conductors 1/0 through 4/0 are run in ladder type trays, the maximum rung spacing must not exceed 9"
NEC Section 318-3(b)(1)
9" max.
Rungs
Single or multiconductor cable marked for use in cable trays
NEC Section 318-3(b)(1)&(2)
Nonmetallic cable tray is permitted for use in corrosive areas and in areas requiring voltage isolation
NEC Section 318-3(e)
Left: Cable tray systems must not be used in hoistways or where subject to severe physical damage
NEC Section 318-4
Steam pipe
Gas pipe
Electrical wiring
Right: Electrical conductors must not be installed in the same raceways or cable tray with steam, water, gas, air, or drainage pipes
NEC Section 300-8

Figure 7-12: *Summary of NEC installation requirements for cable tray*

Cable Installation

NEC Sections 318-8 through 318-12 cover the installation of cables in cable-tray systems.

Section 318-8 covers the general requirements for all conductors used in cable-tray systems; that is, splicing, securing, and running conductors in parallel. For example, cable splices are permitted in cable trays provided they are made and insulated by (*NEC*) approved methods. Furthermore, any splices must be readily accessible and must not project above the side rails of the tray.

In horizontal runs, and in most cases, the cables may be laid in the trays without further securing them in place. However, on vertical runs or any runs other than horizontal, the cables must be secured to transverse members of the cable trays

Cables may enter and leave a cable-tray system in a number of different ways. In general, no junction box is required where such cables are installed in bushed conduit or tubing. Where conduit or tubing is used, either must be secured to the tray with the proper fittings — usually with a Unistrut conduit clamp. Further precautions must be taken to insure that the cable is not bent sharply as it enters or leaves the conduit or tubing.

Conductors Connected In Parallel

Where single-conductor cables comprising each phase or neutral of a circuit are connected in parallel, as permitted in *NEC* Section 310-4, the conductors must be installed in groups. The groups must consist of not more than one conductor per phase or neutral to prevent current unbalance in the paralleled conductors due to inductive reactance. Such conductors must be securely bound in circuit groups to prevent excessive movement due to fault-current magnetic forces.

Number Of Cables Allowed In Cable Tray

The number of multiconductor cables, rated 2000 volts or less, permitted in a single cable tray must not exceed the requirements of *NEC* Section 318-9, which applies to both copper and aluminum conductors.

Where all of the cables installed in the tray are No. 4/0 or larger, the sum of the diameters of all cables shall not exceed the cable tray width, and the cables must be installed in a single layer. For example, if a cable tray installation is to contain three 4/0 multiconductor cables (1.5-inch diameter), two 250 kcmil multiconductor cables (1.85 inches), and two 350 kcmil multiconductor cables (2.5 inches), the minimum width of the cable tray is determined as follows:

$$3(1.5) + 2(1.85) + 2(2.5) = 13.2"$$

The closest standard cable tray size that meets or exceeds 13.2 inches is 18 inches. Therefore, this is the size to use.

Where all of the cables are smaller than No. 4/0, the sum of the cross-sectional area of all cables smaller than 4/0 must not exceed the maximum allowable cable fill area as specified in Column 1 of *NEC* Table 318-9; this gives the appropriate cable tray width. To use *NEC* Table 318-9, however, you must have the manufacturer's data for the cables being used. This will give the cross-sectional area of the cables.

The following shows the steps involved in determining the size of cable tray for multiconductors smaller than No. 4/0 AWG:

1. Calculate the total cross-sectional area of all cables used in the tray. Obtain the area of each from manufacturer's data.
2. Look in Column 1 of *NEC* Table 318-9 and find the smallest number that is at least as large as the calculated number.
3. Look at the number to the left of the row selected in Step 2. This is the minimum width of cable tray that may be used.

For example, let's determine the minimum cable tray width required for the following multiple conductor cables — all less than No. 4/0 AWG:

- Four @1.5-inch diameter
- Five @1.75-inch diameter
- Three @2.15-inch diameter

Step 1. Determine the cross-sectional area of the cables from the equation:

$$A = \frac{\pi \times D^2}{4}$$

where:
A = area
D = diameter

The area of a 1.5-inch diameter cable is:

$$\frac{(3.14159)(1.5)^2}{4} = 1.7671 \text{ sq. inches}$$

The area of the four 1.5-inch cables is:

4 x 1.7671 sq. inches = 7.0685 sq. inches

The area of 1.75-inch diameter cable is:

$$\frac{(3.14159)(1.75)^2}{4} = 2.4053 \text{ sq. inches}$$

The area of the five 1.75-inch cables is:

5 x 2.4053 sq. in. = 12.0264 sq. inches

The area of the three 2.15-inch cables is:

$$\frac{(3.14159)(2.15)^2}{4} = 3.6305 \text{ sq. inches}$$

Therefore, the total area of all three cables is:

3 x 3.6305 sq. inches = 10.8915 sq. inches

The total cross-section area is found by adding the above three totals to obtain:

7.0685 + 12.0264 + 10.8915 = 29.9864 sq. inches

Step 2. Look in Column 1, *NEC* Table 318-9 and find the smallest number that is at least as large as 29.9864 sq. inches. The number is 35.

Step 3. Look to the left of "35" and see inside tray width of 30 inches. Therefore, the minimum tray width that can be used for the given group of conductors is 30 inches.

Combination Cables

Where No. 4/0 or larger cables are installed in the same cable tray with cables smaller than No. 4/0, the sum of the cross-sectional area of all cables smaller than No. 4/0 must not exceed the maximum allowable fill area resulting from computation in Column 2 of *NEC* Table 318-9, for the appropriate cable tray width. The No. 4/0 and larger cables must be installed in a single layer and no other cables can be placed on them.

To determine the size of tray for a combination of cables as discussed in the above paragraph, proceed as follows:

Step 1. Repeat the steps from the procedure used previously to determine the minimum width of tray required for the multiconductor cables having conductors sized No. 4/0 and larger. Call this number *Sd* for the "sum of diameters."

Step 2. Repeat the steps from the procedure used previously to determine the cross-sectional area of all multiconductor cables having conductors smaller than No. 4/0 AWG.

Step 3. Multiply the result of Step 1 by the constant 1.2 and add this product to the results of Step 2. Call this sum A. Search Column 2 of *NEC* Table 318-9 for the smallest number that is at least as large as A. Look to the left in that row to determine the minimum size cable tray required.

To illustrate these steps, let's assume that we desire to find the minimum cable tray width of two multiconductor cables, each with a diameter of 2.54 inches (conductors size 4/0 or larger); three cables, each with a diameter of 3.30 inches (conductors size 4/0 or larger); plus eight cables, each with a diameter of 1.92 inches (conductor size less than No. 4/0).

Step 1. The sum of all the diameters of cable having conductors 4/0 or larger is:

$$2(2.54) + 3(3.30) = Sd = 14.98 \text{ inches}$$

Step 2. The sum of the cross-sectional areas of all cables having conductors smaller than 4/0 is:

$$\frac{(8)(3.14159)(1.92)^2}{4} = 23.1623 \text{ sq. inches}$$

Step 3. Multiply the results of Step 1 by 1.2 and add this product to the results of Step 2:

$$1.2 \times (14.98) + 23.1623 = 41.1383 \text{ sq. inches}$$

The smallest number in Column 2 of *NEC* Table 318-9 that is larger than 41.1383 is 42. The tray width that corresponds to 42 is 36 inches. Select a cable tray width of 36 inches.

Solid-Bottom Tray

Where solid-bottom cable trays contain multiconductor power or lighting cables, or any mixture of multiconductor power, lighting, control, and signal cables, the maximum number of cables must conform to the following:

Where all of the cables are No. 4/0 or larger, the sum of the diameters of all cables must not exceed 90 percent of the cable tray width, and the cables must be installed in a single layer.

Where all of the cables are smaller than No. 4/0, the sum of the cross-sectional areas of all cables must not exceed the maximum allowable cable fill area in Column 3 of Table 318-9 for the appropriate cable tray width.

Where No. 4/0 or larger cables are installed in the same cable tray with cables smaller than No. 4/0, the sum of the cross-sectional areas of all of the smaller cables must not exceed the maximum allowable fill area resulting from the computation in Column 4 of *NEC* Table 318-9 for the appropriate cable tray width. The No. 4/0 and larger cables must be installed in a single layer, and no other cables can be placed on them.

Where a solid-bottom cable tray, having a usable inside depth of 6 inches or less, contains multiconductor control and/or signal cables only, the sum of the cross-sectional areas of all cables at any cross section must not exceed 40 percent of the interior cross-sectional area of the cable tray. A depth of 6 inches must be used to compute the allowable interior cross-sectional area of any cable tray that has a usable inside depth of more than 6 inches.

In a previous example, we determined that the minimum tray size for multiple conductor cables with all conductor sizes 4/0 or larger was 18 inches. The sum of all cable diameters for this example was 13.2 inches. To see if an 18-inch solid-bottom tray can be used, multiply the tray width by .90 (90 percent):

$$18 \times .9 = 16.2 \text{ inches}$$

Therefore, 16.2 inches is the minimum width allowed for solid-bottom tray. Since we are using 18-inch tray, this meets the requirements of *NEC* Section 318-9(c)(1).

When dealing with solid bottom trays, and using *NEC* Table 318-9, use Columns 3 and 4, instead of Columns 1 and 2 as used for ladder and trough type tray.

Single Conductor Cables

Calculating cable tray widths for single conductor cables (2000 volts or under) is similar to the calculations used for multiconductor cables with the following exceptions:

Conductors 1000 kcmil and larger are treated the same as multiconductor cables, having conductor sizes 4/0 or larger.

Conductors that are smaller than 1000 kcmil are treated the same as multiconductor cables having conductors smaller than size 4/0.

NEC Section 318-10 covers the details of installing single conductor cables in cable tray systems.

Ampacity Of Cable-Tray Conductors

NEC Section 318-11 gives the requirements for cables used in tray systems with rated voltages of 2000 volts or less. Section 318-12 covers cables with voltages of 2001 and over, while Section 318-13 deals with Type MV and Type MC cables rated over 2000 volts.

INDUCTION AND DIELECTRIC HEATING EQUIPMENT

The wiring for and connection of induction and dielectric heat generating equipment used in industrial and scientific applications (but not for medical or dental applications) are covered in Article 665 of the *NEC*.

The heating effect of such equipment is accomplished by placing the materials to be heated in the magnetic field of an electric voltage of very high frequency or between two electrodes connected to a source of high frequency voltage. Induction heating is used in heating metals and other conductive materials. Dielectric heating is used in the heating of materials that are poor conductors of electric current.

The equipment used consists either of motor-operated, high-frequency generators, electric tube or solid-state oscillators. Such equipment is supplied by manufacturers or their representatives. Designers, electrical contractors, and electrical workers can benefit by contacting these manufacturers to obtain installation procedures, specifications, and the like.

The size of the supply circuit conductors, overcurrent protection, disconnecting means, type of grounding, and output circuits are covered in *NEC* Article 665. A summary of *NEC* installation requirements for induction heating equipment appears in Figure 7-13. Please check the *NEC* book for further details.

INDUCTION HEATING EQUIPMENT		
Application	**Requirements**	***NEC* Reference**
Access	Doors or detachable panels must be provided for access.	Section 665-22
Control enclosures	DC or low-frequency ac is permitted in the control portion of heating equipment.	Section 665-28
Enclosures	All circuits must be contained within an enclosure.	Section 665-20
Grounding	Must comply with *NEC* Article 250.	Section 665-26
Hazardous locations	Must not be installed in hazardous locations.	Section 665-4
Induction coils operating at 30 V ac or more	Must be enclosed in a nonmetallic or split metallic enclosure and must be isolated or made inaccessible by location.	Section 427-36
Panel controls	Must be of dead-front construction.	Section 665-21

Figure 7-13: *Summary of NEC installation requirements for induction heating equipment*

ELECTRONIC COMPUTER/DATA-PROCESSING SYSTEMS

The use of computers more than doubles each year. Consequently data-processing rooms or areas are becoming more common. *NEC* Article 645 covers the installation of power supply wiring, grounding of equipment, and other such provisions that will insure a safe installation.

In many computer installations, the equipment operates continuously. For this reason, *NEC* Section 645-5(a) requires that branch circuits supplying one or more computers have an ampacity not less than 125 percent of the total connected load.

Therefore, if a computer system is rated at, say, 12 amperes, the branch circuit feeding this equipment should be rated at 12 x 1.25 = 15 amperes. The supply conductors are normally flexible cables or cords for easy moving or when adding new equipment. *NEC* Article 645(b) lists the following as permissible:

- Computer/data processing cable and attachment plug cap.
- Flexible cord and an attachment plug cap.
- Cord-set assembly. When run on the surface of the floor, they shall be protected against physical damage.

Separate data processing units shall be permitted to be interconnected by means of cables and cable assemblies listed for the purpose. Where run on the surface of the floor, they shall be protected against physical damage.

Since there are so many interconnections for power, control, and communications, computer equipment is sometimes installed on raised floors with the cables running underneath the floor. The branch-circuit conductors in that case must be either rigid conduit, intermediate metal conduit, electrical metallic tubing, metal wireway, surface metal raceway with metal cover, flexible metal conduit, liquidtight flexible metal or nonmetallic conduit, Type MI cable, Type MC cable, or Type AC cable. Whichever wiring method is used, it must be installed in accordance with Section 300-11.

This underfloor area can also be used to circulate air for computer equipment cooling. Panels in the floor are removed under the equipment to allow air to enter the cabinets that house the equipment. In such an arrangement, a separate heating/ventilating/air conditioning (HVAC) system should be provided solely for the computer area. If this system services other areas, fire/smoke dampers must be provided at the point of penetration of the room boundary. See *NEC* Section 645-10.

A disconnecting means must be provided that will allow the operator to disconnect all computer equipment in the area. A disconnecting means must also be provided to disconnect the ventilation system serving the computer area.

Figures 7-14 and 7-15 summarize the *NEC* installation requirements for electronic computer/data-processing areas.

Cranes And Hoists

Crane and hoist equipment is usually furnished and mechanically installed by crane manufacturing companies or their representatives. When working on such equipment, refer to *NEC* Article 610.

Crane and hoist wiring consists of the control and operating circuits on the equipment itself and the contact conductors or flexible conductors supplying electric power to the equipment. Electricians on the job normally install the feeder circuit at a point of connection to the contact conductor or flexible cables, including overcurrent protection and disconnecting means.

The motor-control equipment, control and operating circuits, and the bridge contact conductors are usually furnished and installed by the manufacturer; sometimes, it is the responsibility of the owner or electrical contractor.

SWIMMING POOLS

The *NEC* recognizes the potential danger of electric shock to persons in swimming pools, wading pools, and therapeutic pools or near decorative pools or fountains. This shock could occur from electric potential in the water itself or as a result of a person in the water or a wet area touching an enclosure that is not at ground potential. Accordingly, the *NEC* provides rules for the safe installation of electrical equipment and wiring in

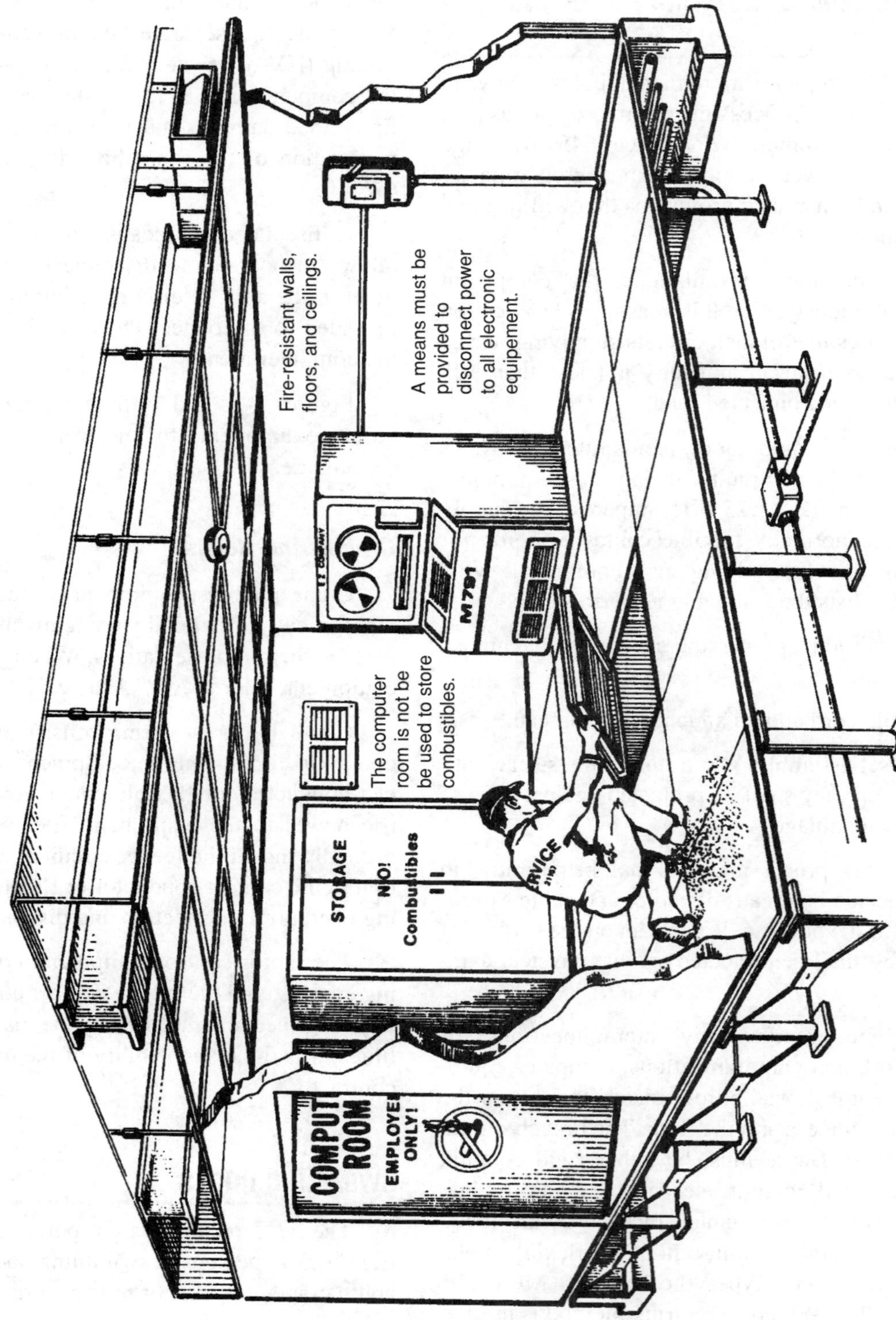
Fire-resistant walls, floors, and ceilings.
A means must be provided to disconnect power to all electronic equipement.
The computer room is not be used to store combustibles.
STORAGE
NO! Combustibles
M791
COMPUT ROOM
EMPLOYEE ONLY!

Figure 7-14: *NEC installation requirements for electronic computer/data-processing areas*

ELECTRONIC COMPUTER/DATA PROCESSING EQUIPMENT		
Application	**Requirements**	***NEC* Reference**
Disconnect	Must comply with *NEC* Section 645-10.	Section 645-2(a)
HVAC system	A separate system must be provided that is dedicated for the electronic computer equipment.	Section 645-2(b)
Listed	Equipment must be listed to comply with *NEC* Section 645-2.	Section 645-2(c)
Occupants	Only those personnel needed for the maintenance and functional operation of the installed equipment.	Section 645-2(d)
Separation	Room must be separated from other occupancies by fire-resistant-rated walls, floors, etc.	Section 645-2(e)

Figure 7-15: *Summary of NEC installation requirements for electronic computer/data-processing areas*

or adjacent to swimming pools and similar locations. *NEC* Article 680 covers the specific rules governing the installation and maintenance of swimming pools and similar installations.

The general requirements for the installation of outlets, overhead conductors and other equipment are summarized in Figures 7-16 and 7-17.

Other installations falling under the category of special equipment are listed below along with the appropriate *NEC* Article for further reference.

- Electrically driven or controlled irrigation machines, Article 675
- Electrolytic cells, Article 668
- Electroplating, Article 669
- Elevators, dumbwaiters, escalators and moving walks, Article 620
- Integrated electrical systems, Article 685
- Manufactured wiring systems, Article 604
- Office furnishings, Article 605
- Pipe organs, Article 650
- Solar photovoltaic systems, Article 690
- Audio signal processing, amplification, and reproduction equipment, Article 640
- X-ray equipment, Article 660

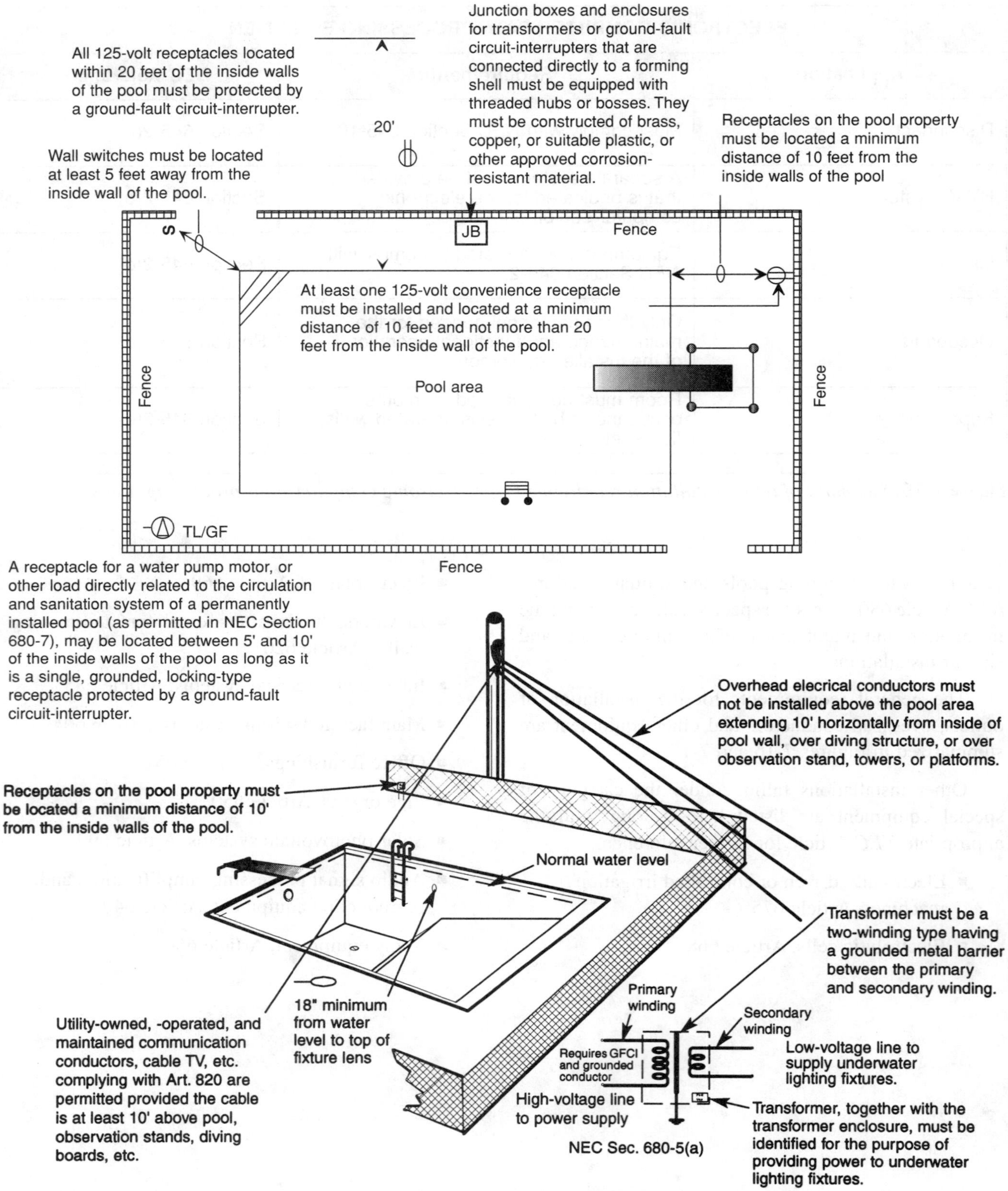

All 125-volt receptacles located within 20 feet of the inside walls of the pool must be protected by a ground-fault circuit-interrupter.
20'
Junction boxes and enclosures for transformers or ground-fault circuit-interrupters that are connected directly to a forming shell must be equipped with threaded hubs or bosses. They must be constructed of brass, copper, or suitable plastic, or other approved corrosion-resistant material.
Receptacles on the pool property must be located a minimum distance of 10 feet from the inside walls of the pool
Wall switches must be located at least 5 feet away from the inside wall of the pool.
S
JB
Fence
At least one 125-volt convenience receptacle must be installed and located at a minimum distance of 10 feet and not more than 20 feet from the inside wall of the pool.
Fence
Pool area
Fence
TL/GF
Fence
A receptacle for a water pump motor, or other load directly related to the circulation and sanitation system of a permanently installed pool (as permitted in NEC Section 680-7), may be located between 5' and 10' of the inside walls of the pool as long as it is a single, grounded, locking-type receptacle protected by a ground-fault circuit-interrupter.
Overhead elecrical conductors must not be installed above the pool area extending 10' horizontally from inside of pool wall, over diving structure, or over observation stand, towers, or platforms.
Receptacles on the pool property must be located a minimum distance of 10' from the inside walls of the pool.
Normal water level
Transformer must be a two-winding type having a grounded metal barrier between the primary and secondary winding.
Primary winding
Secondary winding
18" minimum from water level to top of fixture lens
Requires GFCI and grounded conductor
Low-voltage line to supply underwater lighting fixtures.
Utility-owned, -operated, and maintained communication conductors, cable TV, etc. complying with Art. 820 are permitted provided the cable is at least 10' above pool, observation stands, diving boards, etc.
High-voltage line to power supply
NEC Sec. 680-5(a)
Transformer, together with the transformer enclosure, must be identified for the purpose of providing power to underwater lighting fixtures.

Figure 7-16: *Summary of NEC installation requirements for swimming pools*

SWIMMING POOLS		
Application	**Requirements**	***NEC* Reference**
Equipment	Must comply with NEC Article 680.	Section 680-2
Receptacles	Must be located at least 10 ft away from inside walls of pool.	Section 680-6(a)
Receptacles, for water circulation	Shall be permitted betwen 5 ft and 10 ft from the inside of the walls of the pool.	Section 680-6(a)(1)
Switches	Must be located at least 5 ft away from inside walls of pool.	Section 680-6(c)
Transformers	Must be identified for use to supply power for underwater lighting.	Section 680-5(a)
Water heaters	Each circuit supply loads must not exceed 48 A.	Section 680-9
Water heaters, protection of	Each circuit supply overcurrent device must not exceed 60 A.	

Figure 7-17: *More NEC installation requirements for swimming pools*

Chapter 8
Signaling Systems

A signaling circuit is any electric circuit that energizes signaling equipment. Circuits falling in this category include residential doorbells, security, fire-alarm, and similar systems — employing a wide variety of techniques and often involving special types of equipment and materials designed for specific applications.

Electric signaling and communications is a very broad field, covering everything from simple residential door chimes to elaborate building alarm and detector systems.

NEC Article 725 covers remote-control, signaling, and power-limited circuits that are not an integral part of a device or appliance. *NEC* Section 725-1 states:

> *"The circuits described herein are characterized by usage and electrical power limitations that differentiate them from electric light and power circuits; therefore, alternative requirements to those of Chapters 1 through 4 are given with regard to minimum wire sizes, derating factors, overcurrent protection, insulation requirements, and wiring methods and materials."*

Workers installing security/fire-alarm systems should become thoroughly familiar with *NEC* Article 725 and also *NEC* Article 760 — Fire Alarm Systems.

Other *NEC* Articles and Sections that apply directly to signaling circuits include:

- Article 90 — Introduction
- Article 100-A — Definitions
- Section 225-2 — Other Articles
- Section 230-94 — Relative Location of Overcurrent Device and Other Service Equipment
- Section 240-2 — Protection of Equipment
- Section 240-3 — Protection of Conductors
- Article 250 — Grounding
- Section 310-5 — Minimum Size of Conductors
- Section 310-10 — Temperature Limitations of Conductors
- Section 310-11 — Marking
- Section 310-13 — Conductor Constructions and Applications
- Table 310-13 — Conductor Application and Insulations
- Section 320-12 — Clearance from Piping, Exposed Conductors, etc.
- Article 336 — Nonmetallic-Sheathed Cable
- Article 352 — Surface Metal Raceways and Surfaçe Nonmetallic Raceways
- Article 362 — Metal Wireways and Nonmetallic Wireways
- Article 374 — Auxiliary Gutters
- Article 384 — Switchboards and Panelboards

CLASSIFICATION OF SIGNAL CIRCUITS

A signal circuit used for security or fire-alarm systems may be classified as *open circuit* or *closed circuit*. An open circuit is one in which current flows only when a signal is being sent. A closed circuit is one in which current flows continuously, except when the circuit is opened to allow a signal to be sent.

All security systems have three functions in common:

- Detection
- Control
- Annunciation (or alarm) signaling

Many systems incorporate switches or relays that operate when entry, movement, pressure, infrared-beam interruption, and other intrusions occur. The control senses the operation of the detector with a relay and produces an output that may operate a bell, siren, silent alarm such as telephone dialers to law enforcement agencies, or other signals. The controls frequently contain ON/OFF switches, test meters, time delays, power supplies, standby batteries, and terminals for connecting the system together. The control output usually provides power on alarms to operate signaling devices or switch contacts for silent alarms. See Figure 8-1.

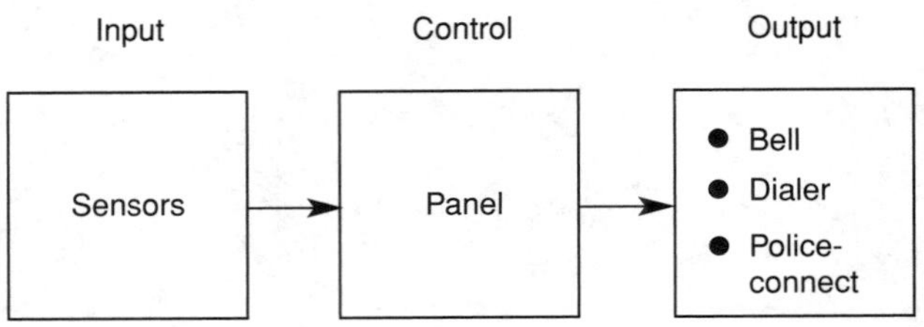
Input
Control
Output
Sensors
Panel
Bell
Dialer
Police-connect

Figure 8-1: *Basic subdivisions of an alarm system*

An example of a basic closed-circuit security system is shown in Figure 8-2. The detection (or input) subdivision in this drawing shows exit/entry door or window

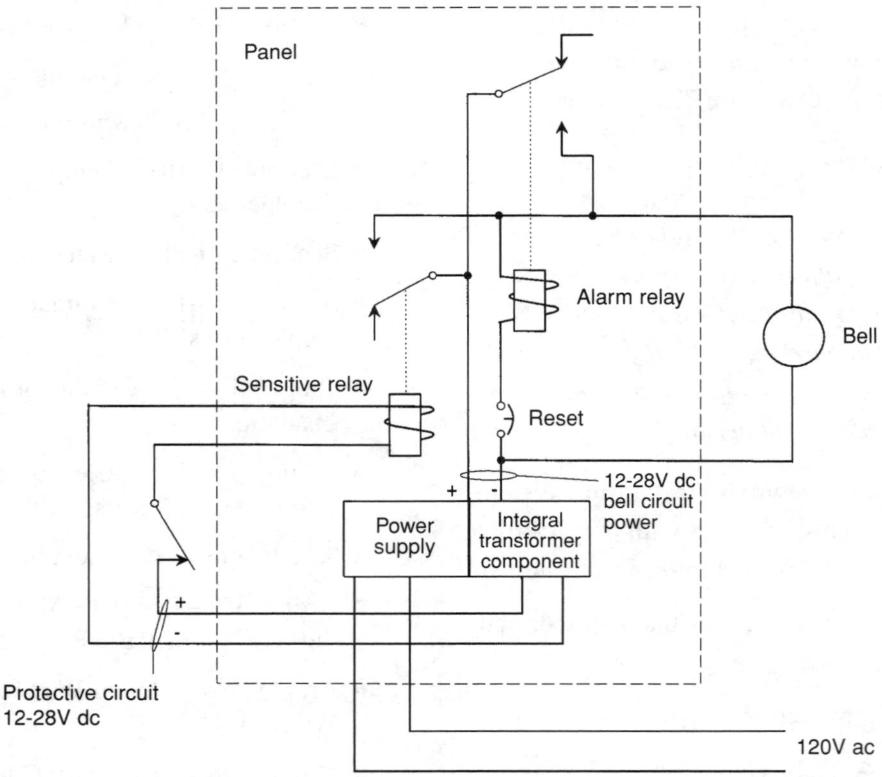
Panel
Alarm relay
Bell
Sensitive relay
Reset
12-28V dc bell circuit power
Power supply
Integral transformer component
Protective circuit 12-28V dc
120V ac

Figure 8-2: *Basic closed-circuit security alarm system*

contacts. However, the detectors could just as well be smoke or heat detectors, switch mats, ultrasonic detectors, and the like.

The control subdivision for the system in Figure 8-2 consists of switches, relays, a power supply, a reset button, and related wiring. The power supply shown is a 12-28V dc battery system that is kept charged by the integral transformer unit. Terminals are provided on the battery housing to furnish the 12-28V dc power for the detection (protective) circuit and power to operate the alarm or output subdivision.

Figure 8-3, on page 212, shows another closed-circuit system. The protective circuit consists of a dc energy source, any number of normally closed intrusion-detection contacts (wired in series), a sensitive relay (R_1) and interconnecting wiring. In operation, the normally closed intrusion contacts are connected to the coil of the sensitive relay. This keeps the relay energized, holding its normally closed contacts open against spring pressure — the all-clear condition of the protective circuit. The opening of any intrusion contact breaks the circuit, which deenergizes the sensitive relay and allows spring force to close the relay contacts. This action initiates the alarm.

The key-operated switch shown in the circuit in Figure 8-3 is provided for opening the protective circuit for test purposes. A meter (M) is activated when the switch is set to CIRCUIT TEST. The meter gives a current reading only if all intrusion contacts are closed. All three sections of the switch (S_1, S_2, S_3) make contact simultaneously as the key is turned.

Opening of intrusion contacts is not the only event that causes the alarm to activate. Any break in protective-circuit wiring or loss of output from the energy source has the same effect. The circuit is broken which deenergizes the sensitive relay and allows spring force to

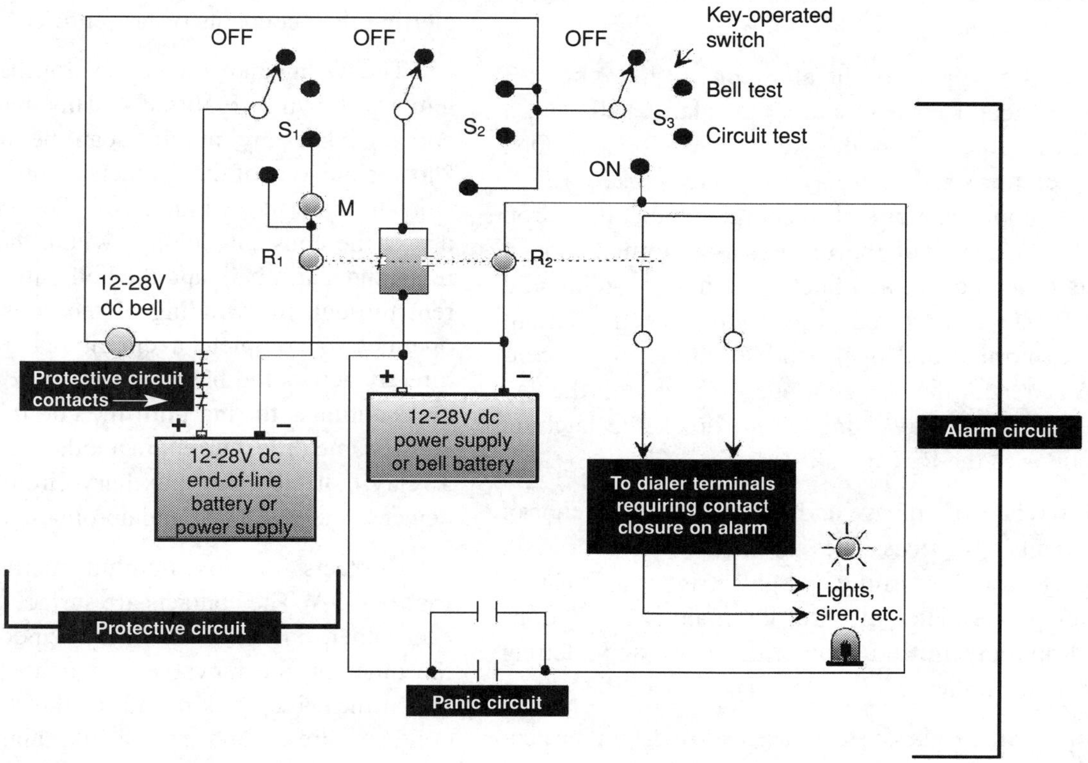

OFF
OFF
OFF
Key-operated switch
Bell test
Circuit test
S1
S2
S3
ON
M
R1
R2
12-28V dc bell
Protective circuit contacts
+
−
12-28V dc end-of-line battery or power supply
12-28V dc power supply or bell battery
To dialer terminals requiring contact closure on alarm
Lights, siren, etc.
Alarm circuit
Protective circuit
Panic circuit

Figure 8-3: *Closed-circuit security alarm system*

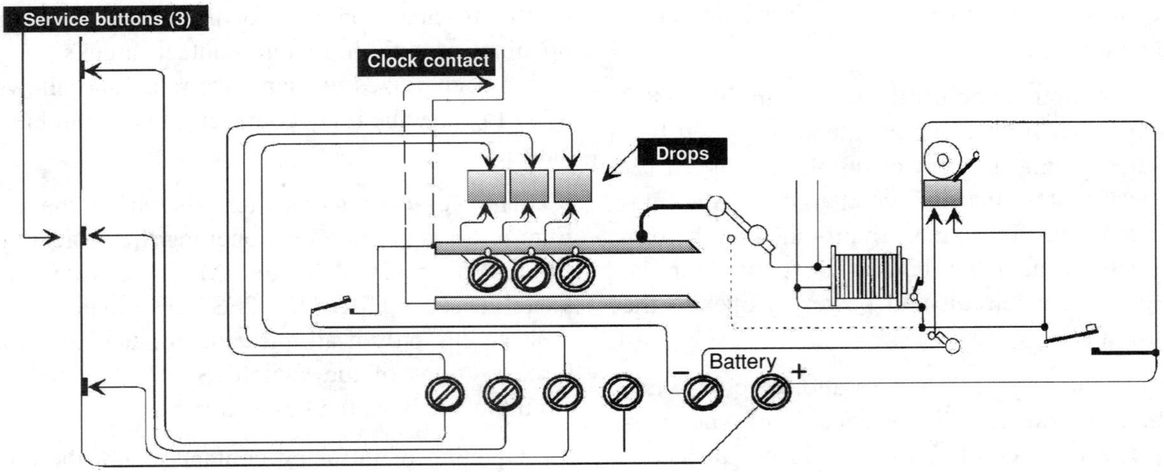
Service buttons (3)
Clock contact
Drops
Battery
–
+

Figure 8-4: *Open-circuit security alarm system*

close the relay contacts, thus sounding the alarm. Any cross or short circuit between the positive and negative wires of the protective circuit also keeps current from reaching the relay coil and causes dropout which again sounds the alarm.

Other components of the alarm circuit in Figure 8-3 include a second energy source, an alarm bell, and a drop relay (R_2). When the keyed switch is at ON, dropout of the sensitive relay (R_1), and closing of its contacts completes a circuit to energize the coil of drop relay (R_2). Closing of the drop relay's normally open contacts rings the bell and latches in the drop-relay coil so that R_2 stays energized even if the protective circuit returns to normal and opens the sensitive relay's contacts. As a result, the bell continues to ring until the key switch is turned away from ON to break the latching connections to the R_2 coil.

Drop relays often have additional contacts to control other circuits or devices. The extra contacts in the circuit in Figure 8-3 are for turning on lights, triggering an automatic telephone dialer, etc. But the main two functions of the drop relay are actuation of the alarm and latching the coil to keep the circuit in the alarm condition.

Almost all burglar systems use a closed-loop protective circuit. In general, the system consists of an annunciator connected to a special design contact on each door and window and a relay so connected that when any window or door is opened it will cause current to pass through the relay. The relay, in turn, will operate to close a circuit on a bell, horn, or other type of annunciator which will continue to sound until it is shut off, thereby alerting the occupants or law enforcement agencies.

The wiring and connections for the open-circuit system are shown in Figure 8-4. This diagram shows three contacts, but any number can be added as needed. Closing any one of the contacts completes the power circuit through the winding of the proper annunciator drops, the constant-ringing switch, the constant-ringing relay, the alarm bell, and the bell-cutoff switch. The current through the winding of the constant-ringing relay operates to complete a circuit placing the alarm bell directly across the battery or other power source so the bell continues to ring until the cutoff switch is opened. At the same time, current in another set of wires operates a relay that closes an auxiliary circuit to operate other devices, such as lights and automatic telephone dialer.

Contacts for closed-circuit operation are shown in Figure 8-5A. The contacts are surface-mounted opposite each other, one on a stationary window or door frame; the other on the movable part of the window or door. When the window is raised, or the door is opened, the contacts break and sound the alarm. Contacts for recessed mounting are shown in Figure 8-5B and operate the same way as described for the surface-mounted contacts.

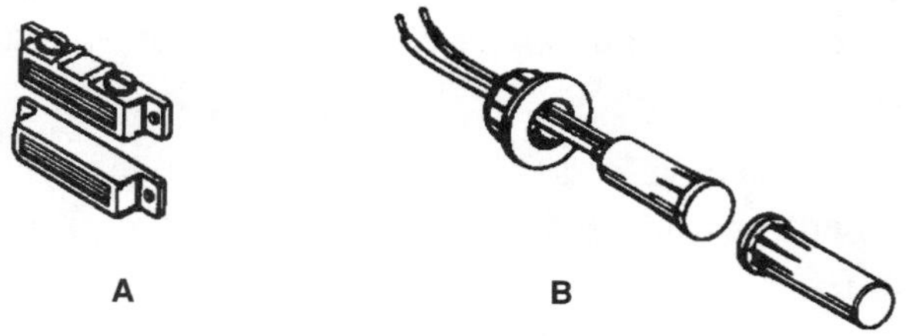
A
B

Figure 8-5: *Magnetic-type contacts for closed-circuit operation*

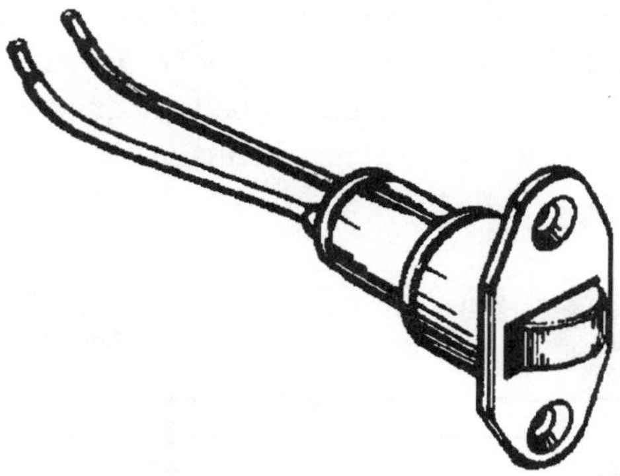

Figure 8-6: *Spring-type contact for open-circuit operation*

A spring-type contact for open-circuit operation is shown in Figure 8-6. This device is recessed in the window frame or a door jamb so that the cam projects outward. When the window is raised, the cam pivots and is pressed in and makes contact with a spring that is insulated from the plate. The contact is connected in series with the power source and the annunciator; that is, one wire is connected to the plate and the other to the spring.

NEC Requirements For Basic Signaling Circuits

A summary of *NEC* requirements for signaling and communications systems is listed in Figure 8-7. In general, the power supply of nonpower-limited signaling circuits must comply with *NEC* Chapters 1 through 4 and the output voltage must not exceed 600 volts, nominal. Signaling circuits and equipment must be grounded in accordance with *NEC* Article 250, except for dc power-limited fire-protection signaling circuit having a maximum current of 0.03 amperes.

FIRE-ALARM SYSTEMS

A fire-alarm system consists of the following:

- Sensors
- Control panel

SIGNAL AND COMMUNICATION CIRCUITS		
Application	**Requirements**	***NEC* Reference**
Conductors, installation of	The provisions in *NEC* Section 780-6 apply.	Section 800-3
Conductors on power poles	Must be located below the power conductors where practical.	Section 800-10(a)(1)
Circuit protection	As per (a) through (c) of this *NEC* Section.	Section 800-30
Entering building	As per (a) and (b) of this *NEC* Section.	Section 800-10
Grounding	As per (a) through (d) of this *NEC* Section.	Section 800-40
Insulation	As per (a) through (i) of this *NEC* Section.	Section 800-51
Power conductors run with communication conductors	Open communication conductors must be separated at least 2 inches from power conductors.	Section 800-52(a)(2)

Figure 8-7: *Summary of NEC installation requirements for basic signal and communications circuits*

- Annunciator
- Related wiring

They are generally divided into the following four types:

- Noncoded
- Master-coded
- Selective-coded
- Dual-coded

Each of these four types has several functional features so designed that a specific system may meet practically any need to comply with local and state codes, statutes, and regulations.

In a noncoded system, an alarm signal is sounded continuously until manually or automatically turned off.

In a master-coded system, a common-coded alarm signal is sounded for not less than three rounds. The same code is sounded regardless of the alarm-initiating device activated.

In a selective-coded system, a unique coded alarm is sounded for each firebox or fire zone on the protected premises.

In a dual-coded system, a unique coded alarm is sounded for each firebox or fire zone to notify the building's personnel of the location of the fire, while noncoded or common-coded alarm signals are sounded separately to notify other occupants to evacuate the building.

Figure 8-8 represents a riser diagram of a fire-alarm system. If the detector senses smoke or if any manual striking station is operated, all bells within the building will ring. At the same time, the magnetic door switches

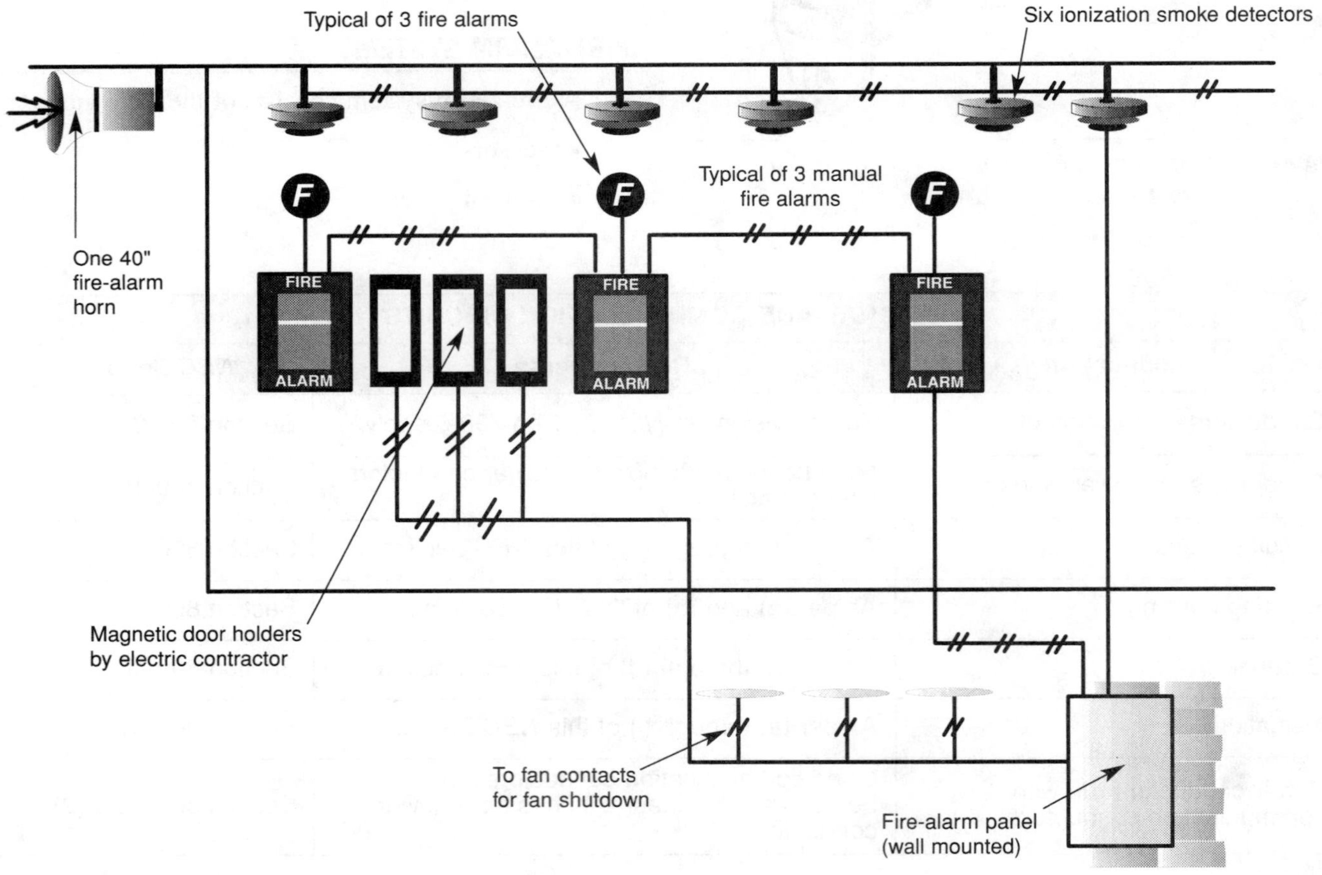
Typical of 3 fire alarms
Six ionization smoke detectors
Typical of 3 manual fire alarms
One 40" fire-alarm horn
FIRE
ALARM
Magnetic door holders by electric contractor
To fan contacts for fan shutdown
Fire-alarm panel (wall mounted)

Figure 8-8: *Riser diagram of a fire-alarm system*

will release the smoke doors to help block smoke and/or drafts. This system is also connected to a water-flow switch on the sprinkler system. If the sprinkler valves are activated causing a flow of water in the system, the fire-alarm system will again go into operation energizing all bells and closing smoke doors.

Smoke And Fire Detectors

Any product of a fire that changes the ambient conditions is called a *fire signature* and is potentially useful for detection purposes. The principal fire signature used in residential smoke detectors is aerosol. *Aerosols* are particles suspended in air. The process of combustion releases into the atmosphere large numbers of such solid and liquid particles that may range in size from 10 µm [a micron (µm) is one thousandth of a millimeter] down to 0.001 µms. Aerosols resulting from a fire represent two different fire signatures. Those particles less than 0.3 µms do not scatter light efficiently and are classified as visible. The invisible aerosol signature is usually referred to as the "products of combustion" and the visible aerosol signature as "smoke." Invisible aerosol is the earliest appearing fire signature.

Types Of Fire Detection Devices

Thermal Detectors: Thermal detectors are devices that operate on high heat — typically 135°F. These units consist of a bimetallic element which bends to complete a circuit under high heat conditions. Since these units do not detect smoke or products of combustion, they are not recommended for living areas of a residence. They do have value for use in attics, unheated garages and furnace rooms.

Flame Detectors: Flame detectors detect actual flames by sensing the ultraviolet emissions. These devices would not be used in residential applications.

Gas Detectors: These units respond to certain gases (propane, carbon monoxide, liquid petroleum, butane, gasoline vapors, etc.) that would not be detected by a smoke and fire detector. While these detectors do have some uses, they should not be a substitute for a smoke and fire detector. They will not respond to aerosols produced by the majority of residential fires.

Ionization Detectors: Inside the ionization chamber, the radioactive source emits radiation, main alpha particles, which bombard the air and ionize the air particles, which, in turn, are attracted by the voltage on the collector electrodes. This action results in a minute current flow. If aerosols, such as products of combustion or smoke, enter the chamber, the ionized air particles attach themselves to the aerosols and the resultant particles, being of larger mass than ionized air, move more slowly, and thus, per unit of time, fewer reach the electrodes. A decrease in current flow, therefore, takes place within the chamber whenever aerosols enter. The decrease in current flow is electronically converted into an alarm signal output. See Figure 8-9.

An ionization type of detector responds best to invisible aerosols in which the particles from burning materials are in the size range of 1.0 µm down to 0.01 µm. A tremendous number of these particles are produced by a flaming fire as opposed to a smoldering fire, which produces large and small particles. But, because of low heat, the low thermal lift tends to allow particles to agglomerate into larger particles if the detector is some distance from the fire.

High air flows will affect the operation of this type of unit by reducing the ion concentration in the detector chamber. In fact, with a high enough air flow, the unit will respond and alarm even though a fire does not exist. For this reason, locations near windows, direct air flows from air vents and comparable areas should be avoided.

Ionization smoke detectors (Figure 8-10) may be used in place of conventional smoke detectors or may be used in combination with standard smoke detectors. They are more sensitive than the conventional smoke detectors.

Photoelectric Detectors: A beam from the light source (Figure 8-11) is projected across a chamber into a light catcher. The chamber is designed to permit access of smoke, but not access of external light. A photo-resistive cell or light sensitive device is located in a recessed area perpendicular to the light beam. When smoke enters the chamber, smoke particles will scatter or reflect a small portion of the light beam to the light receiving device, which, in turn, will provide a signal for amplification to the alarm. This description of operation is the basic operating principle of photoelectric detectors. Some variations in design are used.

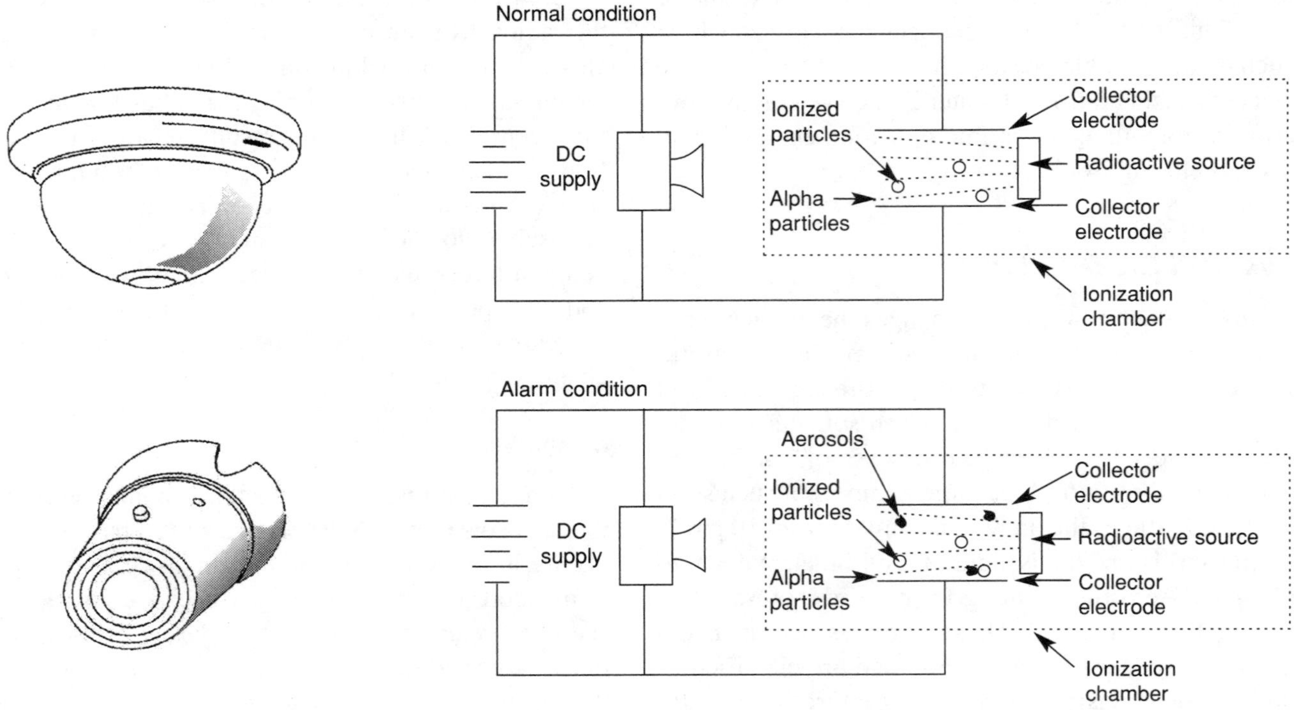
Normal condition
DC supply
Ionized particles
Alpha particles
Collector electrode
Radioactive source
Collector electrode
Ionization chamber
Alarm condition
DC supply
Aerosols
Ionized particles
Alpha particles
Collector electrode
Radioactive source
Collector electrode
Ionization chamber

Figure 8-9: *Diagrams of ionization detectors. The top diagram shows normal conditions. The bottom diagram shows aerosols, such as products of combustion or smoke, entering the sensor. In this latter condition, the alarm is activated.*

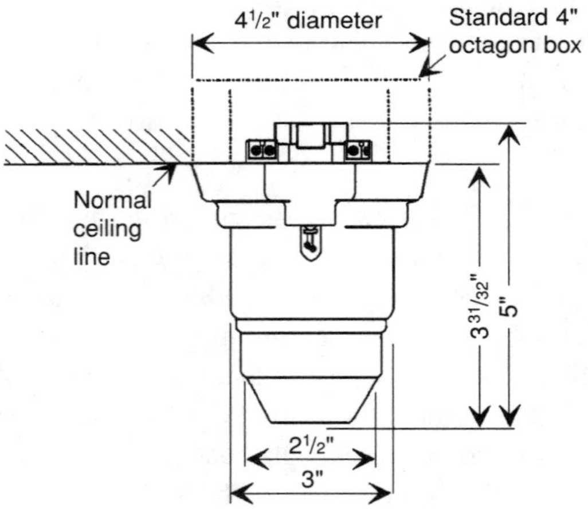
4 1/2" diameter
Standard 4" octagon box
Normal ceiling line
3 31/32"
5"
2 1/2"
3"

Figure 8-10: *Ionization smoke detector*

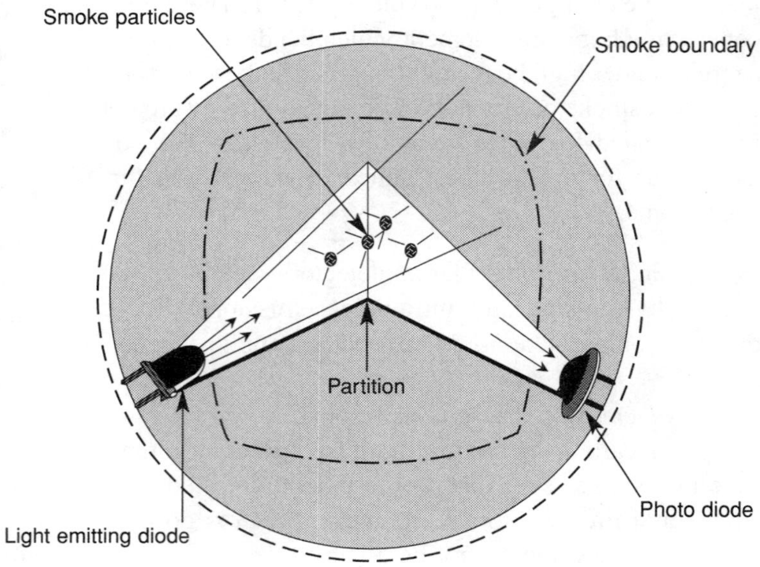
Smoke particles
Smoke boundary
Partition
Light emitting diode
Photo diode

Figure 8-11: *Basic operating principles of photoelectric detectors*

Some photoelectric detectors are adversely affected by dirt films. Any accumulated dirt, dust, film, or foreign matter collecting on either or both lenses of the light source or the photocell will cause an opaque effect and the detector will then become less and less sensitive. It, therefore, will require more smoke in order to respond.

While latest photoelectric models utilize solid state light emitting and receiving devices which have a longer life than previous light devices, the problem of failure of the light source still exists. Underwriters' Laboratories requires an audible alarm if light failure occurs.

Photoelectric units respond best to fires producing visible aerosols where the particles range from 10 μm down to 0.3μm. These particles would be produced by a smoldering fire where very little heat is produced.

Ionization and Photoelectric Devices: The diagram in Figure 8-12 can be used to illustrate both types of devices — the difference is the use of either an ionization sensor or a photoelectric sensor in the reference chamber and detector portions of the circuit. Under normal conditions, the voltage across the reference chamber and the detection chamber is the same. However, when fire occurs, the detection chamber then functions as described in the previous explanation. Thus, when there is sufficient voltage difference between the two chambers, the alarm is activated through the switching circuit.

COMPONENTS OF SECURITY/FIRE-ALARM SYSTEMS

Wire sizes for the majority of low-voltage systems range from No. 22 to No. 16 AWG. However, there are some situations where it may be necessary to use larger wire sizes (such as where larger-than-normal currents are required and longer distance between outlets) to prevent excessive voltage drop.

NOTE: Voltage-drop calculations should be made to determine the correct wire size for a given application.

Most closed systems use two-wire No. 22 or No. 24 AWG conductors and are color-coded to identify them. A No. 18 pair normally is adequate for connecting bells or sirens to controls if the run is 40 feet or less. Many, however, prefer to use No. 16 or even No. 14 nonmetallic cable.

A summary of the various components for a typical security/fire-alarm system is shown in the riser diagram in Figure 8-13. Notice the varying types of sensors or detectors in this system.

Control Station

The control station is the heart of any security system since it is the circuitry in these control panels that senses a broken contact and then either sounds a local bell or horn or omits the bell for a silent alarm. Most modern control panels use computer circuit board con-

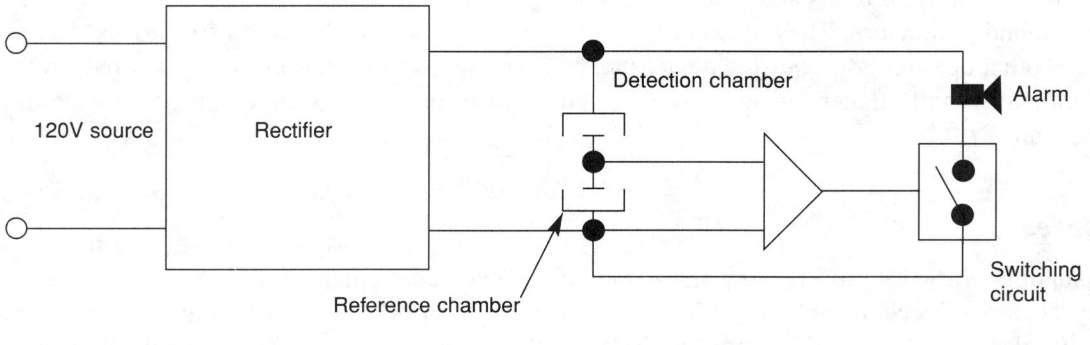
120V source
Rectifier
Detection chamber
Alarm
Reference chamber
Switching
circuit

Figure 8-12: *Diagram of ionization and photoelectric devices*

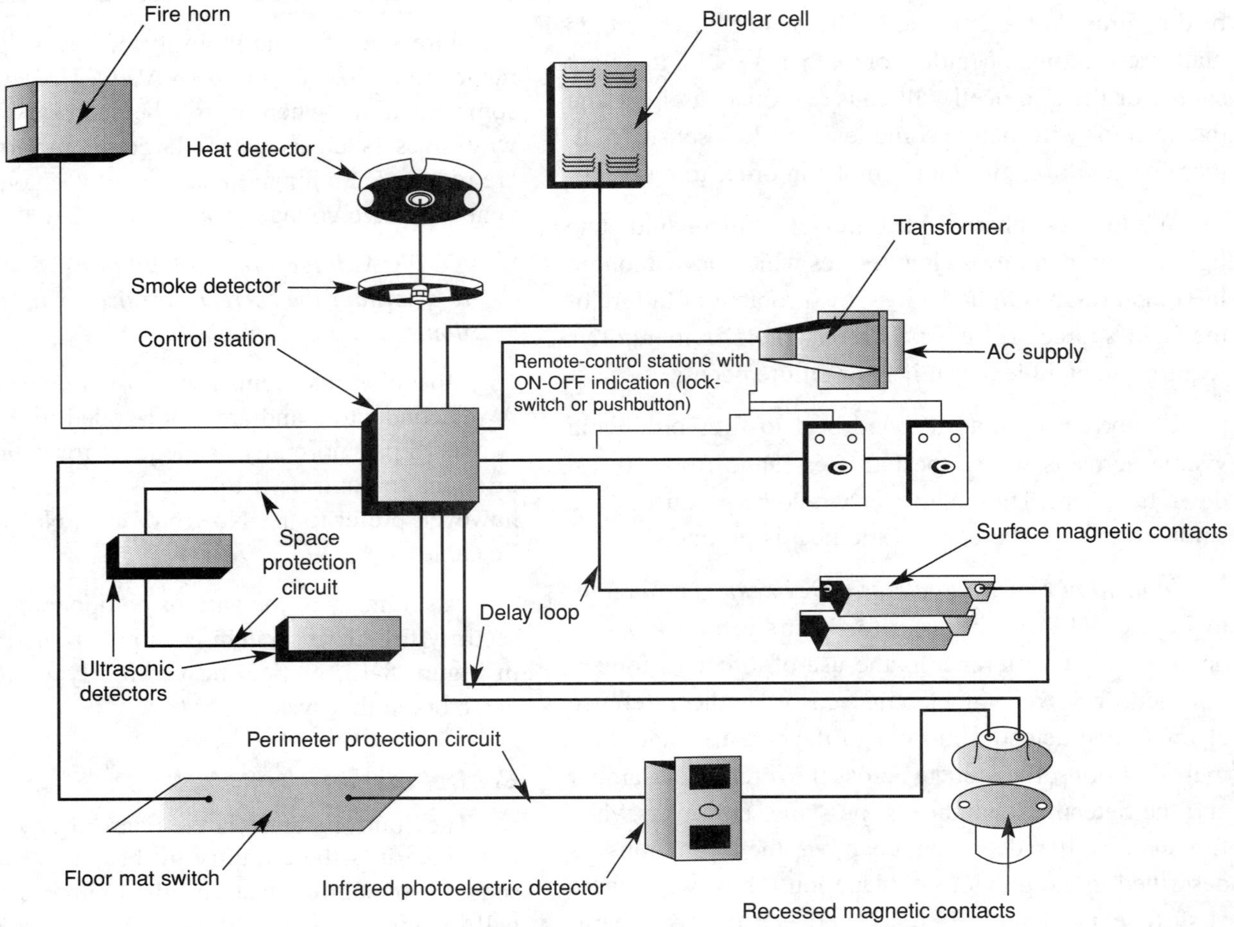
Fire horn
Burglar cell
Heat detector
Smoke detector
Transformer
Control station
Remote-control stations with ON-OFF indication (lock-switch or pushbutton)
AC supply
Space protection circuit
Surface magnetic contacts
Delay loop
Ultrasonic detectors
Perimeter protection circuit
Floor mat switch
Infrared photoelectric detector
Recessed magnetic contacts

Figure 8-13: *Components for a typical security/fire-alarm system*

trols to sense the protective circuits and regulate the output for alarm-sounding devices. They also contain contacts to actuate other deterrent or reporting devices and a silent holdup alarm with dialer or police-connected reporting mechanism.

Power Supplies

Power supplies vary for different systems, but in general they consist of rechargeable 12-28V dc power supplies for burglar alarm systems. The power packs are a new variety: more powerful and longer-life batteries that are charged and monitored by the control system itself. The better systems can maintain an armed system for an extended period of time, and have the capability to operate the alarm for a minimum of two hours, and up to 10 or 12 hours on a high-grade system.

NEC Requirements For Fire-Alarm Systems

Figure 8-14 on the next page summarizes the basic *NEC* requirements for fire-alarm systems, while the illustrations beginning with Figure 8-15 summarize *NEC* requirements for all types of signaling systems. The code book, however, should be consulted for details in all situations.

FIRE-ALARM SYSTEMS		
Application	**Requirements**	***NEC* Reference**
Access to electrical equipment	Must not be denied by an accumulation of wires and cables.	Section 760-5
Circuits extending beyond one building	Must meet the requirements of either *NEC* Articles 800 or 225.	Section 760-7
Classification	Must comply with (a) and (b) of this *NEC* Section.	Section 760-15
Connections at services	May be connected to the supply side.	Section 230-82(4)
	A separate means of overcurrent protection must be provided.	Section 230-94 Exception No. 4
Grounding	Must be grounded where system grounding is required.	Section 250-112(i)
	Must be grounded as per *NEC* Article 250.	Section 760-6
Health care facilities	As specified.	Section 517 F
Location	Must comply with (a) through (f) of this *NEC* Section.	Section 760-3
Nonpower-limited circuits	As specified.	Section 760 B
Power-limited circuits	As specified.	Section 760 C

Figure 8-14: *Summary of NEC installation requirements for fire-alarm systems*

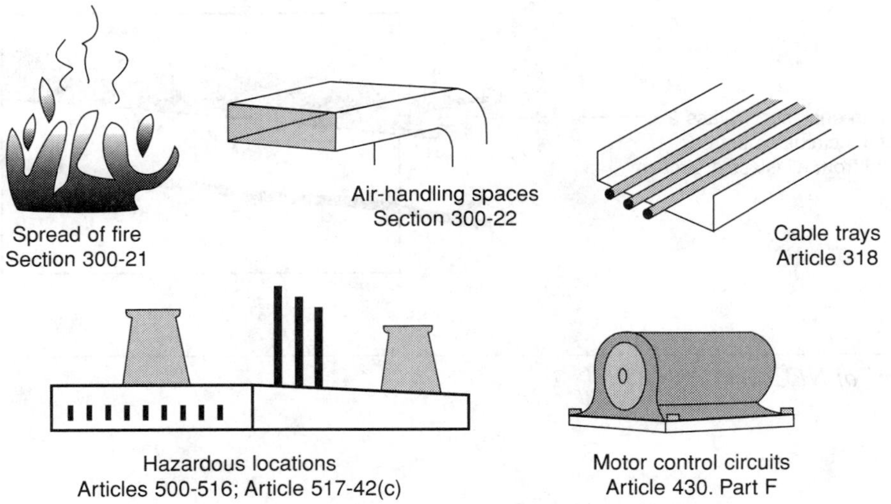
Spread of fire
Section 300-21
Air-handling spaces
Section 300-22
Cable trays
Article 318
Hazardous locations
Articles 500-516; Article 517-42(c)
Motor control circuits
Article 430. Part F

Figure 8-15: *Applicable locations and other NEC Articles that should be consulted*

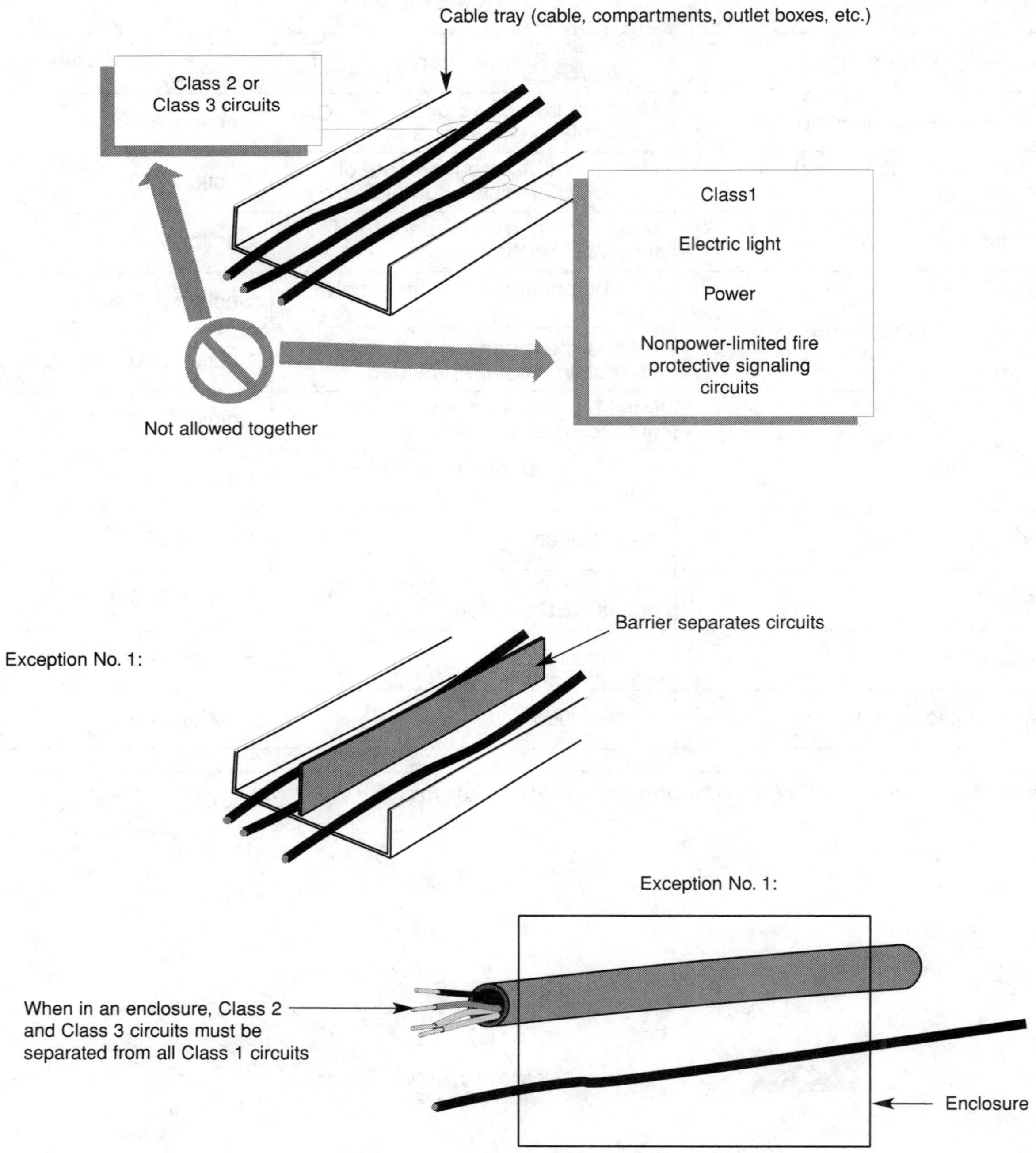
Cable tray (cable, compartments, outlet boxes, etc.)
Class 2 or
Class 3 circuits
Class1
Electric light
Power
Nonpower-limited fire
protective signaling
circuits
Not allowed together
Exception No. 1:
Barrier separates circuits
Exception No. 1:
When in an enclosure, Class 2
and Class 3 circuits must be
separated from all Class 1 circuits
Enclosure

Figure 8-16: *Summary of NEC Section 725-54*

Communications Cable		
Cable Marking / Type		**Reference**
MPP	Multipurpose plenum cable	*NEC* Sections 800-51(g) 800-53(a)
CMP	Communications plenum cable	*NEC* Sections 800-51(a) 800-53(a)
MPR	Multipurpose riser cable	*NEC* Sections 800-51(g) 800-53(b)
CMR	Communications riser cable	*NEC* Sections 800-51(b) 800-53(b)
MPG	Multipurpose general-purpose cable	*NEC* Sections 800-51(g) 800-53(d)
CMG	Communications general-purpose cable	*NEC* Sections 800-51(c) 800-53(d)
MP	Multipurpose general-purpose cable	*NEC* Sections 800-51(c) 800-53(d)
CM	Communications, general-purpose cable	*NEC* Sections 800-51(d) 800-53(d)
CMX	Communications cable, general use	*NEC* Sections 800-51(e) 800-53(d) Exceptions No. 1, 2, 3, 4
CMUC	Undercarpet communications wire and cable	*NEC* Sections 800-51(f)EX 800-53(d)EX

Figure 8-17: *Summary of NEC Table 800-50 — Cable Markings*

Coaxial Cables		
Cable Type	**Permitted Substitutions**	
CATVP	MPP CMP	
CATVR	CATVP MPP CMP	MPR CMTR
CATV	CATVP MPP CMP CATVR MPR	CMR MPG MP CMG CM
CATVX	CATVP MPP CMP CATVR MPR CMR	CATV MPG MP CMG CM

Figure 8-18: *Summary of NEC Section 800-53 — Cable Substitution*

Chapter 9
Electric Motors and Controllers

The principal means of changing electrical energy into mechanical energy or power is the electric motor — ranging in size from small fractional-horsepower, low-voltage motors to the very large high-voltage synchronous motors.

Electric motors are classified according to the following:

- Size (horsepower)
- Type of application
- Electrical characteristics
- Speed, starting, speed control and torque characteristics
- Mechanical protection
- Method of cooling

In basic terms, electric motors convert electric energy into the productive power of rotary mechanical force. This capability finds many applications in unlimited ways in commercial establishments for powering elevators in public buildings, HVAC and refrigeration fans and compressors, gasoline and water pumps, power tools, and a host of other applications.

All of these and more represent the scope of electric motor participation in powering and controlling machines and equipment used in electrical systems.

Electric motors are machines that change electrical energy into mechanical energy. They are rated in horsepower. The attraction and repulsion of the magnetic poles produced by sending current through the armature and field windings cause the armature to rotate. The armature rotation produces a twisting power called torque.

Single-Phase Motors

Single-phase ac motors are usually limited in size to about two or three horsepower, but sometimes single-phase motors are encountered as high as 7.5 horsepower or more — especially on some farm-machinery applications. For residential and small commercial applications, these motors will be found in both central and individual room air-conditioning units, fans, ventilating units, and refrigeration units such as household refrigerators and the larger units used to cool produce and other foods in market places.

Since there are so many applications of electric motors, there are many types of single-phase motors in use. Some of the more common types are repulsion start, universal, and single-phase induction motors. This latter type includes split-phase, capacitor-start, shaded-pole, and repulsion-induction motors.

Figure 9-1 shows the basic parts of a motor to familiarize you with its makeup.

Split-Phase Motors

Split-phase motors are fractional-horsepower units that use an auxiliary winding on the stator to aid in starting the motor until it reaches its proper rotation speed. This type of motor finds use in small pumps, oil burners, and similar applications.

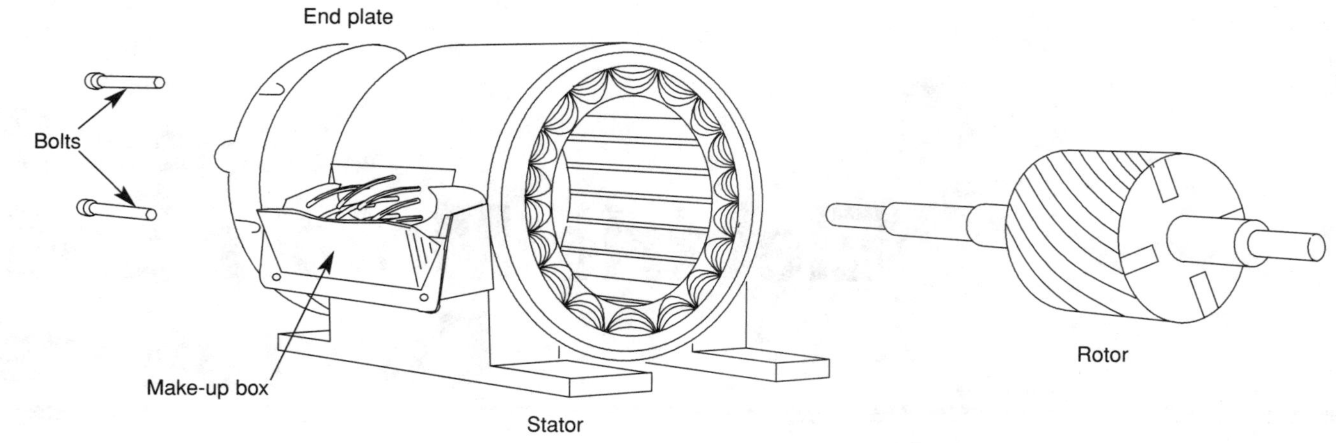
End plate
Bolts
Make-up box
Stator
Rotor

Figure 9-1: *Basic parts of a motor*

In general, the split-phase motor consists of a housing, a laminated iron-core stator with embedded windings forming the inside of the cylindrical housing, a rotor made up of copper bars set in slots in an iron core and connected to each other by copper rings around both ends of the core, plates that are bolted to the housing and contain the bearings that support the rotor shaft, and a centrifugal switch inside the housing. This type of rotor is often called a squirrel-cage rotor since the configuration of the copper bars resembles an actual cage. These motors have no windings as such, and a centrifugal switch is provided to open the circuit to the starting winding when the motor reaches running speed.

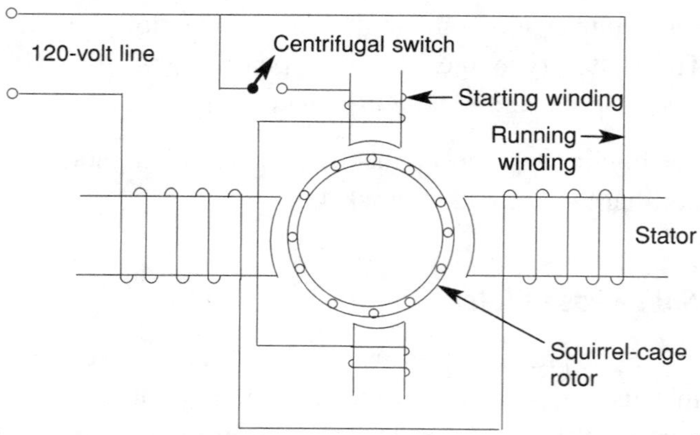
120-volt line
Centrifugal switch
Starting winding
Running winding
Stator
Squirrel-cage rotor

Figure 9-2: *Diagram of a split-phase motor*

To understand the operation of a split-phase motor, look at the wiring diagram in Figure 9-2. Current is applied to the stator windings, both the main winding and the starting winding, which is in parallel with it through the centrifugal switch. The two windings set up a rotating magnetic field, and this field sets up a voltage in the copper bars of the squirrel-cage rotor. Because these bars are shortened at the ends of the rotor, current flows through the rotor bars. The current-carrying rotor bars then react with the magnetic field to produce motor action. When the rotor is turning at the proper speed, the centrifugal switch cuts out the starting winding since it is no longer needed.

Capacitor Motors

Capacitor motors are single-phase ac motors ranging in size from fractional horsepower (hp) to perhaps as high as 15 hp. This type of motor is widely used in all types of single-phase applications such as powering air compressors, refrigerator compressors, and the like. This type of motor is similar in construction to the split-phase motor, except a capacitor is wired in series with the starting winding, as shown in Figure 9-3.

The capacitor provides higher starting torque, with lower starting current, than does the split-phase motor, and although the capacitor is sometimes mounted inside the motor housing, it is more often mounted on top of the motor, encased in a metal compartment to enable easier servicing.

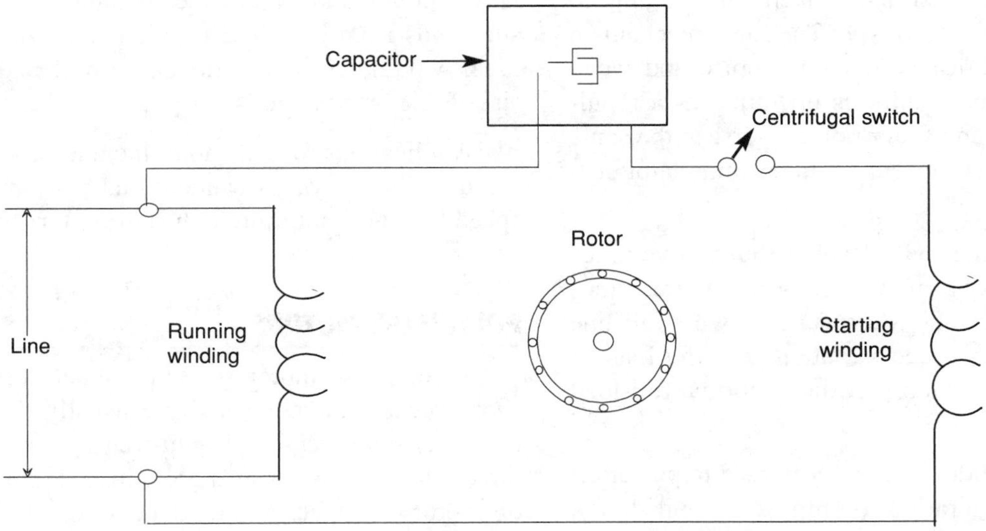
Capacitor
Centrifugal switch
Rotor
Line
Running winding
Starting winding

Figure 9-3: *Diagram of a capacitor-start motor*

In general, two types of capacitor motors are in use: the capacitor-start motor and the capacitor start-and-run motor. As the name implies, the former utilizes the capacitor only for starting; it is disconnected from the circuit once the motor reaches running speed, or at about 75 percent of the motor's full speed. Then the centrifugal switch opens to cut the capacitor out of the circuit.

The capacitor start-and-run motor keeps the capacitor and starting winding in parallel with the running winding, providing a quiet and smooth operation at all times.

Capacitor split-phase motors require the least maintenance of all single-phase motors, but they have a very low starting torque, making them unsuitable for many applications. Their high maximum torque, however, makes them especially useful in HVAC systems to power slow-speed, direct-connected fans.

Repulsion-Start Motors

Repulsion-type motors are divided into several groups, including (1) repulsion-start, induction-run motors, (2) repulsion motors, and (3) repulsion-induction motors. The repulsion-start, induction-run motor is of the single-phase type, ranging in size from about 1/10 hp to as high as 20 hp. It has high starting torque and a constant-speed characteristic, which makes it suitable for such applications as commercial refrigerators, compressors, pumps, and similar applications requiring high starting torque.

The repulsion motor is distinguished from the repulsion-start, induction-run motor by the fact that it is made exclusively as a brush-riding type and does not have any centrifugal mechanism. Therefore, this motor both starts and runs on the repulsion principle. This type of motor has high starting torque and a variable-speed characteristic. It is reversed by shifting the brush holder to either side of the neutral position. Its speed can be decreased by moving the brush holder farther away from the neutral position.

The repulsion-induction motor combines the high starting torque of the repulsion-type and the good speed regulation of the induction motor. The stator of this motor is provided with a regular single-phase winding, while the rotor winding is similar to that used on a dc motor. When starting, the changing single-phase stator flux cuts across the rotor windings and induces currents in them; thus, when flowing through the commutator, a continuous repulsive action on the stator poles is present.

This motor starts as a straight repulsion-type and accelerates to about 75 percent of normal full speed when a centrifugally operated device connects all the commutator bars together and converts the winding to an equivalent squirrel-cage type. The same mechanism usually raises the brushes to reduce noise and wear. Note that, when the machine is operating as a repulsion-type, the rotor and stator poles reverse at the same instant, and that the current in the commutator and brushes is ac.

This type of motor will develop four to five times normal full-load torque and will draw about three times normal full-load current when starting with full-line voltage applied. The speed variation from no load to full load will not exceed 5 percent of normal full-load speed.

The repulsion-induction motor is used to power air compressors, refrigeration (compressor and fans), pumps, stokers, and the like. In general, this type of motor is suitable for any load that requires a high starting torque and constant-speed operation. Most motors of this type are less than 5 hp.

Universal Motors

This type of motor is a special adaptation of the series-connected dc motor, and it gets its name "universal" from the fact that it can be connected on either ac or dc and operates the same. All are single-phase motors for use on 120 or 240 volts.

In general, the universal motor contains field windings on the stator within the frame, an armature with the ends of its windings brought out to a commutator at one end, and carbon brushes that are held in place by the motor's end plate, allowing them to have a proper contact with the commutator.

When current is applied to a universal motor, either ac or dc, the current flows through the field coils and the armature windings in series. The magnetic field set up by the field coils in the stator reacts with the current-carrying wires on the armature to produce rotation. Universal motors are frequently used on small fans.

Shaded-Pole Motors

A shaded-pole motor (Figure 9-4) is a single-phase induction motor provided with an uninsulated and permanently short-circuited auxiliary winding displaced in magnetic position from the main winding. The auxiliary winding is known as the shading coil and usually surrounds from one-third to one-half of the pole. The main winding surrounds the entire pole and may consist of one or more coils per pole.

Applications for this motor include small fans, timing devices, relays, instrument dials, or any constant-speed load not requiring high starting torque.

POLYPHASE MOTORS

Three-phase motors offer extremely efficient and economical application and are usually the preferred type for commercial and industrial applications when three-phase service is available. In fact, the great bulk of motors sold are standard ac three-phase motors. These motors are available in ratings from fractional horsepower up to thousands of horsepower in practically every standard voltage and frequency. In fact, there are few applications for which the three-phase motor cannot be put to use.

Three-phase motors are noted for their relatively constant speed characteristic and are available in designs giving a variety of torque characteristics; that is, some have a high starting torque and others a low starting torque. Some are designed to draw a normal starting current, others a high starting current.

There are three basic types of three-phase motors:

- The squirrel cage induction motor.
- The wound rotor induction motor.
- The synchronous motor.

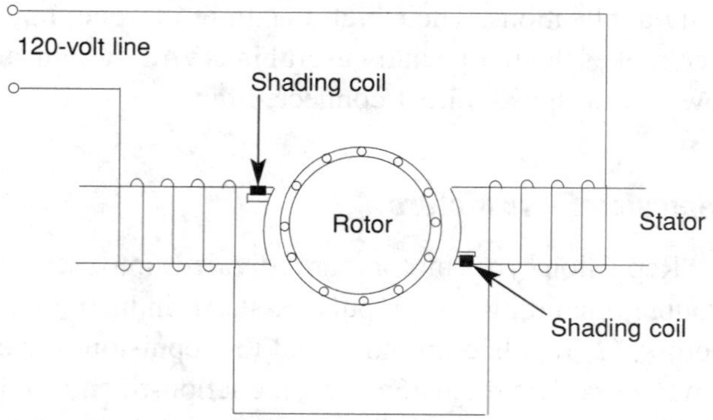
120-volt line
Shading coil
Rotor
Stator
Shading coil

Figure 9-4: *Wiring diagram of a shaded-pole motor*

The type of three-phase motor is determined by the rotor or rotating member (refer again to Figure 9-1). The stator winding for any of these motors is the same.

The principle of operation for all three-phase motors is the rotating magnetic field. There are three factors that cause the magnetic field to rotate:

- The voltages of a three-phase electrical system are 120° out of phase with each other.
- The three voltages change polarity at regular intervals.
- The arrangement of the stator windings around the inside of the motor.

The *NEC* also plays an important role in the installation of electric motors. *NEC* Article 430 covers application and installation of motor circuits and motor control connections — including conductors, short-circuit and ground-fault protection, controllers, disconnects, and overload protection.

NEC Article 440 contains provisions for motor-driven air conditioning and refrigerating equipment including the branch circuits and controllers for the equipment. It also takes into account the special considerations involved with sealed (hermetic-type) motor compressors (Figure 9-5), in which the motor operates under the cooling effect of the refrigeration. In referring to *NEC* Article 440, be aware that the rules in this *NEC* Article are *in addition to*, or *are amendments to*, the rules given in *NEC* Article 430.

Motors are also covered to some degree in *NEC* Articles 422 and 424.

THREE-PHASE MOTOR BASICS

The rotor of an ac squirrel-cage induction motor (Figure 9-6) consists of a structure of steel laminations mounted on a shaft. Embedded in the rotor is the rotor winding, which is a series of copper or aluminum bars that are all short-circuited at each end by a metallic end

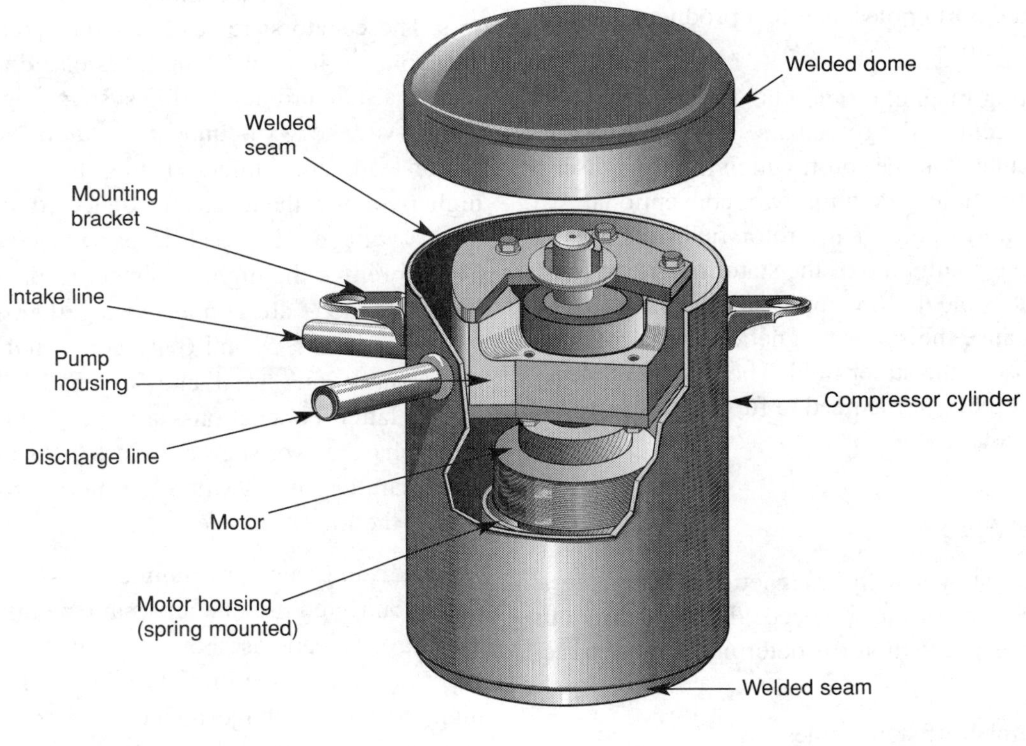
Welded dome
Welded seam
Mounting bracket
Intake line
Pump housing
Compressor cylinder
Discharge line
Motor
Motor housing (spring mounted)
Welded seam

Figure 9-5: *Basic components of a hermetically-sealed compressor*

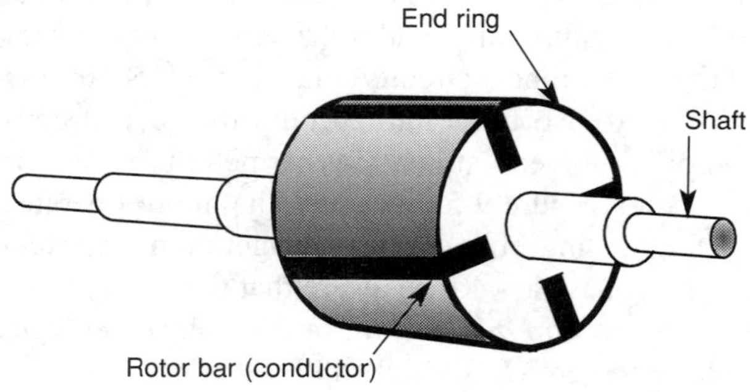
End ring
Shaft
Rotor bar (conductor)

Figure 9-6: *Squirrel-cage rotor*

ring. The stator consists of steel laminations mounted in a frame. Slots in the stator hold stator windings that can be either copper or aluminum wire coils or bars. These are connected to form a circuit.

Energizing the stator coils with an ac supply voltage causes current to flow in the coils. The current produces an electromagnetic field that, in turn, causes magnetic poles to be created in the stator iron. The strength and polarity of these poles vary as the ac current flows in one direction, then the other. This change causes the poles around the stator to alternate between being south and north poles, in effect producing a rotating magnetic field.

The rotating magnetic field cuts through the rotor, inducing a current in the rotor bars. This induced current only circulates in the rotor, which in turn causes a rotor magnetic field. As with two conventional bar magnets, the north pole of the rotor field attempts to line up with the south pole of the stator magnetic field, and the south pole to line up with the north pole. However, because the stator magnetic field is rotating, the rotor "chases" the stator field. The rotor field never quite catches up due to the need to furnish torque to the mechanical load.

Synchronous Speed

The speed at which the magnetic field rotates is known as the *synchronous* speed. The synchronous speed of a three-phase motor is determined by two factors:

- The number of stator poles.
- The frequency of the ac line.

Since 60 Hz is the standard frequency throughout the United States and Canada, the following gives the synchronous speeds for motors with different numbers of poles.

2 Poles	3600 RPM
4 Poles	1800 RPM
6 Poles	1200 RPM
8 Poles	900 RPM

From the above, the RPM of any three-phase, 60 Hz motor can be determined by counting the number of poles in the stator.

Stator Windings

The stator windings of three-phase motors are connected in either wye or delta. See Figure 9-7. Some motor stators are designed to operate both ways; that is, some motors are started as a wye-connected motor to help reduce starting current, and then changed to a delta connection for running.

Many three-phase motors have dual-voltage stators. These stators are designed to be connected to, say, 240 volts or 480 volts. The leads of a dual-voltage stator use a standard numbering system. Figure 9-8 shows a dual-voltage wye-connected stator. Note that the 9 motor leads are numbered in a spiral. For use on the higher voltage, the leads are connected in series; for the lower voltage, the leads are connected in parallel. Therefore, for the higher voltage, leads 4 and 7, 5 and 8, and 6 and 9 are connected together. For the lower voltage, leads 4, 5, and 6 are connected together; further connections are 1 and 7, 2 and 8, and 3 and 9. These latter connections are then connected to the three-phase power source. Figure 9-9 shows the equivalent parallel circuit when the motor is connected for use on the lower voltage.

The same standard numbering system is used for delta-connected motors, and many delta-wound motors also have 9 leads as shown in Figure 9-10. However, there are only three circuits of three leads each. The high- and low-voltage connections for a three-phase, delta-wound, 9-lead, dual-voltage motor are shown in Figure 9-11.

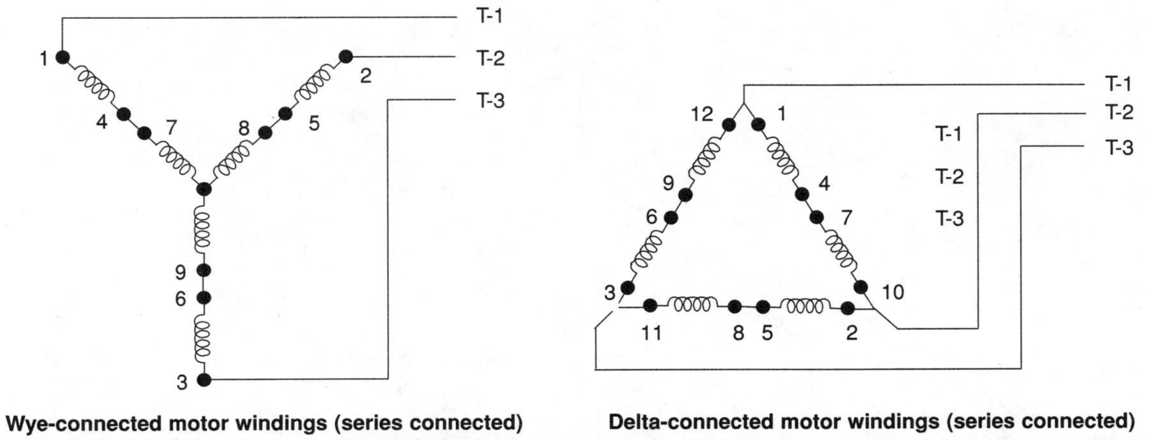
T-1
T-2
T-3
1
2
3
4
5
6
7
8
9
10
11
12
Wye-connected motor windings (series connected)
Delta-connected motor windings (series connected)

Figure 9-7: *Two types of windings found in three-phase motors*

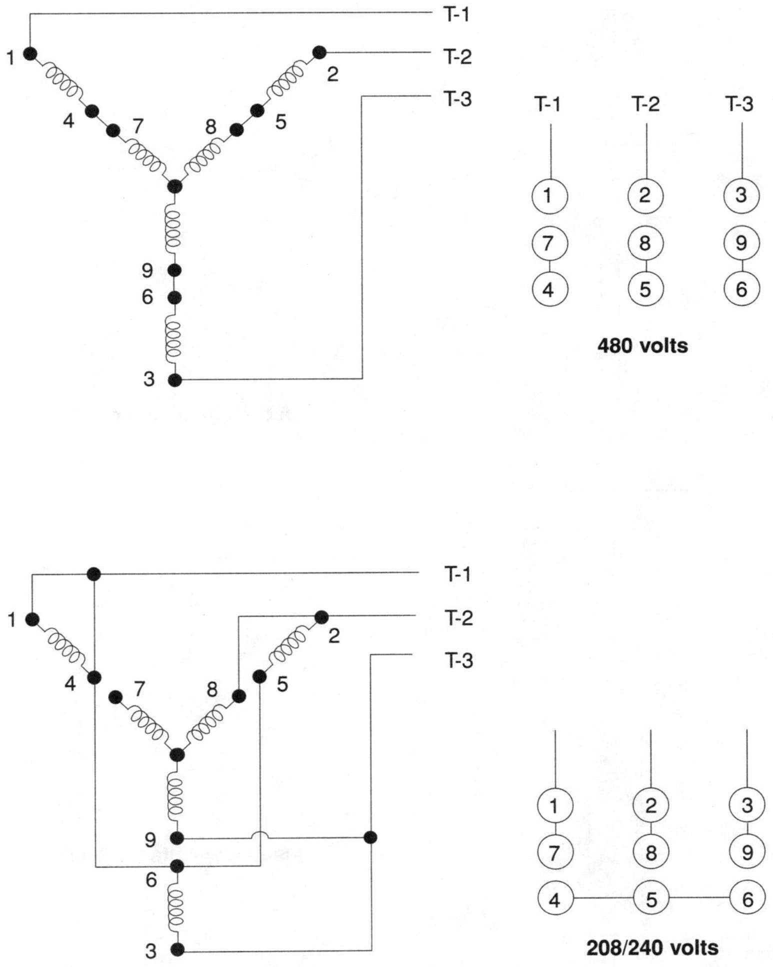
T-1
T-2
T-3
1
2
3
4
5
6
7
8
9
480 volts
208/240 volts

Figure 9-8: *High- and low-voltage connections for wye-connected three-phase motors*

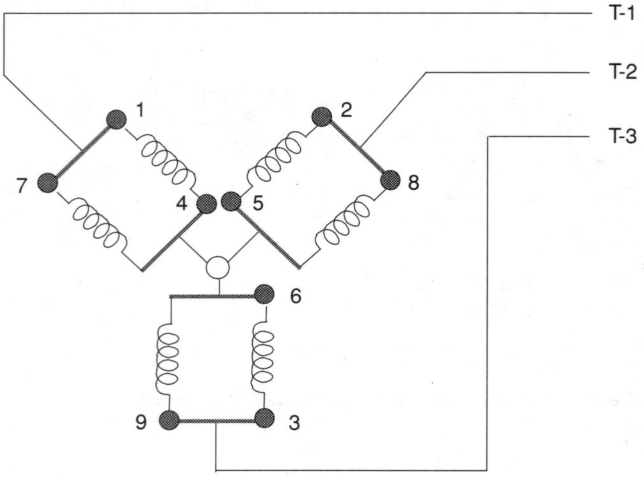
T-1
T-2
T-3
1
2
7
4
5
8
6
9
3

Figure 9-9: *Equivalent parallel circuit*

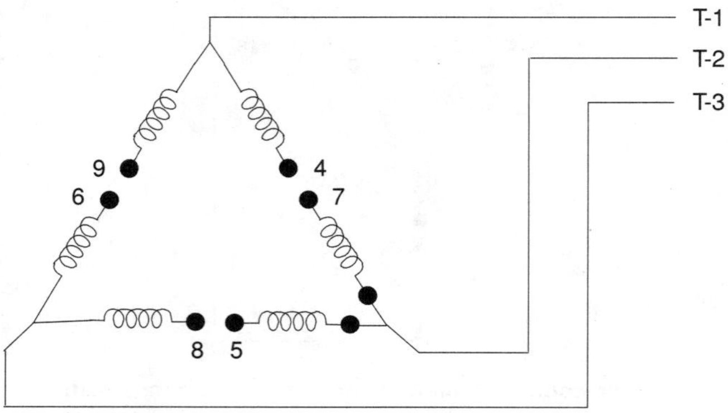
T-1
T-2
T-3
9
4
6
7
8
5

Figure 9-10: *Arrangement of motor leads in a 9-lead, delta-wound, dual-voltage motor*

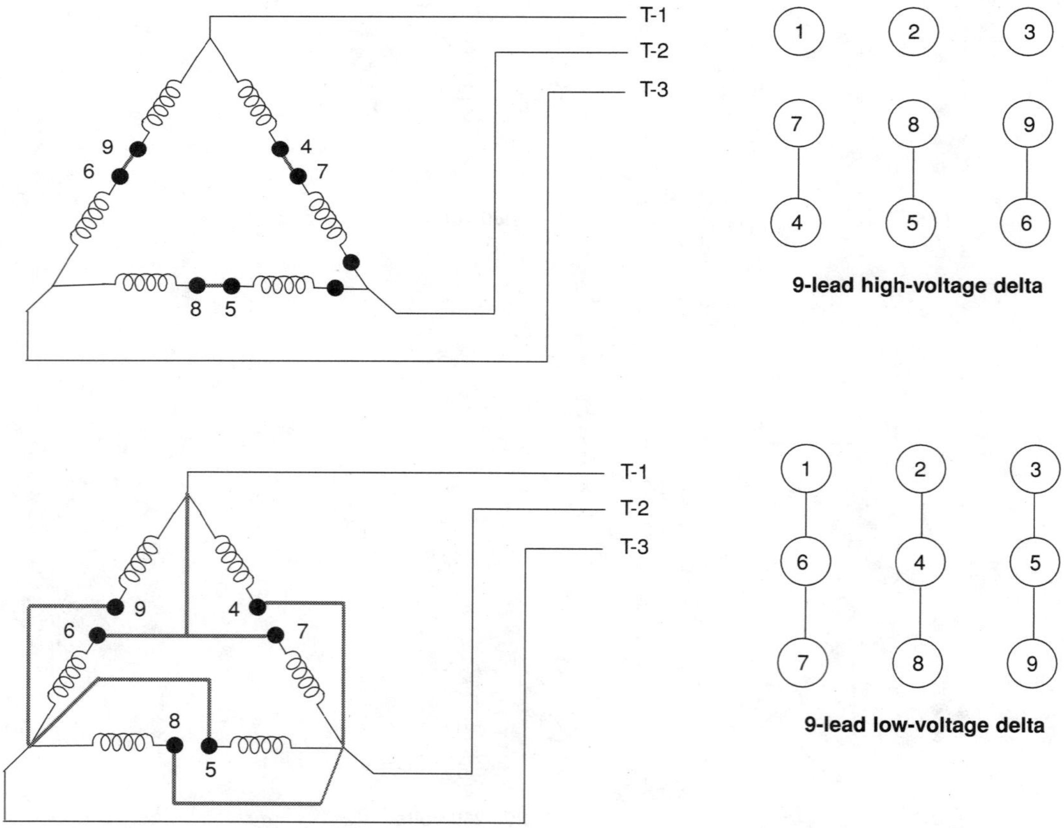
T-1
T-2
T-3
9
4
6
7
8
5
1
2
3
7
8
9
4
5
6
9-lead high-voltage delta
T-1
T-2
T-3
9
4
6
7
8
5
1
2
3
6
4
5
7
8
9
9-lead low-voltage delta

Figure 9-11: *Lead connections for a three-phase, dual-voltage, delta-wound motor*

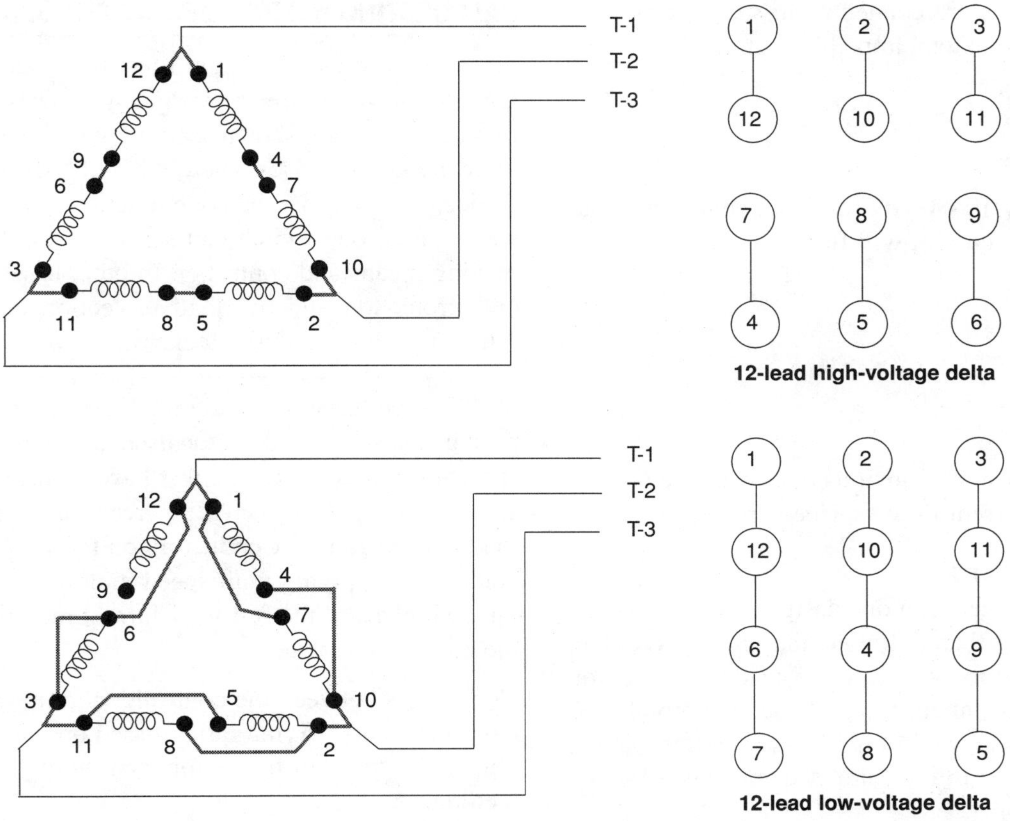
T-1
T-2
T-3
12
1
9
4
6
7
3
10
11
8
5
2
1
2
3
12
10
11
7
8
9
4
5
6
12-lead high-voltage delta
T-1
T-2
T-3
12
1
9
4
7
6
3
5
10
11
8
2
1
2
3
12
10
11
6
4
9
7
8
5
12-lead low-voltage delta

Figure 9-12: *Connections for a 12-lead, dual-voltage, delta-wound motor*

In some instances, a dual-voltage motor connected in delta will have 12 leads instead of 9. Figure 9-12 shows the high-voltage and low-voltage connections for dual-voltage, 12-lead, delta-wound motors.

Principles Of Dual-Voltage Connections

When a motor is operated at 240 volts, the current draw of the motor is double the current draw of a 480-volt connection. For example, if a motor draws 10 amperes of current when connected to 240 volts, it will draw only 5 amperes when connected to 480 volts. The reason for this is the difference of impedance in the windings between a 240-volt connection and a 480-volt connection. Remember that the low-voltage windings are always connected in parallel, while the high-voltage windings are connected in series.

For instance, let's assume that the stator windings of a motor have an impedance of 48 ohms. If the stator windings are connected in parallel, the total impedance may be found as follows:

$$Rt = \frac{R1 \times R2}{R1 + R2}$$

$$Rt = \frac{48 \times 48}{48 + 48}$$

$$Rt = \frac{2304}{96}$$

$$Rt = 24 \text{ ohms}$$

Therefore, the total impedance of the motor winding connected in parallel is 24 ohms, and if 240 volts is applied to this connection, the following current will flow:

$$I = \frac{E}{R}$$

$$I = \frac{240}{24}$$

$$I = 10 \text{ amperes}$$

If the windings are connected in series for operation on 480 volts, the total impedance of the winding is:

Rt = R1 + R2

Rt = 48 + 48

Rt = 96 ohms

Consequently, if 480 volts are applied to this winding, the following current will flow:

$$I = \frac{E}{R}$$

$$I = \frac{480}{96}$$

I = 5 amperes

From the above, it is obvious that twice the voltage means half the current flow, or vice versa.

Special Connections

Some three-phase motors designed for operation on voltages higher than 600 volts may have more than 9 or 12 leads. Motors with 15 or 18 leads are common in high-voltage installations. A 15-lead motor has 3 coils per phase as shown in Figure 9-13. Notice that the leads are numbered in the same sequence as a 9-lead, wye-wound motor.

CALCULATING MOTOR CIRCUIT CONDUCTORS

The basic elements that must be accounted for in any motor circuit are shown in Figure 9-14. Although these elements are shown separately in this illustration, there are certain cases where the *NEC* permits a single device to serve more than one function. For example, in some cases, one switch can serve as both the disconnecting means and controller. In other cases, short-circuit protection and overload protection can be combined in a single circuit breaker or set of fuses.

NOTE: The basic *NEC* rule for sizing conductors supplying a single-speed motor used for continuous duty specifies that conductors must have a current-carrying capacity of not less than 125 percent of the motor full-load current rating. Conductors on the line side of the controller supplying multispeed motors must be based on the highest of the full-load current ratings shown on the motor nameplate.

Conductors between the controller and the motor must have a current-carrying rating based on the current rating for the speed of the motor each set of conductors is feeding.

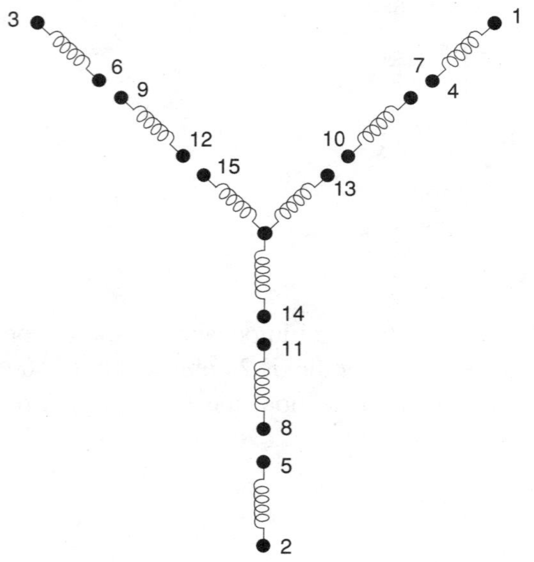
3
6
9
12
15
1
7
4
10
13
14
11
8
5
2

Figure 9-13: *Fifteen-lead motor connection diagram*

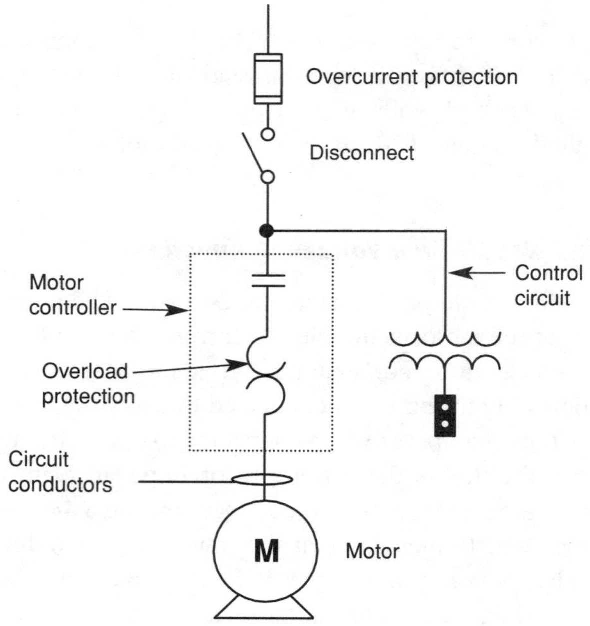
Overcurrent protection
Disconnect
Control circuit
Motor controller
Overload protection
Circuit conductors
M
Motor

Figure 9-14: *Basic elements of any motor circuit*

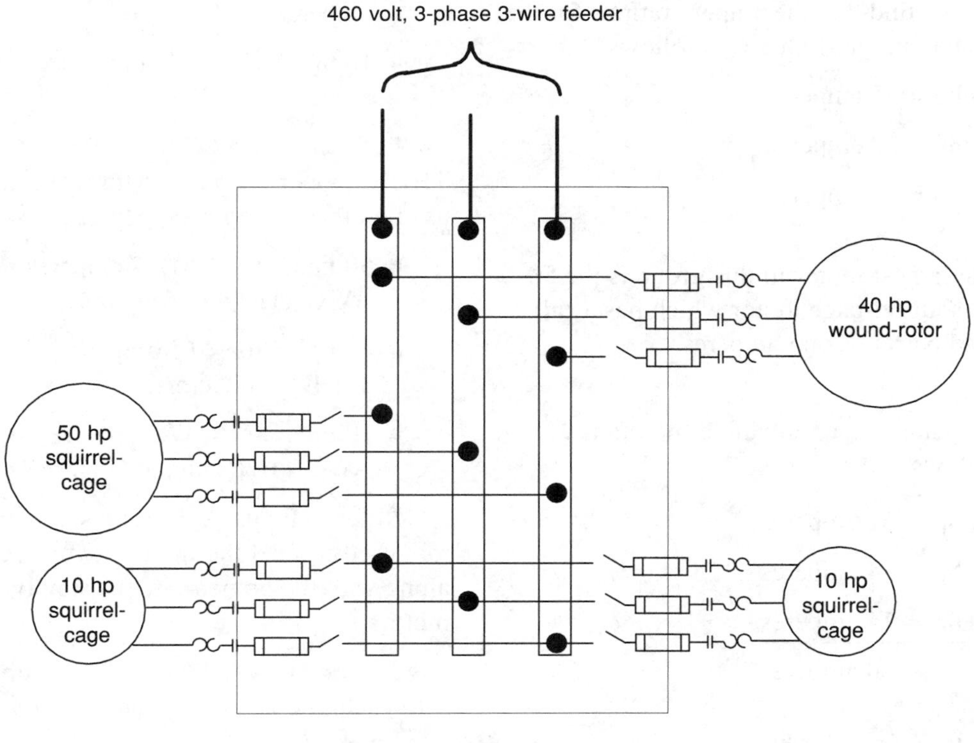
460 volt, 3-phase 3-wire feeder
40 hp
wound-rotor
50 hp
squirrel-
cage
10 hp
squirrel-
cage
10 hp
squirrel-
cage
Motor-control center

Figure 9-15: *Motor branch circuits from motor-control center*

A typical motor-control center and branch circuits feeding four different motors are shown in Figure 9-15. Let's see how the feeder and branch-circuit conductors are sized for these motors, using standard *NEC* installation requirements.

Step 1. Refer to NEC Table 430-150 for the full-load current of each motor.

Step 2. Determine the full-load current of the largest motor in the group.

Step 3. Calculate the sum of the full-load current ratings for the remaining motors in the group.

Step 4. Multiply the full-load current of the largest motor by 1.25 (125 percent) and then add the sum of the remaining motors to the answer.

Step 5. The combined total of Step 4 will give the *minimum* feeder size.

When sizing feeder conductors for motors, be aware that the procedure described in the five steps above will give the *minimum* conductor rating based on temperature rise only. Consequently, it is often necessary to increase the size of conductors to compensate for voltage drop and power loss in the circuit.

Now let's complete the conductor calculations for the motor circuits in Figure 9-15.

Step 1. Refer to *NEC* Table 430-150 in the code book. The motor horsepower is shown in the very left-hand column. Follow across the appropriate row until you come to the column titled "460V" — the voltage of the motor circuits in question. In doing so, we find that the ampere ratings for the motors in question are as follows:

50 hp = 65 amperes

40 hp = 52 amperes

10 hp = 14 amperes

Step 2. The largest motor in this group is the 50 hp squirrel-cage motor which has a full-load current of 65 amperes.

Step 3. The sum of the remaining motors is as follows:

40 hp = 52 amperes

10 hp = 14 amperes

10 hp = 14 amperes

80 amperes

Step 4. Multiplying the full-load current of the largest motor and then adding the total amperage of the remaining motors results in the following:

$$(1.25)(65) + 80 = 161.25 \text{ amperes}$$

Step 5. Therefore, the minimum feeder size for the 460V, 3-phase, 3-wire motor control center will be 161.25 amperes. Referring to NEC Table 310-16, under the column headed "90°C" the closest conductor size is 1/0 copper (rated at 170 amperes) or 3/0 aluminum (rated at 175 amperes).

The branch-circuit conductors feeding the individual motors are calculated somewhat differently. *NEC* Section 430-22(a) requires that the ampacity of branch-circuit conductors supplying a single continuous-duty motor must not be less than 125 percent of the motor's full-load current rating. Therefore, the current-carrying capacity of the branch-circuit conductors feeding the four motors in question are calculated as follows:

- 50 hp motor = 65 amperes x 1.25 = 81.25 amperes
- 40 hp motor = 52 amperes x 1.25 = 65.00 amperes
- 10 hp motor = 14 amperes x 1.25 = 17.5 amperes

Referring to *NEC* Table 310-16, the closest size THHN copper conductors that will be permitted to be used on these various branch circuits are as follows:

- 50 hp motor = 81.25 amperes requires No. 4 AWG THHN conductors.
- 40 hp motor = 65 amperes requires No. 6 AWG THHN conductors.
- 10 hp motor = 17.5 amperes requires No. 12 AWG THHN conductors.

Refer to Figure 9-16 for a summary of the conductors used to feed the motor-control center in question, along with the branch-circuits supplying the individual motors.

For motors with other voltages (up to 2300 volts) or for synchronous motors, refer to *NEC* Table 430-150.

Branch-circuit conductors serving motors used for short-time, intermittent, or other varying duty, must have an ampacity not less than the percentage of the motor nameplate current rating shown in *NEC* Table 430-22(b), *Exception*. However, to qualify as a short-time, intermittent motor, the nature of the apparatus that the motor drives must be arranged so that the motor cannot operate continuously with load under any condition of use. Otherwise, the motor must be considered continuous duty. Consequently, the majority of motors encountered in the electrical trade must be rated for continuous duty, and the branch-circuit conductors sized accordingly.

Wound-Rotor Motors

The primary full-load current of wound-rotor motors is listed in *NEC* Table 430-150 and is the same as squirrel-cage motors. Conductors connecting the secondary leads of wound-rotor induction motors to their controllers must have a current-carrying capacity at least equal to 125 percent of the motor's full-load secondary current if the motor is used for continuous duty. If the motor is used for less than continuous duty, the conductors must have a current-carrying capacity of

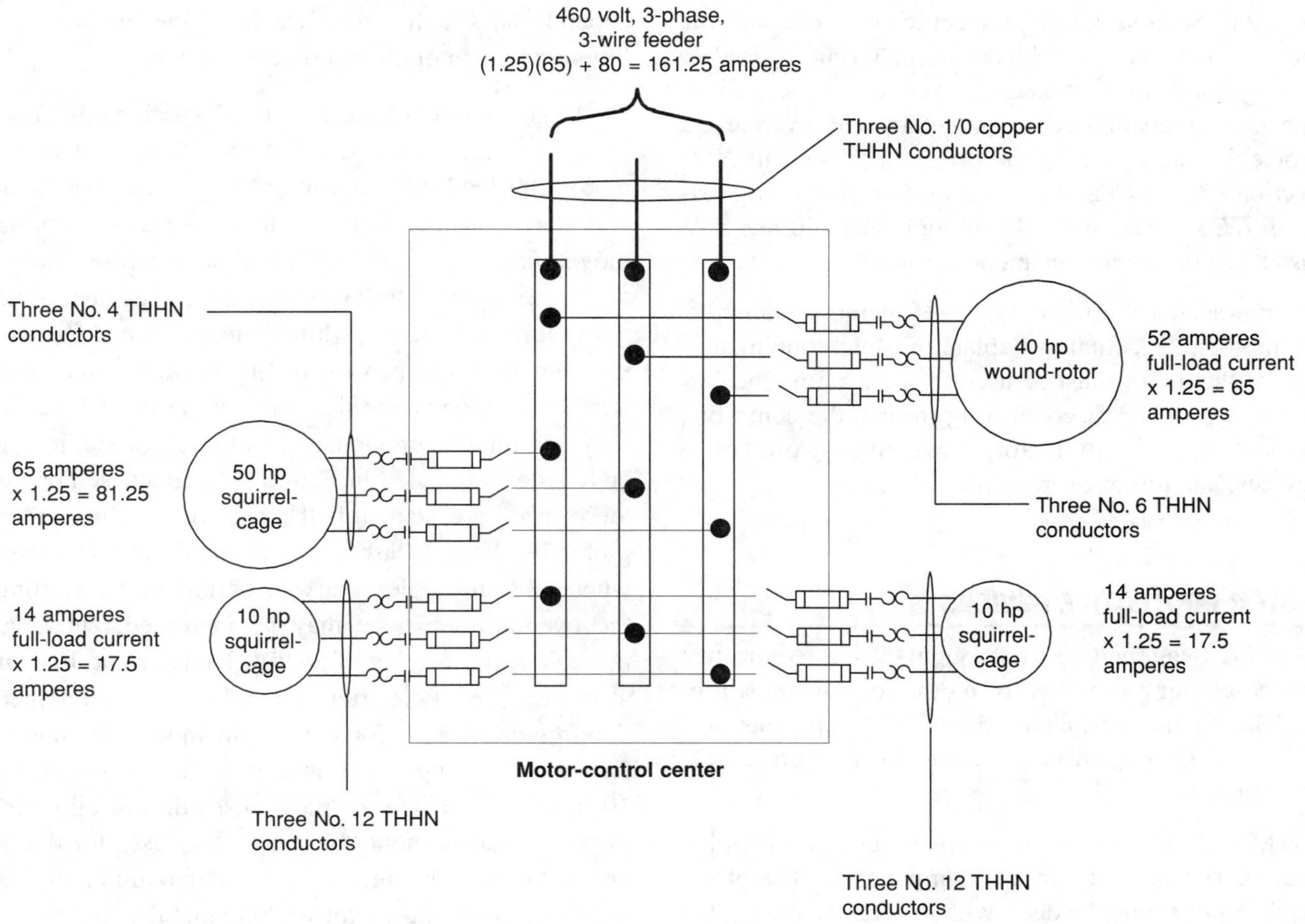
460 volt, 3-phase,
3-wire feeder
(1.25)(65) + 80 = 161.25 amperes
Three No. 1/0 copper
THHN conductors
Three No. 4 THHN
conductors
40 hp
wound-rotor
52 amperes
full-load current
x 1.25 = 65
amperes
65 amperes
x 1.25 = 81.25
amperes
50 hp
squirrel-
cage
Three No. 6 THHN
conductors
14 amperes
full-load current
x 1.25 = 17.5
amperes
10 hp
squirrel-
cage
10 hp
squirrel-
cage
14 amperes
full-load current
x 1.25 = 17.5
amperes
Motor-control center
Three No. 12 THHN
conductors
Three No. 12 THHN
conductors

Figure 9-16: *Sized branch-circuit and feeder conductors*

not less than the percentage of the full-load secondary nameplate current given in *NEC* Table 430-22(a). Conductors from the controller of a wound-rotor induction motor to its starting resistors must have an ampacity in accordance with *NEC* Table 430-23(c).

NOTE: *NEC* Section 430-6(a)(1) specifies that for general motor applications (excluding applications of torque motors and sealed hermetic-type refrigeration compressor motors), the values given in *NEC* Tables 430-147, 430-148, 430-149, and 430-150 should be used instead of the actual current rating marked on the motor nameplate when sizing conductors, switches, and overcurrent protection. Overload protection, however, is based on the marked motor nameplate.

Conductors For DC Motors

NEC Section 430-29 covers the rules governing the sizing of conductors from a dc motor controller to separate resistors for power accelerating and dynamic braking. This section, with its table of conductor ampacity percentages, assures proper application of dc constant-potential motor controls and power resistors. However, when selecting overload protection, the actual motor nameplate current rating must be used.

Conductors For Miscellaneous Motor Applications

Refer to *NEC* Section 430-6 for torque motors, shaded-pole motors, permanent split-capacitor motors and ac adjustable-voltage motors.

NEC Section 430-6(b) specifically states that the motor's nameplate full-load current rating is used to size ground-fault protection for a torque motor. However, branch-circuit conductors and overcurrent protection are sized by the provisions listed in *NEC* Section 430-52(b) and the full-load current rating listed in *NEC* Tables 430-147 through 430-150 are used instead of the motor's nameplate rating.

For sealed (hermetic-type) refrigeration compressor motors, the actual nameplate full-load running current of the motor must be used in determining the current rating of the disconnecting means, the controller, branch-circuit conductor, overcurrent-protective devices, and motor overload protection.

MOTOR PROTECTIVE DEVICES

NEC Sections 430-51 through 430-58 require that branch-circuit protection for motor controls must protect the circuit conductors, the control apparatus, and the motor itself against overcurrent due to short circuits or ground faults.

Motors and motor circuits have unique operating characteristics and circuit components. Therefore, these circuits must be dealt with differently from other types of loads. Generally, two levels of overcurrent protection are required for motor branch circuits:

- *Overload protection* — Motor running overload protection is intended to protect the system components and motor from damaging overload currents.
- *Short-circuit protection (includes ground-fault protection)* — Short-circuit protection is intended to protect the motor circuit components such as the conductors, switches, controllers, overload relays, motor, etc. against short-circuit currents or grounds. This level of protection is commonly referred to as motor branch-circuit protection applications. Dual-element fuses are designed to provide this protection provided they are sized correctly.

There are a variety of ways to protect a motor circuit — depending upon the user's objective. The ampere rating of a fuse selected for motor protection depends on whether the fuse is of the dual-element time-delay type or the nontime-delay type.

In general, *NEC* Table 430-152 specifies that nontime-delay fuses can be sized at 300 percent of the motor full-load current for ordinary motors with no code letter or code letters F to V so that the normal motor starting current does not affect the fuse. Motors with code letters B to E can be sized 250 percent of the motor full-load current, while motors with code letter A may be sized 150 percent of the motor full-load current. Consequently, the sizes of nontime-delay fuses for the four motors previously mentioned (assuming no code letter) are listed in Figure 9-17. Because none of these sizes are standard, the size of the fuses must preferably be decreased to a standard size. However, where absolutely necessary to permit motor starting, the overcurrent device may be increased, but never more than 400 percent of the full-load current. In actual practice, most electricians would use a 200-ampere nontime-delay fuse for the 50 hp motor; 50-ampere fuse for the 40 hp motor, and 40-ampere fuses for the 10 hp motors. If any of these fuses do not allow the motor to start without "blowing," the fuses for the 50 hp motor may be increased to a maximum of 260 amperes; the 40 hp motor to 208 amperes, and the 10 hp motors to 56 amperes.

Dual-element, time-delay fuses are able to withstand normal motor-starting current and can be sized closer to the actual motor rating than can nontime-delay fuses. If necessary for proper motor operation, dual (time-delay) fuses may be sized up to 175 percent of the motor's full-load current for all standard motors, with the exception of those with code letter A. Code A motors must not have fuses sized for more than 150 percent of the motor's full-load current rating. Where absolutely necessary for proper operation, the rating of dual-element (time-delay) fuses may be increased, but never more than 225 percent of the motor's full-load current rating. Sizing dual-element fuses at 175 percent for the four motors in Figure 9-17 will be:

50 hp motor: 65 amps x 175% = 113.75 amps

40 hp motor: 52 amps x 175% = 91 amps

10 hp motors: 14 amps x 175% = 24.5 amps

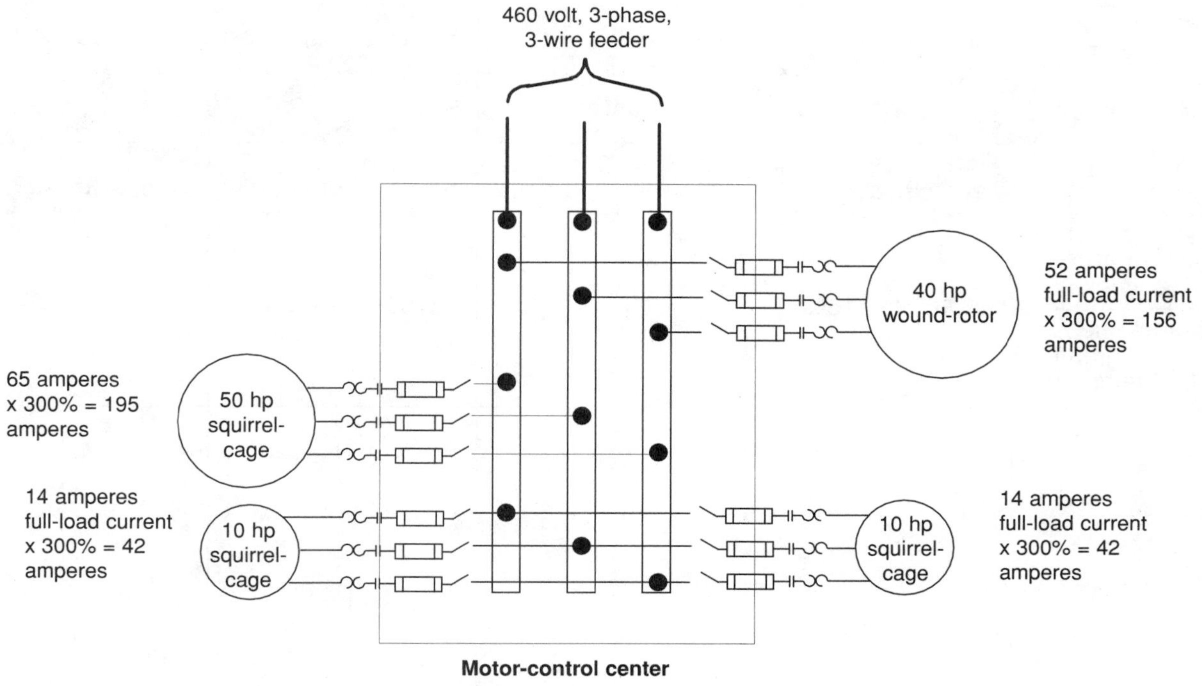
460 volt, 3-phase,
3-wire feeder
40 hp
wound-rotor
52 amperes
full-load current
x 300% = 156
amperes
65 amperes
x 300% = 195
amperes
50 hp
squirrel-
cage
14 amperes
full-load current
x 300% = 42
amperes
10 hp
squirrel-
cage
10 hp
squirrel-
cage
14 amperes
full-load current
x 300% = 42
amperes
Motor-control center

Figure 9-17: *Ratings of nontime-delay fuses for typical motor circuits*

The table in Figure 9-18 gives generalized fuse application guidelines for motor branch circuits. In using this table, bear in mind that in many cases the maximum fuse size depends on the type of motor code letter, motor type, and starting method.

Practical Application

Often, for various reasons, motors are oversized for applications. For instance, a 5 hp motor is installed when the load demand is only 3 hp. In these cases, a much higher degree of overload protection can be obtained by sizing the overload relay elements and/or Fusetron® and Low-Peak® dual-element time-delay fuses based on the actual full-load current draw. In existing installations, here's the procedure for providing the maximum overcurrent protection for oversized motors.

Step 1. With a clamp-on ammeter, determine running RMS current when the motor is at normal full-load as shown in Figure 9-19. (Be sure this current does not exceed nameplate current rating.) The advantage of this method is realized when a lightly loaded motor (especially those over 50 hp) experiences a single-phase condition. Even though the relays and fuses may be sized correctly based on the motor nameplate, circulating currents within the motor may cause damage.

If unable to meter the motor current, then take the current rating off the motor's nameplate.

Step 2. Size the overload relay elements and/or overcurrent protection based on this current. The table in Figure 9-20 may be used to assist in sizing dual-element fuses.

Step 3. Use a labeling system to mark the type and ampere rating of the fuse that should be in the fuse clips. This simple system makes it easy to run spot checks for proper fuse replacements.

Type of Motor	Dual-Element, Time Delay Fuses		Nontime-Delay Fuses	
	Desired Level of Protection			
	Motor Overload and Short-Circuit	Backup Overload and Short-Circuit	Short-Circuit Only	Short-Circuit Only
Service factor 1.15 or greater or 40°C temp. rise or less	100 to 115%	115% or next standard size	150% to 175%	300%
Service factor less than 1.14 or greater than 40°C temp. rise				
		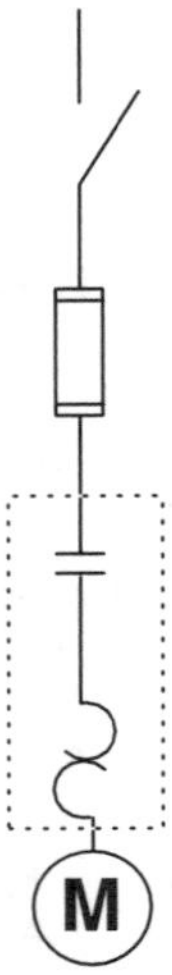	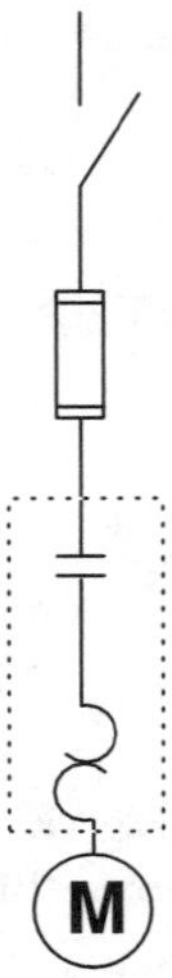	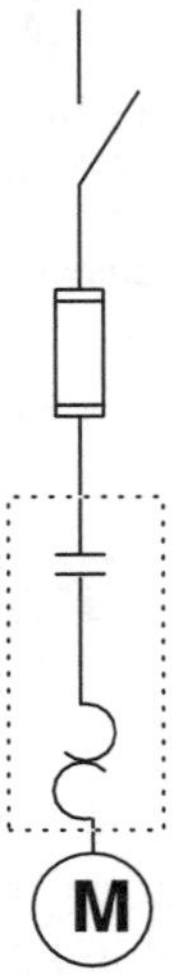
	Fuses give overload and short-circuit protection	Overload relay gives overload protection and fuses provide backup overload protection	Overload relay provides overload protection and fuses provide only short-circuit protection	Overload relay provides overload protection and fuses provide only short-circuit protection

Figure 9-18: *Sizing of fuses as a percentage of motor full-load current*

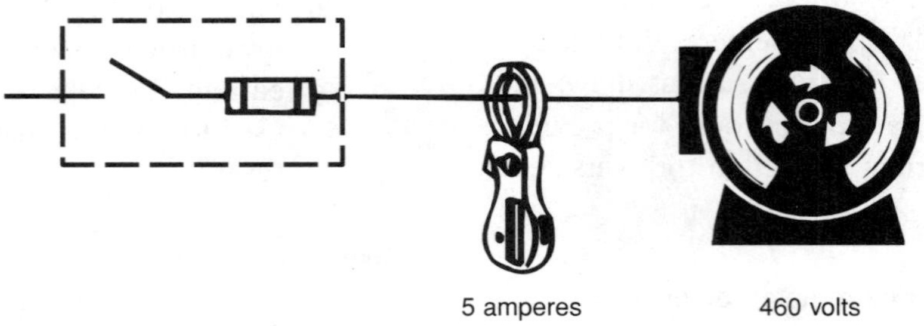
5 amperes
460 volts

Figure 9-19: *Determining running RMS current with an ammeter*

Dual-Element Fuse Size	Motor Protection (Used without properly sized overload relays). Motor Full-Load Amps		Back-up Motor Protection (Used with properly sized overload relays). Motor Full-Load Amps	
	Motor Service Factor of 1.15 or Greater or With Temp. Rise Not Over 40° C.	Motor Service Factor of Less Than 1.15 or With Temp. Rise Not Over 40° C.	Motor Service Factor of 1.15 or Greater or With Temp. Rise Not Over 40° C.	Motor Service Factor of Less Than 1.15 or With Temp. Rise Not Over 40° C.
1/10	0.08 - 0.09	0.09 - 0.10	0 - 0.08	0 - 0.09
1/8	0.10 - 0.11	0.11 - 0.125	0.09 - 0.10	0.10 - 0.11
5/100	0.12 - 0.15	0.14 - 0.15	0.11 - 0.12	0.12 - 0.13
2/10	0.16 - 0.19	0.18 - 0.20	0.13 - 0.16	0.14 - 0.17
1/4	0.20 - 0.23	0.22 - 0.25	0.17 - 0.20	0.18 - 0.22
3/10	0.24 - 0.30	0.27 - 0.30	0.21 - 0.24	0.23 - 0.26
4/10	0.32 - 0.39	0.35 - 0.40	0.25 - 0.32	0.27 - 0.35
1/2	0.40 - 0.47	0.44 - 0.50	0.33 - 0.40	0.36 - 0.43
6/10	0.48 - 0.60	0.53 - 0.60	0.41 - 0.48	0.44 - 0.52
8/10	0.64 - 0.79	0.70 - 0.80	0.49 - 0.64	0.53 - 0.70
1	0.80 - 0.89	0.87 - 0.97	0.65 - 0.80	0.71 - 0.87
1 1/8	0.90 - 0.99	0.98 - 1.08	0.81 - 0.90	0.88 - 0.98
1 1/4	1.00 - 1.11	1.09 - 1.21	0.91 - 1.00	0.99 - 1.09
1 4/10	1.12 - 1.19	1.22 - 1.30	1.01 - 1.12	1.10 - 1.22
1 1/2	1.20 - 1.27	1.31 - 1.39	1.13 - 1.20	1.23 - 1.30
1 6/10	1.28 - 1.43	1.40 - 1.56	1.21 - 1.28	1.31 - 1.39
1 8/10	1.44 - 1.59	1.57 - 1.73	1.29 - 1.44	1.40 - 1.57
2	1.60 - 1.79	1.74 - 1.95	1.45 - 1.60	1.58 - 1.74
2 1/4	1.80 - 1.99	1.96 - 2.17	1.61 - 1.80	1.75 - 1.96
2 1/2	2.00 - 2.23	2.18 - 2.43	1.81 - 2.00	1.97 - 2.17

Figure 9-20: *Selection of dual-element fuses for motor protection*

Dual-Element Fuse Size	Motor Protection (Used without properly sized overload relays). Motor Full-Load Amps		Back-up Motor Protection (Used with properly sized overload relays). Motor Full-Load Amps	
	Motor Service Factor of 1.15 or Greater or With Temp. Rise Not Over 40° C.	Motor Service Factor of Less Than 1.15 or With Temp. Rise Not Over 40° C.	Motor Service Factor of 1.15 or Greater or With Temp. Rise Not Over 40° C.	Motor Service Factor of Less Than 1.15 or With Temp. Rise Not Over 40° C.
2 6/10	2.24 -2.39	2.44 -2.60	2.01 -2.24	2.18 - 2.43
3	2.40 - 2.55	2.61 - 2.78	2.25 - 2.40	2.44 - 2.60
3 2/10	2.56 - 2.79	2.79 - 3.04	2.41 - 2.56	2.61 - 2.78
3 1/2	2.80 - 3.19	3.05 - 3.47	2.57 - 2.80	2.79 - 3.04
4	3.20 - 3.59	3.48 - 3.91	2.81 - 3.20	3.05 - 3.48
4 1/2	3.60 - 3.99	3.92 - 4.34	3.21 - 3.60	3.49 - 3.91
5	4.00 - 4.47	4.35 - 4.86	3.61 - 4.00	3.92 - 4.35
5 6/10	4.48 - 4.79	4.87 - 5.21	4.01 - 4.48	4.36 - 4.87
6	4.80 - 4.99	5.22 - 5.43	4.49 - 4.80	4.88 - 5.22
6 1/4	5.00 - 5.59	5.44 - 6.08	4.81 - 5.00	5.23 - 5.43
7	5.60 - 5.99	6.09 - 6.52	5.01 - 5.60	5.44 - 6.09
7 1/2	6.00 - 6.39	6.53 - 6.95	5.61 - 6.00	6.10 - 6.52
8	6.40 - 7.19	6.96 - 7.82	6.01 - 6.40	6.53 - 6.96
9	7.20 - 7.99	7.83 - 8.69	6.41 - 7.20	6.97 - 7.83
10	8.00 - 9.59	8.70 - 10.00	7.21 - 8.00	7.84 - 8.70
12	9.60 - 11.99	10.44 - 12.00	8.01 - 9.60	8.71 - 10.43
15	12.00 - 13.99	13.05 - 15.00	9.61 - 12.00	10.44 - 13.04
17 1/2	14.00 - 15.99	15.22 - 17.39	12.01 - 14.00	13.05 - 15.21
20	16.00 - 19.99	17.40 - 20.00	14.01 - 16.00	15.22 -17.39
25	20.00 - 23.99	21.74 - 25.00	16.01 - 20.00	17.40 - 21.74
30	24.00 - 27.99	26.09 - 30.00	20.01 - 24.00	21.75 - 26.09
35	28.00 - 31.99	30.44 - 34.78	24.01 - 28.00	26.10 - 30.43

Figure 9-20: *Selection of dual-element fuses for motor protection (continued)*

Dual-Element Fuse Size	Motor Protection (Used without properly sized overload relays). Motor Full-Load Amps		Back-up Motor Protection (Used with properly sized overload relays). Motor Full-Load Amps	
	Motor Service Factor of 1.15 or Greater or With Temp. Rise Not Over 40° C.	Motor Service Factor of Less Than 1.15 or With Temp. Rise Not Over 40° C.	Motor Service Factor of 1.15 or Greater or With Temp. Rise Not Over 40° C.	Motor Service Factor of Less Than 1.15 or With Temp. Rise Not Over 40° C.
40	32.00 - 35.99	34.79 - 39.12	28.01 - 32.00	30.44 - 37.78
45	36.00 - 39.99	39.13 - 43.47	32.01 - 36.00	37.79 - 39.13
50	40.00 - 47.99	43.48 - 50.00	36.01 - 40.00	39.14 - 43.48
60	48.00 - 55.99	52.17 - 60.00	40.01 - 48.00	43.49 - 52.17
70	56.00 - 59.99	60.87 - 65.21	48.01 - 56.00	52.18 - 60.87
75	60.00 - 63.99	65.22 - 69.56	56.01 - 60.00	60.88 - 65.22
80	64.00 - 71.99	69.57 - 78.25	60.01 - 64.00	65.23 - 69.57
90	72.00 - 79.99	78.26 - 86.95	64.01 - 72.00	69.58 - 78.26
100	80.00 - 87.99	86.96 - 95.64	72.01 - 80.00	78.27 - 86.96
110	88.00 - 99.99	95.65 - 108.69	80.01 - 88.00	86.97 - 95.65
125	100.00 - 119.99	108.70 - 125.00	88.01 - 100.00	95.66 - 108.70
150	120.00 - 139.99	131.30 - 150.00	100.01 - 120.00	108.71 - 130.43
175	140.00 - 159.99	152.17 - 173.90	120.01 - 140.00	130.44 - 152.17
200	160.00 - 179.99	173.91 - 195.64	140.01 - 160.00	152.18 - 173.91
225	180.00 - 199.99	195.65 - 217.38	160.01 - 180.00	173.92 - 195.62
250	200.00 - 239.99	217.39 - 250.00	180.01 - 200.00	195.63 - 217.39
300	240.00 - 279.99	260.87 - 300.00	200.01 - 240.00	217.40 - 260.87
350	280.00 - 319.99	304.35 - 347.82	240.01 - 280.00	260.88 - 304.35
400	320.00 - 359.99	347.83 - 391.29	280.01 - 320.00	304.36 - 347.83
450	360.00 - 399.99	391.30 - 434.77	320.01 - 360.00	347.84 - 391.30
500	400.00 - 479.99	434.78 - 500.00	360.01 - 400.00	391.31 - 434.78
600	480.00 - 600.00	521.74 - 600.00	400.01 - 480.00	434.79 - 521.74

Figure 9-20: *Selection of dual-element fuses for motor protection (continued)*

HINT! When installing the proper fuses in the switch to give the desired level of protection, it is often advisable to leave spare fuses on top of the disconnect, starter enclosure or in a cabinet adjacent to the motor-control center. Should the fuses open, the problem can be corrected and the proper size of fuses readily reinstalled.

Disconnect switches must have an ampere rating of at least 115 percent of the motor full-load ampere rating (*NEC* Section 430-110a). The next larger size switches with fuse reducers may sometimes be required.

Abnormal installations may require dual-element fuses of a larger size than shown in the table in Figure 9-20 providing only short-circuit protection. These applications include:

- Dual-element fuses in high ambient temperature environments.
- A motor started frequently or rapidly reversed.
- Motor is directly connected to a machine that cannot be brought up to full speed quickly. Some examples include centrifugal machines (extractors and pulverizers) and machines having large fly wheels (large punch presses), etc.
- Motor has a high Code letter with full-voltage start.

Motor Overload Protection

A high-quality electric motor, properly cooled and protected against overloads, can be expected to have a long life. The goal of proper motor protection is to prolong motor life and postpone the failure that ultimately takes place. Good electrical protection consists of providing both proper overload protection and current-limiting, short-circuit protection. AC motors and other types of high inrush loads require protective devices with special characteristics. Normal, full-load, running currents of motors are substantially less than the currents that result when motors start or are subjected to temporary mechanical overloads. This characteristic is illustrated by the typical motor-starting current curve shown in Figure 9-21.

At the moment an ac motor circuit is energized, the starting current rapidly rises to many times normal current and the rotor begins to rotate. As the rotor accelerates and reaches running speed, the current declines to the normal running current. Thus, for a period of time, the overcurrent protective devices in the motor circuit must be able to tolerate the rather substantial temporary

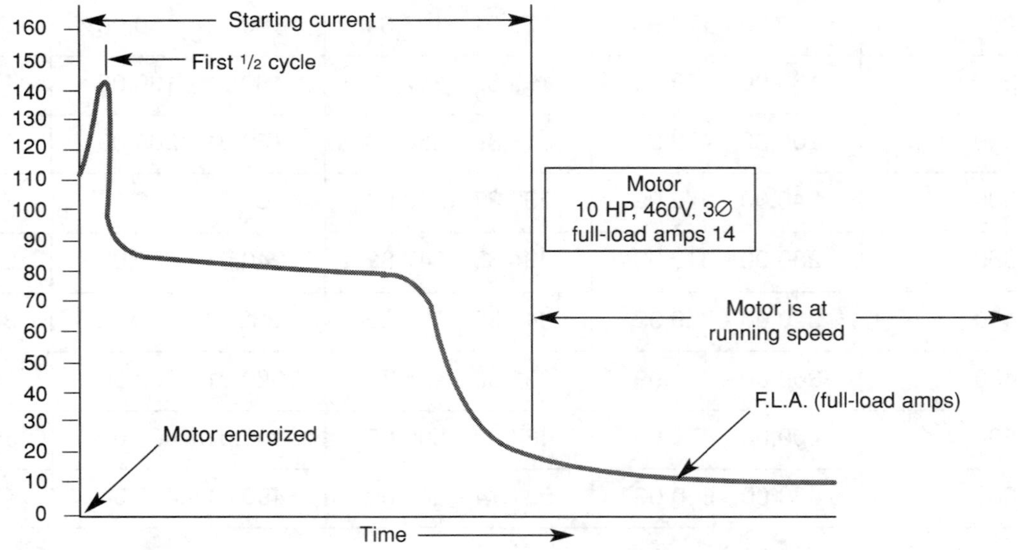
160
150
140
130
120
110
100
90
80
70
60
50
40
30
20
10
0
Starting current
First 1/2 cycle
Motor
10 HP, 460V, 3Ø
full-load amps 14
Motor is at
running speed
F.L.A. (full-load amps)
Motor energized
Time

Figure 9-21: *Motor starting current characteristics*

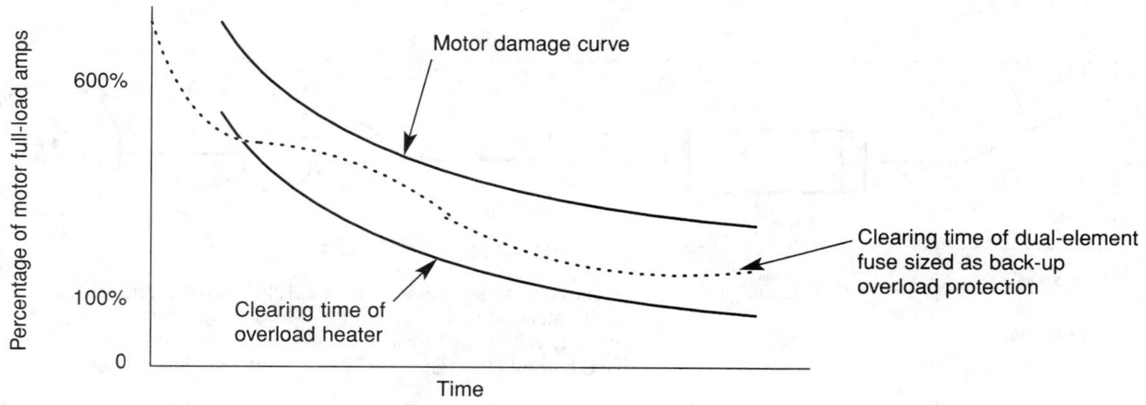
Motor damage curve
Percentage of motor full-load amps
600%
100%
0
Clearing time of dual-element fuse sized as back-up overload protection
Clearing time of overload heater
Time

Figure 9-22: *Clearing times of overload heaters, fuses, and motor damage curves*

overload. Motor starting currents can vary substantially depending on the motor type, load type, starting methods, and other factors. For the initial first 1/2 cycle, the momentary transient RMS current can be as high as 11 times or more. After this first half-cycle, the starting current subsides to 4 to 8 times (typically 6 times) the normal current for several seconds. This current is called the locked rotor current. When the motor reaches running speed, the current then subsides to its normal running level.

In summary, the special requirements for protection of motors require that the motor overload protective device withstand the temporary overload caused by motor starting currents, and, at the same time, protect the motor from continuous or damaging overloads. The main types of devices used to effectively provide overload protection include:

- Overload relays
- Fuses
- Circuit breakers

There are numerous causes of overloads, but if the overload protective devices are properly responsive, such overloads can be removed before damage occurs. To insure this protection, the motor running protective devices should have time-current characteristics similar to motor damage curves but should be slightly faster. This is illustrated in Figure 9-22.

Let's take a 10 hp motor, for example, and determine the proper circuit components that should be employed. Refer to Figure 9-23.

To begin, select the proper size overload relays. Typically, the overload relay is rated to trip at about 115 percent (average). The correct starter size (using NEMA standards) is a NEMA 1. The switch size that should be used is 30 amperes. Switch sizes are based on *NEC* requirements; dual-element, time-delay fuses allow the use of smaller switches.

On large motors with currents in excess of 600 amperes, LOW-PEAK® time-delay fuses are recommended. Most motors of this size will have reduced voltage starters and the inrush currents are not as rigorous. LOW-PEAK®, KRP-C® fuses should be sized at approximately 150 to 225 percent of the motor's full-load current.

Motor controllers with overload relays commonly used on motor circuits provide motor running overload protection. The overload relay setting or selection must comply with *NEC* Section 430-32. On overload conditions, the overload relays should operate to protect the motor. For motor back-up protection, size dual-element fuses at the next ampere rating greater than the overload relay trip setting. This can typically be achieved by sizing dual-element fuses at 125 percent for 1.15 service factor motors and 115 percent for 1.0 service factor motors.

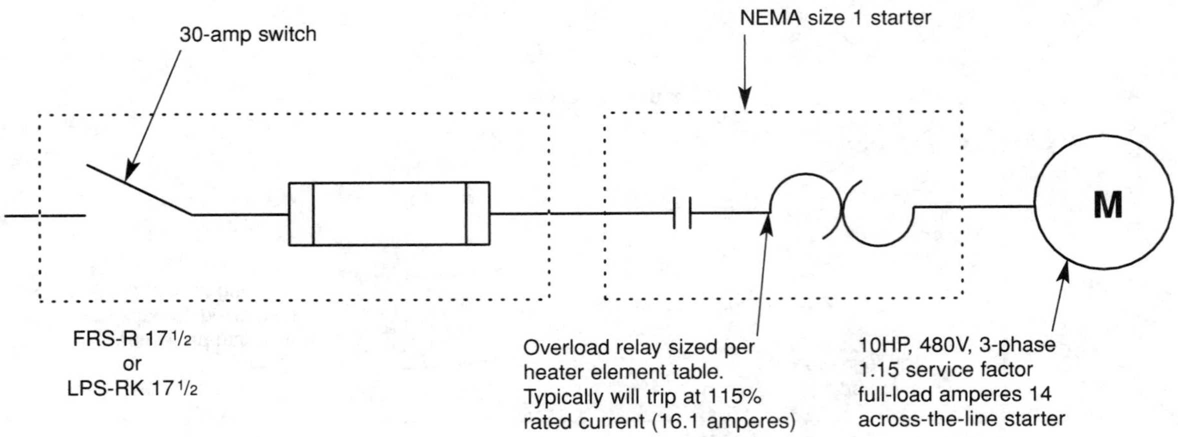
30-amp switch
NEMA size 1 starter
M
FRS-R 17 1/2
or
LPS-RK 17 1/2
Overload relay sized per
heater element table.
Typically will trip at 115%
rated current (16.1 amperes)
10HP, 480V, 3-phase
1.15 service factor
full-load amperes 14
across-the-line starter

Figure 9-23: *Circuit components of a typical 10 hp motor*

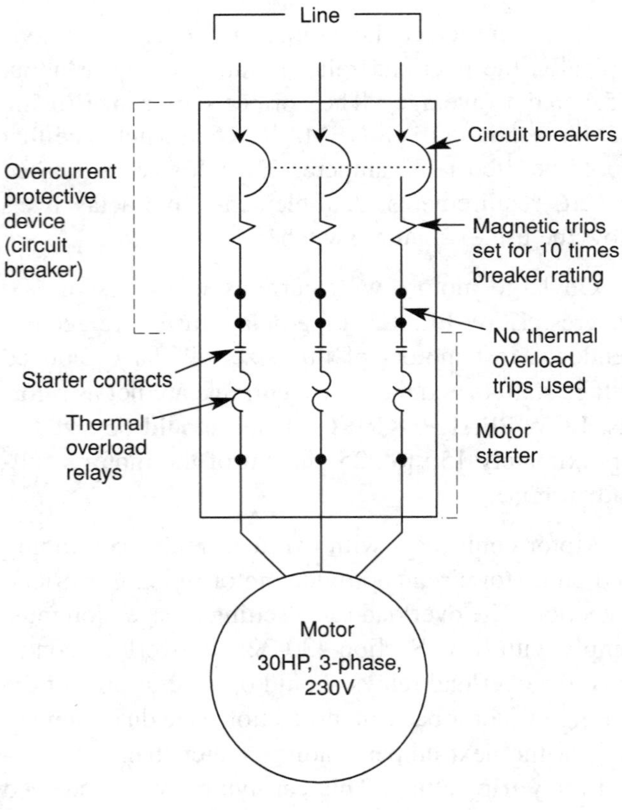
Line
Circuit breakers
Overcurrent
protective
device
(circuit
breaker)
Magnetic trips
set for 10 times
breaker rating
No thermal
overload
trips used
Starter contacts
Thermal
overload
relays
Motor
starter
Motor
30HP, 3-phase,
230V

Figure 9-24: *Listed combination starter arranged as per NEC*

CIRCUIT BREAKERS

The *NEC* recognizes the use of instantaneous-trip circuit breakers (without time delay) for short-circuit protection of motor branch circuits. Such breakers are acceptable only if they are adjustable and are used in combination motor starters. Such starters must have overload protection for each conductor and must be approved for the purpose. This permits the use of smaller circuit breakers than would be allowed if a standard thermal-magnetic circuit breaker was used. Smaller circuit breakers, in this case, offer faster operation for greater protection against grounds and short circuits. Figure 9-24 shows a schematic diagram of magnetic-only circuit breakers used in a combination motor starter.

The use of magnetic-only circuit breakers in motor branch circuits requires careful consideration due to the absence of overload protection up to the short-circuit trip rating that is normally available in thermal elements in circuit breakers. However, heaters in the motor starter protect the entire circuit and all equipment against overloads up to, and including, locked-rotor current. Heaters (thermal overload relays) are commonly set at 115 to 125 percent of the motor's full-load current.

In dealing with such circuits, an adjustable circuit breaker can be set to take over the interrupting task at currents above locked rotor and up to the short-circuit duty of the supply system at the point of the installation. The magnetic trip in such breakers typically can be adjusted from 3 to 13 times the breaker current rating. For example, a 100 ampere circuit breaker can be adjusted to trip anywhere between 300 and 1300 amperes. Consequently, the circuit breaker serves as motor short-circuit protection.

Application Of Magnetic-Only Circuit Breakers

Let's compare the use of both thermal-magnetic and magnetic-only circuit breakers to the motor circuit in Figure 9-24. In doing so, our job is to select a circuit breaker that will provide short-circuit protection and also qualify as the motor circuit disconnecting means.

Step 1. Determine the motor's full-load current from *NEC* Table 430-150. This is found to be 80 amperes.

Step 2. A circuit breaker suitable for use as a motor disconnecting means must have a current rating at least 115 percent of motor's full-load current. Thus,

1.15 x 80 = 92 amperes

NOTE: *NEC* Table 430-152 permits the use of an inverse-time circuit breaker rated not more than 250 percent of the motor's full-load current. However, a circuit breaker could be rated as high as 400 percent of the motor's full-load current if necessary to "hold" the motor-starting current without opening.

Step 3. From the above note, assuming that a circuit breaker rated at 250 percent of the motor's full-load current will be used, perform the following calculation:

2.5 x 80 amperes = 200 amperes

Step 4. Select a regular thermal-magnetic circuit breaker with a 225-ampere frame and set to trip at 200 amperes.

Step 5. Determine the initial starting current of the motor.

Step 6. Refer to *NEC* Table 430-7(b) and note that the highest starting (locked-rotor) current for this motor (letter M) is 11.19 kVA per horsepower; therefore,

$$\frac{30\text{ (horsepower)} \times 11.19\text{ (kVA)}}{230\text{V} \times 1.73} = 843.68\text{ amperes}$$

NOTE: When performing the above calculation, be sure to multiply the "11.19" factor by 1000 to convert kVA to VA.

This thermal-magnetic circuit breaker will provide protection for grounds and short-circuits without interfering with motor-overload protection. Note, however, that the instantaneous trip setting of a 200 ampere circuit breaker will be about 10 times the current rating, or:

200 x 10 = 2000 amperes

Now consider the use of a 100-ampere circuit breaker with thermal and adjustable magnetic trips. The instantaneous trip setting at 10 times the normal current rating would be:

100 x 10 = 1000 amperes

Although this "1000-amperes instantaneous trip setting" is above the 843-ampere locked-rotor current of the 30 hp motor in question, starting current would probably trip the thermal element and open the circuit breaker.

This problem can be solved by removing the circuit breaker's thermal element and leaving only the magnetic element in the circuit breaker. Then the conditions of overload can be cleared by the overload devices (heaters) in the motor starter. To determine the instantaneous trip rating of such a breaker, refer to *NEC* Section 430-52(c)(3) and note *Exception No. 1*:

> *Where the setting specified in NEC Table 430-152 is not sufficient for the starting current of the motor, the setting of an instantaneous trip circuit breaker shall be permitted to be increased but shall in no case exceed 1300 percent of the motor's full-load current.*

Therefore, since it has been determined that the 30 hp motor in question has a full-load ampere rating of 80 amperes, the maximum trip must not be set higher than:

80 x 13 = 1040 or 1000 amperes

This circuit breaker would qualify as the circuit disconnect because it has a rating higher than 115 percent of the motor's full-load current (92 amperes). However, the use of a magnetic-only circuit breaker does not protect against low-level grounds and short-circuits in the branch-circuit conductors on the line side of the motor starter overload relays. Such an application must be made only where the circuit breaker and motor starter are installed as a combination motor starter in a single enclosure.

Motor Short-Circuit Protectors

Motor short-circuit protectors (MSCP) are fuse-like devices designed for use only in their own type of fusible-switch combination motor starter. The combination offers short-circuit protection, overload protection, disconnecting means, and motor control — all with assured coordination between the short-circuit interrupter and the overload devices.

The *NEC* recognizes MSCPs in *NEC* Section 430-40 and 430-52, provided the combination is identified for the purpose. This means that a combination motor starter equipped with an MSCP and listed by UL or another nationally recognized third-party testing lab as a package, is called an MSCP starter.

MULTIMOTOR BRANCH CIRCUITS

NEC Sections 430-53(a) and 430-53(b) permit the use of more than one motor on a branch circuit provided the following conditions are met:

> *Two or more motors, each rated not more than 1 hp, and each drawing a full-load current not exceeding 6 amperes, may be used on a branch circuit protected at not more than 20 amperes at 125 volts or less, or 15 amperes at 600 volts or less. The rating of the branch circuit protective device marked on any of the controllers must not be exceeded. Individual overload protection is necessary in such circuits unless the motor is not permanently installed, or is manually started and is within sight from the controller location, or has sufficient winding impedance to prevent overheating due to locked-rotor current, or is part of an approved assembly which does not subject the motor to overloads and which incorporates protection for the motor against locked-rotor, or the motor cannot operate continuously under load.*

Two or more motors of any rating, each having individual overload protection, may be connected to a single branch circuit that is protected by a short-circuit protective device (MSCP). The protective device must be selected in accordance with the maximum rating or setting that could protect an individual circuit to the motor of the smallest rating. This may be done only where it can be determined that the branch-circuit device so selected will not open under the most severe normal conditions of service that might be encountered. The permission of this *NEC* section offers wide application of more than one motor on a single circuit, particularly in the use of small integral-horsepower motors installed on 208-volt, 240-volt, and 480-volt, three-phase industrial and commercial systems. Only such three-phase motors have full-load operating currents low enough to permit more than one motor on circuits fed from 15-ampere protective devices.

Using these *NEC* rules, let's take a typical branch circuit (Figure 9-25) with more than one motor connected and see how the calculations are made.

Step 1. The full-load current of each motor is taken from *NEC* Table 430-150 as required by *NEC* Section 430-6(a).

Step 2. A circuit breaker must be chosen that does not exceed the maximum value of short-circuit protection (250 percent) required by *NEC* Section 430-52 and *NEC* Table 430-152 for the smallest motor in the group. In this case: 1.5 hp. Since the listed full-load current for the smallest motor (1.5 hp) is 2.6 amperes, the calculation is made as follows:

2.6 amperes x 2.5 (250%) = 6.5A

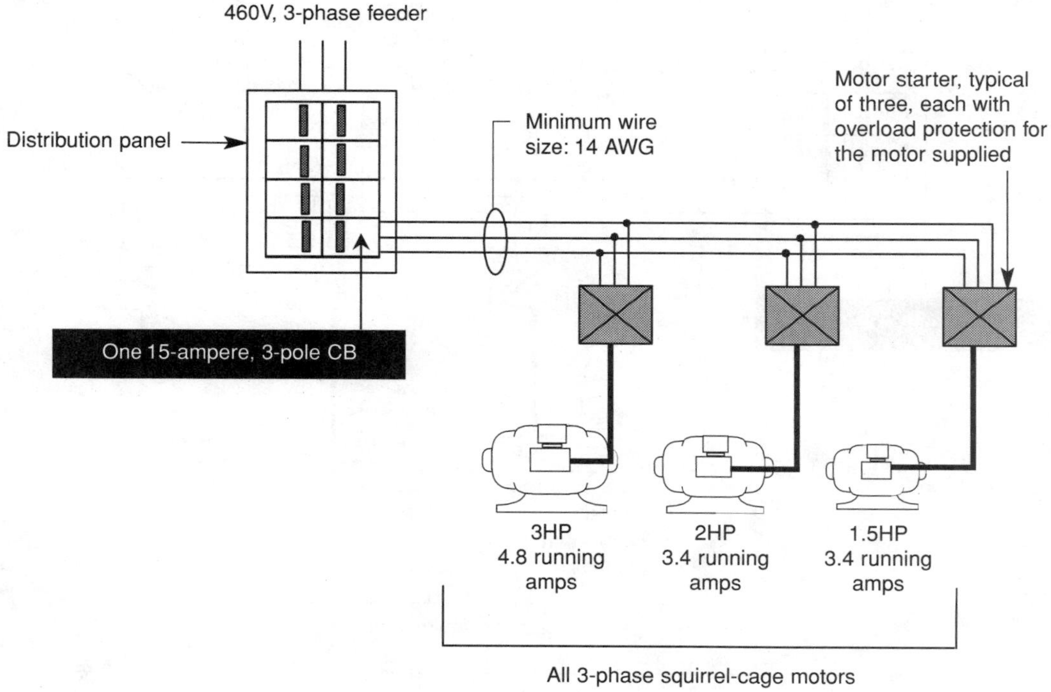
460V, 3-phase feeder
Distribution panel
Minimum wire size: 14 AWG
Motor starter, typical of three, each with overload protection for the motor supplied
One 15-ampere, 3-pole CB
3HP
4.8 running amps
2HP
3.4 running amps
1.5HP
3.4 running amps
All 3-phase squirrel-cage motors

Figure 9-25: *Several motors supplied by one branch circuit*

NOTE: NEC Section 430-52(c), *Exception No. 1*, allows the next higher size, rating or setting for a standard circuit breaker. Since a 15-ampere circuit breaker is the smallest standard rating recognized by *NEC* Section 240-6, a 15-ampere, 3-pole circuit breaker may be used.

Step 3. The total load of the motor currents must be calculated as follows:

4.8 + 3.4 + 2.6 = 10.8 amperes

The total full-load current for the three motors (10.8 amperes) is well within the 15-ampere circuit breaker rating, which has sufficient time delay in its operation to permit starting of any one of these motors with the other two already operating. Torque characteristics of the loads on starting are not high. Therefore, the circuit breaker will not open under the most severe normal service.

Step 4. Make certain that each motor is provided with the properly-rated individual overload protection in the motor starter.

Step 5. Branch-circuit conductors are sized in accordance with *NEC* Section 430-24. In this case:

4.8 + 3.4 + 2.6 + (25% the largest motor — 4.8 amperes) = 12 amperes

No. 14 AWG conductors rated at 75°C will fully satisfy this application.

Another multimotor situation is shown in Figure 9-26 on the next page. In this case, smaller motors are used. In general, *NEC* Section 430-53(b) requires branch-circuit protection to be no greater than the maximum amperes permitted by *NEC* Section 430-52 for the lowest rated motor of the group, which, in our case, is 1.1 amperes for the 0.5 hp motor. With this information in mind, let's size the circuit components for this application.

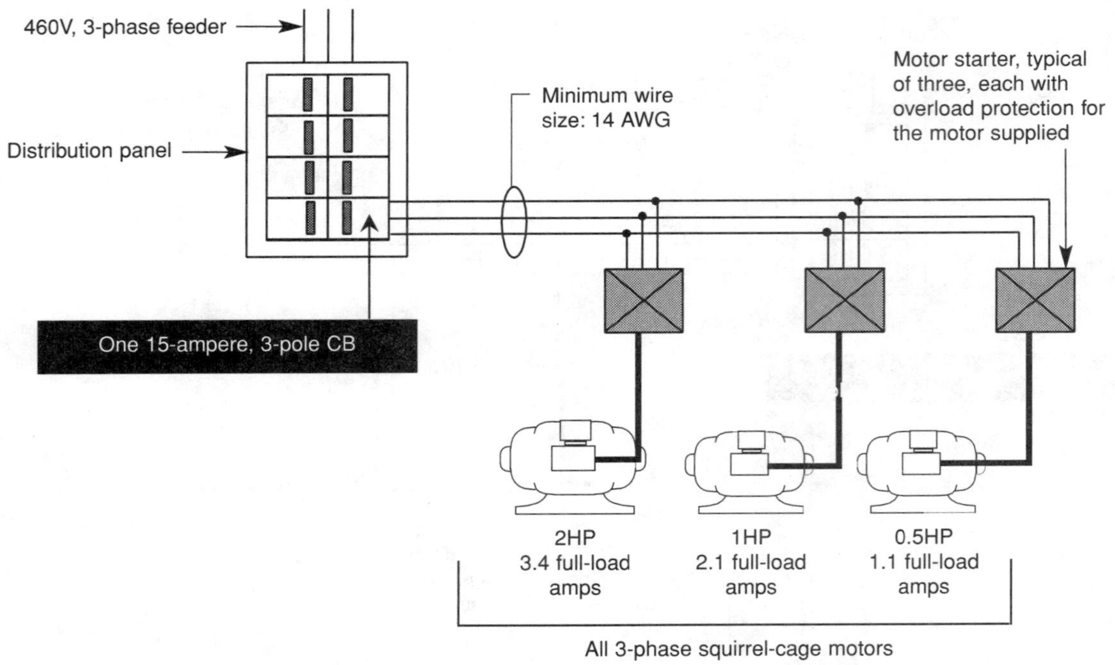
460V, 3-phase feeder
Distribution panel
One 15-ampere, 3-pole CB
Minimum wire
size: 14 AWG
Motor starter, typical
of three, each with
overload protection for
the motor supplied
2HP
3.4 full-load
amps
1HP
2.1 full-load
amps
0.5HP
1.1 full-load
amps
All 3-phase squirrel-cage motors

Figure 9-26: *Several smaller motors supplied by one branch circuit*

Step 1. From *NEC* Section 430-52 and *NEC* Table 430-152, the maximum protection rating for a circuit breaker is 250 percent of the lowest rated motor. Since this rating is 1.1 amperes, the calculation is performed as follows:

2.5 x 1.1 = 2.75 amperes

NOTE: Since 2.75 amperes is not a standard rating for a circuit breaker, according to *NEC* Section 240-6, *NEC* Section 430-52 (*Exception 1*) permits the use of the next higher rating.

Step 2. Because 15 amperes is the lowest standard rating of circuit breakers, it is the next higher device rating above 2.5 amperes and satisfies *NEC* rules governing the rating of the branch-circuit protection.

These two previous applications permit the use of several motors up to the circuit capacity, based on *NEC* Sections 430-24 and 430-53(b) and on starting torque characteristics, operating duty cycles of the motors and their loads, and the time-delay of the circuit breaker. Such applications greatly reduce the number of circuit breakers, number of panels and the amount of wire used in the total system. One limitation, however, is placed on this practice in *NEC* Section 430-52(c)(2):

- Where maximum branch-circuit short-circuit and ground-fault protective device ratings are shown in the manufacturer's overload relay table for use with a motor controller or are otherwise marked on the equipment, they shall not be exceeded even if higher values are allowed as shown in the preceding examples.

POWER-FACTOR CORRECTION AT MOTOR TERMINALS

Generally, the most effective method of power-factor correction is the installation of capacitors at the source of poor power factor — the induction motor. This not only increases power factor, but also releases system capacity, improves voltage stability and reduces power losses. See Chapter 11 for a thorough explanation of power factor.

When power factor correction capacitors are used, the total corrective Kvar on the load side of the motor controller should not exceed the value required to raise the no-load power factor to unity. Corrective Kvar in excess of this value may cause over excitation that results in high transient voltages, currents and torques that can increase safety hazards to personnel and possibly damage the motor or driven equipment.

Do not connect power factor correction capacitors at motor terminals on elevator motors, multispeed motors, plugging or jogging applications or open transition, wye-delta, autotransformer starting and some part-winding start motors.

If possible, capacitors should be located at position No. 2 in Figure 9-27. Placing the capacitor in this position does not change the current flowing through motor overload protectors.

Connection of capacitors at position No. 3 requires a change of overload protectors. Capacitors should be located at position No. 1 for the following:

- Elevator motors
- Multispeed motors
- Plugging or jogging applications

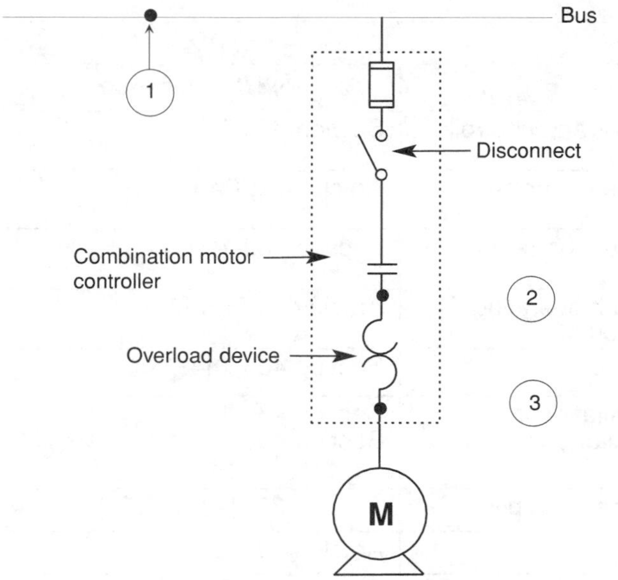
Bus
1
Disconnect
Combination motor controller
2
Overload device
3
M

Figure 9-27: *Placement of capacitors in motor circuits*

- Open transition, wye-delta, autotransformer starting
- Some part-winding motors

The table in Figure 9-28 allows the determination of corrective Kvar required where capacitors are individually connected at motor leads. These values should be considered the maximum capacitor rating when the motor and capacitor are switched as a unit. The figures given are for 3-phase, 60 Hz, NEMA Class B motors to raise full-load power factor to 95 percent. For other types of motors, obtain catalogues and charts from manufacturers of capacitors. These charts are usually available at local electrical suppliers.

Summary of Electric Motors

The *NEC* plays an important role in the selection and application of motors — including branch-circuit conductors, disconnects, controller, overcurrent protection, and overload protection. For example, *NEC* Article 430 covers application and installation of motor circuits and motor-control connections — including conductors, short-circuit and ground-fault protection, controllers, disconnects, and overload protection.

NEC Article 440 contains provisions for motor-driven air conditioning and refrigerating equipment — including the branch circuits and controllers for the equipment. It also takes into account the special considerations involved with sealed (hermetic-type) motor compressors, in which the motor operates under the cooling effect of the refrigeration. In referring to *NEC* Article 440, be aware that the rules in this *NEC* Article are *in addition to*, or *are amendments to*, the rules given in *NEC* Article 430. See Figure 9-29.

MOTOR CONTROLLERS

Starting and stopping or otherwise regulating motors is the function of motor controllers and drives. There are as many different types of controllers (motor starters) and drives as there are types of motors. However, the most common types are those used to control squirrel-cage induction motors and dc motors. The following are those that will be encountered by most electricians:

- Manual full-voltage motor starters

Nominal Motor Speed in RPM												
	3600		1800		1200		900		720		600	
Induction Motor HP Rating	Capacitor Rating KVAR	Line Current Reduction %	Capacitor Rating KVAR	Line Current Reduction %	Capacitor Rating KVAR	Line Current Reduction %	Capacitor Rating KVAR	Line Current Reduction %	Capacitor Rating KVAR	Line Current Reduction %	Capacitor Rating KVAR	Line Current Reduction %
3	1.5	14	1.5	15	1.5	20	2	27	2.5	35	3.5	41
5	2	12	2	13	2	17	3	25	4	32	4.5	37
$7^1/_2$	2.5	11	2.5	12	3	15	4	22	5.5	30	6	34
10	3	10	3	11	3.5	14	5	21	6.5	27	7.5	31
15	4	9	4	10	5	13	6.5	18	8	23	9.5	27
20	5	9	5	10	6.5	12	7.5	16	9	21	12	25
25	6	9	6	10	7.5	11	9	15	11	20	14	23
30	7	8	7	9	9	11	10	14	12	18	16	22
40	9	8	9	9	11	10	12	13	15	16	20	20
50	12	8	11	9	13	10	15	12	19	15	24	19
60	14	8	14	8	15	10	18	11	22	15	27	19
75	17	8	16	8	18	10	21	10	26	14	32.5	18
100	22	8	21	8	25	9	27	10	32.5	13	40	17
125	27	8	26	8	30	9	32.5	10	40	13	47.5	16
150	32.5	8	30	8	35	9	37.5	10	47.5	12	52.5	15
200	40	8	37.5	8	42.5	9	47.5	10	60	12	65	14
250	50	8	45	7	52.5	8	57.5	9	70	11	77.5	13

Figure 9-28: *Motor power factor correction table*

MOTORS		
Application	Requirements	*NEC* Reference
Disconnecting means	Required to disconnect motor and controller from power supply.	Section 430-74
	Must be within sight from the controller location.	Article 430, Part J
	If safety switch is used, it must be rated in horsepower.	Section 430-109
Grounding, stationary motors	Must have frames grounded if operating at voltages over 150 V to ground.	Article 430 Part M
Grounding, motor controllers	Enclosures must be grounded.	Section 430-144
Location	Must be located so that ventilation is provided and maintenance can be readily accomplished.	Section 430-14 Section 430-16
Over 600 V	Special installation requirements apply.	Article 430 Part K
Protection from live parts	Exposed live parts must be guarded.	Article 430 Part L

Figure 9-29: *Summary of NEC installation requirements for electric motors*

- Drum switches
- AC magnetic starters
- AC magnetic reversing starters
- AC combination starters
- AC reduced-voltage starters
- Part-winding starters
- Wye-delta magnetic starters
- Magnetic primary resistor starters
- Electronic solid-state devices
- Miscellaneous control devices such as limit switches, solenoids, etc.

It is the duty of electricians to install, maintain, troubleshoot, and repair such starters to keep the installation in service. Consequently, a thorough knowledge of motors and motor controls is essential for anyone involved in the electrical industry in any capacity.

Solid-state devices have become increasingly popular over the past decade for use in motor controllers — mainly because solid-state components have no moving parts, they are resistant to shock and vibration, and are sealed against dirt and moisture. The greatest advantage, however, is that the control voltage of many solid-state control devices is isolated from the line voltage and also from the motor that it is intended to control. See Figure 9-30. Consequently, all personnel associated with the electrical industry in any capacity should have a good working knowledge of solid-state motor controllers, their individual components, the circuit connections, and the troubleshooting methods used when problems arise.

CONTROLLER COMPONENTS AND THEIR RELATED GRAPHIC SYMBOLS

The electrical symbols, shown in Figure 9-31 on the next page, are those which will most likely be encountered in motor-control schematic and wiring diagrams. All electricians should be thoroughly familiar with these symbols and the function of the components in which they represent. The purpose of this section is to acquaint the reader with the motor-control circuit symbols in all types of motor-control applications. Furthermore, device designations (usually in the form of letters or abbreviations) on electrical drawings help to identify functions and characteristics of motor-starter components. See Figure 9-32.

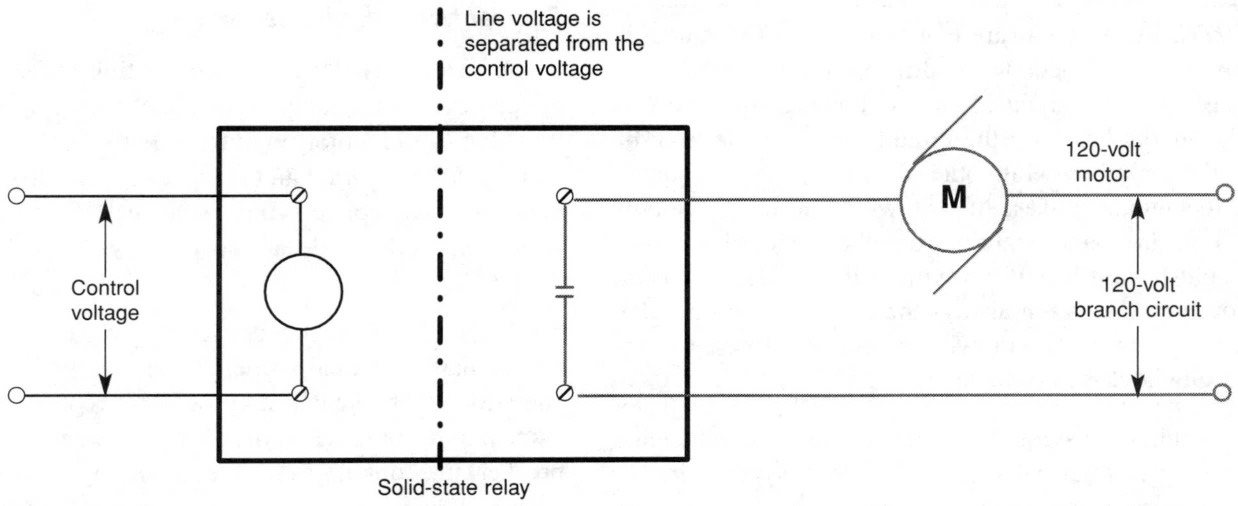
Line voltage is separated from the control voltage
120-volt motor
M
Control voltage
120-volt branch circuit
Solid-state relay

Figure 9-30: *The control input voltage is separated from the higher line voltage on this solid-state control*

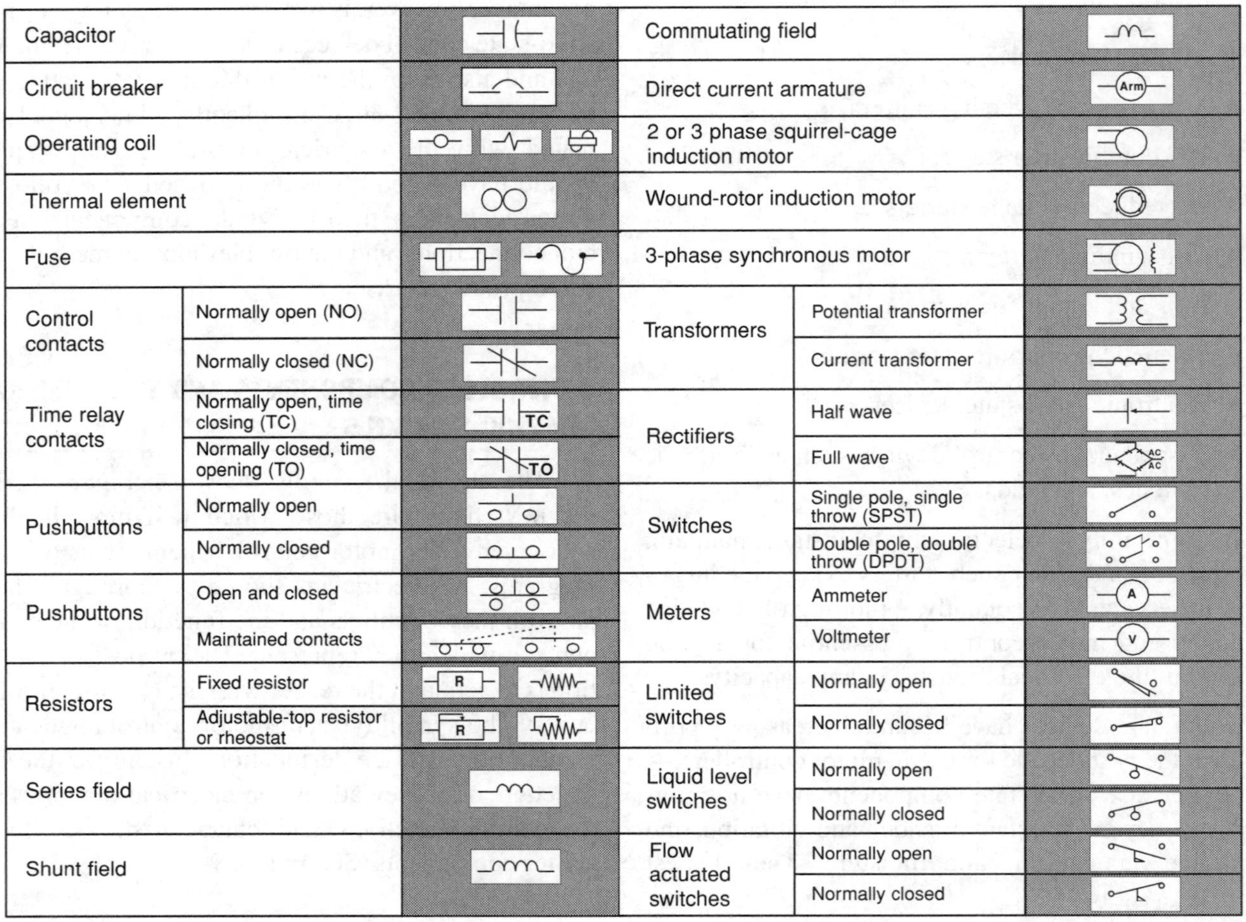

Capacitor
Circuit breaker
Operating coil
Thermal element
Fuse
Control contacts
Normally open (NO)
Normally closed (NC)
Time relay contacts
Normally open, time closing (TC)
Normally closed, time opening (TO)
TC
TO
Pushbuttons
Normally open
Normally closed
Pushbuttons
Open and closed
Maintained contacts
Resistors
Fixed resistor
Adjustable-top resistor or rheostat
R
Series field
Shunt field
Commutating field
Direct current armature
Arm
2 or 3 phase squirrel-cage induction motor
Wound-rotor induction motor
3-phase synchronous motor
Transformers
Potential transformer
Current transformer
Rectifiers
Half wave
Full wave
AC
Switches
Single pole, single throw (SPST)
Double pole, double throw (DPDT)
Meters
Ammeter
A
Voltmeter
V
Limited switches
Normally open
Normally closed
Liquid level switches
Normally open
Normally closed
Flow actuated switches
Normally open
Normally closed

Figure 9-31: *Symbols used on motor-control schematic drawings*

NOTE: Please be aware that any group of "standard" symbols is subject to modification. For example, a relay may be designated on one drawing with two vertical marks breaking the circuit line; a circle may be used on another, while other drawings may use square or rectangular boxes. Any drawings utilizing symbols that deviate significantly from the standard symbols should have a legend or symbol list to identify them. However, this is not always the case; so examine each new style of control drawing carefully before performing any hook-ups or maintenance.

Additional symbols, as they are used on schematic drawings, are usually identified in a symbol list or legend appearing on the drawing itself, the manufacturers shop drawings, or else in the project's written specifications

Special NEC Requirements

A switch used as a disconnecting means for a motor circuit must have a horsepower rating not less than that of the motor it controls. For example, a 5-hp motor must be started and stopped by a control device that has a nameplate rating of at least 5 hp. Certain exceptions to this rule are covered by Section 430 of the *NEC*.

If a magnetic switch is used as a motor controller, it must have a manually-operated disconnecting switch ahead of it. The switch may be a conventional safety switch, or in some cases, the overcurrent device (circuit breaker) may qualify as the disconnecting means. If the switch is more than 50 feet away from the motor, another switch must be provided in the circuit and within sight of the machine as shown in Figure 9-33.

Accelerating contact(s)	**A**	Forward	**F**	Potential transformer	**PT**
Armature shunt	**AS**	Full field	**FF**	Power factor	**PF**
Auxiliary devices	**X, Y, Z**	Fuse	**FU**	Pressure switch	**PS**
Brake contact(s)	**B**	Jog (inch)	**J**	Pushbutton	**PB**
Capacitor	**C, CAP**	Limit switch	**LS**	Rectifier	**REC**
Circuit breaker	**CB**	Line contactor	**M**	Resistor	**RES**
Closing coil	**CC**	Line switch	**ISW**	Start	**S**
Control switch	**CSW**	Master switch	**MS**	Switch	**SW**
Current transformer	**CT**	Neutral	**N**	Time closing	**TC**
Emergency stop	**ES**	Overcurrent	**OC**	Time opening	**TO**
Field contact(s)	**FC**	Overload	**OL**	Time relay	**TR**
Field accelerating	**FA**	Overspeed	**OS**	Time switch	**TS**
Field decelerating	**FD**	Plugging	**P**	Trip coil	**TC**
Field loss (failure)	**FL**	Plugging forward	**PF**	Undervoltage	**UV**
Field protective	**FP**	Plugging reverse	**PR**	Unloader coil	**UC**
Field weakening	**FW**	Potential interlocking	**PI**	Voltage	**V**

Figure 9-32: *Device designations are used on schematic diagrams*

Switches used as disconnecting means for motor circuits should be rated in horsepower for all motors in excess of 2 hp. A switch should carry a rating of at least 125 percent of the full-load nameplate current rating of the motor and be manually operable in a readily accessible location. It must indicate whether it is in the open (off) or closed (on) position. When closed, the switch must disconnect both the controller and the motor from all ungrounded supply conductors.

Protective devices, such as overload relays, low-voltage protection devices, and low-voltage release devices, are an important part of a motor controller. An overload relay will open the contactors in motor circuits when current is too high; a low-voltage protective device will prevent the motor from starting as long as the full-rated voltage is not available, and manual restarting of the motor is necessary after the low-voltage protective device has operated; a low-voltage release device will disconnect the motor during a voltage dip, but the motor will start automatically when the normal voltage returns.

Controllers also contain braking arrangements, accelerators, and reversing switches which reverse the rotation of the motor.

Disconnect
50' maximum
START
Equipment

Figure 9-33: *Disconnect switches for motors must be located within sight of the motor and not more than 50 feet away*

Chapter 10
Transformers

It is important for anyone working with electricity to become familiar with transformer operation; that is, how they work, how they are connected into circuits, their practical applications and precautions to take during the installation or while working on them.

The electric power produced by alternators in a generating station is transmitted to locations where it is utilized and distributed to users. Many different types of transformers play an important role in the distribution of electricity. Power transformers are located at generating stations to step up the voltage for more economical transmission. Substations with additional power transformers and distribution equipment are installed along the transmission line. Finally, distribution transformers are used to step down the voltage to a level suitable for utilization.

Transformers are also used quite extensively in all types of control work, to raise and lower ac voltage on control circuits. For example, in 480Y/277-volt electrical systems, transformers are used to reduce the voltage for operating 208Y/120-volt lighting and other electrically-operated equipment. Buck-and-boost transformers are used for maintaining appropriate voltage levels in certain electrical systems in all types of commercial applications.

This chapter is designed to cover these items as well as *National Electrical Code* (*NEC*) installation requirements and transformer overcurrent protection and grounding. Specialty transformers suitable for use in all types of wiring systems are also covered.

TRANSFORMER BASICS

A very basic transformer consists of two coils or windings formed on a single magnetic core as shown in Figure 10-1. The source of alternating current and voltage is connected to the primary winding of the transformers; the load is connected to the secondary winding. Such an arrangement will allow transforming a large alternating current at low voltage into a small alternating current at high voltage, or vice versa. But let's start at the beginning. What makes a transformer work?

Mutual Inductance

The term *mutual induction* refers to the condition in which two circuits are sharing the energy of one of the circuits. It means that energy is being transferred from one circuit to another.

Consider the diagram in Figure 10-2. Coil A is the primary circuit which obtains energy from the battery. When the switch is closed, the current starts to flow and a magnetic field expands out of coil A. Coil A then changes electrical energy of the battery into the magnetic energy of a magnetic field. When the field of coil A is expanding, it cuts across coil B, the secondary circuit, inducing a voltage in coil B. The indicator (a galvanometer) in the secondary circuit is deflected, and

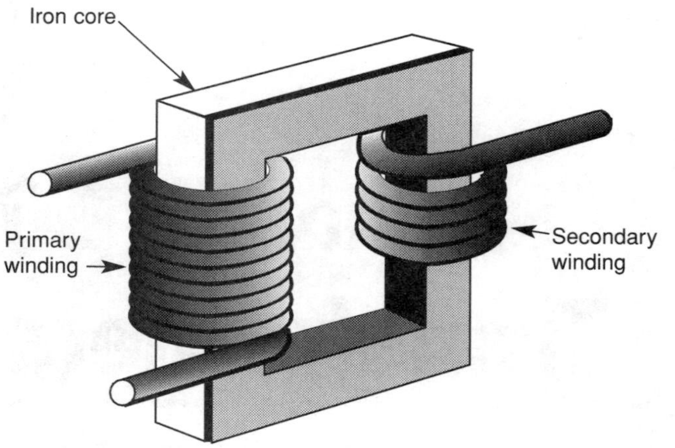
Iron core
Primary winding
Secondary winding

Figure 10-1: *Basic parts of a transformer*

shows that a current, developed by the induced voltage, is flowing in the circuit.

The induced voltage may be generated by moving coil B through the flux of coil A. However, this voltage is induced without moving coil B. When the switch in the primary circuit is open, coil A has no current and no field. As soon as the switch is closed, current passes through the coil and the magnetic field is generated. This expanding field moves or "cuts" across the wires of coil B, thus inducing a voltage without the movement of coil B.

The magnetic field expands to its maximum strength and remains constant as long as full current flows. Flux lines stop their cutting action across the turns of coil B because expansion of the field has ceased. At this point the indicator needle on the meter reads zero because no induced voltage exists anymore. If the switch is opened, the field collapses back to the wires of coil A. As it does so, the changing flux cuts across the wires of coil B, but in the opposite direction. The current present in the coil causes the indicator needle to deflect, showing this new direction. The indicator, then, shows current flow only when the field is changing, either building up or collapsing. In effect, the changing field produces an induced voltage exactly as does a magnetic field moving across a conductor. This principle of inducing voltage by holding the coils steady and forcing the field to change is used in innumerable applications. The transformer is particularly suitable for operation by mutual induction. Transformers are perfect components for transferring and changing ac voltages as needed.

Transformers are generally composed of two coils placed close to each other but not connected together. Refer again to Figure 10-1. The coil that receives energy from the line voltage source, etc., is called the "primary" and the coil that delivers energy to a load is called the "secondary." Even though the coils are not physically connected together they manage to convert and transfer energy as required by a process known as mutual induction.

Transformers, therefore, enable changing or converting power from one voltage to another. For example, generators that produce moderately large alternating currents at moderately high voltages utilize transformers to convert the power to very high voltage and proportionately small current in transmission lines, permitting the use of smaller cable and providing less power loss.

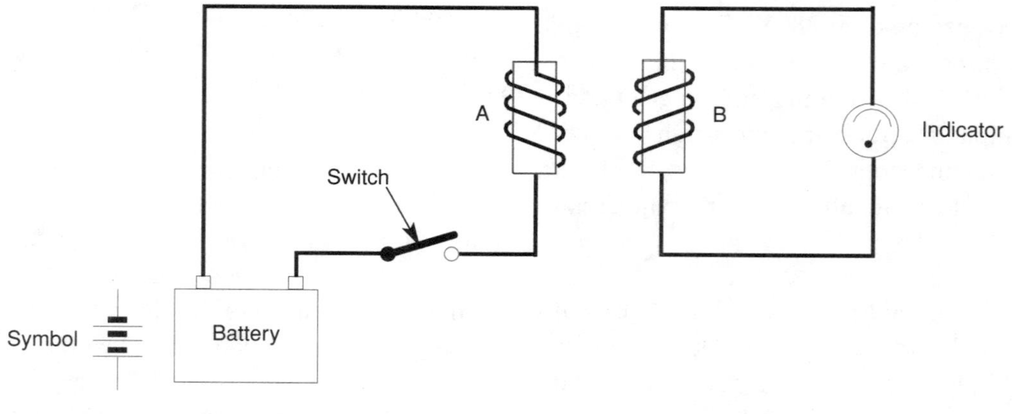
A
B
Indicator
Switch
Symbol
Battery

Figure 10-2: *Circuits demonstrating mutual inductance*

When alternating current (ac) flows through a coil, an alternating magnetic field is generated around the coil. This alternating magnetic field expands outward from the center of the coil and collapses into the coil as the ac through the coil varies from zero to a maximum and back to zero again. Since the alternating magnetic field must cut through the turns of the coil, a self-inducing voltage occurs in the coil which opposes the change in current flow.

If the alternating magnetic field generated by one coil cuts through the turns of a second coil, voltage will be generated in this second coil just as voltage is induced in a coil which is cut by its own magnetic field. The induced voltage in the second coil is called the "voltage of mutual induction," and the action of generating this voltage is called "transformer action." In transformer action, electrical energy is transferred from one coil (called the primary) to another (the secondary) by means of a varying magnetic field.

Induction In Transformers

A simple transformer consists of two coils located very close together and electrically insulated from each other. Alternating current is applied to the *primary*. In doing so, a magnetic field is generated which cuts through the turns of the other coil, and generates a voltage in the secondary. The coils are not physically connected to each other. They are, however, magnetically coupled to each other. Consequently, a transformer transfers electrical power from one coil to another by means of an alternating magnetic field.

Assuming that all the magnetic lines of force from the primary cut through all the turns of the secondary, the voltage induced in the secondary will depend on the ratio of the number of turns in the primary to the number of turns in the secondary. For example, if there are 100 turns in the primary and only 10 turns in the secondary, the voltage in the primary will be 10 times the voltage in the secondary. Since there are more turns in the primary than there are in the secondary, the transformer is called a "step-down transformer." Figure 10-3 shows a diagram of a step-down transformer with a ratio of 100:10 or 10:1. Therefore, if the primary has a potential of 120 volts, then the secondary may be calculated as follows:

$$\frac{10 \text{ turns}}{100 \text{ turns}} = 0.10 = .10 \times 120 = 12 \text{ volts}$$

With fewer turns on the secondary than on the primary, the secondary voltage will not only be proportionately lower than that in the primary, but the secondary current will be that much larger, again in proportion to the current on the primary.

If there are more turns on the secondary than on the primary winding, the secondary voltage will be higher than that in the primary and by the same proportion as the number of turns in the winding. This configuration results in a "step-up" transformer. In this type of transformer, the secondary current, in turn, will be proportionately smaller than the primary current.

Since alternating current continually increases and decreases in value, every change in the primary winding of the transformer produces a similar change of flux in the core. Every change of flux in the core along with every corresponding movement of magnetic field around the core produces a similarly changing voltage in the secondary winding, causing an alternating current to flow in the circuit that is connected to the secondary.

For example, if there are, say, 100 turns in the secondary and only 10 turns in the primary, the voltage induced in the secondary will be 10 times the voltage applied to the primary. See Figure 10-4 on the next page. Since there are more turns in the secondary than in the primary, the transformer is called a "step-up transformer."

$$\frac{100}{10} = 10 \times 12 = 120 \text{ volts}$$

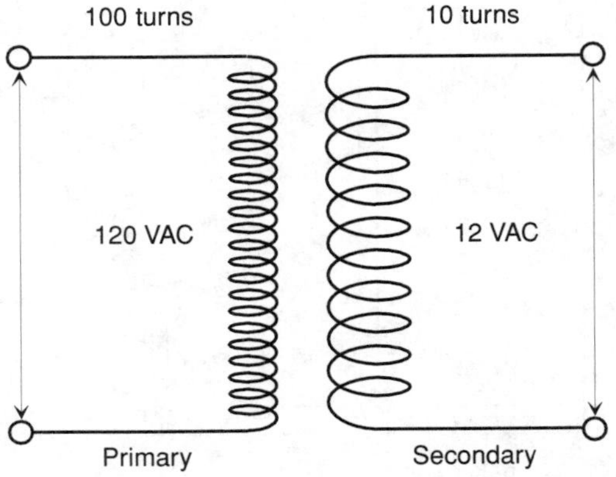
100 turns
10 turns
120 VAC
12 VAC
Primary
Secondary

Figure 10-3: *Step-down transformer with 10:1 ratio*

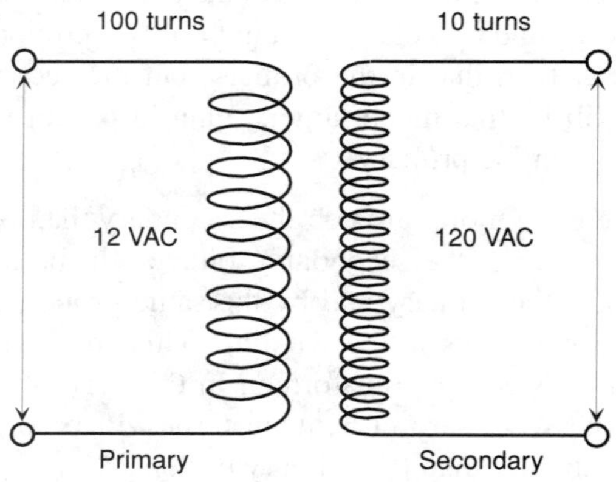
100 turns
10 turns
12 VAC
120 VAC
Primary
Secondary

Figure 10-4: *Step-up transformer with 1:10 ratio*

NOTE : A transformer does not generate electrical power. It simply transfers electric power from one coil to another by magnetic induction. Transformers are rated in either volt-amperes (VA) or kilo-volt-amperes (kVA).

Magnetic Flux In Transformers

Figure 10-5 shows a cross section of what is known as a high-leakage flux transformer. In applications, if no load were connected to the secondary or output winding, a voltmeter would indicate a specific voltage reading across the secondary terminals. If a load were applied, the voltage would drop, and if the terminals were shorted, the voltage would drop to zero. During these circuit changes, the flux in the core of the transformer would also change — it is forced out of the transformer core and is known as leakage flux. Leaking flux can actually be demonstrated with iron filings placed close to the transformer core. As the changes take place, the filings will shift their position — clearly showing the change in the flux pattern.

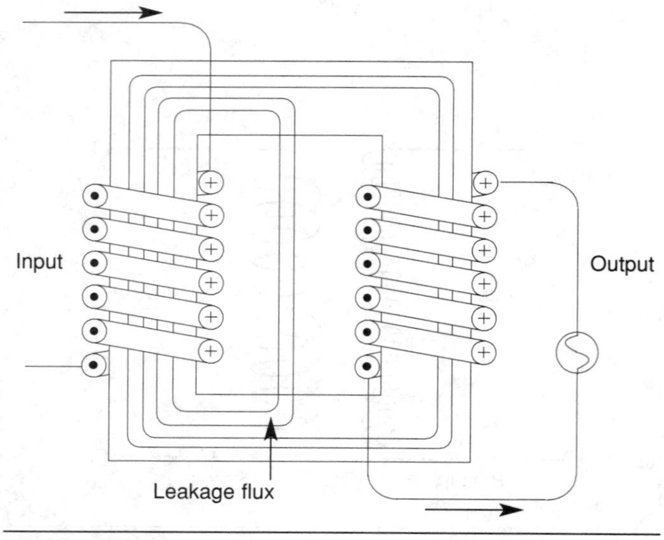
Input
Output
Leakage flux

Figure 10-5: *Transformer with high-leakage flux*

What actually happens is that as the current flows in the secondary, it tries to create its own magnetic field which is in opposition to the original flux field. This action, like a valve in a water system, restricts the flux flow which forces the excess flux to find another path — either through air or in adjacent structural steel as in transformer housings or supporting clamps.

Note that the coils in Figure 10-5 are wrapped on the same iron core, but separated from each other, while the transformer in Figure 10-6 has its coils wrapped around each other which results in a low-leakage transformer design.

TRANSFORMER CONSTRUCTION

Transformers designed to operate on low frequencies have their coils, called "windings," wound on iron cores. Since iron offers little resistance to magnetic lines, nearly all the magnetic field of the primary flows through the iron core and cuts the secondary.

Iron cores of transformers are constructed in three basic types — the open core, the closed core and the shell type. See Figure 10-7. The open core is the least expensive to manufacture as the primary and secondary are wound on one cylindrical core. The magnetic path, as shown in Figure 10-7, is partially through the core and partially through the surrounding air. The air path opposes the magnetic field, so that the magnetic interaction or "linkage" is weakened. The open core transformer, therefore, is highly inefficient.

The closed core improves the transformer efficiency by offering more iron paths and less air path for the magnetic field. The shell type core further increases the magnetic coupling and therefore the transformer efficiency is greater due to two parallel magnetic paths for the magnetic field — providing maximum coupling between the primary and secondary.

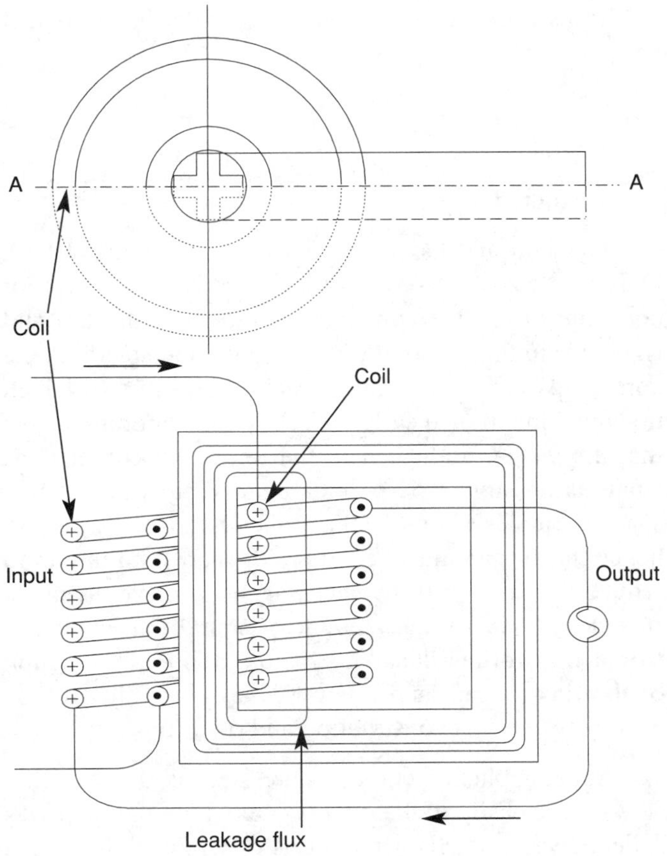
A
A
Coil
Coil
Input
Output
Leakage flux

Figure 10-6: *Transformer with low-leakage design*

Cores

Special core steel is used to provide a controlled path for the flow of magnetic flux generated in a transformer. In most practical applications, the transformer core is not a solid bar of steel, but is constructed of many layers of thin sheet steel called laminations.

While the specifications of the core steel are primarily of interest to the transformer design engineer, the electrical worker should at least have a conversational knowledge of the materials used.

Steel used for transformer core laminations will vary with the manufacturer, but a popular size is .014 inch thick and is called 29-gauge steel. It is processed from silicon iron alloys containing approximately $3^1/_4$ percent silicon. The addition of silicon to the iron increases its ability to be magnetized and also renders it essentially non-aging.

The most important characteristic of electrical steel is core loss. It is measured in watts per pound at a specified frequency and flux density. The core loss is responsible for the heating in the transformer and it also contributes to the heating of the windings. Much of the core loss is a result of eddy currents which are induced in the laminations when the core is energized. To hold this loss to a minimum, adjacent laminations are coated with an inorganic varnish.

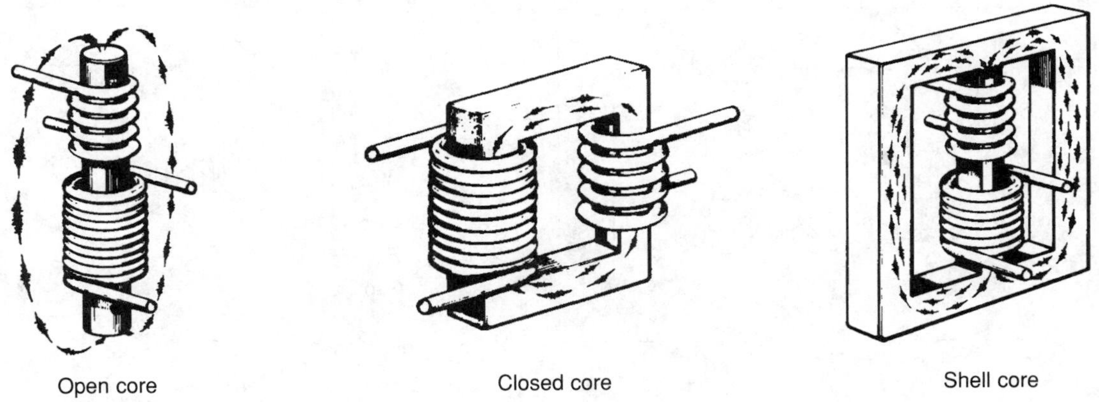
Open core
Closed core
Shell core

Figure 10-7: *Three types of iron-core transformers*

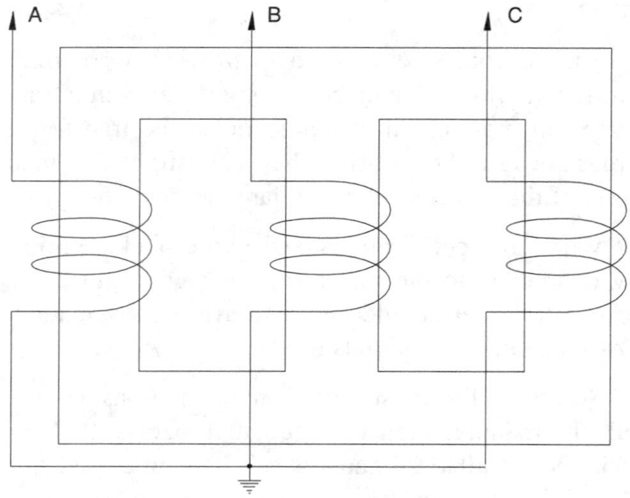
A
B
C

Figure 10-8: *Core type transformer construction*

Cores on modern power transformers may be of either the "core type" as shown in Figure 10-8 or the "shell type" as shown in Figure 10-9. Of the two, the core type is favored for dry-type transformers because:

- Only three core legs require stacking; thus, reducing cost.
- Steel does not encircle the two outer coils; this provides better cooling.
- Floor space is reduced.

Types Of Cores

Transformer cores normally fall under three distinct types:

- Butt
- Wound
- Mitered

The butt-and-lap core is shown in Figure 10-10. Only two sizes of core steel are needed in this type of core due to the lap construction shown at the top and right side in Figure 10-10. For ease of understanding, the core strips are shown much thicker than the .014 inch thickness mentioned earlier. Each strip is carefully cut so that the air gap indicated in the lower left corner is as small as possible. The permeability of steel to the passage of flux is about 10,000 times as effective as air, hence the air gap must be held to the barest minimum to reduce the ampere turns necessary to achieve adequate flux density. Also, the amount of sound that emanates from a transformer due to magneto-striction is a function of the flux density and this poses an interesting difference between this construction and other types.

Another phenomenon in core steel is that the flux flows more easily in the direction in which the steel was rolled. Even this characteristic is different in hot-rolled versus cold-rolled steel. For example, the core loss due to flux passing at right angles to the rolling direction is

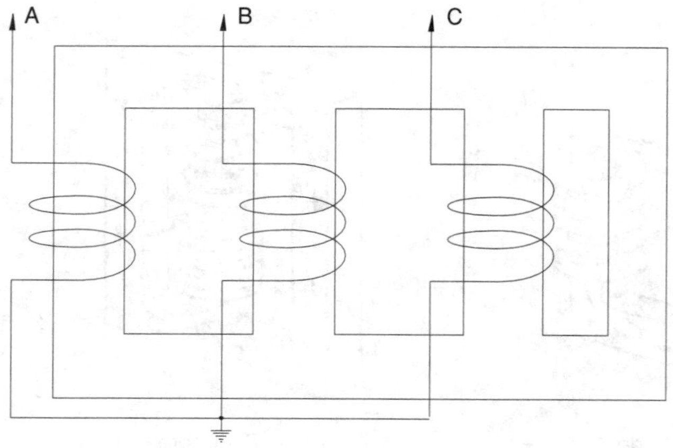
A
B
C

Figure 10-9: *Shell type transformer core*

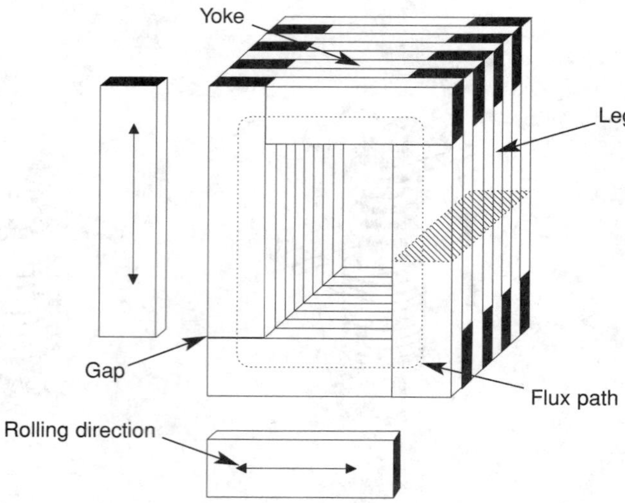
Yoke
Legs
Gap
Flux path
Rolling direction

Figure 10-10: *Butt-and-lap transformer core*

almost $1^1/2$ times as great in hot rolled and $2^1/2$ times as great in cold-rolled when compared with the core loss in the direction of rolling. The difference in exciting current is more dramatic, with ratios of two to one in hot-rolled and almost 40 to 1 in cold-rolled. These are primarily the designer's concern, but at least you will know now that there is a difference.

Eddy currents are restricted from passage from one lamination to another due to the inorganic insulating coating. However, the magnetic lines of flux easily transfer at adjacent laminations in the lap area but in so doing, are forced to cross at an angle to the preferred direction.

Wound Cores

Because of the unique characteristics of core steel, some core designs were made that took advantage of these differences. One such type is shown in Figure 10-11. The core loops are cut to predetermined lengths so that the gap locations do not coincide. These cuts permit assembling the core around a prewound coil which passes through both openings. Another design, now discontinued because of unfavorable cost, used a continuous core with no cuts. Separate coils had to be wound on each of the vertical legs of the completed core. You may encounter transformers of this type in existing installations.

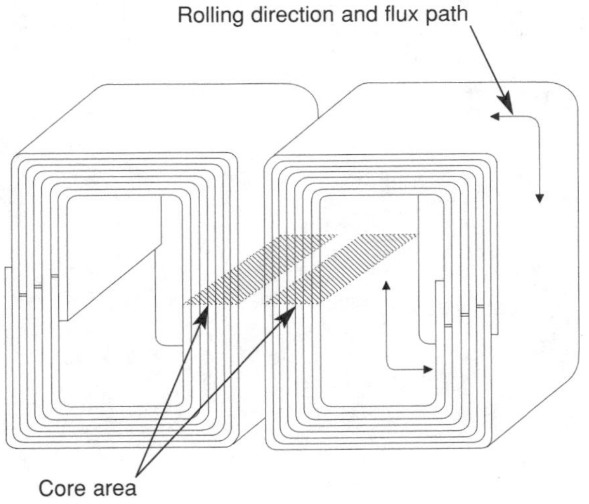
Rolling direction and flux path
Core area

Figure 10-11: *Wound transformer core*

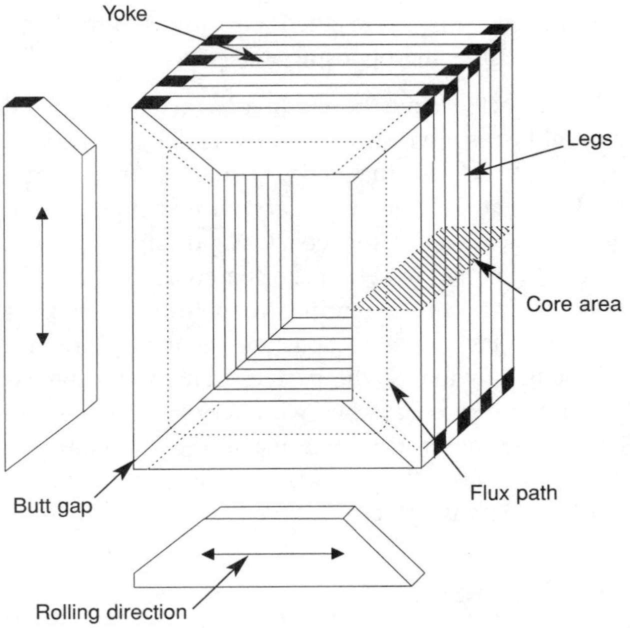
Yoke
Legs
Core area
Flux path
Butt gap
Rolling direction

Figure 10-12: *Mitered transformer core*

Mitered Cores

Figure 10-12 shows a mitered-core design. In effect, it is a butt-lap core with the joints made at 45-degree angles. There are two benefits derived from this type of joint:

- It eliminates all cross grain flux thereby improving the core loss and exciting current values.
- It reduces the flux density in the air gap — resulting in lower sound levels.

This type of core is normally used only with cold-rolled, grain oriented steel which permits this steel to be used to its fullest capability in this type of transformer core.

Transformer Characteristics

In a well-designed transformer, there is very little magnetic leakage. The effect of the leakage is to cause a decrease of secondary voltage when the transformer is loaded. When a current flows through the secondary in phase with the secondary voltage, a corresponding current flows through the primary in addition to the magnetizing current. The magnetizing effects of the two currents are equal and opposite.

In a perfect transformer, that is, one having no eddy-current losses, no resistance in its windings, and no magnetic leakage, the magnetizing effects of the primary load current and the secondary current neutralize each other, leaving only the constant primary magnetizing current effective in setting up the constant flux. If supplied with a constant primary pressure, such a transformer would maintain constant secondary pressure at all loads. Obviously, the perfect transformer has yet to be built; the closest is one with very small eddy-current loss where the drop in pressure in the secondary windings is not more than 1 to 3 percent, depending on the size of the transformer.

TRANSFORMER TAPS

If the exact rated voltage could be delivered at every transformer location, transformer taps would be unnecessary. However, this is not possible, so taps are provided on the secondary windings to provide a means of either increasing or decreasing the secondary voltage.

Generally, if a load is very close to a substation or power plant, the voltage will consistently be above normal. Near the end of the line the voltage may be below normal.

In large transformers, it would naturally be very inconvenient to move the thick, well-insulated primary leads to different tap positions when changes in source-voltage levels make this necessary. Therefore, taps are used, such as shown in the wiring diagram in Figure 10-13. In this transformer, the permanent high-voltage leads would be connected to H_1 and H_2, and the secondary leads, in their normal fashion, to X_1 and X_2, X_3, and X_4. Note, however, the tap arrangements available at taps 2 through 7. Until a pair of these taps is interconnected with a jumper wire, the primary circuit is not completed. If this were, say, a typical 7200-volt primary, the transformer would have a normal 1620 turns. Assume 810 of these turns are between H_1 and H_6 and another 810 between H_3 and H_2. Then, if taps 6 and 3 were connected together with a flexible jumper on which lugs have already been installed, the primary circuit is completed, and we have a normal ratio transformer that could deliver 120/240 volts from the secondary.

Between taps 6 and either 5 or 7, 40 turns of wire exist. Similarly, between taps 3 and either 2 or 4, 40 turns are present. Changing the jumper from 3 to 6 to 3 to 7 removes 40 turns from the left half of the primary. The same condition would apply on the right half of the winding if the jumper were between taps 6 and 2. Either connection would boost secondary voltage by $2^1/_2$ percent. Had taps 2 and 7 been connected, 80 turns would

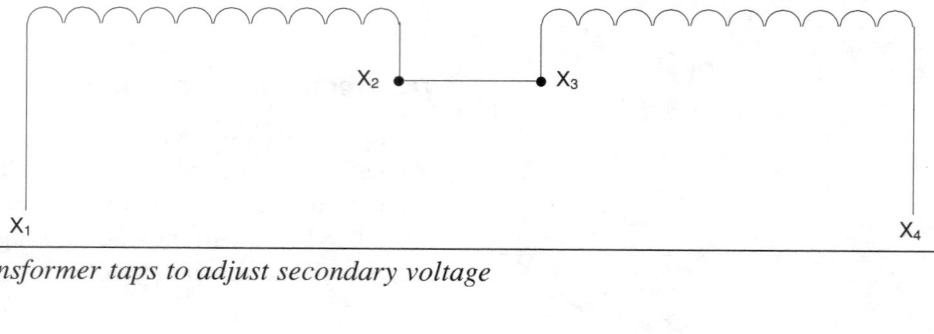
H1
7
6
5
4
3
2
H2
X2
X3
X1
X4

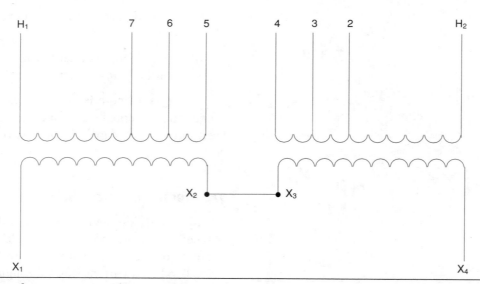

Figure 10-13: *Transformer taps to adjust secondary voltage*

have been omitted and a 5 percent boost would result. Placing the jumper between taps 6 and 4 or 3 and 5 would reduce the output voltage by 5 percent.

Caution! Before changing any transformer taps, make sure that the primary is de-energized and that the circuit has been "tagged out."

TRANSFORMER CONNECTIONS — BASIC

Transformer connections are many, and space does not permit the description of all of them here. However, an understanding of a few will give the basic requirements and make it possible to use manufacturer's data for others should the need arise on any commercial electrical installation.

Single-Phase For Light And Power

The diagram in Figure 10-14 is a transformer connection used quite extensively for residential and small commercial applications. It is the most common single-phase distribution system in use today. It is known as the 120/240-volt, single-phase, three-wire system and is used where 120 and 240 volts are used simultaneously.

In application, a single transformer is mounted on a power pole close to the premises where the power is to be used. In some instances, the transformer is mounted on a concrete pad (called *padmount*) so that the electric service may be routed underground to the point of utilization. Underground services, as discussed in Chapter 2, are called *service laterals*.

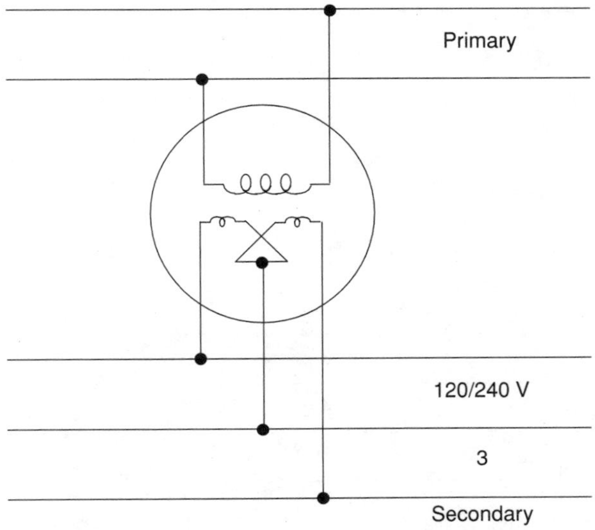
Primary
120/240 V
3
Secondary

Figure 10-14: *120/240V, single-phase, three-wire*

In either case, the transformer connections are identical, and both provide 120/240-volt, single-phase, three-wire power.

Y-Y For Light And Power

The primaries of the transformer connection in Figure 10-15 are connected in wye — sometimes called *star* connection. When the primary system is 2400/4160Y volts, a 4160-volt transformer is required when the system is connected in delta-Y. However, with a Y-Y system, a 2400-volt transformer can be used, offering a saving in transformer cost. It is necessary that a primary neutral be available when this connection is used, and the neutrals of the primary system and the transformer bank are tied together as shown in the diagram. If the three-phase load is unbalanced, part of the load current flows in the primary neutral. For these reasons, it is essential that the neutrals be tied together as shown. If this tie were omitted, the line-to-neutral voltages on the secondary would be very unstable. That is, if the load on one phase were heavier than on the other two, the voltage on this phase would drop excessively and the voltage on the other two phases would rise. Also, varying voltages would appear between lines and neutral, both in the transformers and in the secondary system, in addition to the 60-hertz component of voltage. This means that for a given value of rms (root-mean-square) voltage, the peak voltage would be much higher than for a pure 60-hertz voltage. This overstresses the insulation both in the transformers and in all apparatus connected to the secondaries.

Delta-Connected Transformers

The delta-connected system in Figure 10-16 operates a little differently from the previously described Y-Y system. While the wye-connected system is formed by connecting one terminal from three equal voltage transformer windings together to make a common terminal, the delta-connected system has its windings connected in series, forming a triangle or the Greek delta symbol Δ. Note in Figure 10-16 that a center-tap terminal is used on

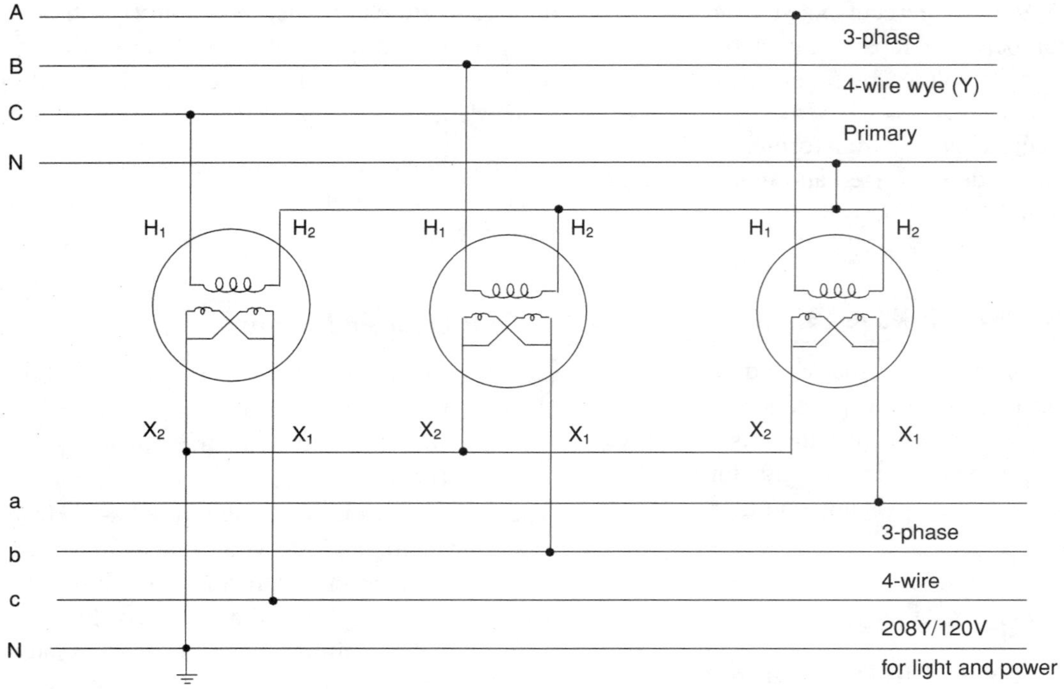
A
B
C
N
3-phase
4-wire wye (Y)
Primary
H1
H2
H1
H2
H1
H2
X2
X1
X2
X1
X2
X1
a
b
c
N
3-phase
4-wire
208Y/120V
for light and power

Figure 10-15: *Three-phase, four-wire, Y-Y connected transformer system*

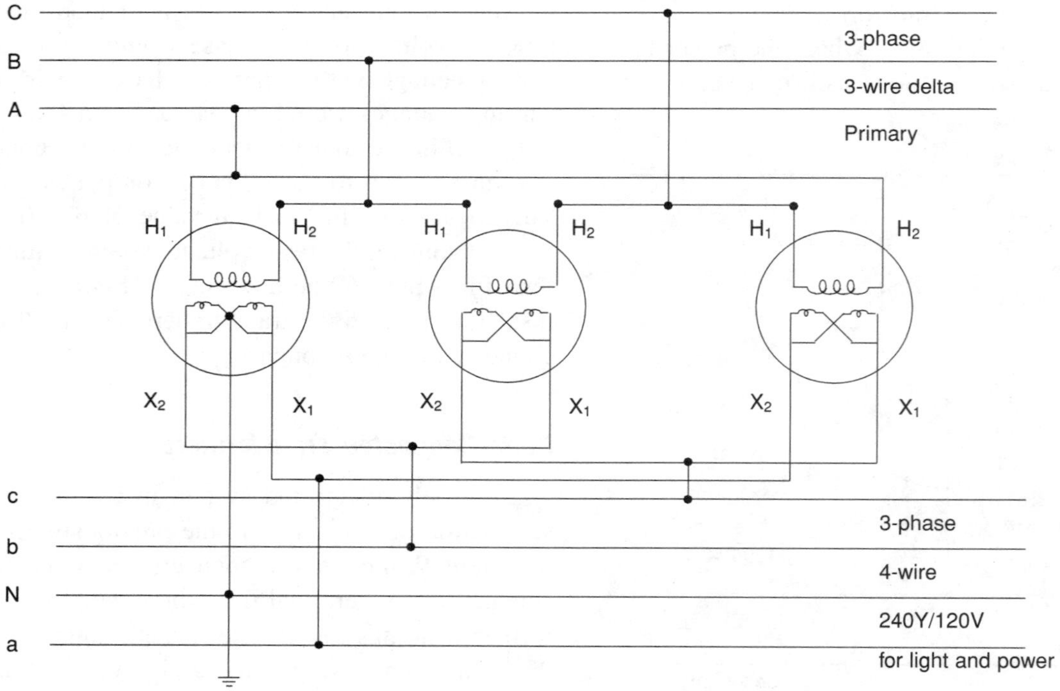
C
B
A
3-phase
3-wire delta
Primary
H1
H2
H1
H2
H1
H2
X2
X1
X2
X1
X2
X1
c
b
N
a
3-phase
4-wire
240Y/120V
for light and power

Figure 10-16: *Three-phase, four-wire, delta-connected transformer system*

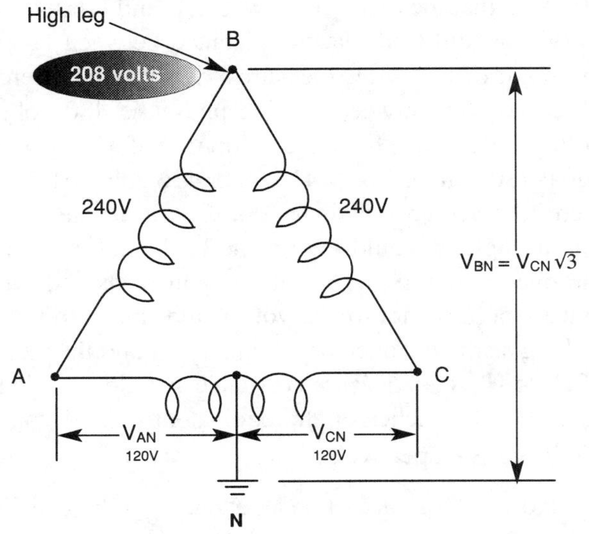
High leg
B
208 volts
240V
240V
VBN = VCN √3
A
C
VAN
120V
VCN
120V
N

On a 3-phase, 4-wire, 120/240V delta-connected system, the midpoint of one phase winding is grounded to provide 120V between phase A and ground; also between phase C and ground. Between phase B and ground, however, the voltage is higher and may be calculated by multiplyng the voltage between C and ground (120V) by the square root of 3 or 1.73. Consequently, the voltage between phase B and ground is approximately 208 volts. Thus, the name "high leg."

The *NEC* requires that conductors connected to the high leg of a 4-wire delta system be color-coded with orange insulation or tape.

Figure 10-17: *Characteristics of a center-tap, delta-connected system*

one winding to ground the system. On a 240/120-volt system, there are 120 volts between the center-tap terminal and each ungrounded terminal on either side; that is, phases A and C. There are 240 volts across the full winding of each phase.

Refer to Figure 10-17 and note that a high leg results at point "B." This is known in the trade as the "high leg," "red leg," or "wild leg." This high leg has a higher voltage to ground than the other two phases. The voltage of the high leg can be determined by multiplying the voltage to ground of either of the other two legs by the square root of 3. Therefore, if the voltage between phase A to ground is 120 volts, the voltage between phase B to ground may be determined as follows:

$$120 \times \sqrt{3} = 207.84 = 208 \text{ volts}$$

From this, it should be obvious that no single-pole breakers should be connected to the high leg of a center-tapped, four-wire delta-connected system. In fact, *NEC* Section 215-8 requires that the phase busbar or conductor having the higher voltage to ground to be permanently marked by an outer finish that is orange in color. By doing so, this will prevent future workers from connecting 120-volt single-phase loads to this high leg which will probably result in damaging any equipment connected to the circuit. Remember the color ***orange***; no 120-volt loads are to be connected to this phase.

Open Delta

Three-phase, delta-connected systems may be connected so that only two transformers are used; this arrangement is known as open *delta* as shown in Figure 10-18. This arrangement is frequently used on a delta system when one of the three transformers becomes damaged. The damaged transformer is disconnected from the circuit and the remaining two transformers carry the load. In doing so, the three-phase load carried by the open delta bank is only 86.6 percent of the combined rating of the remaining two equal sized units. It is only 57.7 percent of the normal full-load capability of a full bank of transformers. In an emergency, however, this capability permits single- and three-phase power at a location where one unit has burned out and a replacement was not readily available. The total load must be curtailed to avoid another burnout.

Tee-Connected Transformers

When a delta-wye transformer is used, we would usually expect to find three primary and three secondary coils. However, in a tee-connected three-phase transformer, only two primary and two secondary windings are used as shown in Figure 10-19. If an equilateral triangle is drawn as indicated by the dotted lines in Figure

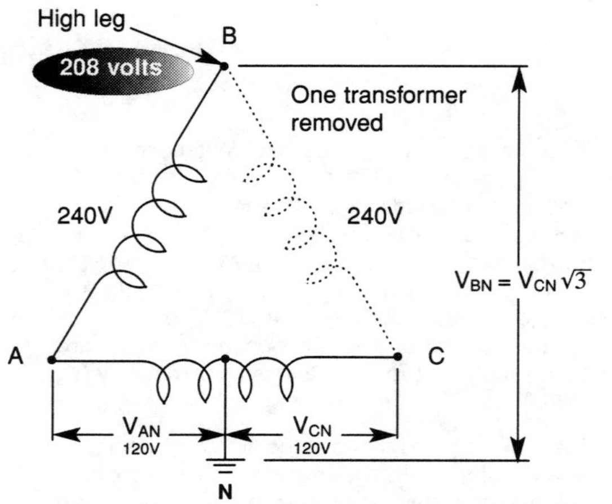
High leg
208 volts
B
One transformer
removed
240V
240V
$V_{BN} = V_{CN}\sqrt{3}$
A
C
V_{AN}
120V
V_{CN}
120V
N

Figure 10-18: *Open delta system*

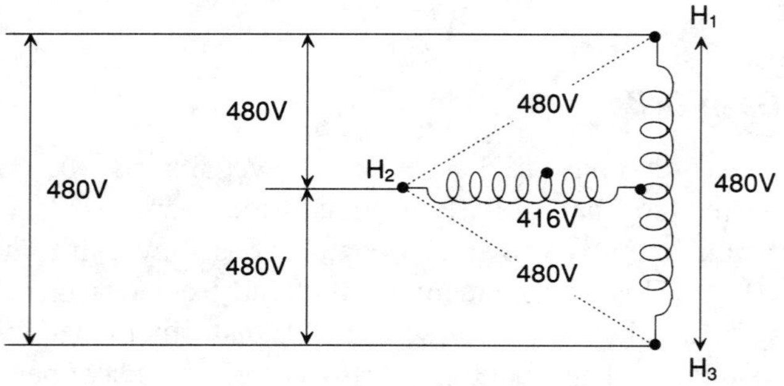
H_1
480V
480V
480V
H_2
480V
416V
480V
480V
H_3

Figure 10-19: *Typical tee-connected transformer*

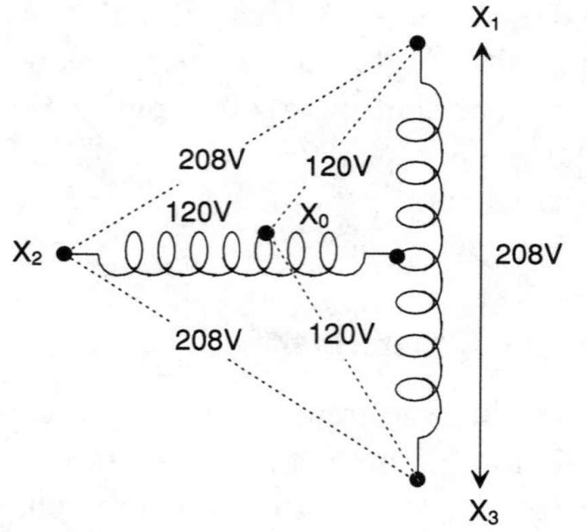
X_1
208V
120V
120V
X_0
X_2
208V
208V
120V
X_3

Figure 10-20: *Secondary voltage on tee-connected system*

10-19 so that the distance between H_1 and H_3 is 4.8 inches, you would find that the distance between H_2 to the midpoint of H_1 — H_3 measures 4.16 inches. Therefore, if the voltage between outside phases is 480 volts, the voltage between H_2 to the midpoint of H_1 — H_3 will equal 480 volts x .866 = 415.68 or 416 volts. Also, if you were to place an imaginary dot exactly in the center of this triangle it would lay on the horizontal winding — the one containing 416 volts. If you measured the distance from this dot to H_2, you would find it to be twice as long as the distance between the dot and the midpoint of H_1 to H_3. The measured distances would be 2.77 inches and 1.385 inches or the equivalent of 277 volts and $138\frac{1}{2}$ volts respectively.

Now, let's look at the secondary winding in Figure 10-20. By placing a neutral tap X_0 so that $\frac{1}{3}$ the number of turns exist between it and the midpoint of X_1 and X_3, as exist between it and X_2, we then can establish X_0 as a neutral point which may be grounded. This provides 120 volts between X_0 and any of the three secondary terminals and the three-phase voltage between X_1, X_2, and X_3, will be 208 volts.

AUTOTRANSFORMERS

An autotransformer is a transformer whose primary and secondary circuits have part of a winding in common and therefore the two circuits are not isolated from each other. See Figure 10-21. The application of an autotransformer is a good choice for some users where a 480Y/277- or 208Y/120-volt, three-phase, four-wire distribution system is utilized. Some of the advantages are as follows:

- Lower purchase price
- Lower operating cost due to lower losses
- Smaller size; easier to install
- Better voltage regulation
- Lower sound levels

For example, when the ratio of transformation from the primary to secondary voltage is small, the most economical way of stepping down the voltage is by using autotransformers as shown in Figure 10-22. For this application, it is necessary that the neutral of the autotransformer bank be connected to the system neutral.

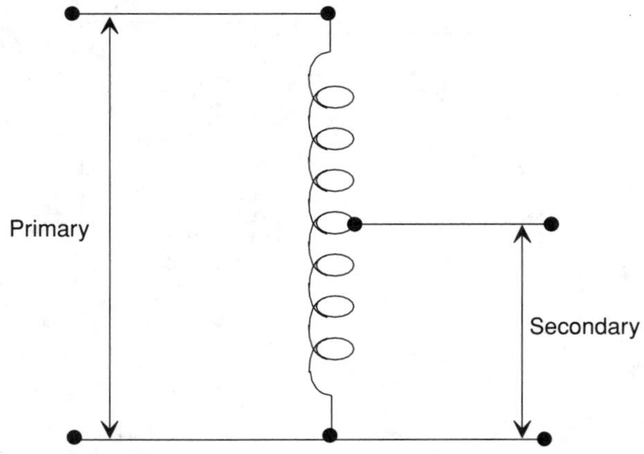
Primary
Secondary

Figure 10-21: *Step-down autotransformer*

An autotransformer, however, cannot be used on a 480- or 240-volt, three-phase, three-wire delta system. A grounded neutral phase conductor must be available in accordance with *NEC* Article 210-9, which states:

NEC Section 210-9: Circuits Derived from Autotransformers. Branch circuits shall not be derived from autotransformers . . .

Exception: Where the system supplied has a grounded conductor that is electrically connected to a grounded conductor of the system supplying the autotransformer.

Another Exception: An autotransformer used to extend or add an individual branch circuit in an existing installation for an equipment load without the connection to a similar grounded conductor when transforming from a nominal 208 volts to a nominal 240 volt supply or similarly from 240 volts to 208 volts.

The *NEC*, in general, requires that separately derived alternating-current systems be grounded. The secondary of a two-winding insulated transformer is a separately derived system. Therefore, it must be grounded in accordance with *NEC* Section 250-26.

A typical drawing to illustrate this *NEC* requirement is shown in Figure 10-23. In the case of an autotransformer, the grounded conductor of the supply is brought into the transformer to a common terminal, and the ground is established to satisfy the *NEC*.

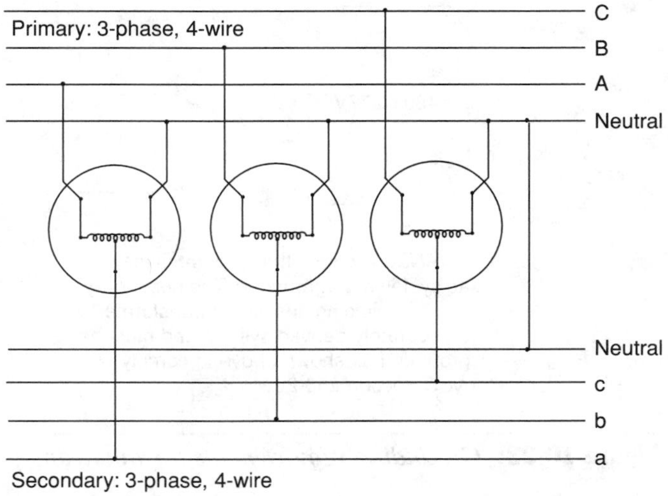
Primary: 3-phase, 4-wire
C
B
A
Neutral
Neutral
c
b
a
Secondary: 3-phase, 4-wire

Figure 10-22: *Autotransformer supplying power from a three-phase, four-wire system*

PARALLEL OPERATION OF TRANSFORMERS

Transformers will operate satisfactorily in parallel on a single-phase, three-wire system if the terminals with the same relative polarity are connected together. However, the practice is not very economical because the individual cost and losses of the smaller transformers are greater than one larger unit giving the same output. Therefore, paralleling of smaller transformers is usually done only in an emergency. In large transformers, however, it is often practical to operate units in parallel as a regular practice. See Figure 10-24.

In connecting large transformers in parallel, especially when one of the windings is for a comparatively low voltage, the resistance of the joints and interconnecting leads must not vary materially for the different transformers, or it will cause an unequal division of load.

Two three-phase transformers may also be connected in parallel provided they have the same winding arrangement, are connected with the same polarity, and have the same phase rotation. If two transformers — or two banks of transformers — have the same voltage ratings, the same turn ratios, the same impedances, and the same ratios of reactance to resistance, they will divide the load current in proportion to their kVA ratings, with no phase difference between the currents in the two transformers. However, if any of the preceding conditions

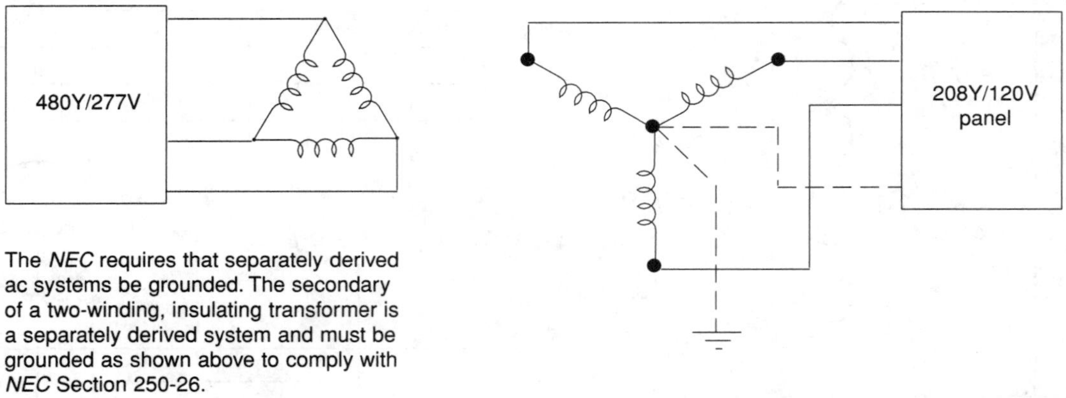
480Y/277V
208Y/120V panel
The *NEC* requires that separately derived ac systems be grounded. The secondary of a two-winding, insulating transformer is a separately derived system and must be grounded as shown above to comply with *NEC* Section 250-26.

Figure 10-23: *Grounding requirements for autotransformers*

are not met, then it is possible for the load current to divide between the two transformers in proportion to their kVA ratings. There may also be a phase difference between currents in the two transformers or banks of transformers.

Some three-phase transformers cannot be operated properly in parallel. For example, a transformer having its coils connected in delta on both high-tension and low-tension sides cannot be made to parallel with one connection. However, if the transformers are connected in delta on the high-tension side, and in Y on the low-tension side, they can be made to parallel provided the transformers have their coils joined in accordance with certain approved schemes. That is, connected in star or Y on the high-tension side, and in delta on the low-tension side or vice versa.

To determine whether or not three-phase transformers will operate in parallel, connect them as shown in Figure 10-25, leaving two leads on one of the transformers unjoined. Test with a voltmeter across the unjoined leads. If there is no voltage between the points shown in the drawing, the polarities of the two transformers are the same, and the connections may then be made and put into service.

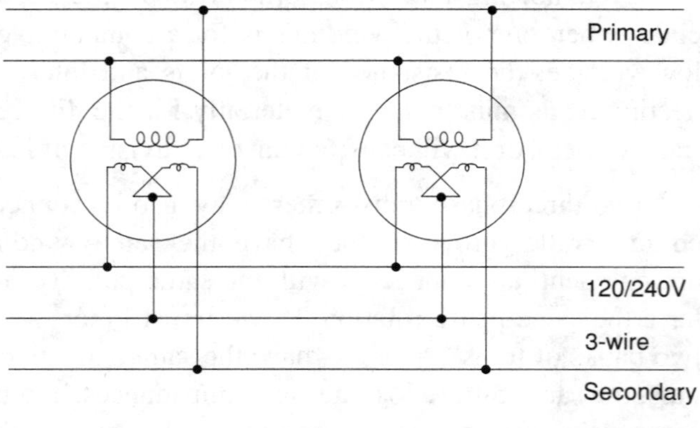
Primary
120/240V
3-wire
Secondary

Figure 10-24: *Parallel operation of single-phase transformers*

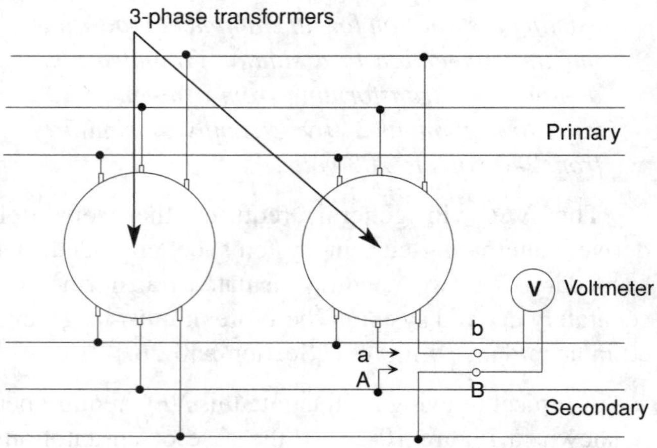
3-phase transformers
Primary
V
Voltmeter
b
a
A
B
Secondary

Figure 10-25: *Testing three-phase transformers for parallel operation*

If a reading indicates a voltage between the points indicated in the drawing (either one of the two or both), the polarity of the two transformers are different. Should this occur, disconnect transformer lead A successively to mains 1, 2, and 3 as shown in Figure 10-25 and at each connection test with the voltmeter between b and B and the legs of the main to which lead A is connected. If with any trial connection the voltmeter readings between b and B and either of the two legs is found to be zero, the transformer will operate with leads b and B connected to those two legs. If no system of connections can be discovered that will satisfy this condition, the transformer will not operate in parallel without changes in its internal connections, and there is a possibility that it will not operate in parallel at all.

In parallel operation, the primaries of the two or more transformers involved are connected together, and the secondaries are also connected together. With the primaries so connected, the voltages in both primaries and secondaries will be in certain directions. It is necessary that the secondaries be so connected that the voltage from one secondary line to the other will be in the same direction through both transformers. Proper connections to obtain this condition for single-phase transformers of various polarities are shown in Figure 10-26. In Figure 10-26(a), both transformers A and B have additive polarity; in Figure 10-26(b), both transformers have subtractive polarity; in Figure 10-26(c), transformer A has additive polarity and B has subtractive polarity.

Transformers, even when properly connected, will not operate satisfactorily in parallel unless their transformation ratios are very close to being equal and their impedance voltage drops are also approximately equal. A difference in transformation ratios will cause a circulating current to flow, even at no load, in each winding of both transformers. In a loaded parallel bank of two transformers of equal capacities, for example, if there is a difference in the transformation ratios, the load circuit will be superimposed on the circulating current. The result in such a case is that in one transformer the total circulating current will be added to the load current, whereas in the other transformer the actual current will be the difference between the load current and the circulating current. This may lead to unsatisfactory operation. Therefore, the transformation ratios of transformers for parallel operation must be definitely known.

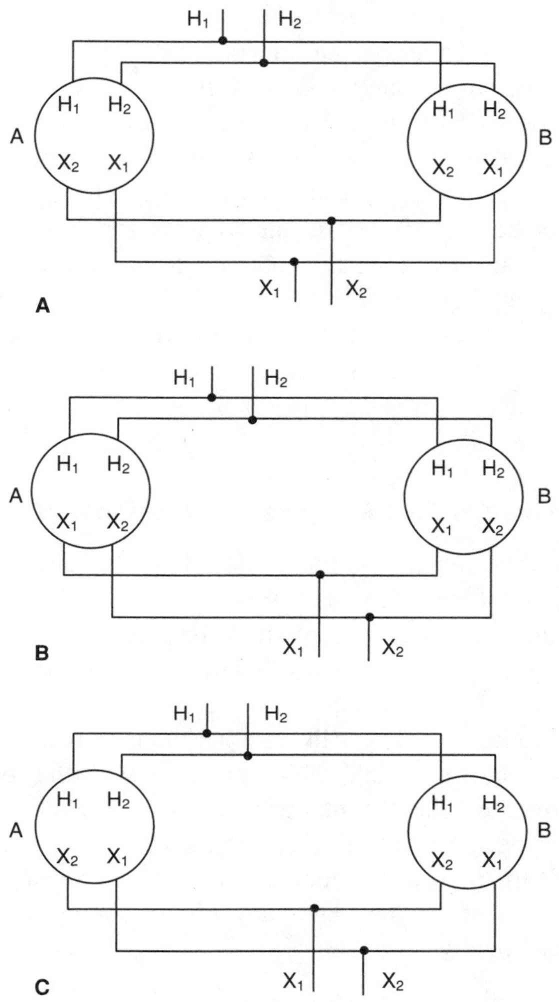
H1 H2
A
B
X1 X2
A
B
C

Figure 10-26: *Transformers connected in parallel*

When two transformers are connected in parallel, the circulating current caused by the difference in the ratios of the two is equal to the difference in open-circuit voltage divided by the sum of the transformer impedances, because the current is circulated through the windings of both transformers due to this voltage difference. To illustrate, let I represent the amount of circulating current — in percent of full-load current — and the equation will be:

$$I = \frac{\%\text{ voltage difference} \times 100}{\text{Sum of percent impedances}}$$

Let's assume an open-circuit voltage difference of 3 percent between two transformers connected in parallel. If each transformer has an impedance of 5 percent, the circulating current, in percent of full-load current, is I = (3 x 100)/5 + 5) = 30 percent. A current equal to 30 percent full-load current therefore circulates in both the high-voltage and low-voltage windings. This current adds to the load current in the transformer having the higher induced voltage and subtracts from the load current of the other transformer. Therefore, one transformer will be overloaded, while the other may or may not be — depending on the phase-angle difference between the circulating current and the load current.

Impedance In Parallel-Operated Transformers

Impedance plays an important role in the successful operation of transformers connected in parallel. The impedance of the two or more transformers must be such that the voltage drop from no load to full load is the same in all transformer units in both magnitude and phase. In most applications, you will find that the total resistance drop is relatively small when compared with the reactance drop and that the total percent impedance drop can be taken as approximately equal to the percent reactance drop. If the percent impedances of the given transformers at full load are the same, they will, of course, divide the load equally.

The following equation may be used to obtain the division of loads between two transformer banks operating in parallel on single-phase systems. In this equation, it can be assumed that the ratio of resistance to reactance is the same in all units since the error introduced by differences in this ratio is usually so small as to be negligible:

$$\text{Power} = \frac{(\text{kVA} - 1) / (Z - 1)}{[(\text{kVA} - 1) / (Z - 1)] + [(\text{kVA} - 2) / (Z - 2)]} \times \text{total kVA}$$

where

kVA − 1 = kVA rating of transformer 1

kVA − 2 = kVA rating of transformer 2

Z − 1 = percent impedance of transformer 1

Z − 2 = percent impedance of transformer 2

The preceding equation may also be applied to more than two transformers operated in parallel by adding, to the denominator of the fraction, the kVA of each additional transformer divided by its percent impedance.

Parallel Operation Of Three-Phase Transformers

Three-phase transformers, or banks of single-phase transformers, may be connected in parallel provided each of the three primary leads in one three-phase transformer is connected in parallel with a corresponding primary lead of the other transformer. The secondaries are then connected in the same way. The corresponding leads are the leads which have the same potential at all times and the same polarity. Furthermore, the transformers must have the same voltage ratio and the same impedance voltage drop.

When three-phase transformer banks operate in parallel and the three units in each bank are similar, the division of the load can be determined by the same method previously described for single-phase transformers connected in parallel on a single-phase system.

In addition to the requirements of polarity, ratio, and impedance, paralleling of three-phase transformers also requires that the angular displacement between the voltages in the windings be taken into consideration when they are connected together.

Phasor diagrams of three-phase transformers that are to be paralleled greatly simplify matters. With these, all that is required is to compare the two diagrams to make sure they consist of phasors that can be made to coincide; then connect together terminals corresponding to coinciding voltage phasors. If the diagram phasors can be made to coincide, leads that are connected together will have the same potential at all times. This is one of the fundamental requirements for paralleling. Phasor diagrams are covered in most basic electrical theory books. Such books should be reviewed to gain a basic understanding of how phasor diagrams are developed.

TRANSFORMER CONNECTIONS — DRY TYPE

Electricians performing work on commercial installations will more often be concerned with the installation and connection of dry-type transformers (Figure 10-27) as opposed to oil-filled ones. Dry-type transformers are

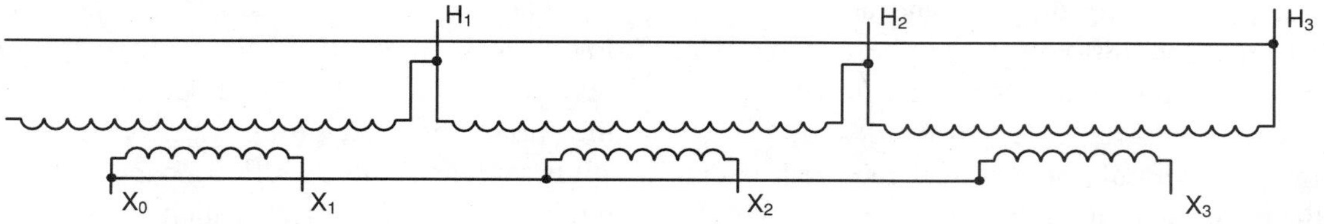

H_1
H_2
H_3
X_0
X_1
X_2
X_3

Primary Volts	Connect Primary Lines to	Connect Secondary Lines to
480	H_1, H_2, H_3	—
Secondary Volts		
208	—	X_1, X_2, X_3
120 1 Phase	—	X_1 to X_0 X_2 to X_0 X_3 to X_0

Figure 10-27: *Typical transformer manufacturer's wiring diagram — delta-wye*

available in both single-and three-phase with a wide range of sizes from the small control transformers to those rated at 500 kVA or more. Such transformers have wide application in electrical systems of all types.

NEC Section 450-11 requires that each transformer be provided with a nameplate giving the manufacturer; rated kVA; frequency; primary and secondary voltage; impedance of transformers 25 kVA and larger; required clearances for transformers with ventilating openings; and the amount and kind of insulating liquid where used. In addition, the nameplate of each dry-type transformer must include the temperature class for the insulation system.

In addition, most manufacturers include a wiring diagram and a connection chart as shown in Figure 10-27 for a 480-volt delta primary to 208Y/120-volt secondary and Figure 10-28 for a 480-volt delta to 240-volt delta connection. It is recommended that all transformers be connected as shown on the manufacturer's nameplate.

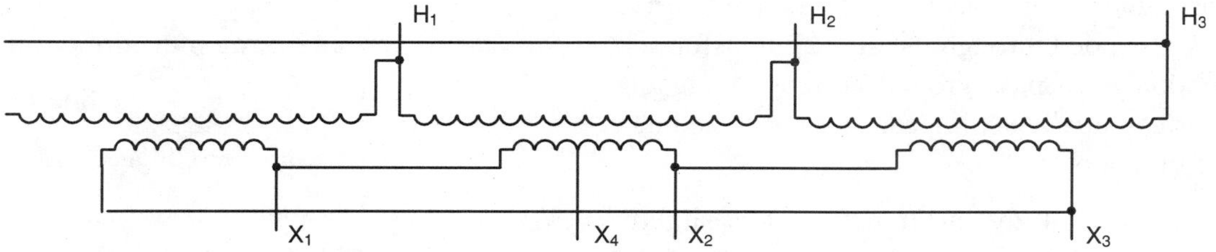

H_1
H_2
H_3
X_1
X_4
X_2
X_3

Primary Volts	Connect Primary Lines to	Connect Secondary Lines to
480	H_1, H_2, H_3	—
Secondary Volts		
240	—	X_1, X_2, X_3
120	—	X_1, X_4 or X_2, X_4

Figure 10-28: *480-volt to 240-volt delta transformer connections*

In general, this wiring diagram and accompanying table indicate that the 480-volt, three-phase, three-wire primary conductors are connected to terminals H_1, H_2, and H_3, respectively — regardless of the desired voltage on the primary. A neutral conductor, if required, is carried from the primary through the transformer to the secondary. Two variations are possible on the secondary side of this transformer: 208-volt, three-phase, three- or four-wire or 120-volt, single-phase, two-wire. To connect the secondary side of the transformer as a 208-volt, three-phase, three-wire system, the secondary conductors are connected to terminals X_1, X_2, and X_3; the neutral is carried through with conductors usually terminating at a solid-neutral bus in the transformer.

Zig-Zag Connections

There are many occasions where it is desirable to upgrade a building's lighting system from 120-volt fixtures to 277-volt fluorescent lighting fixtures. Oftentimes these buildings have a 480/240-volt, three-phase, four-wire delta system. One way to obtain 277 volts from a 480/240-volt system is to connect 480/240-volt transformers in a zig-zag fashion as shown in Figure 10-29. In doing so, the secondary of one phase is connected in series with the primary of another phase, thus changing the phase angle.

The zig-zag connection may also be used as a grounding transformer where its function is to obtain a neutral point from an ungrounded system. With a neutral being available, the system may then be grounded. When the system is grounded through the zig-zag transformer, its sole function is to pass ground current. A zig-zag transformer is essentially six impedances connected in a zig-zag configuration.

The operation of a zig-zag transformer is slightly different from that of the conventional transformer. We will consider current rather than voltage. While a voltage rating is necessary for the connection to function, this is actually line voltage and is not transformed. It provides only exciting current for the core. The dynamic portion of the zig-zag grounding system is the fault current. To understand its function, the system must also be viewed backward; that is, the fault current will flow into the transformer through the neutral as shown in Figure 10-30.

The zero sequence currents are all in phase in each line; that is, they all hit the peak at the same time. In reviewing Figure 10-30, we see that the current leaves the motor, goes to ground, flows up the neutral, and splits three ways. It then flows back down the line to the motor through the fuses which then open — shutting down the motor.

The neutral conductor will carry full fault current and must be sized accordingly. It is also time rated (0-60 seconds) and can therefore be reduced in size. This should be coordinated with the manufacturer's time/current curves for the fuse.

To determine the size of a zig-zag grounding transformer, proceed as follows:

Step 1. Calculate the system line-to-ground asymmetrical fault current.

Step 2. If relaying is present, consider reducing the fault current by installing a resistor in the neutral.

Step 3. If fuses or circuit breakers are the protective device, you may need all the fault current to quickly open the overcurrent protective devices.

Step 4. Obtain time/current curves of relay, fuses, or circuit breakers.

Step 5. Select zig-zag transformer for:

a. Fault current — the line-to-ground

b. Line-to-line voltage

c. Duration of fault (determined from time/current curves)

d. Impedance per phase at 100 percent; for any other, contact manufacturer

BUCK-AND-BOOST TRANSFORMERS

The buck-and-boost transformer is a very versatile unit for which a multitude of applications exist. Buck-and-boost transformers, as the name implies, are

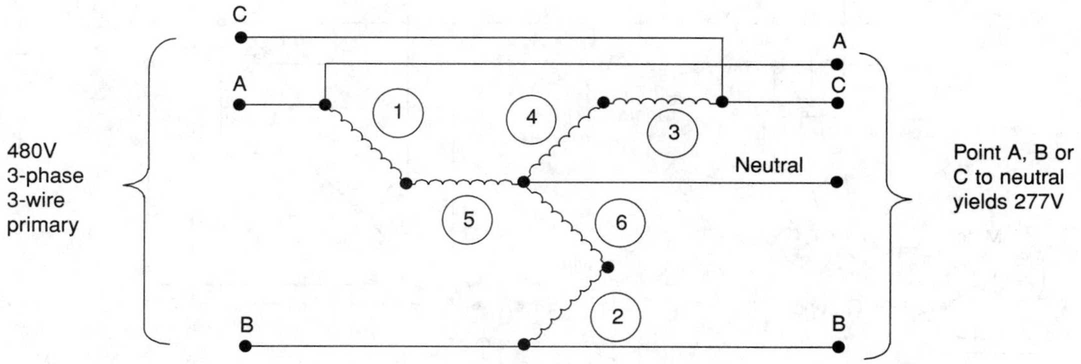
C
A
A
C
1
4
3
480V
3-phase
3-wire
primary
Neutral
Point A, B or
C to neutral
yields 277V
5
6
2
B
B

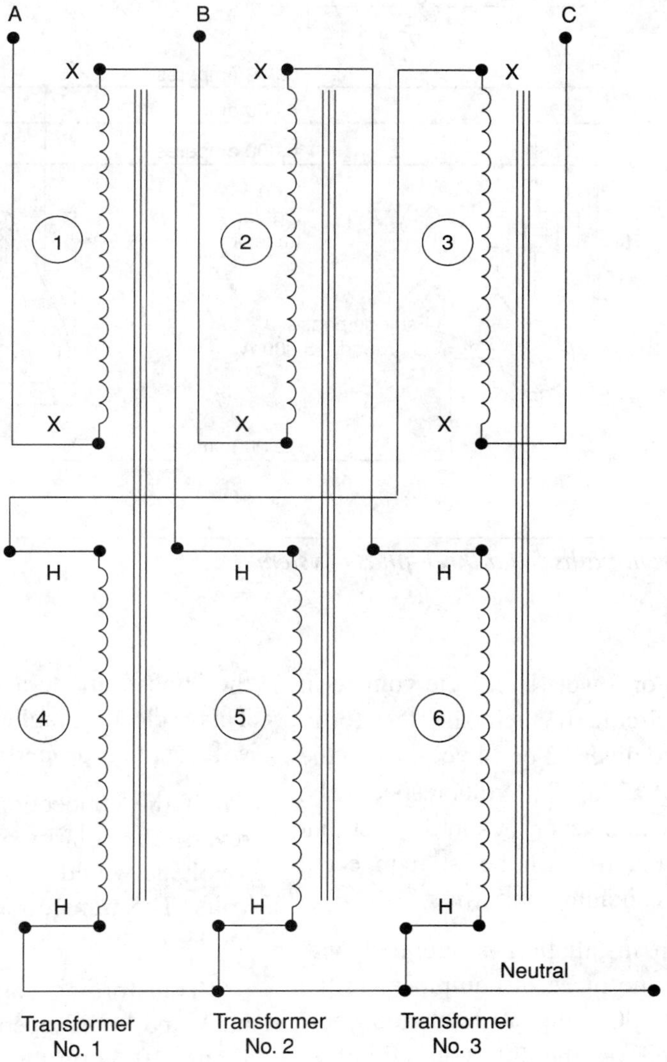
A
B
C
X
X
X
1
2
3
X
X
X
H
H
H
4
5
6
H
H
H
Neutral
Transformer
No. 1
Transformer
No. 2
Transformer
No. 3

Figure 10-29: *Zig-zag connection*

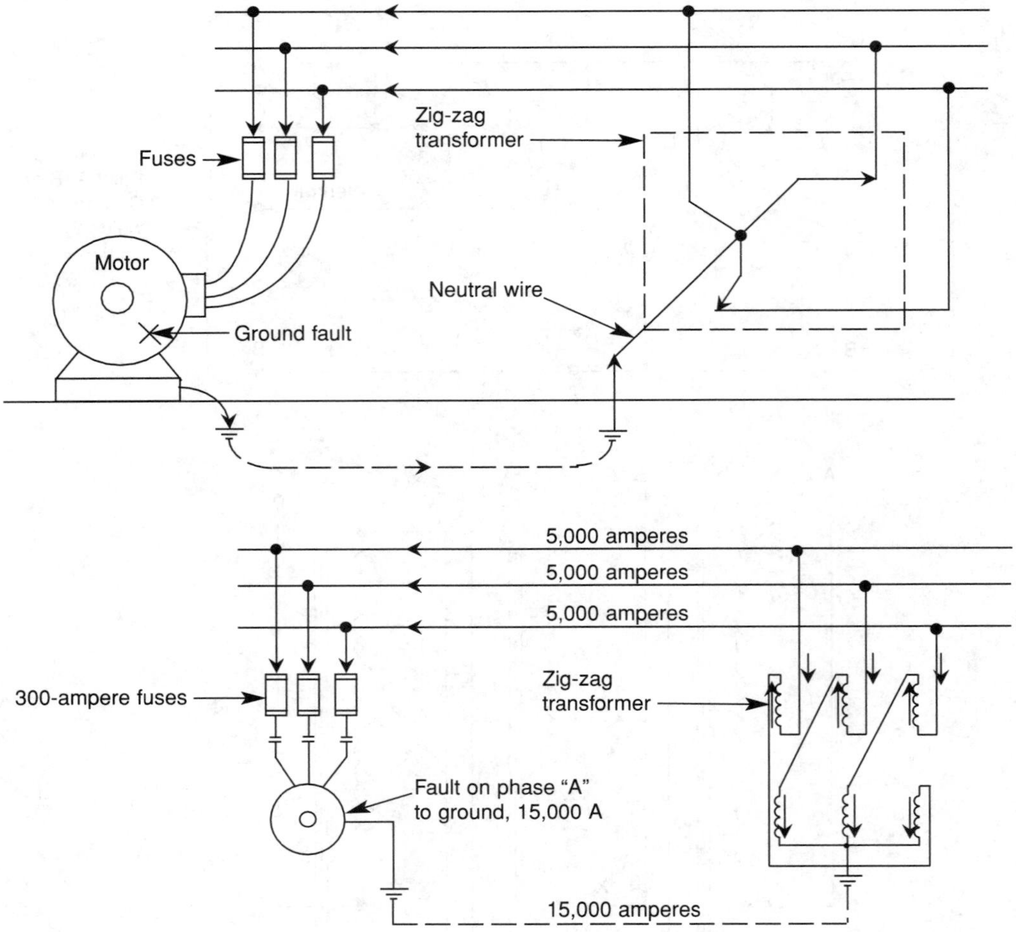
Zig-zag
transformer
Fuses
Motor
Ground fault
Neutral wire
5,000 amperes
5,000 amperes
5,000 amperes
Zig-zag
transformer
300-ampere fuses
Fault on phase "A"
to ground, 15,000 A
15,000 amperes

Figure 10-30: *Fault-current paths for a three-phase system*

designed to raise (boost) or lower (buck) the voltage in an electrical system or circuit. In their simplest form, these insulated units will deliver 12 or 24 volts when the primaries are energized at 120 or 240 volts respectively. Their prime use and value, however, lies in the fact that the primaries and the secondaries can be interconnected — permitting their use as an autotransformer.

Let's assume that an installation is supplied with 208Y/120V service, but one piece of equipment in the installation is rated for 230 volts. A buck-and-boost transformer may be used on the 208-volt circuit to increase the voltage from 208 volts to 230 volts. See Figure 10-31. With this connection, the transformer is in the "boost" mode and delivers 228.8 volts at the load. This is close enough to 230 volts that the load equipment will function properly.

If the connections were reversed, this would also reverse the polarity of the secondary with the result that a voltage would be 208 volts minus 20.8 volts = 187.2 volts. The transformer is now operating in the "buck" mode.

Transformer connections for typical three-phase buck-and-boost open-delta transformers are shown in Figure 10-32 on the next page. The connections shown are in the "boost" mode; to convert to "buck" mode, reverse the input and output.

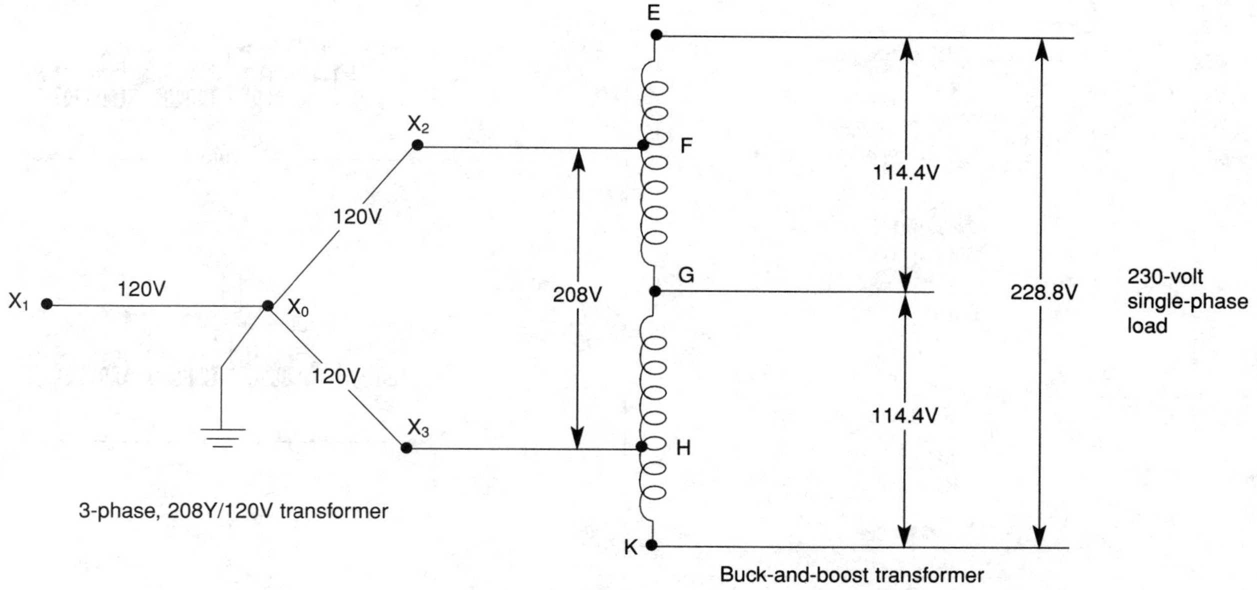
E
F
G
H
K
X1
X2
X3
X0
120V
120V
120V
208V
114.4V
114.4V
228.8V
230-volt single-phase load
3-phase, 208Y/120V transformer
Buck-and-boost transformer

Figure 10-31: *Buck-and-boost transformer connected to a 208-volt system to obtain 230 volts*

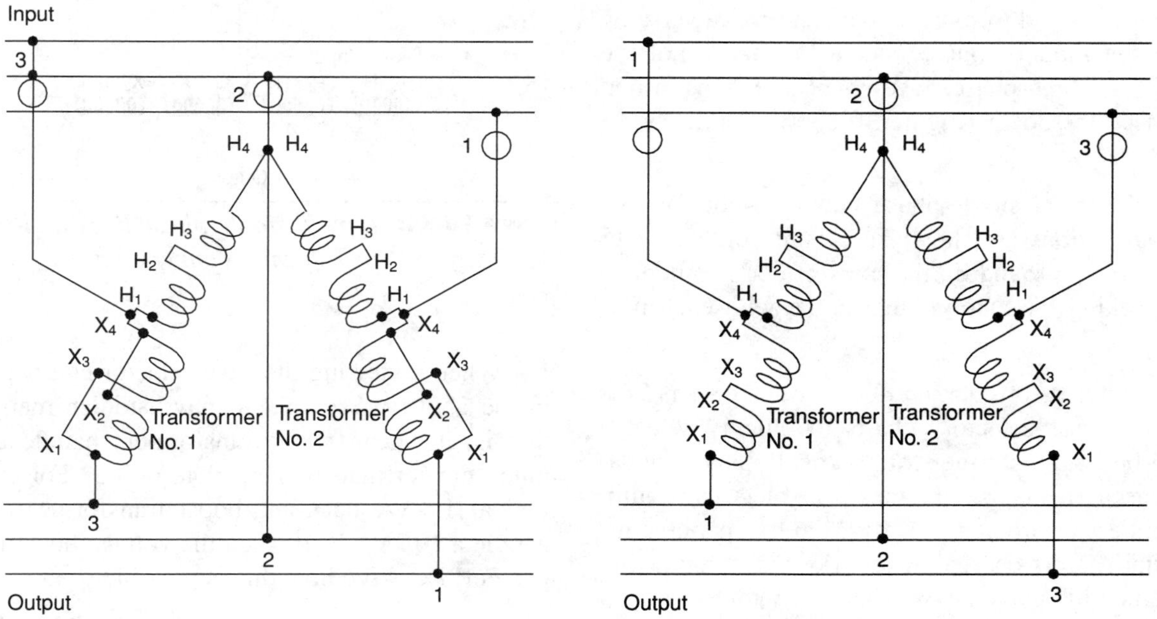
Input
Output
Transformer No. 1
Transformer No. 2
H1
H2
H3
H4
X1
X2
X3
X4
1
2
3

Figure 10-32: *Open delta, three-phase, buck-and-boost transformer connections*

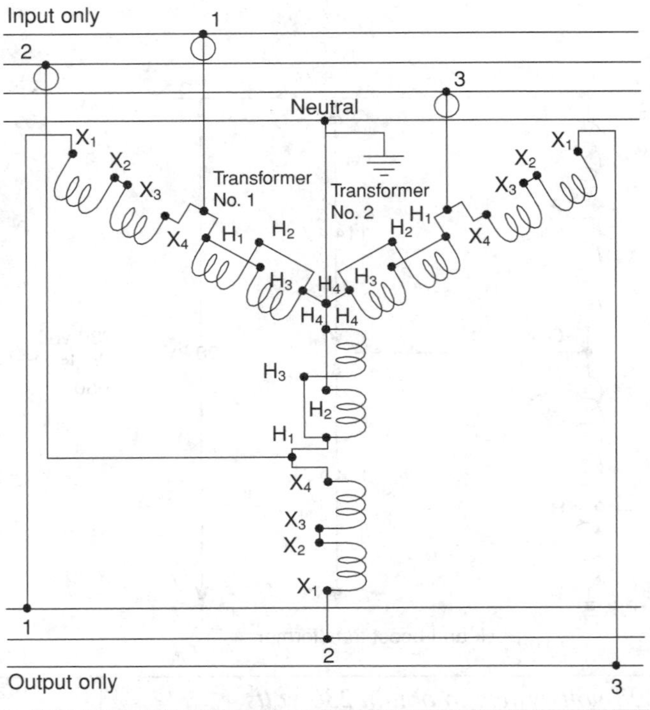
Input only
1
2
3
Neutral
Transformer No. 1
Transformer No. 2
X_1
X_2
X_3
X_4
H_1
H_2
H_3
H_4
1
2
3
Output only

Figure 10-33: *Three-phase, wye-connected*

Another three-phase buck-and-boost transformer connection is shown in Figure 10-33, this time wye-connected. While the open-delta transformers (Figure 10-32) can be converted from buck to boost or vice versa by reversing the input/output connections, this is not the case with the three-phase, wye-connected transformer. The connection shown (Figure 10-33) is for the boost mode only.

Several typical single-phase buck-and-boost transformer connections are shown in Figure 10-34 . Other diagrams may be found on the transformer's nameplate or with packing instructions that come with each new transformer.

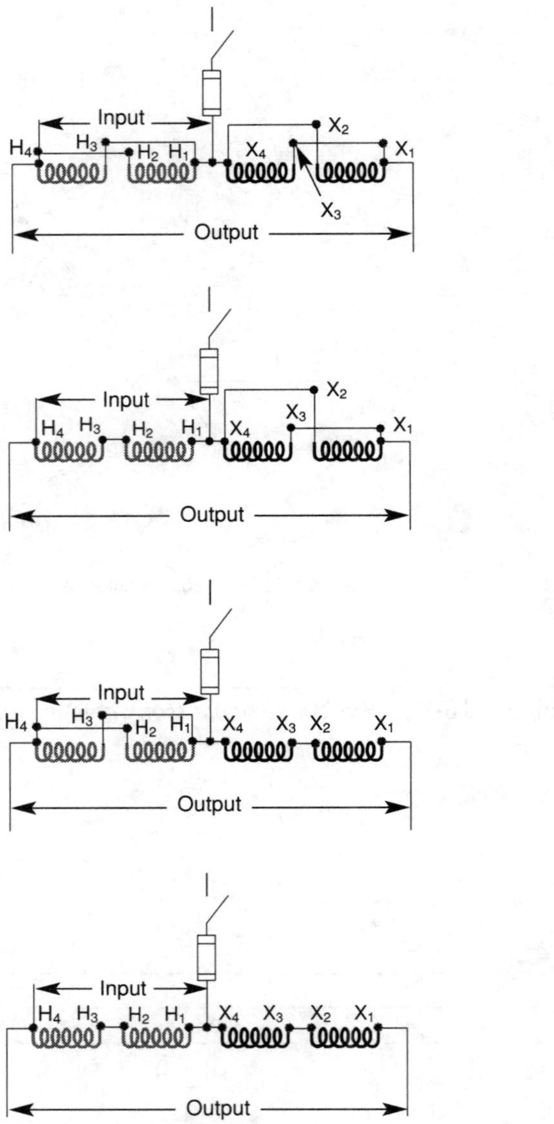
Input
Output
H_4
H_3
H_2
H_1
X_4
X_3
X_2
X_1

Figure 10-34: *Typical single-phase buck-and-boost transformer connections*

Manufacturers of buck-and-boost transformers normally offer "easy-selector" charts to quickly select a buck-and-boost transformer for practically any application. These charts may be obtained from electrical equipment suppliers or ordered directly from the manufacturer — often at no charge. Instructions accompanying these charts will enable anyone familiar with transformers and electrical circuits to use them. However, a brief review of principles involved in using buck-and-boost transformers is in order.

When reviewing the easy-selector charts mentioned in the above paragraph, it may astonish many people; that is, how can these transformers handle a load so much greater that its nameplate rating? For example, a typical 1 kVA buck-and-boost transformer can easily handle a 10 kVA load when the voltage boost is only 10 percent! Let's see how this is possible.

Assume that we have a 1 kVA (1000 VA) insulating transformer designed to transform 120/240 volts to 12/24 volts. This results in a transformer winding with a ratio of 10:1. The primary current may be found by the following equation:

$$\text{Primary current} = \frac{1000 \text{ VA}}{240 \text{ Volts}} = 4.166 \text{ amperes}$$

Therefore, 4.166 amperes rounds off to 4.17 amperes. Because the transformation ratio is 10 to 1, the secondary amperes will be 41.7 amperes; that is, 4.17 multiplied by 10 = 41.7, or the amperage may be determined by the following equation:

$$\text{Secondary current} = \frac{1000 \text{ VA}}{24 \text{ Volts}} = 41.66 \text{ amperes}$$

Figure 10-35 shows a wiring diagram of the transformer under consideration. Note that a 240-volt source delivers 10 kVA, but the secondary winding of the 1 kVA buck-and-boost transformer has been placed in series with the line to the load.

$$P(\text{kVA}) = EI, \text{ kVA} = \text{kV} \times I$$

or

$$I = \frac{\text{kVA}}{\text{kV}}$$

Therefore, by substituting I = 10 over 0.24, the secondary current equals 41.7 amperes. Thus, the current from the source is 41.7 amperes. The 1 kVA buck-and-boost transformer at full load has a secondary current that also equals 41.7 amperes. Consequently, there is no harm in connecting it in the line because its secondary current rating is adequate to handle the load current. Because we started with 240 volts at the source and now add 24 volts to it, the load actually gets 264 volts. To find the kVA ratings of this system at the load, multiply volts divided by 1000 times amperes:

$$\text{kVA} = \frac{\text{Volts}}{1000} \times \text{amperes}$$

$$= \frac{264}{1000} \times 41.7 = 11 \text{ kVA}$$

Ten kVA comes from the source and the extra 1 kVA from the booster secondary. The total current consumed by the circuit is 41.7 amperes plus 4.17 or 45.87 amperes.

In actual practice, four leads would not be brought out and connected to the source as shown in Figure 10-35. Rather, the circuit would probably be simplified as shown in Figure 10-36. Since the two source lines marked A in Figure 10-35 are the same point and the two

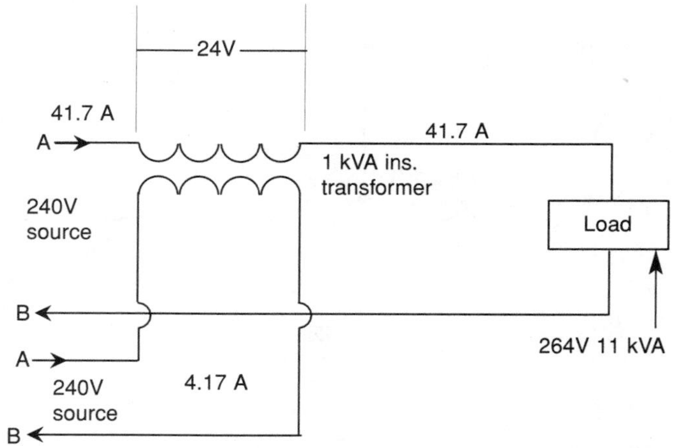
24V
41.7 A
A
41.7 A
1 kVA ins.
transformer
240V
source
Load
B
A
264V 11 kVA
240V
source
4.17 A
B

Figure 10-35: *Transformer circuit under consideration*

Bs are identical, they may be connected together as shown in Figure 10-36. The connections are identical to Figure 10-35 except now there are only two lines running to the power source and the combined current may be shown as 45.87 amperes.

In actual practice, drawing diagrams such as the ones shown in Figures 10-35 and 10-36 are usually simplified even more — in kind of a "ladder" or schematic diagram as shown in Figure 10-37. Actually, all three of these wiring diagrams (Figures 10-35, 10-36, and 10-37) indicate the same thing, and following the connections on any of these drawings will produce the same results at the load.

It should now be evident how a little 1 kVA buck-and-boost transformer, when connected in the circuit as described previously, can actually carry 11 kVA in its secondary winding.

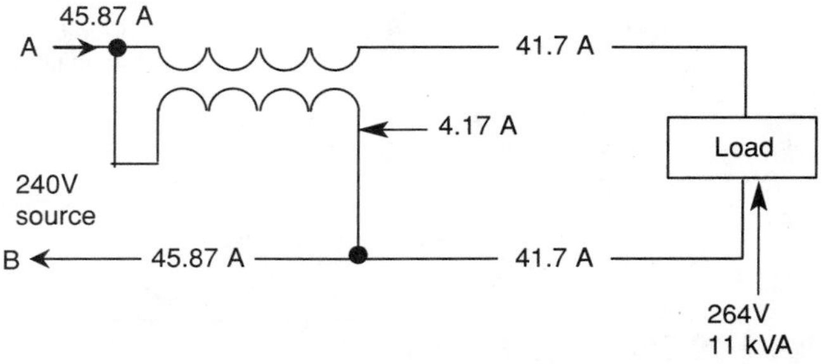
45.87 A
A
41.7 A
4.17 A
Load
240V
source
B
45.87 A
41.7 A
264V
11 kVA

Figure 10-36: *Simplified diagram of Figure 10-35*

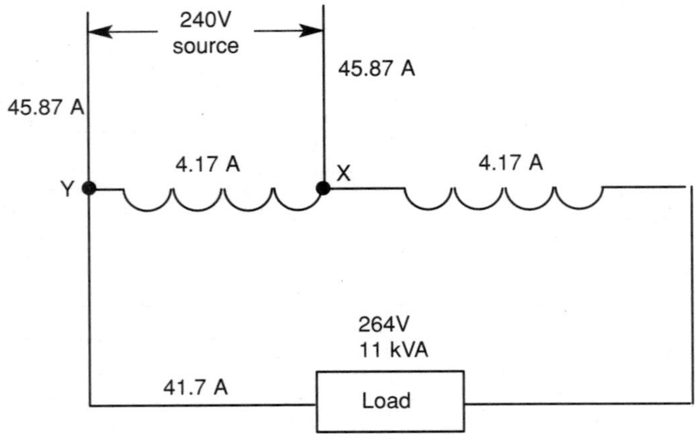
240V
source
45.87 A
45.87 A
4.17 A
4.17 A
Y
X
264V
11 kVA
41.7 A
Load

Figure 10-37: *Further simplification of the transformer circuit under consideration*

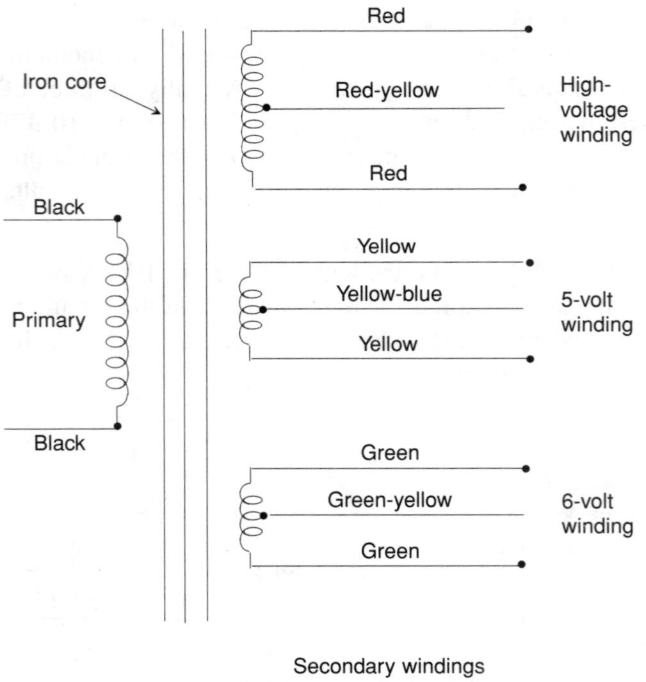
Iron core
Red
Red-yellow
Red
High-
voltage
winding
Black
Primary
Black
Yellow
Yellow-blue
Yellow
5-volt
winding
Green
Green-yellow
Green
6-volt
winding
Secondary windings

Figure 10-38: *Typical control-transformer wiring diagram*

CONTROL TRANSFORMERS

Control transformers are available in numerous types, but most control transformers are dry-type step-down units with the secondary control circuit isolated from the primary line circuit to assure maximum safety. See Figure 10-38. These transformers and other components are usually mounted within an enclosed control box or control panel, which has a pushbutton station or stations independently grounded as recommended by the *NEC*. Industrial control transformers are especially designed to accommodate the momentary current inrush caused when electromagnetic components are energized, without sacrificing secondary voltage stability beyond practical limits.

Other types of control transformers, sometimes referred to as control and signal transformers, normally do not have the required industrial control transformer regulation characteristics. Rather, they are constant-potential, self-air-cooled transformers used for the purpose of supplying the proper reduced voltage for control circuits of electrically operated switches or other equipment and, of course, for signal circuits. Some are of the open type with no protective casing over the winding, while others are enclosed with a metal casing over the winding.

In seeking control transformers for any application, the loads must be calculated and completely analyzed before the proper transformer selection can be made. This analysis involves every electrically energized component in the control circuit. To select an appropriate control transformer, first determine the voltage and frequency of the supply circuit. Then determine the total inrush volt-amperes (watts) of the control circuit. In doing so, do not neglect the current requirements of indicating lights and timing devices that do not have inrush volt-amperes, but are energized at the same time as the other components in the circuit. Their total volt-amperes should be added to the total inrush volt-amperes.

POTENTIAL AND CURRENT TRANSFORMERS

In general, a potential transformer is used to supply voltage to instruments such as voltmeters, frequency meters, power-factor meters, and watt-hour meters. The voltage is proportional to the primary voltage, but it is small enough to be safe for the test instrument. The secondary of a potential transformer may be designed for several different voltages, but most are designed for 120 volts. The potential transformer is primarily a distribution transformer especially designed for good voltage regulation so that the secondary voltage under all conditions will be as nearly as possible a definite percentage of the primary voltage.

Current Transformers

A current transformer (Figure 10-39) is used to supply current to an instrument connected to its secondary, the current being proportional to the primary current, but small enough to be safe for the instrument. The secondary of a current transformer is usually designed for a rated current of five amperes.

A current transformer operates in the same way as any other transformer in that the same relation exists between the primary and the secondary current and voltage. A current transformer is connected in series with the power lines to which it is applied so that line current flows in its primary winding. The secondary of the current transformer is connected to current devices such as ammeters, wattmeters, watt-hour meters, power-factor meters, some forms of relays, and the trip coils of some types of circuit breakers.

When no instruments or other devices are connected to the secondary of the current transformer, a short-circuit device or connection is placed across the secondary to prevent the secondary circuit from being opened while the primary winding is carrying current. There will be no secondary ampere turns to balance the primary ampere turns, so the total primary current becomes exciting current and magnetizes the core to a high flux density. This produces a high voltage across both primary and secondary windings and endangers the life of anyone coming in contact with the meters or leads.

NEC REQUIREMENTS

Transformers must normally be accessible for inspection except for dry-type transformers under certain specified conditions. Certain types of transformers with a high voltage or kVA rating are required to be enclosed in transformer rooms or vaults when installed indoors. The construction of these vaults is covered in *NEC* Sections 450-41 through 450-48.

In general, the *NEC* specifies that the walls and roofs of vaults must be constructed of materials that have adequate structural strength for the conditions with a minimum fire resistance of three hours. However, where

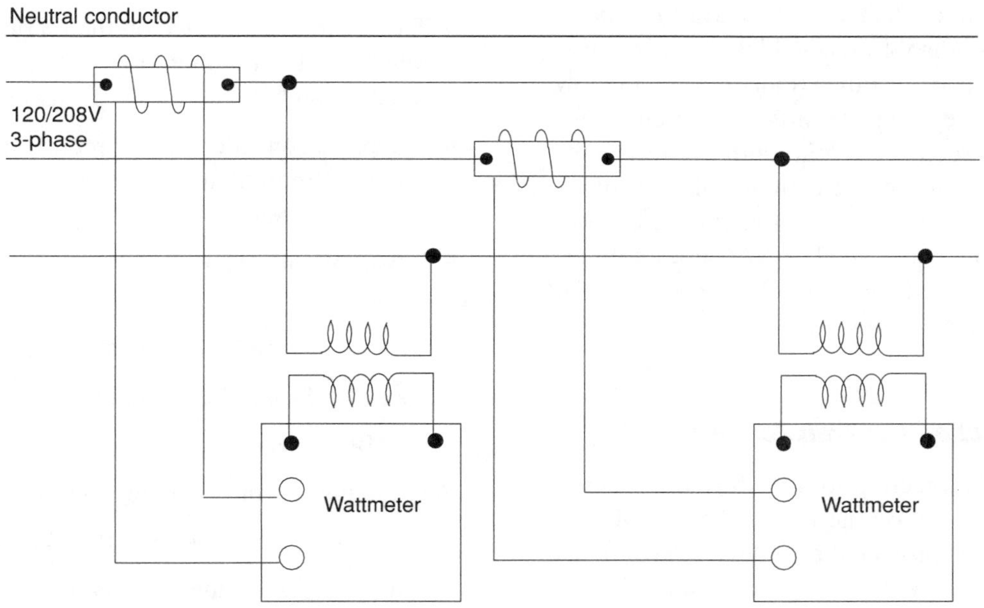
Neutral conductor
120/208V
3-phase
Wattmeter
Wattmeter

Figure 10-39: *Current and potential transformers used in conjunction with watt-hour meter*

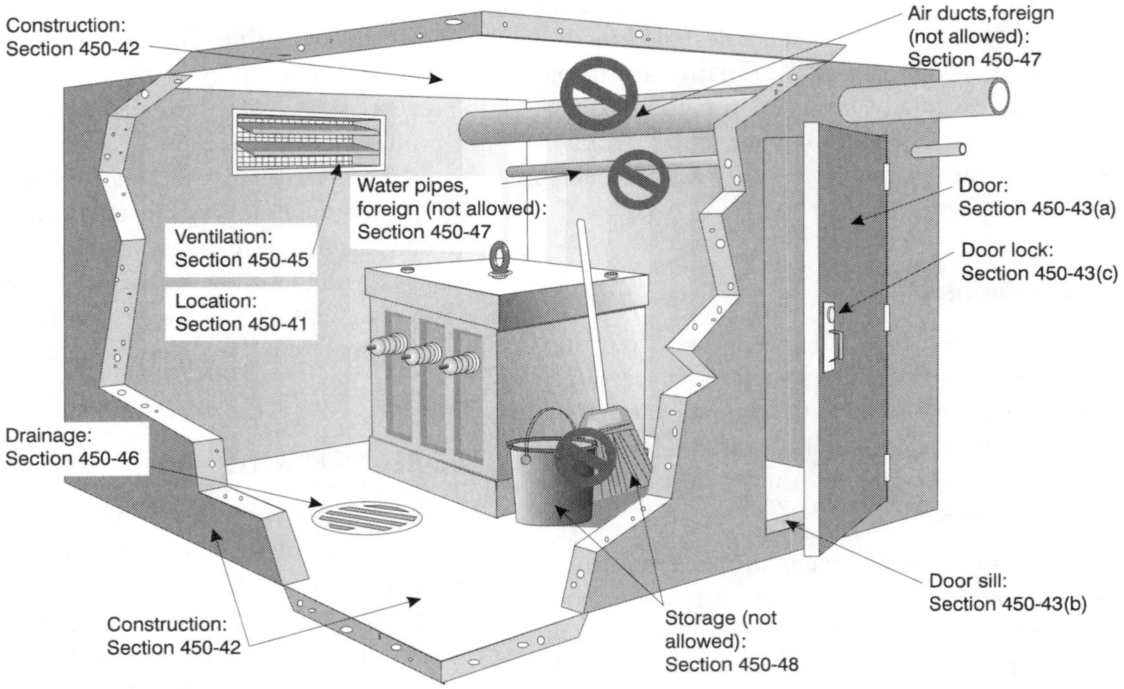
Construction:
Section 450-42
Air ducts,foreign
(not allowed):
Section 450-47
Water pipes,
foreign (not allowed):
Section 450-47
Ventilation:
Section 450-45
Location:
Section 450-41
Door:
Section 450-43(a)
Door lock:
Section 450-43(c)
Drainage:
Section 450-46
Door sill:
Section 450-43(b)
Construction:
Section 450-42
Storage (not
allowed):
Section 450-48

Figure 10-40: *Summary of NEC installation requirements for transformer vaults*

transformers are protected with an automatic sprinkler system, water spray, carbon dioxide, or halon, the fire resistance construction may be lowered to only one hour. The floors of vaults in contact with the earth must be of concrete and not less than 4 inches thick. If the vault is built with a vacant space or other floors (stories) below it, the floor must have adequate structural strength for the load imposed thereon and a minimum fire resistance of three hours. Again, if the fire extinguishing facilities are provided, as outlined above, the fire resistance construction need only be one hour. The *NEC* does not permit the use of studs and wall board construction for transformer vaults. See Figures 10-40 and 10-41.

Overcurrent Protection For Transformers

The overcurrent protection for transformers is based on their rated current, not on the load to be served. The primary circuit may be protected by a device rated or set at not more than 125 percent of the rated primary current of the transformer for transformers with a rated primary current of nine amperes or more.

Instead of individual protection on the primary side, the transformer may be protected only on the secondary side if all the following conditions are met.

- The overcurrent device on the secondary side is rated or set at not more than 125 percent of the rated secondary current.
- The primary feeder overcurrent device is rated or set at not more than 250 percent of the rated primary current.

For example, if a 12 kVA transformer has a primary current rating of:

12,000 watts/480 volts = 25 amperes

and a secondary current rated at:

12,000 watts/120 volts = 100 amperes

the individual primary protection must be set at:

1.25 x 25 amperes = 31.25 amperes

In this case, a standard 30-ampere cartridge fuse rated at 600 volts could be used, as could a circuit breaker approved for use on 480 volts. However, if certain

TRANSFORMER VAULT		
Application	**Requirements**	***NEC* Reference**
Air ducts, foreign	Must not pass through the vault.	Section 450-47
Construction of	Walls, roof, and floor must have adequate structural strength and a minimum fire rating of 3 hours.	Section 450-42
Door	Must be tight-fitting with a minimum fire rating of 3 hours.	Section 450-43(a)
Door lock	Doors must be kept locked and accessible only by qualified personnel.	Section 450-43(c)
Door sill	Must be of sufficient height to confine oil from largest transformer in the vault area.	Section 450-43(b)
Drainage	Where practical, adequate drain must be provided when transformers of more than 100 kVA are used.	Section 450-46
Ducts, foreign	Must not pass through the vault.	Section 450-47
Location	Must be located where they can be ventilated to the outside air.	Section 450-41
Storage	Materials must not be stored in vaults.	Section 450-48
Ventilation	Where required, ventilation openings must comply with (a) through (f) of *NEC* Section 450-45.	Section 450-45
Water pipes, foreign	Must not pass through the transformer vault.	Section 450-47

Figure 10-41: *Description of NEC installation requirements for transformer vaults*

conditions are met, individual primary protection for the transformer is not necessary in this case if the feeder overcurrent-protective device is rated at not more than:

2.5 x 25 amperes = 62.5 amperes

and the protection of the secondary side is set at not more than:

1.25 x 100 amperes = 125 amperes

A 125-ampere circuit breaker could be used.

NOTE: The example cited above is for the transformer only; not the secondary conductors. The secondary conductors must be provided with overcurrent protection as outlined in *NEC* Section 210-20.

The requirements of *NEC* Section 450-3 cover only transformer protection; in practice, other components must be considered in applying circuit overcurrent protection. For circuits with transformers, requirements for conductor protection per *NEC* Articles 240 and 310 and for panelboards per *NEC* Article 384 must be observed. Refer to *NEC* Sections 240-3(f); 240-21(c), and *NEC* Section 384-16(e).

Primary Fuse Protection Only (NEC Section 450-3(b): If secondary fuse protection is not provided, then the primary fuses must not be sized larger than 125 percent of the transformer primary full-load amperes, except if the transformer primary full-load amperes (F.L.A.) is that shown in *NEC* Table 450-3(b) (see Figure 10-42).

Individual transformer primary fuses are not necessary where the primary circuit fuse provides this protection.

Primary and Secondary Protection: In unsupervised locations, with primary over 600 volts, the primary fuse can be sized at a maximum of 300 percent. If the secondary is also over 600 volts, the secondary fuses can be sized at a maximum of 250 percent for transformers with impedances not greater than 6 percent; 225 percent for

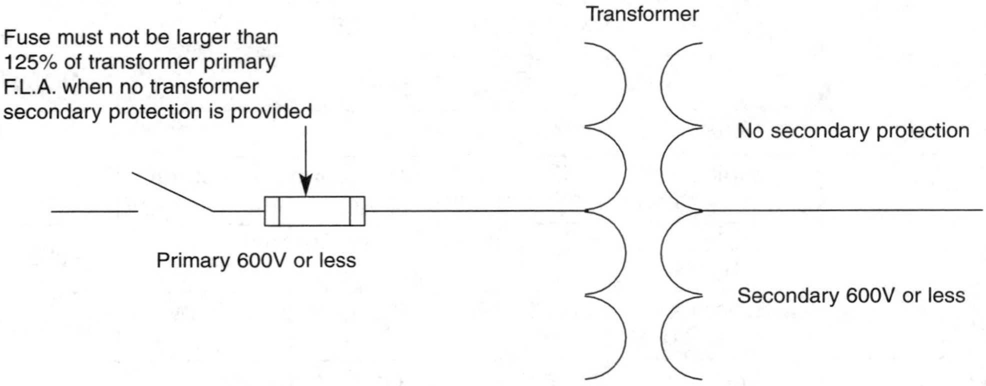
Fuse must not be larger than 125% of transformer primary F.L.A. when no transformer secondary protection is provided
Transformer
No secondary protection
Primary 600V or less
Secondary 600V or less

Primary Current	Primary Fuse Rating
9 amperes or more	125% or next higher standard rating if 125% does not correspond to a standard fuse size
2 amperes or more	167% maximum
Less than 2 amperes	300% maximum

Figure 10-42: *Transformer circuit with primary fuse only*

transformers with impedances greater than 6 percent and not more than 10 percent. If the secondary is 600 volts or below, the secondary fuses can be sized at a maximum of 125 percent. Where these settings do not correspond to a standard fuse size, the next higher standard size is permitted.

In supervised locations, the maximum settings are as shown in Figure 10-43 on the next page except for secondary voltages of 600 volts or below, where the secondary fuses can be sized at a maximum of 250 percent.

Primary Protection Only: In supervised locations, the primary fuses can be sized at a maximum of 250 percent, or the next larger standard size if 250 percent does not correspond to a standard fuse size.

NOTE: The use of "Primary Protection Only" does not remove the requirements for compliance with *NEC* Articles 240 and 384. See (FPN) in *NEC* Section 450-3 which references *NEC* Sections 240-3 and 240-100 for proper protection for secondary conductors.

Protection For Small Power Transformers

Low amperage, E-rated medium voltage fuses are general purpose current limiting fuses. The E rating defines the melting-time-current characteristic of the fuse and permits electrical interchangeability of fuses with the same E rating.

Low amperage, E-rated fuses are designed to provide primary protection for potential, small service, and control transformers. These fuses offer a high level of fault current interruption in a self-contained non-venting package which can be mounted indoors or in an enclosure. As for all current-limiting fuses, the basic application rules found in the *NEC* and manufacturer's literature should be adhered to. In addition, potential transformer fuses must have sufficient inrush capacity to successfully pass through the magnetizing inrush current of the transformer. If the fuse is not sized properly, it will open before the load is energized. The maximum magnetizing inrush currents to the transformer at system voltage and the duration of this inrush current varies with the transformer design. Magnetizing inrush currents are usually denoted as a percentage of the transformer full load current, i.e., 10X, 12X, 15X, etc. The inrush current duration is usually given in seconds. Where this information is available, an easy check can be made on the appropriate minimum-melting curve to verify proper fuse selection. In lieu of transformer inrush data, the rule-of-thumb is to select a fuse size rated at 300 percent of the primary full load current or the next larger standard size.

Example

The transformer manufacturer states that an 800 VA 240-volt, single-phase potential transformer has a magnetizing inrush current of 12X lasting for 0.1 second.

A. I_{FL} = 800VA/2400V = 0.333 ampere

Inrush Current = 12 x 0.333 = 4 amperes

Since the voltage is 2400 volts we can use either a JCW or a JCD fuse. The proper fuse would be a JCW-1E, or JCD-1E.

B. Using the rule-of-thumb — 300 percent of .333 ampere is .999 ampere.

Therefore we would choose a JCW-1E or JCD-1E.

Typical Potential Transformer Connections: The typical potential transformer connections encountered in industry can be grouped into two categories:

- Those connections that require the fuse to pass only the magnetizing inrush of one potential transformer.
- Those connections that must pass the magnetizing inrush of more than one potential transformer.

Fuses for Medium Voltage Transformers and Feeders: E-rated, medium-voltage fuses are general purpose current limiting fuses. The fuses carry either an "E" or an "X" rating which defines the melting-time-current characteristic of the fuse. The ratings are used to allow electrical interchangeability among different manufacturers' fuses.

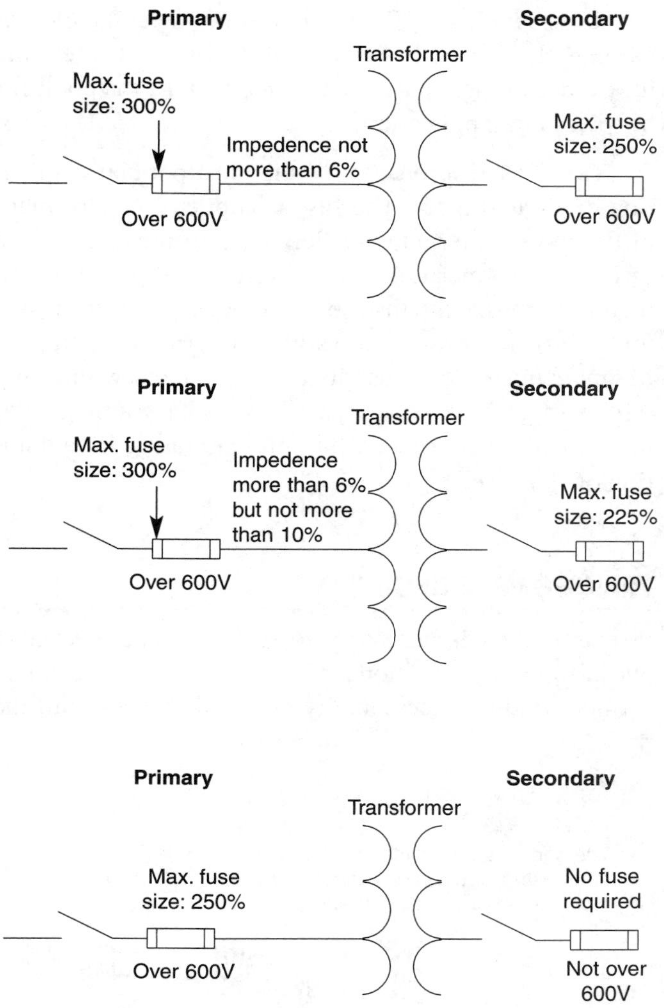
Primary
Secondary
Transformer
Max. fuse size: 300%
Impedence not more than 6%
Max. fuse size: 250%
Over 600V
Over 600V
Primary
Secondary
Transformer
Max. fuse size: 300%
Impedence more than 6% but not more than 10%
Max. fuse size: 225%
Over 600V
Over 600V
Primary
Secondary
Transformer
Max. fuse size: 250%
No fuse required
Over 600V
Not over 600V

Figure 10-43: *Minimum overcurrent protection for transformers in supervised locations*

For a general purpose fuse to have an E rating, the following conditions must be met:

- The current responsive element with ratings 100 amperes or below shall melt in 300 seconds at an rms current within the range of 200 percent to 240 percent of the continuous current rating of the fuse unit. (ANSI C37.46).
- The current responsive element with ratings above 100 amperes shall melt in 600 seconds at an rms current within the range of 220 percent to 264 percent of the continuous current rating of the fuse unit. (ANSI C37.46).

A fuse with an "X" rating does not meet the electrical interchangeability for an E rated fuse but offers the user other ratings that may provide better protection for his particular application.

Transformer protection is the most popular application of E-rated fuses. The fuse is applied to the primary of the transformer and is solely used to prevent rupture of the transformer due to short circuits. It is important, therefore, to size the fuse so that it does not clear on system inrush or permissible overload currents. Magnetizing inrush must also be considered with sizing a fuse. In general, power transformers have a magnetizing inrush current of 12X the full load rating for a duration of 1/10 second.

TRANSFORMER GROUNDING

Grounding is necessary to remove static electricity and also as a precautionary measure in case the transformer windings accidentally come in contact with the core or enclosure. All should be grounded and bonded to meet *NEC* requirements and also local codes, where applicable.

The tank of every power transformer should be grounded to eliminate the possibility of obtaining static shocks from it or being injured by accidental grounding of the winding to the case. A grounding lug is provided on the base of most transformers for the purpose of grounding the case and fittings.

The *NEC* specifically states the requirements of grounding and should be followed in every respect. Furthermore, certain advisory rules recommended by manufacturers provide additional protection beyond that of the *NEC*. In general, the *NEC* requires that separately derived alternating current systems be grounded as stated in *NEC* Section 250-30.

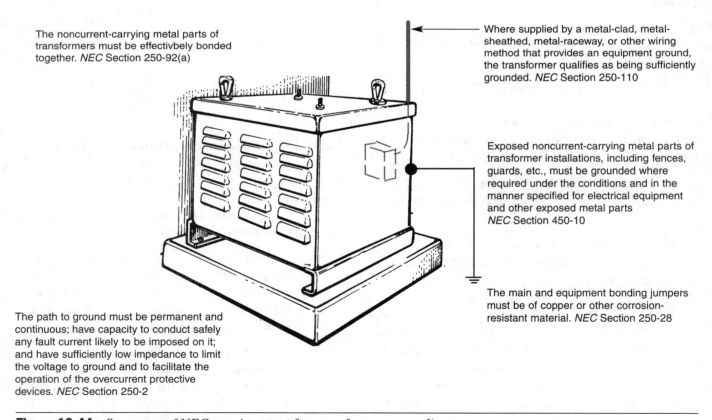

Figure 10-44 summarizes *NEC* regulations governing the grounding of transformers to provide for fault current to trip overcurrent protective devices. This sub-

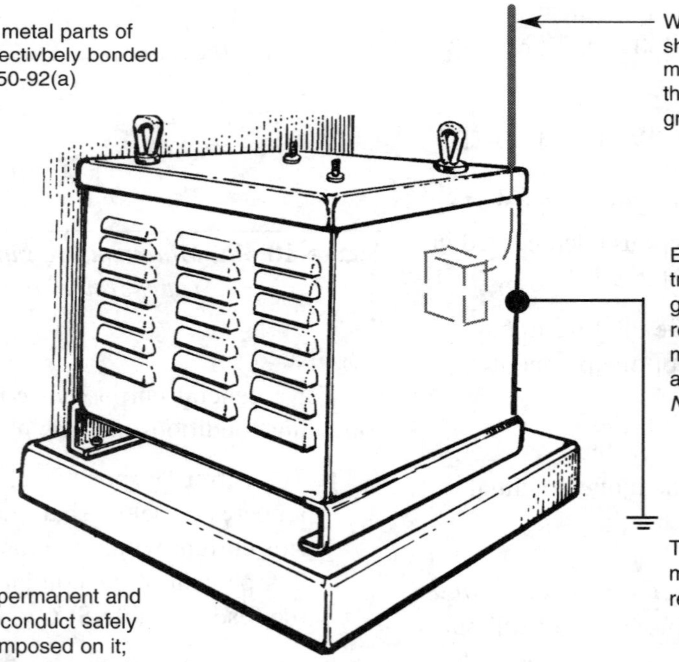
The noncurrent-carrying metal parts of transformers must be effectivbely bonded together. *NEC* Section 250-92(a)
Where supplied by a metal-clad, metal-sheathed, metal-raceway, or other wiring method that provides an equipment ground, the transformer qualifies as being sufficiently grounded. *NEC* Section 250-110
Exposed noncurrent-carrying metal parts of transformer installations, including fences, guards, etc., must be grounded where required under the conditions and in the manner specified for electrical equipment and other exposed metal parts *NEC* Section 450-10
The main and equipment bonding jumpers must be of copper or other corrosion-resistant material. *NEC* Section 250-28
The path to ground must be permanent and continuous; have capacity to conduct safely any fault current likely to be imposed on it; and have sufficiently low impedance to limit the voltage to ground and to facilitate the operation of the overcurrent protective devices. *NEC* Section 250-2

Figure 10-44: *Summary of NEC requirements for transformer grounding*

ject is of the utmost importance, and anyone involved with the design or installation of transformers should thoroughly review all parts of *NEC* Section 250-30.

Rectifier Transformers

Transformers used in connection with rectifiers have to be specially designed to be able to provide single-phase, as well as three-phase, alternating current to the rectifier. The transformation of primary voltage to the secondary voltage makes the rectified current output smoother and more efficient. Because of complicated connections and special requirements, rectifier transformers are considerably more expensive than power transformers of equivalent kVA rating.

Reactors

Reactors are really transformers with only one winding. They are designed so that, for a given current through the reactor, a definite voltage drop exists across the winding. The current flowing through a reactor is all exciting current. The magnetic circuit, therefore, is designed to give the required exciting current at the desired voltage drop.

There are two kinds of reactors, namely, iron-core reactors and air-core reactors. Since the same characteristics can be obtained with either the air-core or the iron-core reactor for a desired range of current and voltage, the decision as to which of the two types to use for a given application is generally made on the basis of comparative costs. Iron-core reactors are usually cheaper wherever the short-time overloads of the reactor are less than three times its continuous-current capacity. For heavier short-time overloads, air-core reactors are ordinarily used.

Typical applications of iron-core reactors are:

- To increase the reactance of one branch of parallel circuits so the load will divide properly among the branches.
- To compensate for the leading charging current of a long transmission line, as when used for shunt reactors operating at nearly constant voltage.
- To limit the current of an arc furnace transformer.

The principal application of air-core reactors is as current-limiting reactors, where the short-circuit current of the system may be many times the normal current. Values from 10 to 20 times the normal current are typical. Air-core reactors are also used for large-capacity shunt reactors. Reactors used to limit the current required for the starting of motors may be of either type, depending on the time required by the motor to come up to speed.

Step-Voltage Regulators

Regulators of the step-voltage type are small transformers (not above 2500-kVA three-phase or 833-kVA single-phase) provided with load tap changers. They are used to raise or to lower the voltage of a circuit in response to a voltage regulating relay or other voltage-control device. Regulators are usually designed to provide secondary voltages ranging from 10 percent below the supply voltage to 10 percent above it, or a total change of 20 percent in 32 steps of $^5/_8$ percent each.

Specialty Transformers

Specialty transformers make up a large class of transformers and autotransformers used for changing line voltage to some particular value best adapted to the load device. The primary voltage is generally 600 volts, or less. Examples of specialty transformers are: sign-lighting transformers, where 120 volts are stepped down to 25 volts for low-voltage tungsten sign lamps; arc-lamp autotransformers, where 240 volts are stepped down to the voltage required for best operation of the arc; and transformers used to change 240 volts power to 120 volts for operating portable tools, fans, welders, and other devices. Also included in this specialty class are neon-sign transformers that step 120 volts up to between 2000 and 15,000 volts for the operation of neon signs.

Summary

When the ac voltage needed for an application is lower or higher than the voltage available from the source, a transformer is used. The essential parts of a transformer are the primary winding, which is connected to the source, and the secondary winding, which is connected to the load, both wound on an iron core. The two windings are not physically connected. The alternating voltage in the primary winding induces an alternating voltage in the secondary winding. The ratio of the primary and secondary voltages is equal to the ratio of the number of turns in the primary and secondary windings. Transformers may step up the voltage applied to the primary winding and have a higher voltage at the secondary terminals, or they may step down the voltage applied to the primary winding and have a lower voltage available at the secondary terminals. Transformers are applied in ac systems only, single-phase and polyphase, and would not work in dc systems since the induction of voltage depends on the rate of change of current.

A transformer is constructed as a single-phase or a three-phase apparatus. A three-phase transformer has three primary and three secondary windings which may be connected in delta (Δ) or wye (Y). Combinations such as Δ-Δ, Δ-Y, Y-Δ, and Y-Y are possible connections of the primary and secondary windings. The first symbol indicates the connection of the primary winding, and the second, that of the secondary winding. A bank of three single-phase transformers can serve the same purpose as one three-phase transformer. See Figures 10-45 and 10-46 for a summary of *NEC* installation requirements for transformers.

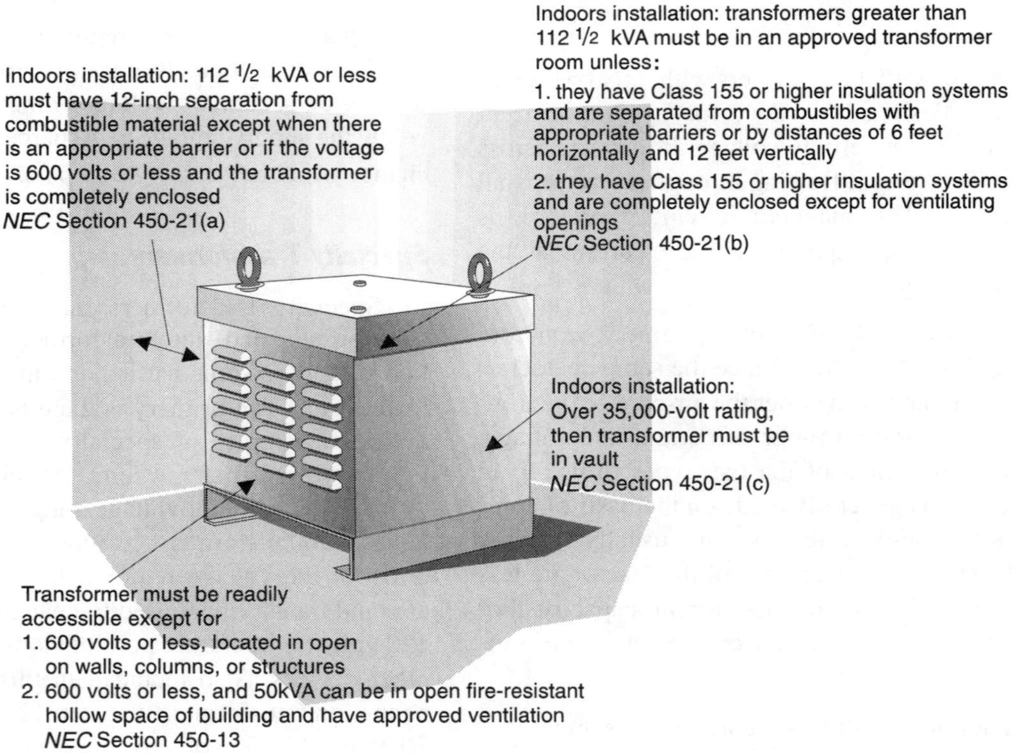
Indoors installation: transformers greater than 112 1/2 kVA must be in an approved transformer room unless:
1. they have Class 155 or higher insulation systems and are separated from combustibles with appropriate barriers or by distances of 6 feet horizontally and 12 feet vertically
2. they have Class 155 or higher insulation systems and are completely enclosed except for ventilating openings
NEC Section 450-21(b)
Indoors installation: 112 1/2 kVA or less must have 12-inch separation from combustible material except when there is an appropriate barrier or if the voltage is 600 volts or less and the transformer is completely enclosed
NEC Section 450-21(a)
Indoors installation: Over 35,000-volt rating, then transformer must be in vault
NEC Section 450-21(c)
Transformer must be readily accessible except for
1. 600 volts or less, located in open on walls, columns, or structures
2. 600 volts or less, and 50kVA can be in open fire-resistant hollow space of building and have approved ventilation
NEC Section 450-13

Figure 10-45: *Summary of NEC installation requirements for dry-type transformers installed indoors*

TRANSFORMERS		
Application	**Requirements**	***NEC* Section**
Accessibility	Transformers must be readily accessible to qualified personnel.	Section 450-13
Location	Must be readily accessible to qualified personnel for maintenance and replacement.	Section 450-13
	Dry-type transformers may be located in the open.	Section 450-13
	Dry-type transformers not exceeding 600 volts and 50 kVA are permitted in fire-resistant hollow spaces of buildings under conditions specified.	Section 450-13
	Liquid-filled transformers must be installed as specified, and usually in vaults when installed indoors or below grade.	Article 450, Part B
Over 600 V	Special regulations apply.	Section 450-3(a)
Overcurrent protection	Primary protection must be rated or set as specified.	Article 450-3

Figure 10-46: *Description of NEC installation requirements for transformers*

Chapter 11
Capacitors, Resistors and Reactors

Capacitors are used in electrical systems to improve the power factor of an electrical installation or an individual piece of electrically-operated equipment. This efficiency, in general, lowers the cost of power.

POWER FACTOR

Power factor is the ratio of useful working current to the total current in an electrical circuit. Since electrical power is the product of current and voltage, the power factor can also be described as the ratio of real power to apparent power as shown in the following equation:

$$PF = \frac{kW}{kVA}$$

where:

PF = power factor

kW = kilowatts

kVA = kilovolt-amperes

Apparent power is made up of two components:

- Real power (expressed in kilowatts).
- The reactive component (expressed in kilovars).

This relationship is shown in Figure 11-1. The horizontal line AB represents the useful real power (kW) in the circuit. The line BC represents the reactive component (kvar) as drawn in a downward direction. Then a line from A to C represents apparent power (kilovolt-amperes or kVA). To the uninitiated, the use of lines representing quantity and direction is often confusing. Lines utilized as such are called vectors. Imagine yourself at point A desiring to reach point C, but, due to obstructions, you must first walk to B, and then turn a right angle and walk to C. The energy you dissipated in reaching C was increased because you could not take the direct course. But, in the final analysis, you ended up at a point the direction and distance of which can be represented by the straight line AC. The power-factor angle shown is called *theta* (θ).

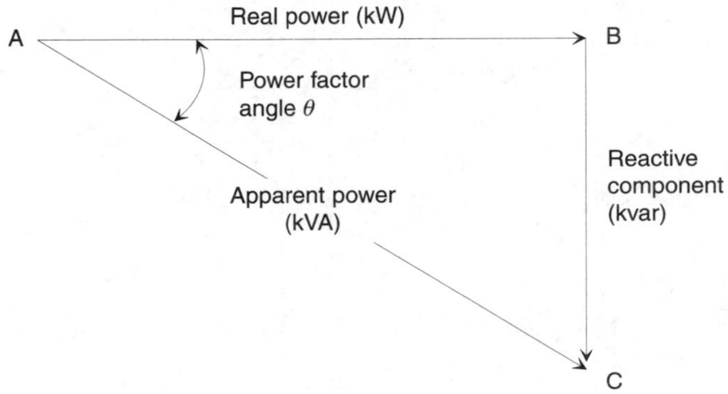

A
Real power (kW)
B
Power factor
angle θ
Reactive
component
(kvar)
Apparent power
(kVA)
C

Figure 11-1: *Relationship of real power and the reactive component (kvar)*

There is interest in power factor because of the peculiarity of certain ac electrical equipment requiring power lines to carry more current than is actually needed to do a specific job. When this effect exists, most power companies penalize the customers, and in a large industrial installation, this penalty can run into lots of wasted money. Consequently, a principle application is used to correct this wasted energy — utilizing the application of capacitors.

Lead And Lag

Real and reactive components cannot be added arithmetically. To understand why, consider the characteristics of electrical circuits. In a pure resistance circuit, the alternating voltage and current curves have the same shape, and the changes occur in perfect step, or phase, with each other. Both are at zero with maximum positive peaks and maximum negative peaks at identical instants. Compare this with a circuit having magnetic characteristics involving units such as induction motors, transformers, fluorescent lights, and welding machines. It is typical that the current needed to establish a magnetic field lags the voltage by 90°.

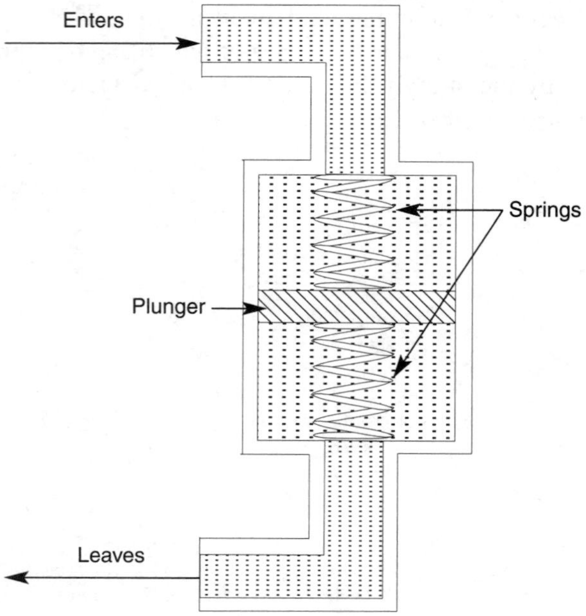
Enters
Springs
Plunger
Leaves

Figure 11-2: *A water system used to illustrate lead and lag in an electrical system*

Visualize a tube of toothpaste. Pressure must be exerted on the tube before the contents ooze out. In other words, there is no flow until pressure is exerted, or, analogously, the flow (current) lags the pressure (voltage). In like manner, the magnetic part of a circuit resists, or opposes, the flow of current through it. In a magnetic circuit, the pressure precedes or leads the current flow, or conversely, the current lags the voltage.

We have the exact opposite in a capacitor. Visualize an empty tank to which a high-pressure air line is attached by means of a valve. At the instant the valve is opened, a tremendous rush of air enters the tank, gradually reducing in rate of flow as the tank pressure approaches the air line pressure. When the tank is up to full pressure, no further flow exists. Accordingly, you must first have a flow of air into the tank before it develops an internal pressure. Consider the tank to be a capacitor and the air line to be the electrical system. In like fashion, current rushes into the capacitor before it builds up a voltage or, in a sense, the current leads the voltage.

A water system provides a better comparison. The system shown in Figure 11-2 is connected to the inlet and outlet of a water pump. For every gallon that enters the upper section, a gallon must flow from the lower section as the plunger is forced down. If the pressure on the upper half is removed, the stored energy in the lower spring will return the plunger to midposition. When the direction of flow is reversed, the plunger travels upward, and if flowmeters were connected to the inlet and outlet, the inlet and outlet flows would prove to be equal. Actually, there is no flow through the cylinder but merely a displacement. The only way we could get a flow through the cylinder would be by leakage around the plunger or if excessive pressure punctured a hole in the plunger.

In a capacitor, we have a similar set of conditions. The electrical insulation can be visualized as the plunger. When a high voltage is applied to one capacitor plate (higher than the other plate), current will rush in. If we remove the pressure and provide an external conducting path, the stored energy will flow to the other plate, discharging the capacitor and bringing it to a balanced condition in much the same fashion as the spring returned the plunger to midposition.

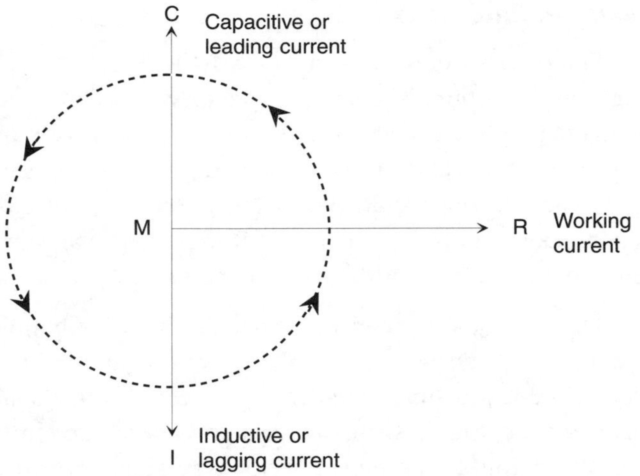
C
Capacitive or
leading current
M
R
Working
current
I
Inductive or
lagging current

Figure 11-3: *Diagram representing three different kinds of current: phase, lagging and leading*

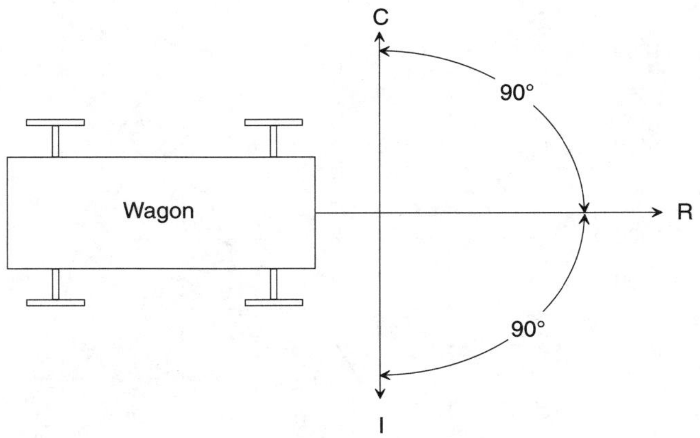
C
90°
Wagon
R
90°
I

Figure 11-4: *Horses and wagon illustrating phase, lagging, and leading currents*

Since no insulator is perfect, some leakage current will flow through it when a voltage differential exists. If we raise the voltage to a point where we break down or puncture the insulation, then the capacitor is damaged beyond repair and must be replaced.

Recognition of the three different kinds of current discussed — in phase, lagging, and leading — permits the drawing of the relationship in Figure 11-3.

The industry accepts a counterclockwise rotation about point M as a means of determining the relative phase position of voltage and current vectors. These may be considered as hands of a clock running in reverse. MC is preceding MR; hence, it is considered leading. MI follows MR; therefore, it is lagging.

The two angles shown are right angles (90°); therefore, it becomes apparent that MC and MI are exactly opposite in direction and will cancel out each other if of equal value.

Consider a wagon to which three horses (*C, I, R*) are hitched, as shown in Figure 11-4. If *C* and *I* pull with equal force, they merely cancel one another's effort, and the wagon will proceed in the direction of *R*, the working horse. If only *R* and *I* are hitched, the wagon will travel in a course or direction between the two horses.

Most electrical circuits contain quantities of working current and lagging current, and one type is just as effective in loading up the circuit as the other. If we know how much inductive current a line or circuit is carrying, then we can connect enough capacitors to that line to cancel out this wasteful and undesired component.

Just as a wattmeter will register the kilowatts in a circuit, a varmeter will register the kvar of reactive power in the line. If an inductive circuit is checked by a meter that reads, say, 150,000 var (150 kvar), then application of a 150-kvar capacitor would completely cancel out the inductive component, leaving only working current in the line.

Power-Factor Correction

In actual practice, full correction to establish the unity power factor is rarely, if ever, recommended. If a system had a constant 24-hour load at a given factor, such correction could be readily approached. Unfortunately, such is not the case, and we are faced with peaks and valleys in the load curves.

If we canceled out, by the addition of capacitors, the inductive (lagging) kvars at peak conditions, the capacitors would continuously pump their full value of leading kilovars into the system. Thus, during early morning hours when inductive kvars are much below peak conditions, a surplus of capacitive kvars would be supplied, and a leading power factor would result. Local conditions may justify such overcorrection, but, in general, overcorrection is not recommended. Figure 11-5 shows how far one should go with fixed capacitors.

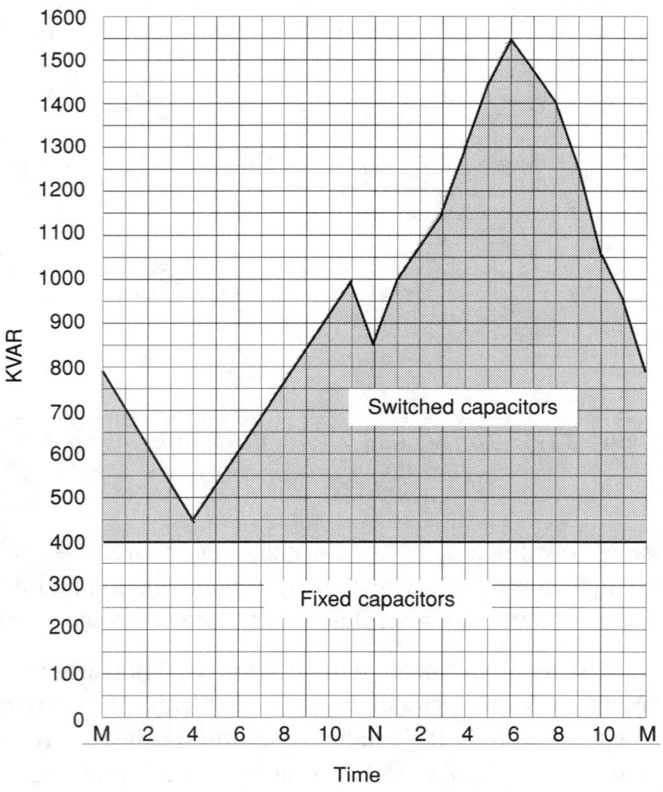
1600
1500
1400
1300
1200
1100
1000
900
800
700
600
500
400
300
200
100
0
KVAR
Switched capacitors
Fixed capacitors
M 2 4 6 8 10 N 2 4 6 8 10 M
Time

Figure 11-5: *Graph depicting relation of kvar to time with fixed capacitors*

A recording kilovarmeter can readily give the curve shown, or else it can be calculated if we know the kW curve and the power factor throughout the day. From 2 to 6 a.m., refrigeration, transformer excitation, high factory loads, and the like result in relatively low readings. When the community comes to life in the morning, televisions, appliances, and factory loads build up a high inductive kvar peak.

The area under the shaded section, which is limited by the lowest 24-hour kvar reading, represents the maximum degree of correction to which switched capacitors are generally applied. It becomes apparent that if we go beyond this point, the power company would be faced with leading power-factor conditions during light-load periods. The shaded area falls into the zone that can be handled effectively with switched capacitors.

Switched Capacitors

There are several ways in which switching of capacitors can be accomplished. A large factory would arrange by manual or automatic operation to switch in a bank of capacitors at the start of the working day and disconnect them when the plant shuts down. The energizing of a circuit breaker control coil can be effected with var, current, voltage, temperature, and time controls.

The application of capacitors to electrical distribution systems has been justified by the overall economy provided. Loads are supplied at reduced cost. The original loads on the first distribution systems were predominantly lighting so the power factor was high. Over the years, the character of loads has changed. Today, loads are much larger and consist of many motor-operated devices that place greater kilovar demands upon electrical systems. Because of the kilovar demand, system power factors have been lower. The results may be threefold:

- Substation and transformer equipment may be taxed to full thermal capacity or overburdened.
- High kilovar demands may, in many cases, cause excessive voltage drops.
- A low power factor may cause an unnecessary increase in system losses.

Capacitors can alleviate these conditions by reducing the kilovar demand from the point of demand all the way back to the generators. Depending on the uncorrected power factor of the system, the installation of capacitors can increase generator and substation capability for an additional load of at least 30 percent, and can increase individual circuit capability such as induction motors, lighting, etc. — from the standpoint of voltage regulation — from 30 percent to a maximum of 100 percent.

CAPACITORS

NEC Article 460 states specific rules for the installation and protection of capacitors other than surge capacitors or capacitors that are part of another apparatus. The chief use of capacitors, as mentioned previously, is to improve the power factor of an electrical installation or an individual piece of electrically-operated equipment. This efficiency, in general, lowers the cost of power.

CAPACITORS		
Application	**Requirements**	***NEC* Reference**
Ampacity of conductors	Must not be less than 135% of rated capacitor current.	Section 460-8(a)
Circuits over 600 V	Special NEC regulations apply.	Article 460, Part B
Disconnecting means	Required for a capacitor unless capacitor is connected to the load side of a motor-running overcurrent device.	Section 460-8(c)
Enclosing and guarding	Must be enclosed, located, or guarded so that persons cannot come into accidental contact with exposed energized parts, terminals, or buses.	Section 460-2(b)
Grounding	Cases must be grounded except when the system is designed to operate at other than ground potential.	Section 460-10
Overcurrent protection	Required in each ungrounded conductor with some exceptions.	Section 460-8(b)
Overcurrent protection for improved power factor	Must be selected on reduced current draw.	Section 460-9
Motor circuits	Capacitor conductors must not be less than 1/3 the size of the motor circuit conductors.	Section 460-8(a)
Stored charge	A means must be provided to drain the stored charge.	Section 460-6

Figure 11-6: *Summary of NEC installation requirements for capacitors*

The *NEC* requirements for capacitors are summarized in Figures 11-6 and 11-7.

Application Of Capacitors

Since capacitors may store an electrical charge and hold a voltage that is present even when a capacitor is disconnected from a circuit, capacitors must be enclosed, guarded, or located so that persons cannot accidentally contact the terminals. In most installations, capacitors are installed out of reach or are placed in an enclosure accessible only to qualified persons. The stored charge of a capacitor must be drained by a discharge circuit either permanently connected to the capacitor or automatically connected when the line voltage of the capacitor circuit is removed. The windings of a motor or a circuit consisting of resistors and reactors will serve to drain the capacitor charge.

Capacitor circuit conductors must have an ampacity of not less than 135 percent of the rated current of the capacitor. This current is determined from the kVA rating of the capacitor as for any load. A 100 kVA (100,000 watts) three-phase capacitor operating at 480 volts has a rated current of

100,000 va / 1.73 x 480 volts = 120.4 amperes

The minimum conductor ampacity is then:

1.35 x 120.4 amperes = 162.5 amperes

When a capacitor is switched into a circuit, a large inrush current results to charge the capacitor to the circuit voltage. Therefore, an overcurrent protective device for the capacitor must be rated or set high enough to allow the capacitor to charge. Although the exact setting is not specified in the *NEC*, typical settings vary between 150 percent and 250 percent of the rated capacitor current.

In addition to overcurrent protection, a capacitor must have a disconnecting means rated at not less than 135 percent of the rated current of the capacitor unless the capacitor is connected to the load side of the motor running overcurrent device. In this case, the motor disconnecting means would serve to disconnect the capacitor and the motor.

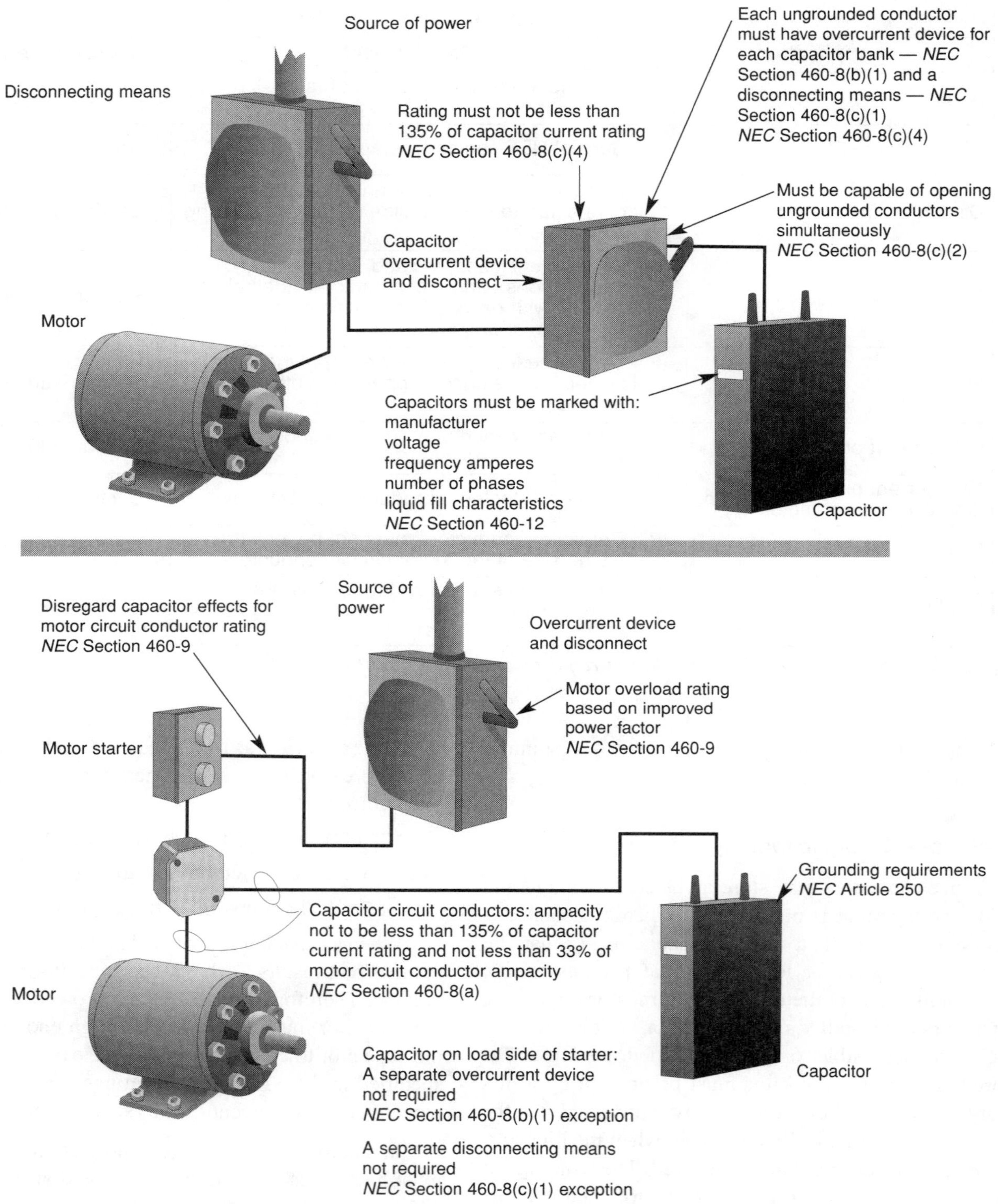
Source of power
Disconnecting means
Each ungrounded conductor must have overcurrent device for each capacitor bank — NEC Section 460-8(b)(1) and a disconnecting means — NEC Section 460-8(c)(1)
NEC Section 460-8(c)(4)
Rating must not be less than 135% of capacitor current rating
NEC Section 460-8(c)(4)
Must be capable of opening ungrounded conductors simultaneously
NEC Section 460-8(c)(2)
Capacitor overcurrent device and disconnect
Motor
Capacitors must be marked with:
manufacturer
voltage
frequency amperes
number of phases
liquid fill characteristics
NEC Section 460-12
Capacitor
Source of power
Disregard capacitor effects for motor circuit conductor rating
NEC Section 460-9
Overcurrent device and disconnect
Motor overload rating based on improved power factor
NEC Section 460-9
Motor starter
Grounding requirements
NEC Article 250
Capacitor circuit conductors: ampacity not to be less than 135% of capacitor current rating and not less than 33% of motor circuit conductor ampacity
NEC Section 460-8(a)
Motor
Capacitor on load side of starter:
A separate overcurrent device not required
NEC Section 460-8(b)(1) exception
A separate disconnecting means not required
NEC Section 460-8(c)(1) exception
Capacitor

Figure 11-7: *Graphic representation of NEC installation requirements for capacitors used to improve the power factor of motor circuits*

A capacitor connected to a motor circuit serves to increase the power factor and reduce the total kVA required by the motor-capacitor circuit. The power factor is defined as the true power in kilowatts divided by the total kVA or

$$PF = \frac{kW}{kVA}$$

where the power factor is a number between .0 and 1.0. A power factor less than one represents a lagging current for motors and inductive devices. The capacitor introduces a leading current that reduces the total kVA and raises the power factor to a value closer to unity. If the inductive load of the motor is completely balanced by the capacitor, a maximum power factor of unity results and all of the input energy serves to perform useful work.

The capacitor circuit conductors for a power factor correction capacitor must have an ampacity of not less than 135 percent of the rated current of the capacitor. In addition, the ampacity must not be less than one-third the ampacity of the motor circuit conductors.

The connection of a capacitor reduces current in the feeder up to the point of connection. If the capacitor is connected on the load side of the motor-running overcurrent device, the current through this device is reduced and its rating must be based on the actual current, not on the full-load current of the motor.

RESISTORS AND REACTORS

NEC Article 470 covers the installation of separate resistors and reactors on electric circuits. However, this Article does not cover such devices that are component parts of other machines and equipment.

In general, the *NEC* requires resistors and reactors to be placed where they will not be exposed to physical damage. Therefore, such devices are normally installed in a protective enclosure such as a controller housing or other type of cabinet. When these enclosures are constructed of metal, they must be grounded as specified in *NEC* Article 250. Furthermore, a thermal barrier must be provided between resistors and/or reactors and any combustible material that is less than 12 inches away from them. A space of 12 inches or more between the devices and combustible material is considered sufficient distance so as not to require a thermal barrier for heat/fire protection.

Insulated conductors used for connections between resistors and motor controllers must be rated at not less than 90°C (194°F) except for motor-starting service. In this latter case, other conductor insulation is permitted provided other Sections of the *NEC* are not violated.

A summary of *NEC* installation requirements for resistors appears in Figure 11-8, while *NEC* installation requirements for reactors are summarized in Figure 11-9.

RESISTORS		
Application	**Requirements**	**NEC Reference**
Clearances	Must be adequate for the voltage involved.	Section 470-18(d)
Conductor insulation	Must have an operating temperature of not less than 90° C.	Section 470-4
Lighting fixtures, direct current	Must be equipped with resistors especially designed for dc operation.	Section 410-74
Lighting fixtures, resistors in	Must be enclosed in an accessible metal cabinet.	Section 410-77
Location	Must be placed where not exposed to physical damage.	Section 470-2
Motors, constant voltage dc motors	Conductor ampacity must be calculated as per *NEC* Table 430-29.	Section 430-29
Space separation	Must be 12" or more between resistor and combustible material.	Section 470-3

Figure 11-8: *Summary of NEC installation requirements for resistors*

REACTORS (600 V or Under)		
Application	**Requirements**	***NEC* Reference**
Conductor insulation	Must have an operating temperature of not less than 90°C.	Section 470-4
Location	Must not be exposed to physical damage.	Section 470-2
Space separation	If distance is less than 12" from any combustible material, a thermal barrier is required.	Section 470-3
REACTORS (Over 600 V)		
Application	**Requirements**	***NEC* Reference**
Clearances	Must be adequate for the voltage involved.	Section 470-18(d)
Combustible materials	Must be installed no less than 12" from combustible materials.	Section 470-18(c)
Grounding	Cases and enclosures must be grounded as per *NEC* Article 250.	Section 470-19
Isolation	Must be isolated by enclosure or elevation.	Section 470-18(b)
Oil-filled reactors	Must comply with applicable requirements of *NEC* Article 450.	Section 470-20
Protection	Must be protected from physical damage.	Section 470-18(a)
Temperature rise	Must be installed so that temperature rise in metallic parts caused by induced circulating current will not be hazardous or constitute a fire hazard.	Section 470-18(e)

Figure 11-9: *Summary of NEC installation requirements for reactors*

Chapter 12
Special Occupancies

Any area in which the atmosphere or a material in the area is such that the arcing of operating electrical contacts, components, and equipment may cause an explosion or fire is considered a hazardous location. In all such cases, special equipment, raceways, and fittings are used to provide a safe wiring system.

INTRODUCTION TO HAZARDOUS LOCATIONS

NEC Articles 500 through 504 cover the requirements of electrical equipment and wiring for all voltages in locations where fire or explosion hazards may exist due to flammable gases or vapor, flammable liquids, combustible dust, or ignitable fibers or flyings. Locations are classified depending on the properties of the flammable vapors, liquids, gases or combustible dusts or fibers that may be present, as well as the likelihood that a flammable or combustible concentration or quantity is present.

Hazardous locations have been classified in the *NEC* into certain class locations. Various atmospheric groups have been established on the basis of the explosive character of the atmosphere for the testing and approval of equipment for use in the various groups. However, it must be understood that considerable skill and judgment must be applied when deciding to what degree an area contains hazardous concentrations of vapors, combustible dusts or easily ignitable fibers and flyings. Furthermore, many factors — such as temperature, barometric pressure, quantity of release, humidity, ventilation, distance from the vapor source, and the like — must be considered. When information on all factors concerned is properly evaluated, a consistent classification for the selection and location of electrical equipment can be developed.

The National Fire Protection Association has compiled a list of the combustible liquids, gases and vapors that will give an area a hazardous location classification. The booklet (NFPA 497) is available for $23.25 for non-members and $21.00 for members. You can call the NFPA at 800-344-3555 or check their Web site at www.nfpa.org for this and other publications you may find helpful.

Class I Locations

Class I atmospheric hazards (Figure 12-1 on the next page) are divided into two Divisions (1 and 2) and also into four groups (A, B, C, and D). The Divisions are summarized in the paragraphs to follow.

Those locations in which flammable gases or vapors may be present in the air in quantities sufficient to produce explosive or ignitable mixtures are classified as Class I locations. If these gases or vapors are present under normal operations, under frequent repair or maintenance operations, or where breakdown or faulty operation of process equipment might also cause simultaneous failure of electrical equipment, the area is designated as Class I, Division 1. Examples of such locations are interiors of paint spray booths where volatile, flammable solvents are used, inadequately ventilated pump rooms where flammable gas is pumped, anesthetizing locations

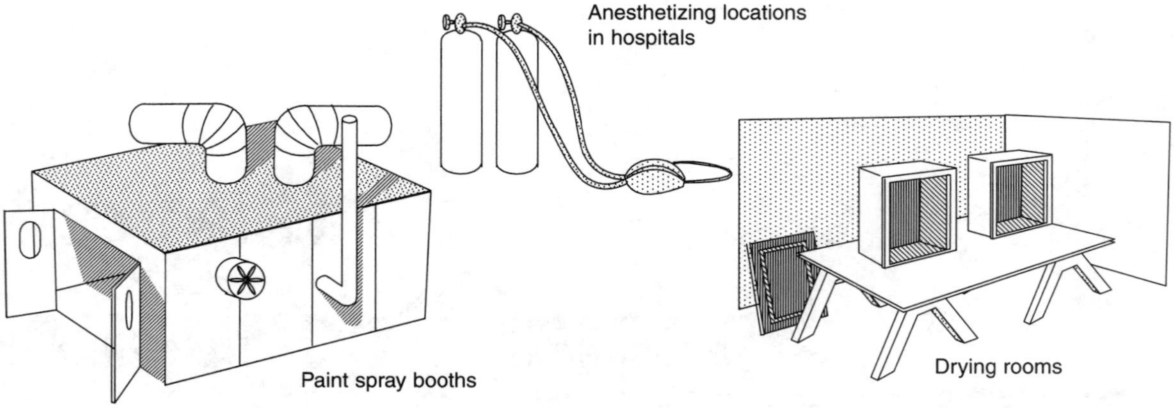
Anesthetizing locations
in hospitals
Paint spray booths
Drying rooms

Figure 12-1: *Typical NEC Class I locations*

CLASS I, DIVISION 1 LOCATIONS		
Application	**Requirements**	***NEC* Reference**
Alarm systems	All components must be suitable for use in Class I, Division 1 locations	Section 501-14
Boxes, fittings	Explosionproof	Section 501-4(a)
Circuit breakers	Class I enclosure	Section 501-6(a)
Dry-type transformers	Class I, Division 1 enclosures	Section 501-7(a)
Flexible connections	Class I, explosionproof	Section 501-4
Generators	Class I, totaly enclosed or submerged	Section 501-8(a)
Lighting fixtures	Explosionproof	Section 501-9
Liquid-filled transformers	Installed in an approved vault	Section 501-2(a)
Motors	Class I, totally enclosed or submerged	Section 501-8(a)
Motor controls	Class I, Division 1	Section 501-10(a)
Panelboards	Explosionproof	Section 501-6(a)
Portable lamps	Explosionproof	Section 501-9(a)
Receptacles	Explosionproof	Section 501-12
Switches	Class I enclosure	Section 501-9(b)(4)
Utilization equipment	Class I, Division 1	Section 501-10(a)
Wiring methods	Rigid metal conduit, steel intermediate metal conduit, or Type MI cable	Section 501-4(a)

Figure 12-2: *Summary of NEC installation requirements in Class I, Division 1 locations*

CLASS I, DIVISION 2 LOCATIONS		
Application	**Requirements**	***NEC* Reference**
Alarm systems	Class I, Division 1	Section 501-14(b)
Boxes, fittings	Need not be explosionproof unless current interrupting contacts are exposed	Section 501-4(b)
Circuit breakers	Class I enclosure	Section 501-6(b)
Dry-type transformers	Class I, general purpose except switching mechanism must be Division 1 enclosures	Section 501-7(b)
Flexible connections	Class I, explosionproof	Section 501-11
Fuses	Class I enclosure	Section 501-6(b)(2)
Generators	Class I, totally enclosed/submerged	Section 501-8(b)
Lighting fixtures	Protected from physical damage	Section 501-9(b)
Liquid-filled transformers	General purpose	Section 501-2(b)
Motors	General purpose unless motor has sliding contacts, switching contacts or integral resistance devices: use Class I	Section 501-8(b)
Motor controls	Class I, Division 1	Section 501-10(b)
Panelboards	General purpose with some exceptions	Section 501-6(b)
Portable lamps	Explosionproof	Section 501-9(b)
Receptacles	Explosionproof	Section 501-12
Switches	Class I enclosure	Section 501-6(b)
Utilization equipment	Class I, Division 1	Section 501-10(b)
Wiring methods	Rigid metal conduit, IMC, or Types MI, MC, MV, TC, SMN, PLTC cables, or enclosed gasketed busways or wireways	Section 501-4(b)

Figure 12-3: *Summary of NEC installation requirements in Class I, Division 2 locations*

of hospitals (to a height of 5 feet above floor level), and drying rooms for the evaporation of flammable solvents. See Figure 12-1.

Class I, Division 2 covers locations where flammable gases, vapors or volatile flammable gases, vapors or volatile liquids are handled either in a closed system, or confined within suitable enclosures, or where hazardous concentrations are normally prevented by positive mechanical ventilation. Areas adjacent to Division 1 locations, into which gases might occasionally flow, would also belong in Division 2. See Figure 12-2 for a summary of *NEC* installation requirements for Class I, Division 1 locations, and Figure 12-3 for requirements in Class I, Division 2 locations.

Class II Locations

Class II locations (Figure 12-4) are those that are hazardous because of the presence of combustible dust. Class II, Division 1 locations are areas where combustible dust, under normal operating conditions, may be present in the air in quantities sufficient to produce explosive or ignitable mixtures; examples are working

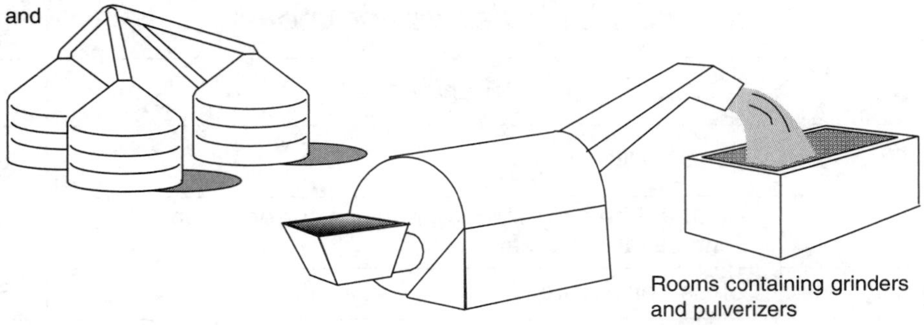
Grain-handling and
storage plants
Rooms containing grinders
and pulverizers

Figure 12-4: *Typical NEC Class II locations*

CLASS II, DIVISION 1 LOCATIONS		
Application	**Requirements**	***NEC* Reference**
Alarm systems	Must comply with (1) through (6) of this *NEC* Section	Section 502-14(a)
Boxes, fittings	Must be provided with threaded bosses for connection to conduit or cable terminations	Section 502-4(a)(1)
Capacitors	Must be installed in vaults as per *NEC* Sections 450-41 through 450-48	Section 502-2(a)
Flexible connections	Liquidtight flexible metal conduit with approved fittings is one method allowed	Section 502-4(a)(2)
Flexible cords	Must be approved for extra-hard usage	Section 502-12
Generators	Must be totally enclosed or approved for Class II, Division 1 locations	Section 501-8
Lighting fixtures	Must comply with (1) through (4) of this *NEC* Section	Section 502-11(a)
Motors	Must be totally enclosed or approved for Class II, Division 1 locations	Section 502-8(a)
Overcurrent protection	Must be protected with dust-ignitionproof enclosures	Section 502-6(a)(1)
Panelboards	Must comply with (1) through (3) of this *NEC* Section	Section 502-6(a)
Portable lamps	Must comply with (1) through (4) of this *NEC* Section	Section 502-11(a)
Receptacles	Must be approved for Class II locations	Section 502-13(a)
Seals	One method that may be used to prevent the entrance of dust into dust-ignitionproof enclosures	Section 502-5
Switches	Must comply with (1) through (3) of this *NEC* Section	Section 502-6(a)
Transformers	Must comply with (1) through (3) of this *NEC* Section	Section 502-2(a)
Utilization equipment	Must be approved for the specific location	Section 502-10(a)
Ventilating piping	Must be dusttight through their length	Section 502-9(a)
Wiring methods	Must use threaded rigid steel conduit, IMC, or Type MI cable	Section 502-4(a)

Figure 12-5: *Summary of NEC installation requirements for Class II, Division 1 locations*

CLASS II, DIVISION 2 LOCATIONS		
Application	**Requirements**	***NEC* Reference**
Alarm systems	Must comply with (1) through (5) of this *NEC* Section	Section 502-14(b)
Boxes, fittings	Must be made to minimize the entrance of dust	Section 502-4(b)(1)
Capacitors	Must comply with (1) through (2) of this NEC Section	Section 502-2(b)
Flexible connections	Liquidtight flexible metal conduit with approved fittings is one method allowed	Section 502-4(b)(2)
Flexible cords	Must be approved for extra-hard usage	Section 502-12
Generators	Must be totally enclosed and approved for the usage	Section 502-8(b)
Lighting fixtures	Must be as specified in (1) through (5) of this *NEC* Section	Section 502-11(b)
Motors	Must be totally enclosed and approved for the usage	Section 502-8(b)
Overcurrent protection	Must be dusttight	Section 502-6(b)
Panelboards	Must be dusttight	Section 502-6(b)
Portable lamps	Must be approved for Class II locations	Section 502-11(b)(1)
Receptacles	Must be designed so that connection to the supply circuit cannot be made or broken while live parts are exposed	Section 502-13(b)
Seals	Must be accessible but are not required to be explosionproof	Section 502-5
Switches	Must be dusttight	Section 502-6(b)
Transformers	Must comply with (1) through (3) of this *NEC* Section	Section 502-2(b)
Utilization equipment	Must comply with (1) through (4) of this *NEC* Section	Section 502-10(b)
Ventilating piping	Must be sufficiently tight to prevent appreciable quantities of dust into the ventilated quipment	Section 502-9(b)
Wiring methods	Rigid steel conduit, IMC, EMT, Type ME or MI cable, and other methods specified in this *NEC* Section	Section 502-4(b)

Figure 12-6: *Summary of NEC installation requirements for Class II, Division 2 locations*

areas of grain-handling and storage plants and rooms containing grinders or pulverizers. Class II, Division 2 locations are areas where dangerous concentrations of suspended dust are not likely, but where dust accumulations might form.

Besides the two Divisions (1 and 2), Class II atmospheric hazards cover three groups of combustible dusts. The groupings are based on the resistivity of the dust. Group E is always Division 1. Group F, depending on the resistivity, and Group G may be either Division 1 or 2. Since the *NEC* is considered the definitive classification tool and contains explanatory data about hazardous atmospheres, refer to *NEC* Section 500-5(b) for exact definitions of Class II, Divisions 1 and 2. See Figure 12-5 for a summary of *NEC* installation requirements in Class II, Division 1 locations and Figure 12-6 for those in Class II, Division 2 locations.

Class III Locations

These locations are those areas that are hazardous because of the presence of easily ignitable fibers or flyings, but such fibers and flyings are not likely to be in suspension in the air in these locations in quantities sufficient to produce ignitable mixtures. Such locations usually include some parts of rayon, cotton, and textile mills, clothing manufacturing plants, and woodworking plants. See Figure 12-7 on the next page. Explosionproof fittings and enclosures are normally not required in Class III locations, but there are still many special wiring techniques employed.

Figure 12-8 summarizes *NEC* installation requirements for Class III locations.

Groups

In Class I and Class II locations, the hazardous materials are further divided into groups; that is, Groups A, B, C, D in Class I and Groups E, F, and G in Class II. For more information on flammable liquids, gases and solids, see *Classification of Flammable Liquids, Gases, or Vapors and Hazardous (Classified) Locations*, NFPA Publication No. 497.

Once the class of an area is determined, the conditions under which the hazardous material may be present determine the division. In Class I and Class II, Division 1 locations, the hazardous gas or dust may be present in the air under normal operating conditions in dangerous concentrations. In Division 2 locations, the hazardous material is not normally in the air, but it might be released if there is an accident or if there is faulty operation of equipment.

PREVENTION OF EXTERNAL IGNITION/EXPLOSION

The main purpose of using explosionproof fittings and wiring methods in hazardous areas is to prevent ignition of flammable liquids or gases and to prevent an explosion. Explosionproof fittings are designed to retain an explosion within the fitting or conduit system.

Sources Of Ignition

In certain atmospheric conditions when flammable gases or combustible dusts are mixed in the proper proportion with air, any source of energy is all that is needed to touch off an explosion.

One prime source of energy is electricity. Equipment such as switches, circuit breakers, motor starters, push-button stations, or plugs and receptacles, can produce arcs or sparks in normal operation when contacts are opened and closed. This could easily cause ignition if proper precautions are not taken.

Other hazards are devices that produce heat, such as lighting fixtures and motors. Here surface temperatures may exceed the safe limits of many flammable atmospheres.

Finally, many parts of the electrical system can become potential sources of ignition in the event of insulation failure. This group would include wiring (particularly splices in the wiring), transformers, impedance coils, solenoids, and other low-temperature devices without make-or-break contacts.

Non-electrical hazards such as sparking metal can also easily cause ignition. A hammer, file or other tool that is dropped on masonry or on a ferrous surface can cause a hazard unless the tool is made of non-sparking material. For this reason, portable electrical equipment is usually made from aluminum or other material that will not produce sparks if the equipment is dropped.

Electrical safety, therefore, is of crucial importance. The electrical installation must prevent accidental ignition of flammable liquids, vapors and dusts released to the atmosphere. In addition, since much of this equipment is used outdoors or in corrosive atmospheres, the material and finish must be such that maintenance costs and shutdowns are minimized.

Combustion Principles

Three basic conditions must be satisfied for a fire or explosion to occur:

- A flammable liquid, vapor or combustible dust must be present in sufficient quantity.
- The flammable liquid, vapor or combustible dust must be mixed with air or oxygen in the proportions required to produce an explosive mixture.
- A source of energy must be applied to the explosive mixture.

In applying these principles, the quantity of the flammable liquid or vapor that may be liberated and its physical characteristics must be recognized.

Vapors from flammable liquids also have a natural tendency to disperse into the atmosphere, and rapidly become diluted to concentrations below the lower explosion limit particularly when there is natural or mechanical ventilation.

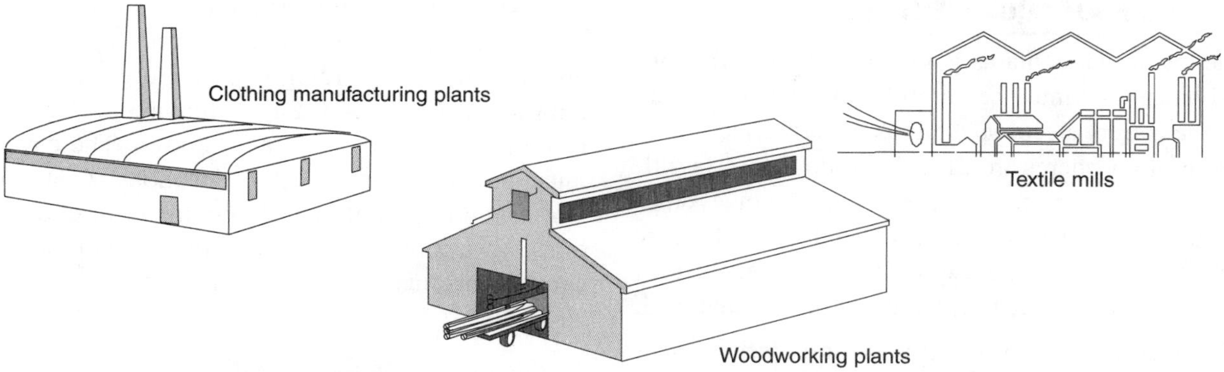
Clothing manufacturing plants
Textile mills
Woodworking plants

Figure 12-7: *Typical NEC Class III locations*

CLASS III LOCATIONS		
Application	**Requirements**	***NEC* Reference**
Alarm systems	Must comply with applicable requirements of *NEC* Section 503	Section 503-12
Boxes, fittings	Must be dusttight	Section 503-3(a)(1)
Capacitors	Must comply with *NEC* Section 502-2(b)	Section 503-2
Cranes	Must comply with (a) through (d) of this *NEC* Section	Section 503-13
Flexible cords	Must be approved for extra-hard usage	Section 502-10
Generators	Must be totally enclosed	Section 503-6
Lighting fixtures	Must comply with (a) through (d) of this *NEC* Section	Section 503-9
Motors	Must be totally enclosed	Section 503-6
Overcurrent protection	Must be provided with dusttight enclosures	Section 503-4
Panelboards	Must be provided with dusttight enclosures	Section 503-4
Portable lamps	As specified in this *NEC* Section	Section 503-9(d)
Receptacles	Must be designed to minimize the entrance of fibers or flyings	Section 503-11
Switches	Must be provided with dusttight enclosures	Section 503-4
Transformers	Must comply with *NEC* Section 502-2(b)	Section 503-2
Utilization equipment	Must be approved for Class III locations	Section 503-8
Ventilating piping	Must be sufficiently tight to prevent appreciable quantities of fibers or flyings	Section 503-7
Wiring methods	Rigid steel conduit, rigid nonmetallic conduit, IMC, EMT, dusttight wireways, or Type MC or MI cable with approved fittings	Section 503-3(a)

Figure 12-8: *Summary of NEC installation requirements for Class III locations*

EXPLOSIONPROOF EQUIPMENT

Each area that contains gases or dusts that are considered hazardous must be carefully evaluated to make certain the correct electrical equipment is selected. Many hazardous atmospheres are Class I, Group D, or Class II, Group G. However, certain areas may involve other groups, particularly Class I, Groups B and C. Conformity with the *NEC* requires the use of fittings and enclosures approved for the specific hazardous gas or dust involved.

The wide assortment of explosionproof equipment now available makes it possible to provide adequate electrical installations under any of the various hazardous conditions. However, the electrician must be thoroughly familiar with all *NEC* requirements and know what fittings are available, how to install them properly, and where and when to use the various fittings.

For example, some workers are under the false belief that a fitting rated for Class I, Division 1 can be used under any hazardous conditions. However, remember the groups! A fitting rated for, say, Class I, Division 1, Group C cannot be used in areas classified as Group A or B. On the other hand, fittings rated for use in Group A may be used for any group beneath A; fittings rated for use in Class I, Division 1, Group B can be used in areas rated as Group B areas or below, but not vice versa.

Explosionproof fittings are rated for both classification and groups. All parts of these fittings, including covers, are rated accordingly. Therefore, if a Class I, Division 1, Group A fitting is required, a Group B (or below) fitting cover must not be used. The cover itself must be rated for Group A locations. Consequently, when working on electrical systems in hazardous locations, always make certain that fittings and their related components match the condition at hand.

Intrinsically Safe Equipment

Intrinsically safe equipment is equipment and wiring that are incapable of releasing sufficient electrical energy under normal or abnormal conditions to cause ignition of a specific hazardous atmospheric mixture in its most easily ignitable concentration.

The use of intrinsically safe equipment is primarily limited to process control instrumentation, since these electrical systems lend themselves to the low energy requirements.

Installation rules for intrinsically safe equipment for Class I, II, and II locations are covered in *NEC* Article 504, and a summary of *NEC* installation requirements is shown in Figure 12-9. In general, intrinsically safe equipment and its associated wiring must be installed so

INTRINSICALLY SAFE EQUIPMENT		
Application	**Requirements**	***NEC* Reference**
Approval	All intrinsically safe equipment must be approved	Section 504-4
Bonding	Must be bonded as per (a) and (b) of this *NEC* Section	Section 504-60
Conductors, open wiring, separation of	Must be separated as per (1) through (3) of this *NEC* Section	Section 504-30(a)
Grounding	Must be grounded	Section 504-50
Identification, at terminals	Must be identified at terminals and junction locations	Section 504-80(a)
Identification, wiring	Raceways, cable trays, and open wiring must be identified with permanently affixed labels	Section 504-80(b)
Sealing	Conduits and cables must be sealed as required	Section 504-70
Wiring methods	Any wiring method suitable for unclassified locations may be used	Section 504-20

Figure 12-9: *NEC installation requirements for intrinsically safe equipment*

they are positively separated from the non-intrinsically safe circuits because induced voltages could defeat the concept of intrinsically safe circuits. Underwriters' Laboratories Inc. and Factory Mutual list several devices in this category.

Explosionproof Condit And Fittings

A floor plan for a hazardous area is shown in Figure 12-10. In hazardous locations where threaded metal conduit is required, the conduit must be threaded with a standard conduit cutting die that provides ¾-inch taper per foot. The conduit should be made up wrench tight in order to minimize sparking in the event fault current flows through the raceway system (*NEC* Section 500-2). Where it is impractical to make a threaded joint tight, a bonding jumper shall be used. All boxes, fittings, and joints shall be threaded for connection to the conduit system and shall be an approved, explosionproof type (Figure 12-11). Threaded joints must be made up with at least five threads fully engaged. Where it becomes necessary to employ flexible connectors at motor or fixture terminals (Figure 12-12), flexible fittings approved for the particular class location shall be used.

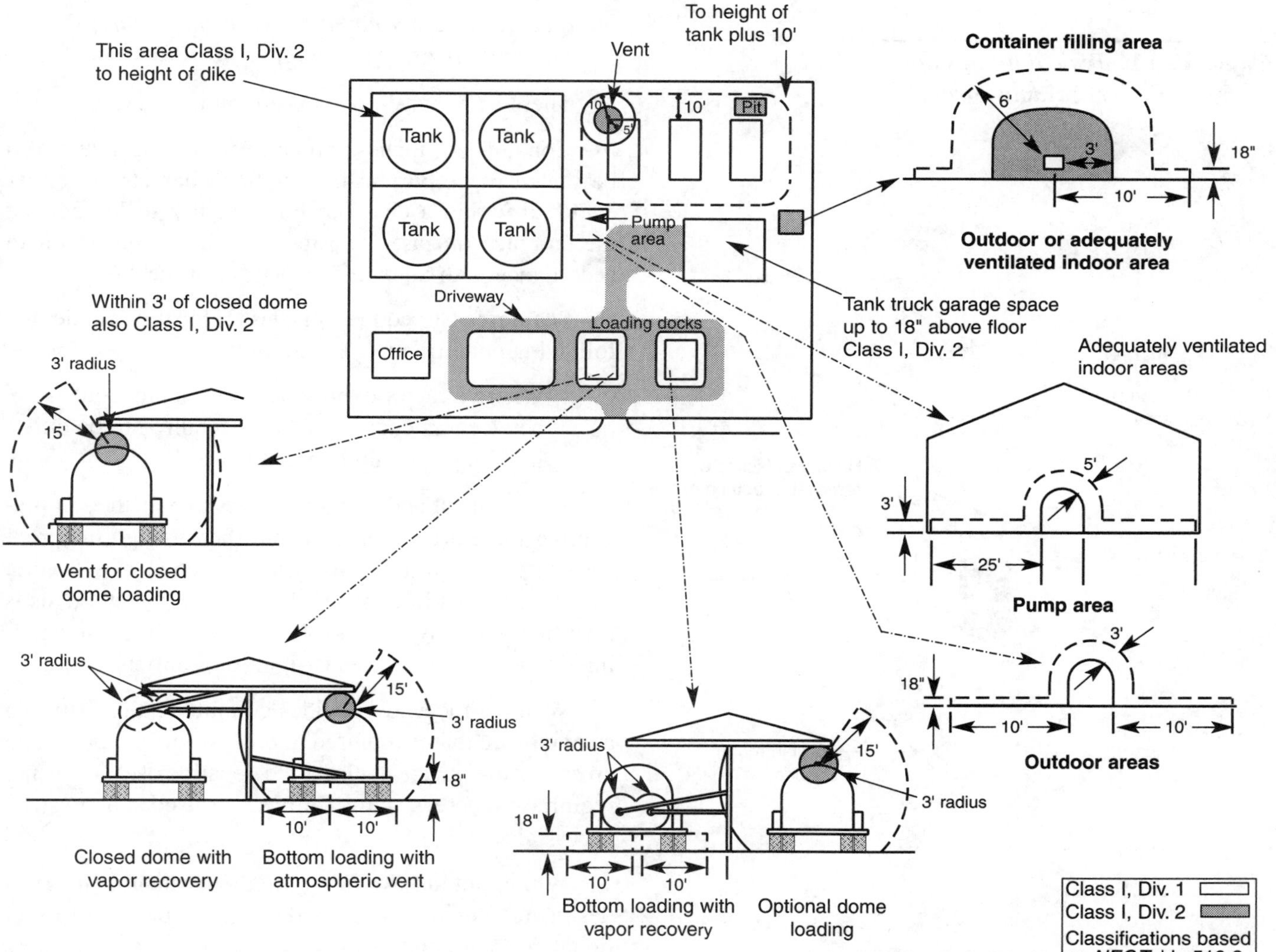

This area Class I, Div. 2 to height of dike
Vent
To height of tank plus 10'
Container filling area
Tank
Pit
Pump area
Outdoor or adequately ventilated indoor area
Tank truck garage space up to 18" above floor Class I, Div. 2
Driveway
Loading docks
Office
Within 3' of closed dome also Class I, Div. 2
3' radius
15'
Vent for closed dome loading
Adequately ventilated indoor areas
Pump area
25'
Outdoor areas
18"
10'
Closed dome with vapor recovery
Bottom loading with atmospheric vent
Bottom loading with vapor recovery
Optional dome loading
Class I, Div. 1
Class I, Div. 2
Classifications based on NEC Table 516-2

Figure 12-10: *Floor plan and elevations of a hazardous location — a bulk storage plant*

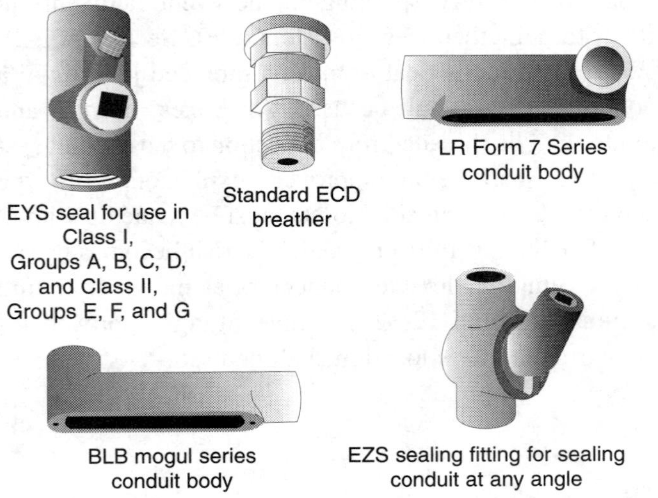
EYS seal for use in Class I, Groups A, B, C, D, and Class II, Groups E, F, and G
Standard ECD breather
LR Form 7 Series conduit body
BLB mogul series conduit body
EZS sealing fitting for sealing conduit at any angle

Figure 12-11: *Typical fittings approved for hazardous locations*

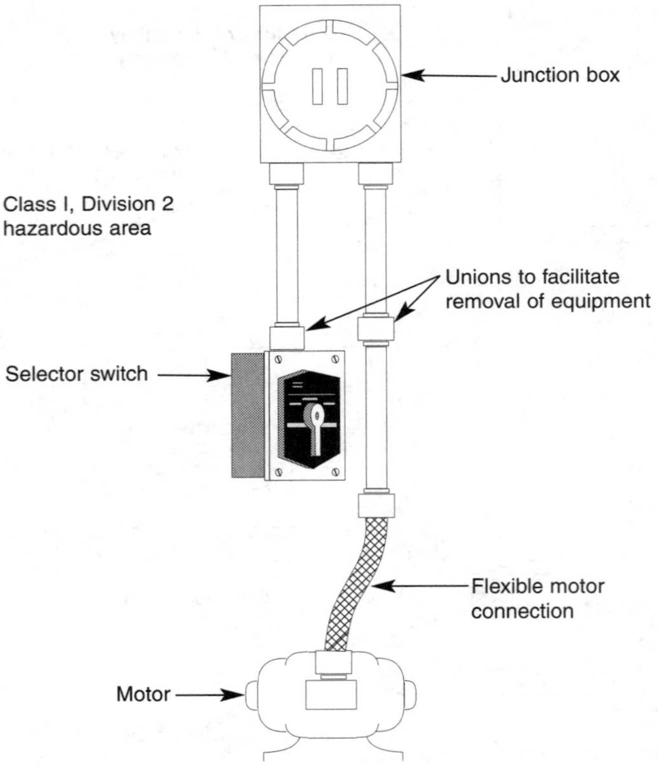
Junction box
Class I, Division 2 hazardous area
Unions to facilitate removal of equipment
Selector switch
Flexible motor connection
Motor

Figure 12-12: *Explosionproof flexible connectors are frequently used for motor terminations*

SEALS AND DRAINS

Seal-off fittings (Figure 12-13 on the next page) are required in conduit systems to prevent the passage of gases, vapors, or flames from one portion of the electrical installation to another at atmospheric pressure and normal ambient temperatures. Furthermore, seal-offs (seals) limit explosions to the sealed-off enclosure and prevent precompression of pressure piling in conduit systems. For Class I, Division 1 locations, *NEC* Section 501-5(a)(1) states:

> *Conduit seals shall be installed within 18 inches from the enclosure. Only explosionproof unions, couplings, reducers, elbows, capped elbows and conduit bodies similar to L, T, and Cross types that are not larger than the trade size of the conduit shall be permitted between the sealing fitting and the explosionproof enclosure.*

There is, however, one exception to this rule:

Conduits 1½ inches and smaller are not required to be sealed if the current-interrupting contacts are either enclosed within a chamber hermetically sealed against the entrance of gases or vapors, or immersed in oil in accordance with Section 501-5(a)(1) of the *NEC*.

Seals are also required in Class II locations under the following condition (*NEC* Section 502-5):

- Where a raceway provides communication between an enclosure that is required to be dust-ignitionproof and one that is not.

A permanent and effective seal is one method of preventing the entrance of dust into the dust-ignitionproof enclosure through the raceway. A horizontal raceway, not less than 10 feet long, is another approved method, as is a vertical raceway not less than 5 feet long and extending downward from the dust-ignitionproof enclosure.

Where a raceway provides communication between an enclosure that is required to be dust-ignitionproof and an enclosure in an unclassified location, seals are not required. Where sealing fittings are used, all must be accessible.

While not an *NEC* requirement, many electrical designers and workers consider it good practice to sectionalize long conduit runs by inserting seals not more than 50 to 100 feet apart, depending on the conduit size, to minimize the effects of "pressure piling."

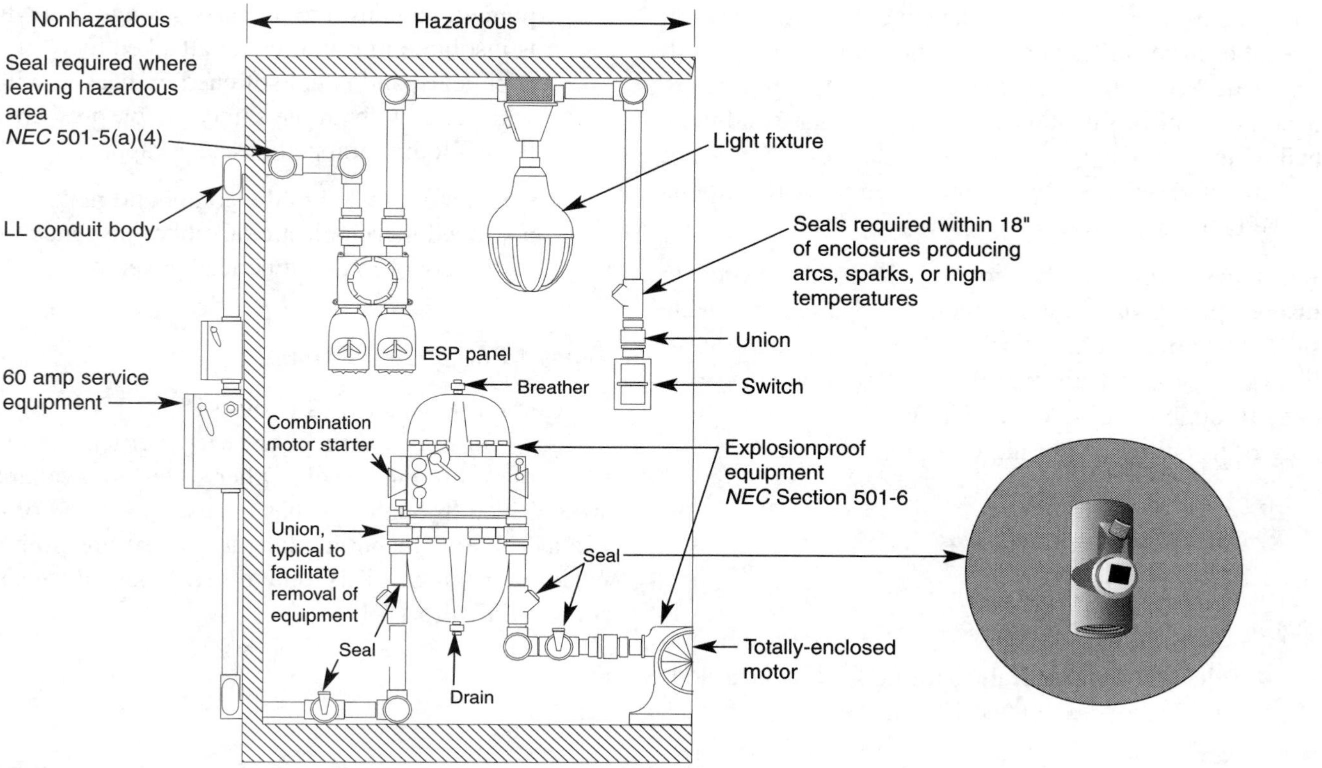
Nonhazardous
Hazardous
Seal required where leaving hazardous area NEC 501-5(a)(4)
LL conduit body
Light fixture
Seals required within 18" of enclosures producing arcs, sparks, or high temperatures
Union
ESP panel
Breather
Switch
60 amp service equipment
Combination motor starter
Explosionproof equipment NEC Section 501-6
Union, typical to facilitate removal of equipment
Seal
Seal
Drain
Totally-enclosed motor

Figure 12-13: *Seals must be installed at various locations in Class I, Division 1 areas*

In general, seals are installed at the same time as the conduit system. However, the conductors are installed after the raceway system is complete and *prior* to packing and sealing the seal-offs.

Drains

In humid atmospheres or in wet locations, where it is likely that water can gain entrance to the interiors of enclosures or raceways, the raceways should be inclined so that water will not collect in enclosures or on seals but will be led to low points where it may pass out through integral drains.

Frequently, the arrangement of raceway runs makes this method impractical — if not impossible. In such instances, special drain/seal fittings should be used, such as Crouse-Hinds Type EZDs as shown in Figure 12-14. These fittings prevent harmful accumulation of water above the seal and meet the requirements of *NEC* Section 501-5(d).

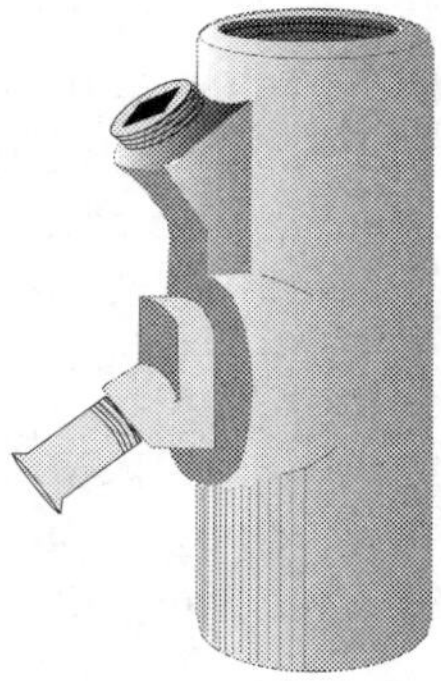

Figure 12-14: *Typical drain seal*

In locations which usually are considered dry, surprising amounts of water frequently collect in conduit systems. No conduit system is airtight; therefore, it may "breathe." Alternate increases and decreases in temperature and/or barometric pressure due to weather changes or due to the nature of the process carried on in the location where the conduit is installed will cause "breathing."

Outside air is drawn into the conduit system when it "breathes in." If this air carries sufficient moisture, it will be condensed within the system when the temperature decreases and chills this air. The internal conditions being unfavorable to evaporation, the resultant water accumulation will remain and be added to by repetitions of the breathing cycle.

In view of this likelihood, it is good practice to insure against such water accumulations and probable subsequent insulation failures by installing drain/seal fittings with drain covers or fittings with inspection covers even through conditions prevailing at the time of planning or installing do not indicate their need.

Selection Of Seals And Drains

The primary considerations for selecting the proper sealing fittings are as follows:

- Select the proper sealing fitting for the hazardous vapor involved; that is, Class 1, Groups A, B, C, or D.
- Select a sealing fitting for the proper use in respect to mounting position. This is particularly critical when the conduit runs between hazardous and nonhazardous areas. Improper positioning of a seal may permit hazardous gases or vapors to enter the system beyond the seal, and permit them to escape into another portion of the hazardous area, or to enter a nonhazardous area. Some seals are designed to be mounted in any position; others are restricted to horizontal or vertical mounting.
- Install the seals on the proper side of the partition or wall as recommended by the manufacturer.
- Installation of seals should be made *only* by trained personnel in strict compliance with the instruction sheets furnished with the seals and sealing compound.
- It should be noted that *NEC* Section 501-5(c)(4) prohibits splices or taps in sealing fittings.
- Sealing fittings are listed by UL for use in Class I hazardous locations with Chico A compound only. This compound, when properly mixed and poured, hardens into a dense, strong mass which is insoluble in water, is not attacked by chemicals, and is not softened by heat. It will withstand, with ample safety factor, pressure of the exploding trapped gases or vapor.
- Conductors sealed in the compound may be approved thermoplastic or rubber insulated type. Both may or may not be lead covered.

Types Of Seals And Fittings

Certain seals, such as Crouse-Hinds EYS seals, are designed for use in vertical or nearly vertical conduit in sizes for ½ through 1 inch. Other styles are available in sizes ½ inch through 6 inches for use in vertical or horizontal conduit. In horizontal runs, these are limited to face-up openings. This and other types of seals are shown in Figure 12-15.

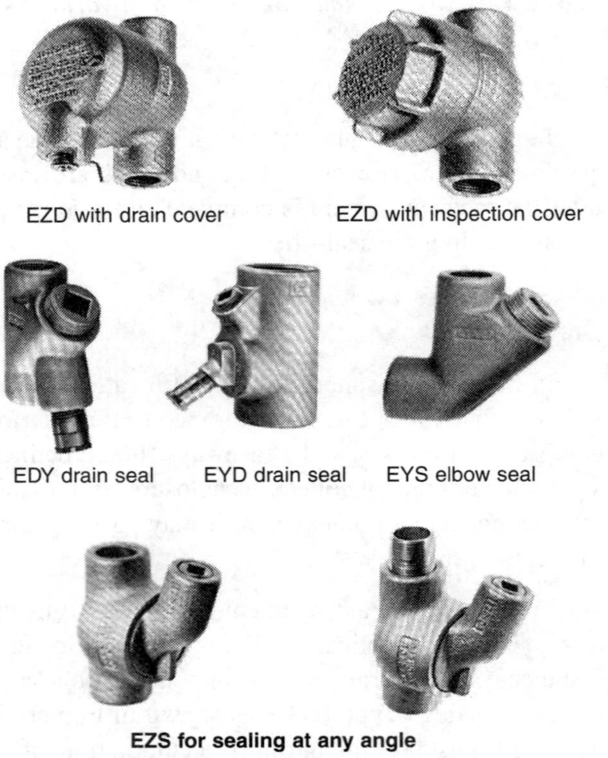
EZD with drain cover
EZD with inspection cover
EDY drain seal
EYD drain seal
EYS elbow seal
EZS for sealing at any angle

Figure 12-15: *Various types of seals and related components*

Seals ranging in sizes from 1¼ inches through 6 inches have extra large work openings, and separate filling holes, so that fiber dams are easy to make. However, the overall diameters of these fittings are scarcely greater than that of unions of corresponding sizes, permitting close conduit spacing.

Crouse-Hinds EZS seals are for use with conduit running at any angle, from vertical through horizontal.

EYD drain seals provide continuous draining and thereby prevent water accumulation. EYD seals are for vertical conduit runs and range in size from ½ inch to 4 inches inclusive. They are provided with one opening for draining and filling, a rubber tube to form drain passage and a drain fitting.

EZD drain seals provide continuous draining and thereby prevent water accumulation. The covers should be positioned so that the drain will be at the bottom. A set screw is provided for locking the cover in this position.

EZD fittings are suitable for sealing vertical conduit runs between hazardous and nonhazardous areas, but must be installed in the hazardous area when it is above the nonhazardous area. They must be installed in the nonhazardous area when it is above the hazardous area.

EZD drain seals are designed so that the covers can be removed readily, permitting inspection during installation or at any time thereafter. After the fittings have been installed in the conduit run and conductors are in place, the cover and barrier are removed. After the dam has been made in the lower hub opening with packing fiber, the barrier must be replaced so that the sealing compound can be poured into the sealing chamber.

EZD inspection seals are identical to EZD drain seals to provide all inspection, maintenance and installation advantages except that the cover is not provided with an automatic drain. Water accumulations can be drained periodically by removing the cover (when no hazards exist). The cover must be replaced immediately.

Sealing Compounds And Dams

Poured seals should be made only by trained personnel in strict compliance with the specific instruction sheets provided with each sealing fitting. Improperly poured seals are worthless.

Sealing compound shall be approved for the purpose; it shall not be affected by the surrounding atmosphere or liquids; and it shall not have a melting point of less than 200° F (93° C). The sealing compound and dams must also be approved for the type and manufacturer of the fitting. For example, Crouse-Hinds Chico A sealing compound is the only sealing compound approved for use with Crouse-Hinds ECM sealing fittings.

To pack the seal off, remove the threaded plug or plugs from the fitting and insert the fiber supplied with the packing kit. Tamp the fiber between the wires and the hub before pouring the sealing compound into the fitting. Then pour in the sealing cement and reset the threaded plug tightly. The fiber packing prevents the sealing compound (in the liquid state) from entering the conduit lines.

Most sealing-compound kits contain a powder in a polyethylene bag within an outer container. To mix, remove the bag of powder, fill the outside container, and pour in the powder and mix.

In practical applications, there may be dozens of seals required for a particular installation. Consequently, after the conductors are pulled, each seal in the system is first packed. To prevent the possibility of overlooking a seal, one color of paint is normally sprayed on the seal hub at this time. This indicates that the seal has been packed. When the sealing compound is poured, a different color paint is once again sprayed on the seal hub to indicate a finished job. This method permits the job supervisor to visually inspect the conduit run, and if a seal is not painted the appropriate color, he or she knows that proper installation on this seal was not done; therefore, action can be taken to correct the situation immediately.

The seal-off fittings in Figure 12-16 are typical of those used. The type in Figure 12-16A is for vertical mounting and is provided with a threaded, plugged opening into which the sealing cement is poured. The seal-off in Figure 12-16B has an additional plugged opening in the lower hub to facilitate packing fiber around the conductors to form a dam for the sealing cement.

The following procedures are to be observed when preparing sealing compound:

- Use a clean mix vessel for every batch. Particles of previous batches or dirt will spoil the seal.

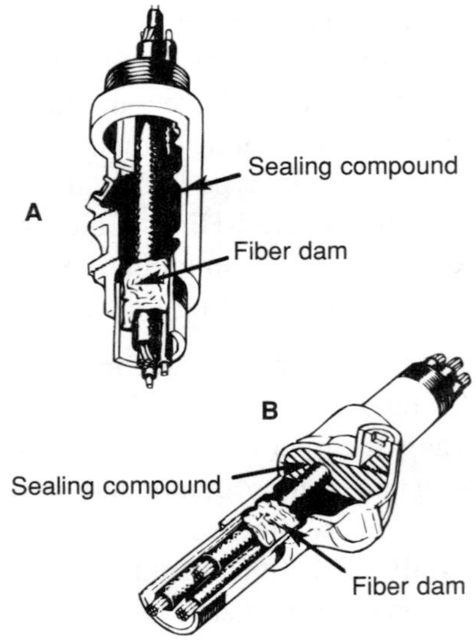
A
Sealing compound
Fiber dam
B
Sealing compound
Fiber dam

Figure 12-16: *Seals made with fiber dams and sealing compound*

- Recommended proportions are by volume — usually two parts powder to one part clean water. Slight deviations in these proportions will not affect the result.
- Do not mix more than can be poured in 15 minutes after water is added. Use cold water. Warm water increases setting speed. Stir immediately and thoroughly.
- If batch starts to set, do not attempt to thin it by adding water or by stirring. Such a procedure will spoil seal. Discard partially set material and make fresh batch. After pouring, close opening immediately.
- Do not pour compound in sub-freezing temperatures, or when these temperatures will occur during curing.
- See that compound level is in accordance with the instruction sheet for that specific fitting.

Most other fittings are provided with threaded hubs for securing the conduit as described previously. All conduit terminations should be drawn up wrench tight. Typical fittings include switch and junction boxes, conduit bodies, union and connectors, flexible couplings, explosionproof lighting fixtures, receptacles, and panelboard and motor starter enclosures. A practical representation of these and other fittings is shown in Figure 12-17.

GARAGES AND SIMILAR LOCATIONS

Garages and similar locations (Figures 12-18 and 12-19) where volatile or flammable liquids are handled or used as fuel in self-propelled vehicles (including automobiles, buses, trucks, and tractors) are not usually considered critically hazardous locations. However, the entire area up to a level of 18 inches above the floor is considered a Class I, Division 2 location, and certain precautionary measures are required by the *NEC*. Likewise, any pit or depression below floor level shall be considered a Class I, Division 2 location, and the pit or depression may be judged as Class I, Division 1 location if it is unvented.

Normal raceway (conduit) and wiring may be used for the wiring method above this hazardous level, except where conditions indicate that the area concerned is more hazardous than usual. In this case, the applicable type of explosionproof wiring may be required.

Approved seal-off fittings should be used on all conduit passing from hazardous areas to nonhazardous areas. The requirements set forth in *NEC* Sections 501-5 and 501-5(b)(2) shall apply to horizontal as well as vertical boundaries of the defined hazardous areas. Raceways embedded in a masonry floor or buried beneath a floor are considered to be within the hazardous area above the floor if any connections or extensions lead into or through such an area. However, conduit systems terminating to an open raceway, in an outdoor unclassified area, shall not be required to be sealed between the point at which the conduit leaves the classified location and enters the open raceway.

Note that space in the immediate vicinity of the gasoline-dispensing island is denoted as Class I, Division 1, to a height of 4 feet above grade. The surrounding area, within a radius of 20 feet of the island, falls under Class I, Division 2, to a height of 18 inches above grade. Bulk storage plants for gasoline are subject to comparable restrictions.

AIRPORT HANGARS

Buildings used for storing or servicing aircraft in which gasoline, jet fuels, or other volatile flammable liquids or gases are used fall under *NEC* Article 513. In general, any depression below the level of the hangar floor is considered to be a Class I, Division 1 location. The entire area of the hangar including any adjacent and

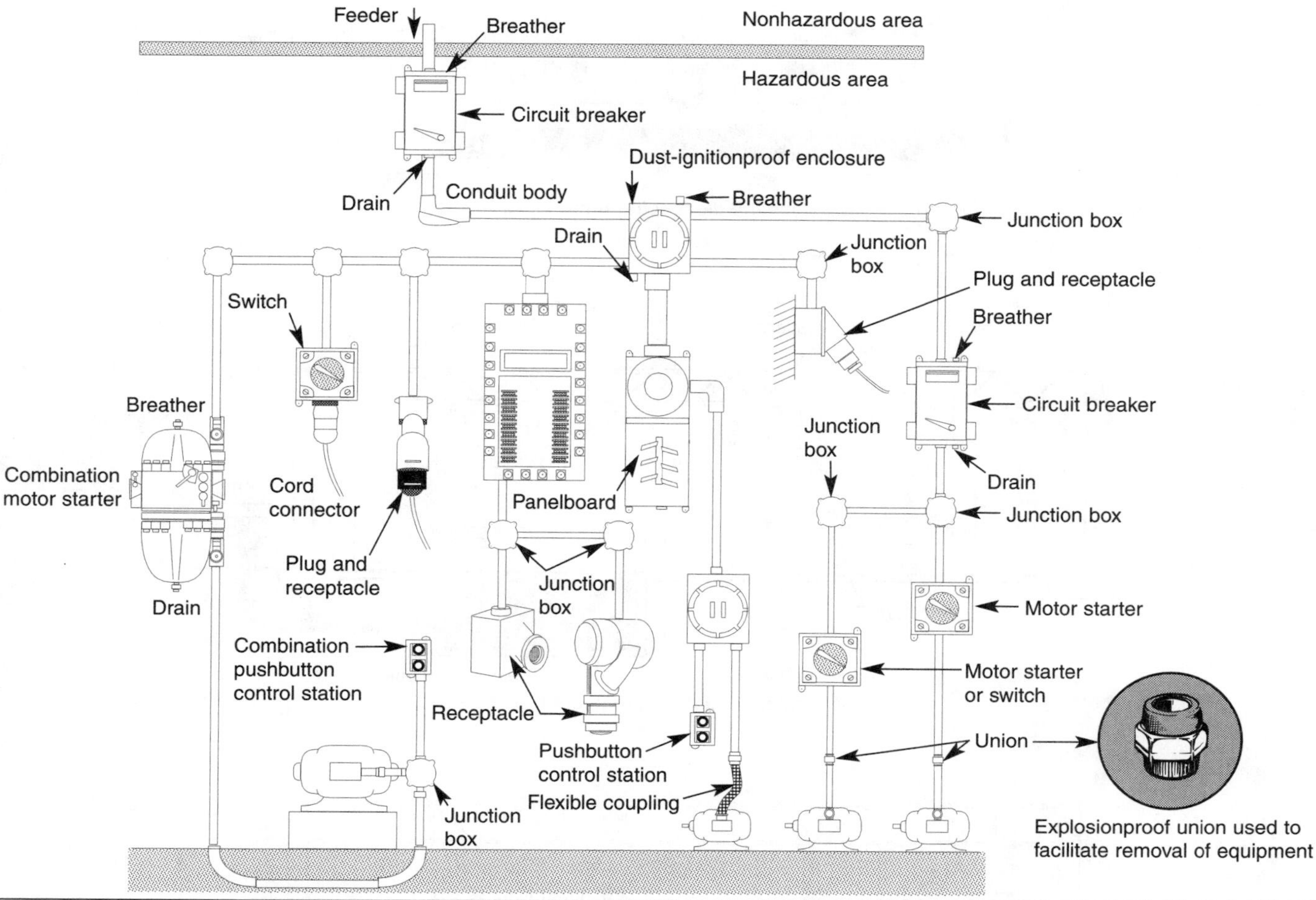
Feeder
Breather
Nonhazardous area
Hazardous area
Circuit breaker
Dust-ignitionproof enclosure
Drain
Conduit body
Breather
Junction box
Drain
Junction box
Plug and receptacle
Switch
Breather
Breather
Circuit breaker
Junction box
Combination motor starter
Cord connector
Panelboard
Drain
Junction box
Plug and receptacle
Junction box
Drain
Motor starter
Combination pushbutton control station
Motor starter or switch
Receptacle
Union
Pushbutton control station
Flexible coupling
Junction box
Explosionproof union used to facilitate removal of equipment

Figure 12-17: *Class II, Division 1 electrical installation*

COMMERCIAL GARAGES		
Application	**Requirements**	***NEC* Reference**
Attachment plug receptacles	Must be located above the level of any defined Class I locations or else be aproved for the location	Section 511-6(d)
Electric vehicle charging	Equipment and wiring must be installed as per *NEC* Article 625 except as noted in (b) and (c) of this *NEC* Section	Section 511-9
Equipment in Class I locations	Must conform to *NEC* Article 501	Section 511-4
Ground-fault circuit-interrupter protector	Must be installed on all 125-volt, single-phase, 15- and 20-amp receptacles	Section 511-10
Grounding	As specified	Section 250-112(f)
Sealing	Approved seals must be used and conform to *NEC* Sections 501-5 and 501-5(b)(2)	Section 511-5
Wiring Class I locations	Must conform to applicable provisions of *NEC* Article 501	Section 511-4

Figure 12-18: *Summary of NEC installation requirements for garages and similar establishments*

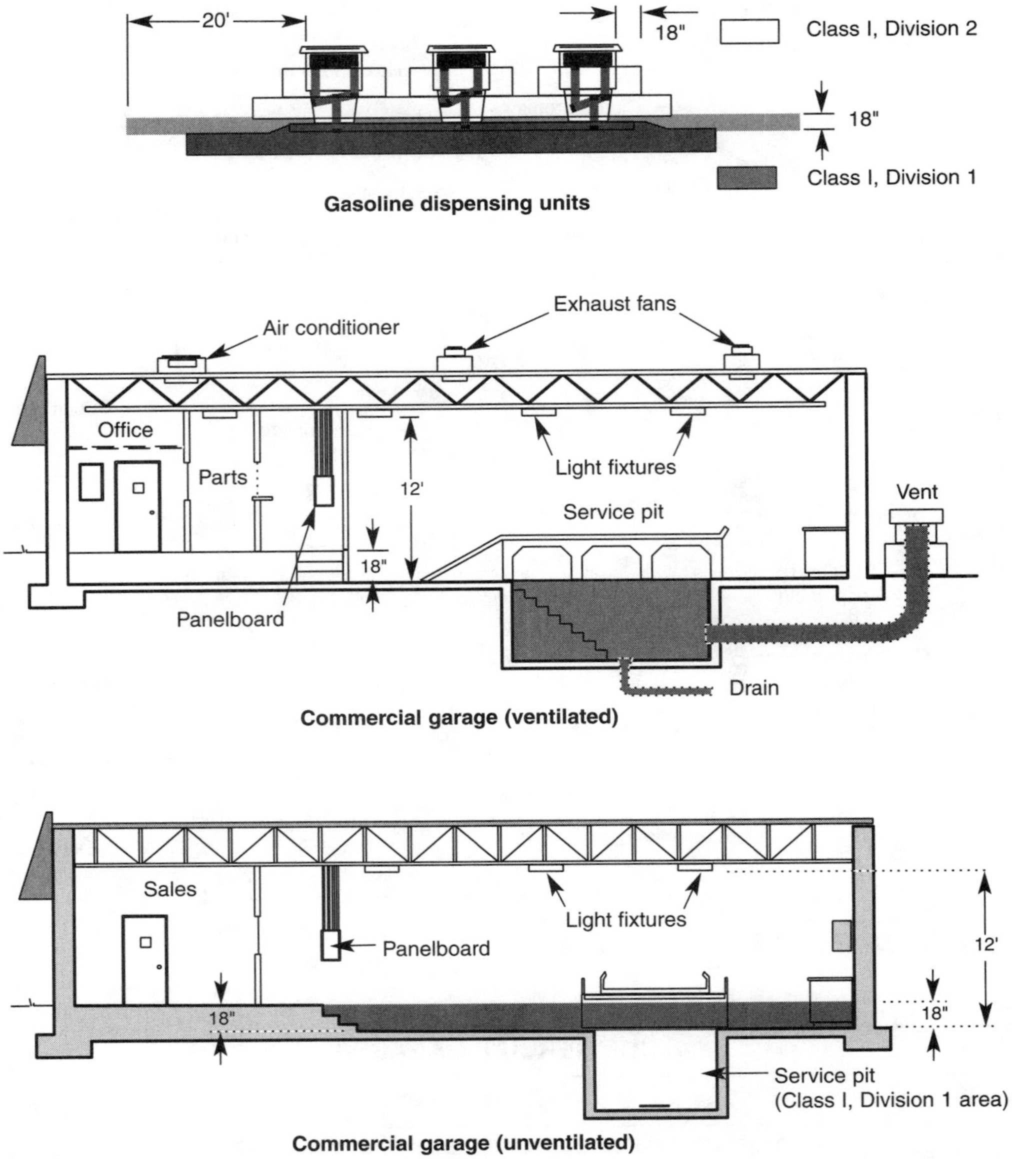
20'
18"
Class I, Division 2
18"
Class I, Division 1
Gasoline dispensing units
Exhaust fans
Air conditioner
Office
Parts
12'
Light fixtures
Vent
Service pit
18"
Panelboard
Drain
Commercial garage (ventilated)
Sales
Light fixtures
Panelboard
12'
18"
18"
Service pit
(Class I, Division 1 area)
Commercial garage (unventilated)

Figure 12-19: *Gasoline dispensing units and commercial garage classifications*

communicating area not suitably cut off from the hangar is considered to be a Class I, Division 2 location up to a level of 18 inches above the floor. The area within 5 feet horizontally from aircraft power plants, fuel tanks, or structures containing fuel is considered to be a Class I, Division 2 hazardous location; this area extends upward from the floor to a level 5 feet above the upper surface of wings and of engine enclosures.

Adjacent areas in which hazardous vapors are not likely to be released, such as stock rooms and electrical control rooms, should not be classed as hazardous when they are adequately ventilated and effectively cut off from the hangar itself by walls or partitions. All fixed wiring in a hangar not within a hazardous area, as defined in Section 513-2, must be installed in metallic raceway or shall be Type MI or Type ALS cable. The only exception is wiring in nonhazardous locations, as defined in Section 513-2(d), which may be of any type recognized in Chapter 3 (Wiring Methods and Materials) of the *NEC*. Figures 12-20 and 12-21 summarize the *NEC* requirements for airport hangars. Please refer to the *NEC* book for additional details.

AIRCRAFT HANGARS		
Application	**Requirements**	***NEC* Reference**
Aircraft batteries	Must be charged outside of hangars	Section 513-10
Aircraft electrical systems	Electrical systems must be deenergized when the aircraft is stored in a hangar	Section 513-9
Aircraft energizers	Not less than 18 inches above the hangar floor level	Section 513-11
Definition	Describes areas that fall under this *NEC* Article	Section 513-1
Equipment	Must comply with applicable provisions of the *NEC*	Section 513-4, 513-6, 513-12
Grounding	Must comply with *NEC* Article 250	Section 513-16
Classification of locations	Classifications as per (a), (b), (c), and (d) under this *NEC* Section	Section 513-3
Sealing	Seals must be installed according to *NEC* Section 501-5	Section 513-8

Figure 12-20: *Summary of NEC installation requirements for airport hangars*

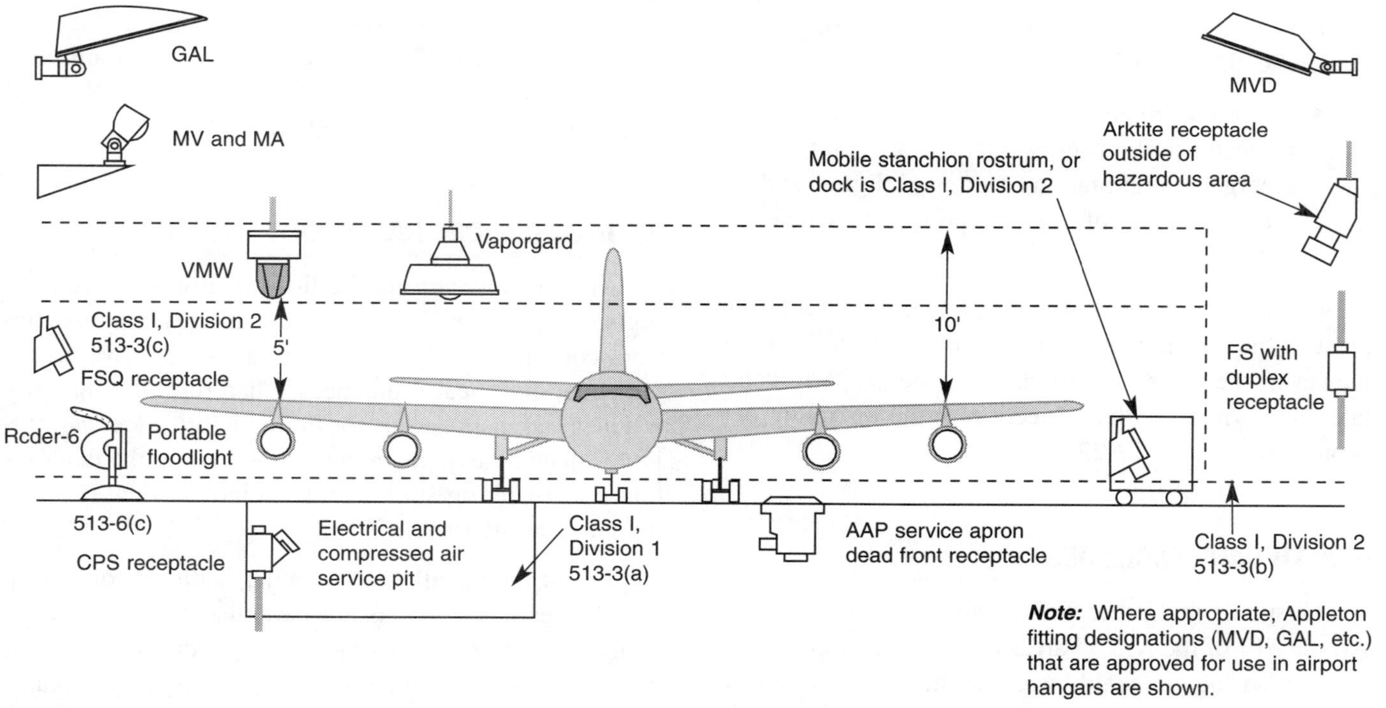
GAL
MVD
MV and MA
Arktite receptacle outside of hazardous area
Mobile stanchion rostrum, or dock is Class I, Division 2
Vaporgard
VMW
Class I, Division 2 513-3(c)
5'
10'
FSQ receptacle
FS with duplex receptacle
Rcder-6
Portable floodlight
513-6(c)
CPS receptacle
Electrical and compressed air service pit
Class I, Division 1 513-3(a)
AAP service apron dead front receptacle
Class I, Division 2 513-3(b)
Note: Where appropriate, Appleton fitting designations (MVD, GAL, etc.) that are approved for use in airport hangars are shown.

Figure 12-21: *Sections of airport hangar showing hazardous locations*

THEATERS

The *NEC* recognizes that hazards to life and property due to fire and panic exist in theaters, cinemas, and the like. The *NEC* therefore requires certain precautions in these areas in addition to those for commercial installations. These requirements include the following:

- Proper wiring of motion picture projection rooms (Article 540).
- Heat-resistant, insulated conductors for certain lighting equipment (Section 520-43(b)).
- Adequate guarding and protection of the stage switchboard and proper control and overcurrent protection of circuits (Section 520-22).
- Proper type and wiring of lighting dimmers (Sections 520-53(e) and 520-25).
- Use of proper types of receptacles and flexible cables for stage lighting equipment (Section 520-45).
- Proper stage flue damper control (Section 520-49).
- Proper dressing-room wiring and control (Sections 520-71, 72, and 73).
- Fireproof projection rooms with automatic projector port closures, ventilating equipment, emergency lighting, guarded work lights, and proper location of related equipment (Article 540).

Outdoor or drive-in motion picture theaters do not present the inherent hazards of enclosed auditoriums. However, the projection rooms must be properly ventilated and wired for the protection of the operating personnel. See Figure 12-22.

HEALTH CARE FACILITIES

Hospitals and other health-care facilities fall under Article 517 of the *NEC*. Part B of Article 517 covers the general wiring of health-care facilities. Part C covers essential electrical systems for hospitals. Part D gives the performance criteria and wiring methods to minimize shock hazards to patients in electrically susceptible patient areas. Part E covers the requirements for electrical wiring and equipment used in inhalation anesthetizing locations.

With the widespread use of X-ray equipment of varying types in health-care facilities, electricians are often required to wire and connect equipment such as discussed in Article 660 of the *NEC*. Conventional wiring methods are used, but provisions should be made for 50- and 60-ampere receptacles for medical X-ray equipment (Section 660-4b).

Anesthetizing locations of hospitals are deemed to be Class I, Division 1, to a height of 5 feet above floor level. Gas storage rooms are designated as Class I, Division 1, throughout. Most of the wiring in these areas, however, can be limited to lighting fixtures only — locating all switches and other devices outside of the hazardous area.

The *NEC* recommends that wherever possible electrical equipment for hazardous locations be located in less hazardous areas. It also suggests that by adequate, positive-pressure ventilation from a clean source of outside air the hazards may be reduced or hazardous locations limited or eliminated. In many cases the installation of dust-collecting systems can greatly reduce the hazards in a Class II area. See Figure 12-23 and also refer to the *Life Safety Code*, available from the National Fire Protection Association, Quincy, MA.

PETRO/CHEMICAL HAZARDOUS LOCATIONS

Most manufacturing facilities involving flammable liquids, vapors, or fibers must have their wiring installations conform strictly to the *NEC* as well as governmental, state, and local ordinances. Therefore, the majority of electrical installations for these facilities are carefully designed by experts in the field — either the plant in-house engineering staff or else an independent consulting engineering firm.

Industrial installations dealing with petroleum or some types of chemicals are particularly susceptible to many restrictions involving many governmental agencies. Electrical installations for petro/chemical plants will therefore have many pages of electrical drawings and specifications which first go through the procedure for approval from all the agencies involved. Once approved,

THEATERS		
Application	**Requirements**	***NEC* Reference**
Arc lamps	Must be listed	Section 520-61
Border lighting fixtures	Must be constructed as per *NEC* Section 520-43 and suitably supported	Section 520-44
Busbars and terminals	Must have ampacity equal to the sum of the circuits connected	Section 520-62(c)
Cable connectors	Tension of the cord must not be transmitted to the connector	Section 520-67
Cable terminals	Must be accessible	Section 520-53(n)
Circuit protection	Must be provided in each ungrounded circuit conductor	Section 520-53(d)
Conductors for portable equipment	Must be listed for extra-hard usage	Section 520-68
Curtain motors	Must be listed	Section 520-48
Dimmers	May be placed in either the grounded or ungrounded conductor of the circuit	Section 520-25(b)
Dressing-room lighting	Pendant lampholders must not be installed in dressing rooms. Guards must be provided for all lamps	Section 520-71 Section 520-72
Dead front switchboards	Stage switchboards must be of the dead-front type	Section 520-21
Enclosures, portable switchboards	Switchboards must be placed inside an enclosure	Section 520-53(a)
Footlights, construction of	As specified	Section 520-43
Grounding	All metallic frames of enclosures, lighting fixtures, conduit and cable must be grounded	Section 520-81
Hood over switchboard	Must protect panel from falling objects	Section 520-24
Lighting fixtures on scenery	Brackets must be wired internally	Section 520-63
Portable switchboard	Must be supplied only from power outlets of sufficient size	Section 520-51
Switches	All lights and receptacles in dressing room must be controlled by wall switches	Section 520-73

Figure 12-22: *Summary of NEC installation requirements for theaters*

these drawings and specifications must be followed exactly because any change whatsoever must once again go through the various agencies for approval.

MANUFACTURERS' DATA

Manufacturers of explosionproof equipment and fittings expend a lot of time, energy, and expense in developing guidelines and brochures to ensure that their products are used correctly and in accordance with the latest *NEC* requirements. The many helpful charts, tables, and application guidelines are invaluable to anyone working on projects involving hazardous locations. Therefore, it is recommended that electrical workers obtain as much of this data as possible. Once obtained, study this data thoroughly. Doing so will enhance anyone's qualifications for working in hazardous locations of any type.

Manufacturers' data is usually available to qualified personnel (electrical workers) at little or no cost and can be obtained from local distributors of electrical supplies, or directly from the manufacturer.

HEALTH CARE FACILITIES		
Application	**Requirements**	***NEC* Reference**
Applicability	Part B of *NEC* Article 517 applies to all health care facilities	Section 517-10
Branch circuits	Each patient bed location must be supplied with a least two branch circuits — one from the emergency system and one from the normal system	Section 517-18(a)
Grounding	All receptacles and fixed electrical equipment operating at 100 volts or more must be grounded	Section 517-13(a)
Grounding conductors	Must be sized as per *NEC* Table 250-122	Section 517-13(a)
Grounding, installation methods	Branch circuits must be installed in metal raceway or cable assembly	Section 517-13(b)
Ground-fault protection, feeders	In addition to requirements of *NEC* Sections 230-95 or 215-10, an additional step of ground-fault protection must be provided in the next level of feeder disconnecting means	Section 517-17(a)
Panelboard bonding	The normal and essential branch-circuit panelboard must be bonded together	Section 517-14
Receptacles	Each patient bed location must be provided with a minimum of four receptacles	Section 517-18(b)
Wiring methods	Except as modified in *NEC* Article 517, wiring methods must comply with *NEC* Chapters 1 through 4	Section 517-12

Figure 12-23: *Summary of NEC installation requirements for health care facilities*

Summary

Any area in which the atmosphere or a material in the area is such that the arcing of operating electrical contacts, components, and equipment may cause an explosion or fire is considered as a hazardous location. In all such cases, explosionproof equipment, raceways, and fittings are used to provide an explosionproof wiring system.

The wide assortment of explosionproof equipment now available makes it possible to provide adequate electrical installations under any of the various hazardous conditions. However, the electrician must be thoroughly familiar with all *NEC* requirements and know what fittings are available, how to install them properly, and where and when to use the various fittings.

Many factors — such as temperature, barometric pressure, quantity of release, humidity, ventilation, distance from the vapor source, and the like — must be considered. When information on all factors concerned is properly evaluated, a consistent classification for the selection and location of electrical equipment can be developed.

Appendix
Electrical Wiring Symbols

The purpose of an electrical drawing is to show the complete design and layout of the electrical systems for lighting, power, signal and communication systems, special raceways, and related electrical equipment. In preparing such drawings, the electrical layout is shown through the use of lines, symbols, and notation which should indicate, beyond any question or any doubt, exactly what is required.

Many engineers, designers, and draftsmen use symbols adapted by the American National Standards Institute (ANSI). However, no definite standard schedule of symbols is always used in its entirety. Consulting engineering firms quite frequently modify these standard symbols to meet their own needs. Therefore, in order to identify the symbols properly, the engineer provides, on one of the drawings or in the written specifications, a list of symbols with a descriptive note for each — clearly indicating the part of the wiring system which it represents.

Figure A2-1 shows a list of electrical symbols which are currently recommended by ANSI, while Figure A2-2 shows another list of symbols which was prepared by the Consulting Engineers Council/U.S. and the Construction Specifications Institute, Inc. Figure A2-3 shows still another list of electrical reference symbols which have been modified for use by the one consulting engineering firm.

It should be evident from these symbols that many have the same basic form, but, because of some slight difference, their meaning changes. For example, the outlet symbols in Figure A2-4 each have the same basic form — a circle — but the addition of a line or an abbreviation gives each an individual meaning. A good way to learn symbols is to first learn the basic form and then apply the variations for obtaining different meanings.

The electrical symbols described in the following paragraphs represent a system of electrical notation whose compactness and clarity may be of assistance to electrical engineers, designers, and draftsmen.

The system should be used in place of standard symbols only if there seems to be a decided advantage in doing so.

Some of the symbols are abbreviated idioms, like "WP" for weatherproof and "AFF" for above finished floor. Other symbols are simplified pictographs, like for a double floodlight fixture or for an infrared electric heater with two lamps.

In some cases there are combinations of idioms and pictographs, as in for fusible safety switch, for nonfusible safety switch, and for double-throw safety switch, where the pictograph of a switch enclosure is combined with the abbreviated idioms of F (fusible), N (nonfusible), and DT (double throw), respectively. The numerals indicate the bus bar capacity in amperes.

This list came about as a result of much discussion with consulting engineers, electrical designers, electrical draftsmen, electrical estimators, electricians, and others who are required to interpret electrical drawings. It is felt that this list represents a good set of symbols in that they are:

- Easy to draw
- Easily interpreted by workers
- Sufficient for most applications

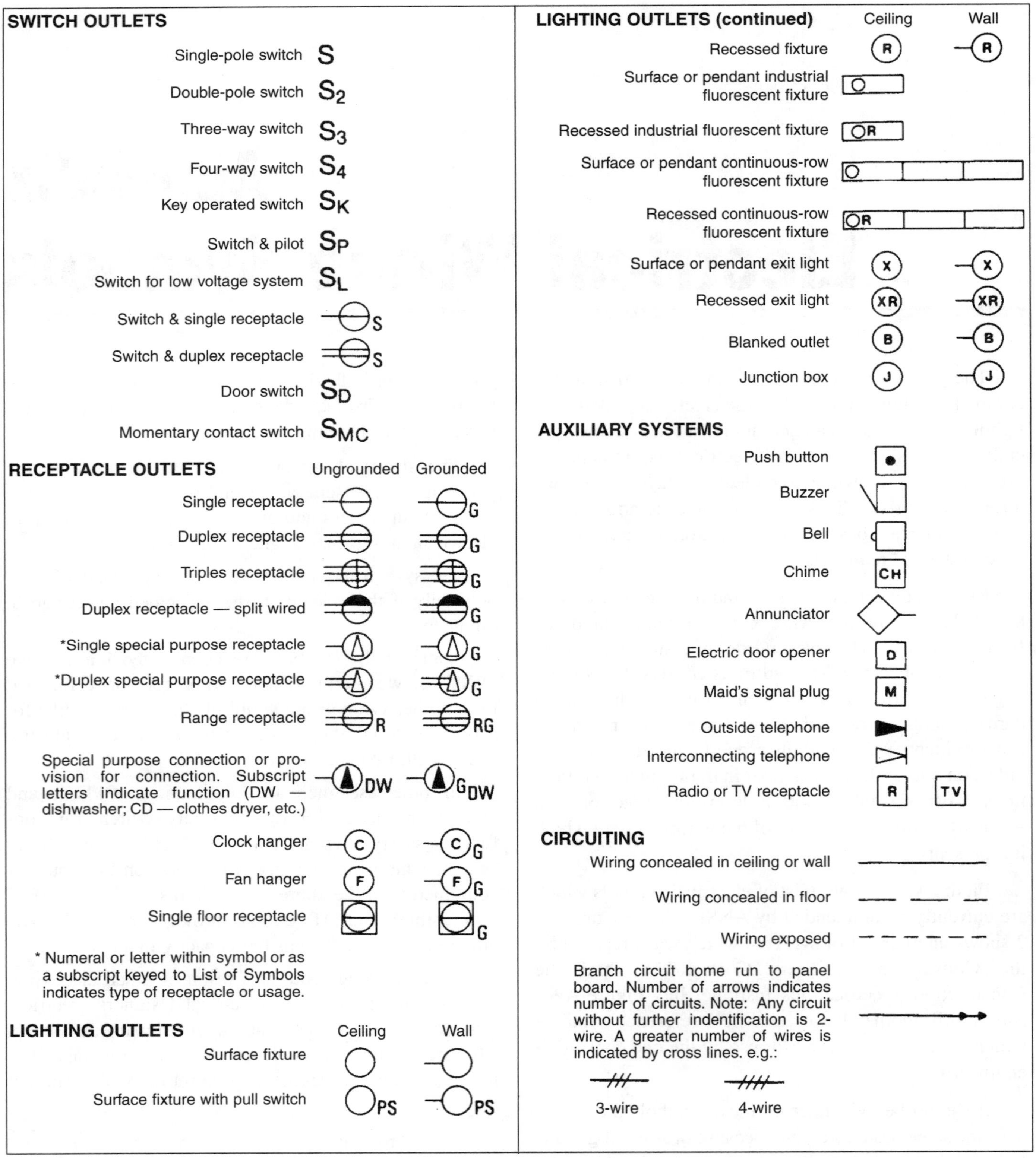
SWITCH OUTLETS
Single-pole switch S
Double-pole switch S2
Three-way switch S3
Four-way switch S4
Key operated switch SK
Switch & pilot SP
Switch for low voltage system SL
Switch & single receptacle S
Switch & duplex receptacle S
Door switch SD
Momentary contact switch SMC
RECEPTACLE OUTLETS
Ungrounded
Grounded
Single receptacle G
Duplex receptacle G
Triples receptacle G
Duplex receptacle — split wired G
*Single special purpose receptacle G
*Duplex special purpose receptacle G
Range receptacle R RG
Special purpose connection or provision for connection. Subscript letters indicate function (DW — dishwasher; CD — clothes dryer, etc.) DW GDW
Clock hanger C G
Fan hanger F G
Single floor receptacle G
* Numeral or letter within symbol or as a subscript keyed to List of Symbols indicates type of receptacle or usage.
LIGHTING OUTLETS
Ceiling
Wall
Surface fixture
Surface fixture with pull switch PS PS
LIGHTING OUTLETS (continued)
Ceiling
Wall
Recessed fixture R R
Surface or pendant industrial fluorescent fixture
Recessed industrial fluorescent fixture R
Surface or pendant continuous-row fluorescent fixture
Recessed continuous-row fluorescent fixture R
Surface or pendant exit light X X
Recessed exit light XR XR
Blanked outlet B B
Junction box J J
AUXILIARY SYSTEMS
Push button
Buzzer
Bell
Chime CH
Annunciator
Electric door opener D
Maid's signal plug M
Outside telephone
Interconnecting telephone
Radio or TV receptacle R TV
CIRCUITING
Wiring concealed in ceiling or wall
Wiring concealed in floor
Wiring exposed
Branch circuit home run to panel board. Number of arrows indicates number of circuits. Note: Any circuit without further indentification is 2-wire. A greater number of wires is indicated by cross lines. e.g.:
3-wire
4-wire

Figure A2-1: *Electrical symbols currently recommended by USASI*

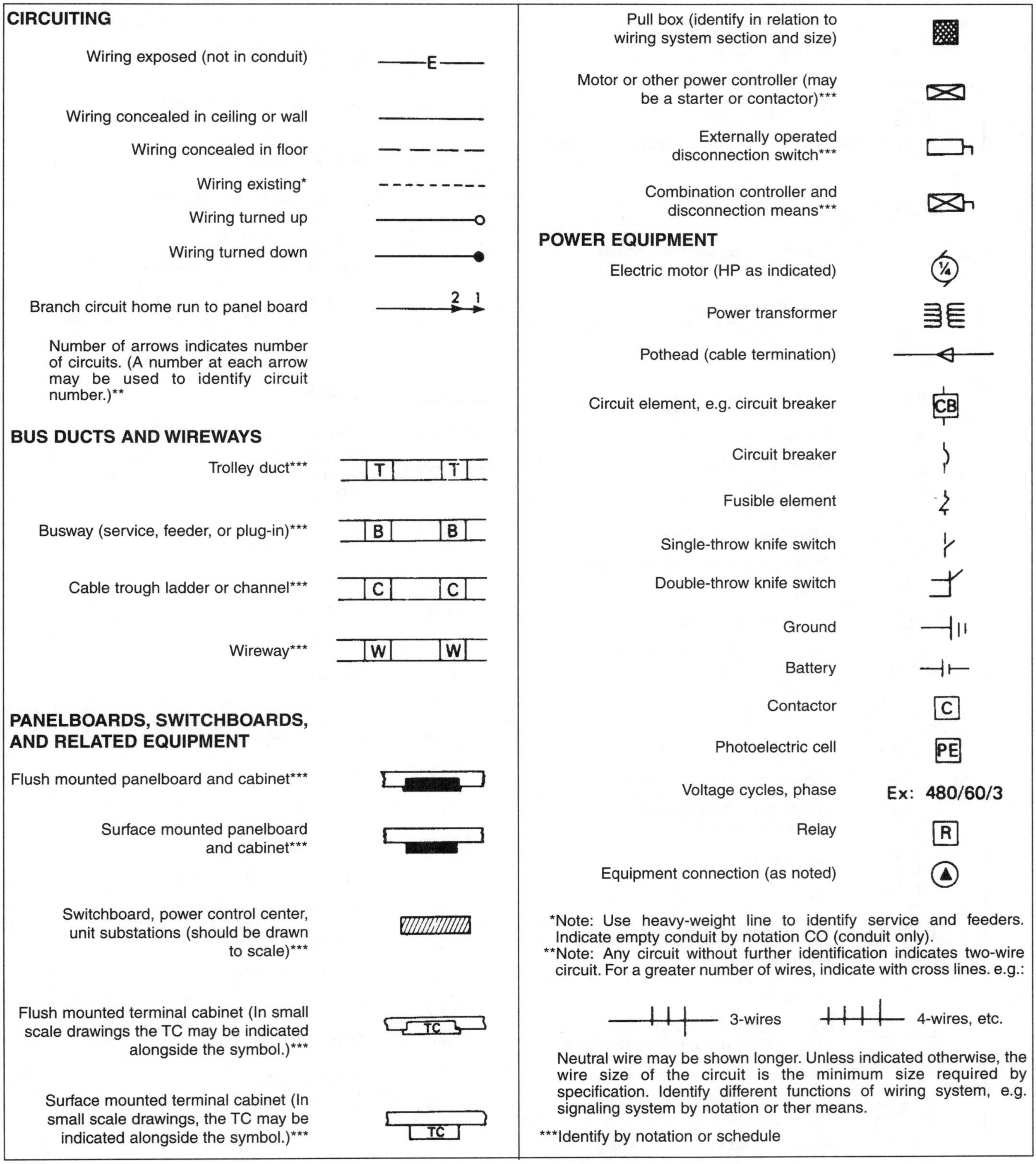
CIRCUITING
Wiring exposed (not in conduit)
E
Wiring concealed in ceiling or wall
Wiring concealed in floor
Wiring existing*
Wiring turned up
Wiring turned down
Branch circuit home run to panel board
2 1
Number of arrows indicates number of circuits. (A number at each arrow may be used to identify circuit number.)**
BUS DUCTS AND WIREWAYS
Trolley duct***
T T
Busway (service, feeder, or plug-in)***
B B
Cable trough ladder or channel***
C C
Wireway***
W W
PANELBOARDS, SWITCHBOARDS, AND RELATED EQUIPMENT
Flush mounted panelboard and cabinet***
Surface mounted panelboard and cabinet***
Switchboard, power control center, unit substations (should be drawn to scale)***
Flush mounted terminal cabinet (In small scale drawings the TC may be indicated alongside the symbol.)***
TC
Surface mounted terminal cabinet (In small scale drawings, the TC may be indicated alongside the symbol.)***
TC
Pull box (identify in relation to wiring system section and size)
Motor or other power controller (may be a starter or contactor)***
Externally operated disconnection switch***
Combination controller and disconnection means***
POWER EQUIPMENT
Electric motor (HP as indicated)
¼
Power transformer
Pothead (cable termination)
Circuit element, e.g. circuit breaker
CB
Circuit breaker
Fusible element
Single-throw knife switch
Double-throw knife switch
Ground
Battery
Contactor
C
Photoelectric cell
PE
Voltage cycles, phase
Ex: 480/60/3
Relay
R
Equipment connection (as noted)
*Note: Use heavy-weight line to identify service and feeders. Indicate empty conduit by notation CO (conduit only).
**Note: Any circuit without further identification indicates two-wire circuit. For a greater number of wires, indicate with cross lines. e.g.:
3-wires
4-wires, etc.
Neutral wire may be shown longer. Unless indicated otherwise, the wire size of the circuit is the minimum size required by specification. Identify different functions of wiring system, e.g. signaling system by notation or ther means.
***Identify by notation or schedule

Figure A2-2: *A list of electrical symbols recommended by the Consulting Engineers Council/US and the Construction Specifications Institute, Inc.*

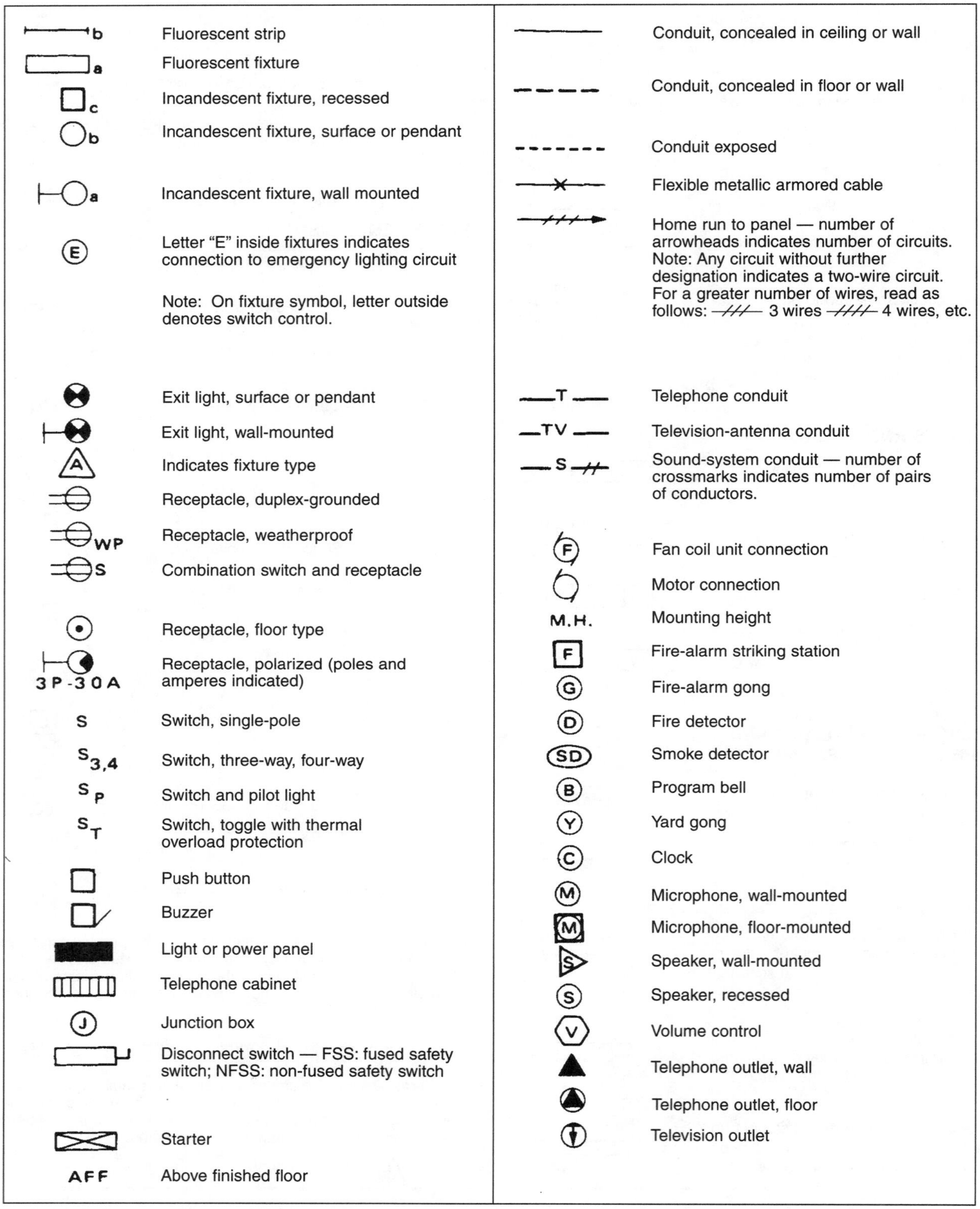
b Fluorescent strip
a Fluorescent fixture
c Incandescent fixture, recessed
b Incandescent fixture, surface or pendant
a Incandescent fixture, wall mounted
E Letter "E" inside fixtures indicates connection to emergency lighting circuit
Note: On fixture symbol, letter outside denotes switch control.
Exit light, surface or pendant
Exit light, wall-mounted
A Indicates fixture type
Receptacle, duplex-grounded
WP Receptacle, weatherproof
S Combination switch and receptacle
Receptacle, floor type
3P-30A Receptacle, polarized (poles and amperes indicated)
S Switch, single-pole
S3,4 Switch, three-way, four-way
SP Switch and pilot light
ST Switch, toggle with thermal overload protection
Push button
Buzzer
Light or power panel
Telephone cabinet
J Junction box
Disconnect switch — FSS: fused safety switch; NFSS: non-fused safety switch
Starter
AFF Above finished floor
Conduit, concealed in ceiling or wall
Conduit, concealed in floor or wall
Conduit exposed
Flexible metallic armored cable
Home run to panel — number of arrowheads indicates number of circuits. Note: Any circuit without further designation indicates a two-wire circuit. For a greater number of wires, read as follows: 3 wires 4 wires, etc.
T Telephone conduit
TV Television-antenna conduit
S Sound-system conduit — number of crossmarks indicates number of pairs of conductors.
F Fan coil unit connection
Motor connection
M.H. Mounting height
F Fire-alarm striking station
G Fire-alarm gong
D Fire detector
SD Smoke detector
B Program bell
Y Yard gong
C Clock
M Microphone, wall-mounted
M Microphone, floor-mounted
S Speaker, wall-mounted
S Speaker, recessed
V Volume control
Telephone outlet, wall
Telephone outlet, floor
Television outlet

Figure A2-3: *A modified list of electrical symbols used by a consulting engineering firm*

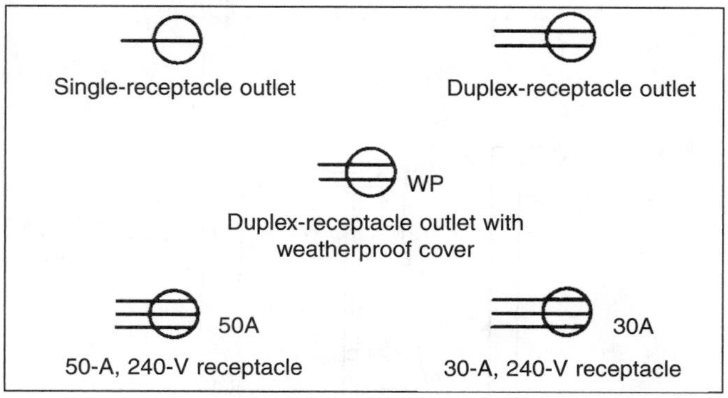
Single-receptacle outlet
Duplex-receptacle outlet
WP
Duplex-receptacle outlet with weatherproof cover
50A
30A
50-A, 240-V receptacle
30-A, 240-V receptacle

Figure A2-4: *Many electrical symbols have the same basic form but a line or note added gives each an individual meaning*

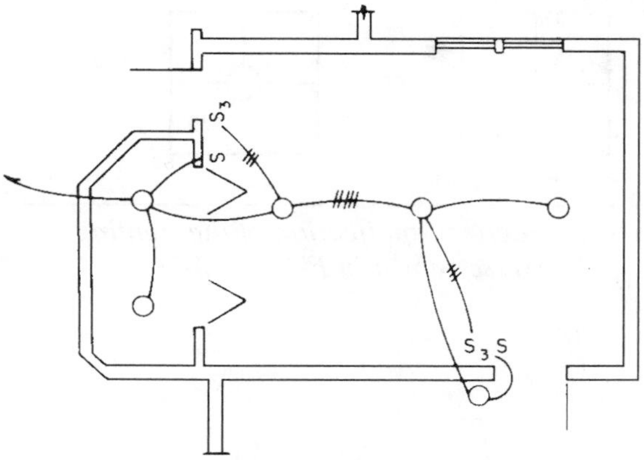

Figure A2-5: *Practical application of surface-mounted incandescent fixtures*

Lighting Outlets

 Ceiling outlet with surface-mounted incandescent lighting fixture

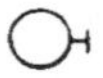 Wall outlet with surface-mounted incandescent fighting fixture

Practical use of these symbols is shown in Figure A2-5.

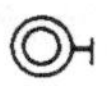 Wall outlet with recessed incandescent lighting fixture

 Ceiling outlet with recessed incandescent lighting fixture

Practical use of these symbols is shown in Figure A2-6.

Sometimes these symbols are modified in order to indicate the physical shape of a particular incandescent fixture. The lighting fixture in Figure A2-7 consists of four 6-in. cubes. This type of lighting fixture may be indicated on the electrical floor plan as shown in Figure A2-8.

A lighting fixture consisting of one cube may be indicated as shown in Figure A2-9. All should be drawn as close to scale as possible.

The type of mounting of all fixtures is usually indicated in the lighting fixture schedule shown on the drawings or in the written specifications. The fixture illustrated in Figure A2-7 is obviously pendant-mounted and should be so indicated in an appropriate column in the lighting fixture schedule, since the floor-plan view in Figure A2-9 does not indicate this fact.

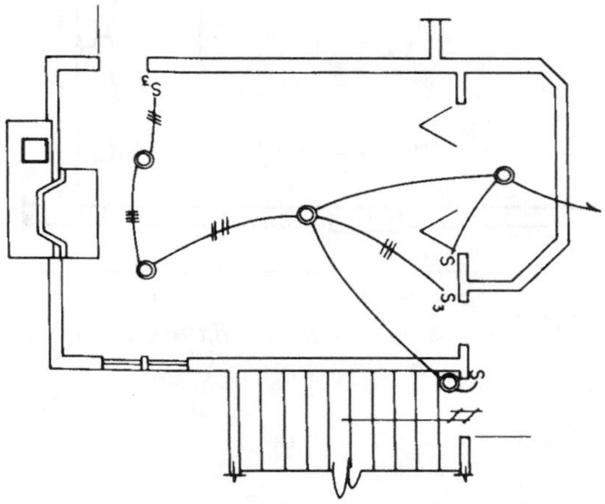

Figure A2-6: *Practical application of recessed incandescent lighting fixtures*

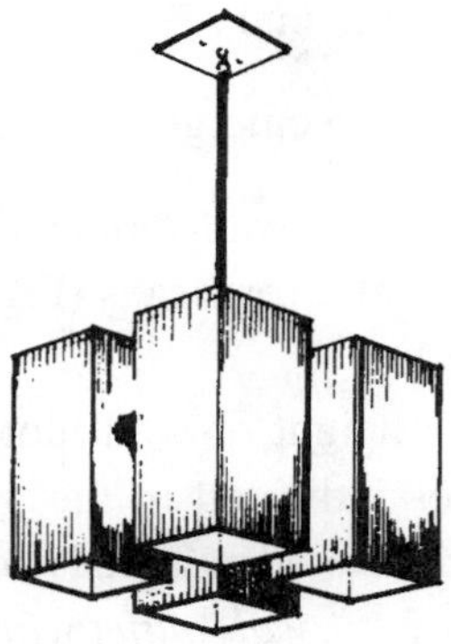

Figure A2-7: *Perspective view of a lighting fixture*

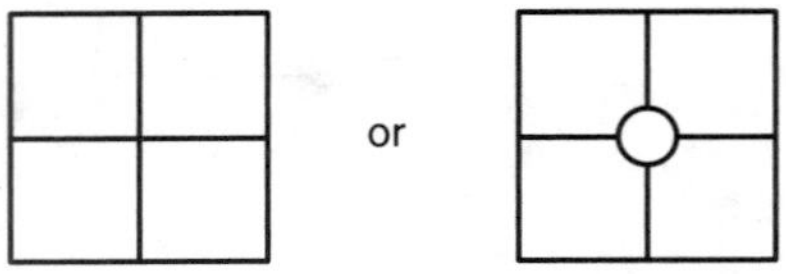
or

Figure A2-8: *Practical application of the lighting fixture shown in Figure A2-7*

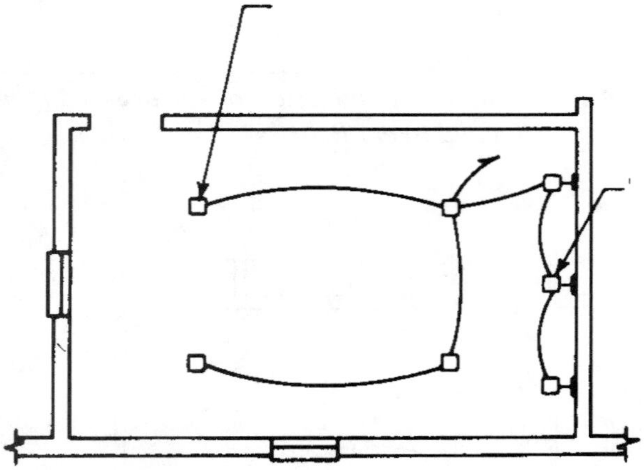

Figure A2-9: *Example showing how to indicate a one-cube lighting fixture on a working drawing*

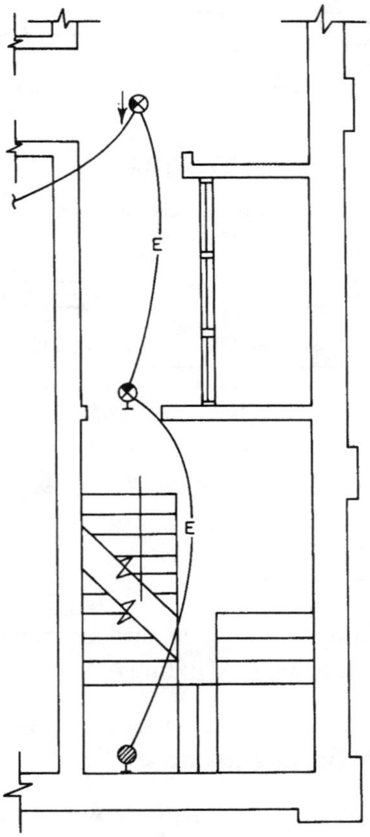
E
E

Figure A2-10: *Practical application of exit lights and emergency white incandescent fixtures*

 Ceiling or wall outlet with incandescent lighting connected to emergency circuit (emergency white light)

 Exit light on emergency circuit, surface- or pendant-mounted

 Exit light on emergency circuit, wall-mounted

 Surface- or wall-mounted exit light with directional arrowheads (Figure A2-10)

The mounting height of wall-mounted lighting fixtures is sometimes indicated in the symbol lists, especially where most are to be mounted at one height. For example, it might read "... *wall outlet with incandescent fixture mounted 6 ft. 6 in. above finished floor to center of outlet box unless otherwise indicated.*" If a few wall-mounted fixtures were to be mounted at 8 ft. 0 in. above finished floor, they could be indicated as shown in Figure A2-11, the letters AFF meaning "above finished floor."

 Ground-mounted incandescent upright (Figure A2-12)

 Post-mounted incandescent fixture

If only one lamp (or more than two lamps) is required on the floodlight outlet, it can be shown as in Figure A2-13 or Figure A2-14.

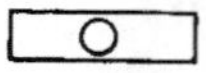 Ceiling outlet with surface- or pendant-mounted fluorescent fixture

 Ceiling outlet with recessed fluorescent fixture

 Ceiling outlet with continuous row of surface or pendant fluorescent fixtures

+ 96 in. AFF
+ 96 in. AFF

Figure A2-11: *Method of indicating mounting height of a few wall-mounted fixtures*

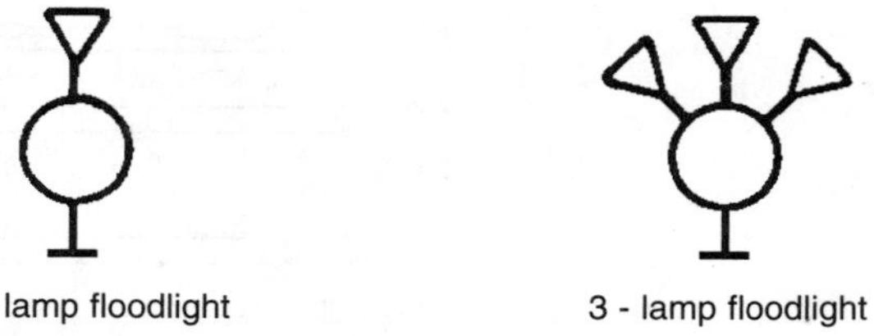
1 - lamp floodlight
3 - lamp floodlight

Figure A2-13: *Practical application showing floodlights on the drawings, either one lamp or more than two lamps*

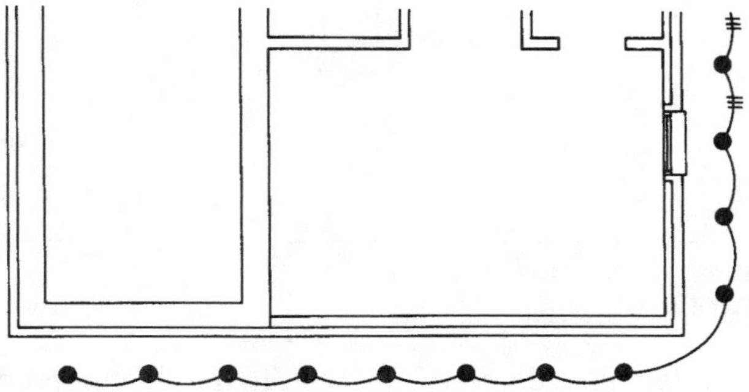

Figure A2-12: *Practical application of ground-mounted incandescent fixtures*

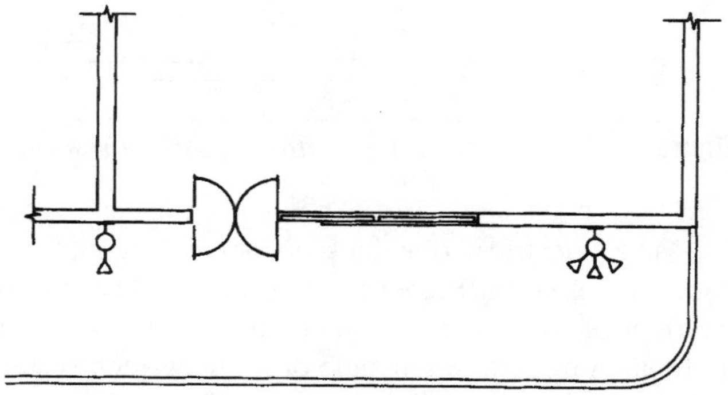

Figure A2-14: *Practical application of floodlights*

 Ceiling outlet with continuous row of recessed fixtures

 Ceiling outlet with bare-lamp fluorescent strip

 Ceiling outlet with continuous row of bare-lamp fluorescent-strip lighting

 Wall outlet with fluorescent fixture

 Fluorescent fixture mounted under cabinet

Modification of the symbols for fluorescent lighting is common. For example, cove lighting with bare-lamp fluorescent strips may be indicated on the drawings as shown in Figure A2-15.

The lighting layout illustrated in Figure A2-16 shows practical applications of all fluorescent symbols covered in this writing.

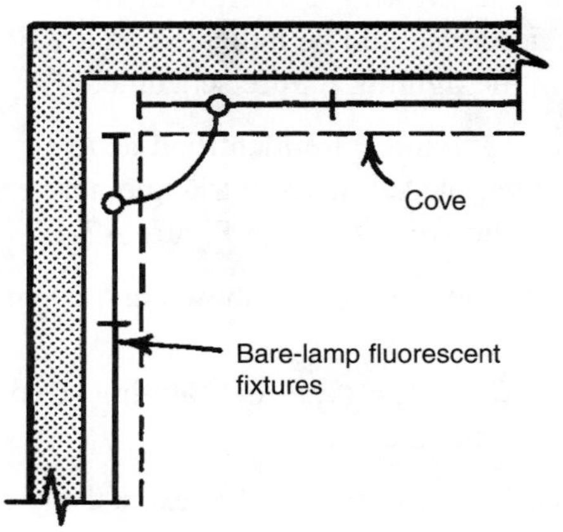
Cove
Bare-lamp fluorescent fixtures

Figure A2-15: *Example of bare-tube fluorescent fixtures shown on a working drawing*

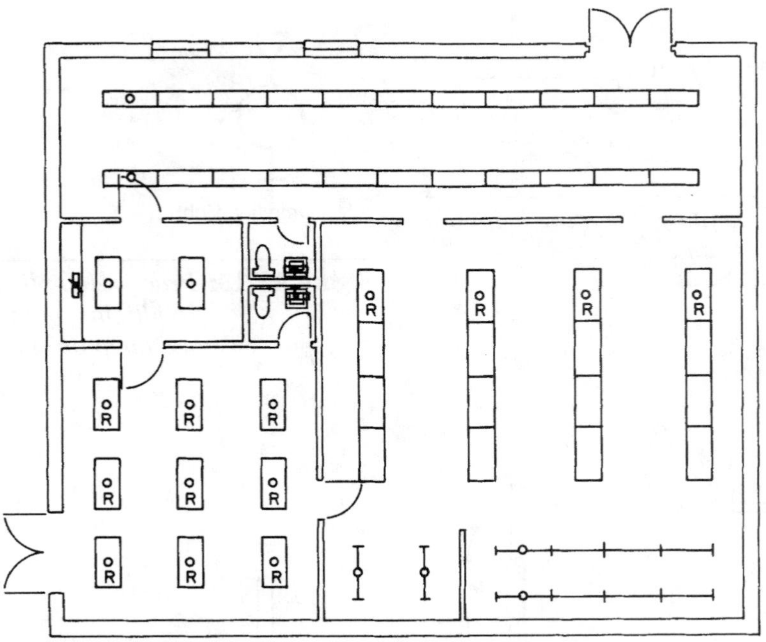

Figure A2-16: *Practical application of all fluorescent lighting symbols covered in this Appendix*

Many electrical drawings do not differentiate between recessed, surface, or pendant-mounted fixtures on the floor plans. Rather, the mounting is indicated either in the lighting fixture schedule or in the written specifications. However, since a major variation in the type of outlet box, outlet supporting means, wiring system arrangement, and outlet connection, plus need of special items such as plaster rings or roughing-in cans, depends upon the way in which a fixture is mounted, the electrician should be able to know the type of mounting in a glance at the drawings. Therefore, the mounting of a lighting fixture should be indicated on the floor plans as well as in the lighting fixture schedule.

Lighting fixtures are identified as to type of fixture by a numeral placed inside a triangle near each lighting fixture, as shown here and in Figure A2-17.

The indicated fixture is shown in the symbol lists as follows:

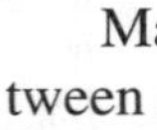
5

Indicates type of lighting fixture — see schedule

If one type of fixture is used exclusively in one room or area, as shown in Figure A2-18, the indicator need only appear once with the word "ALL" lettered at the bottom.

A complete description of the fixture identified by the symbol must be given in the lighting fixture schedule and should include the manufacturer, catalog number, number and type of lamps, voltage, finish, mounting, and any other necessary information needed for a proper installation of the fixtures. Figure A2-19 shows an example of a lighting fixture schedule.

Fluorescent fixtures should be drawn to approximate scale, showing physical size whenever practical.

Mercury vapor and other electric-discharge lighting fixtures are indicated on the drawings in the same way as incandescent fixtures. The type of lamp is indicated in the lighting fixture schedule.

Receptacles and Switches

Every day those who work in the electrical construction industry hear the word "standard." Yes, it would be the ideal situation if all electrical engineers could use one set of standard electrical symbols for all their projects. However, with the present symbols known as "standard" this is not practical.

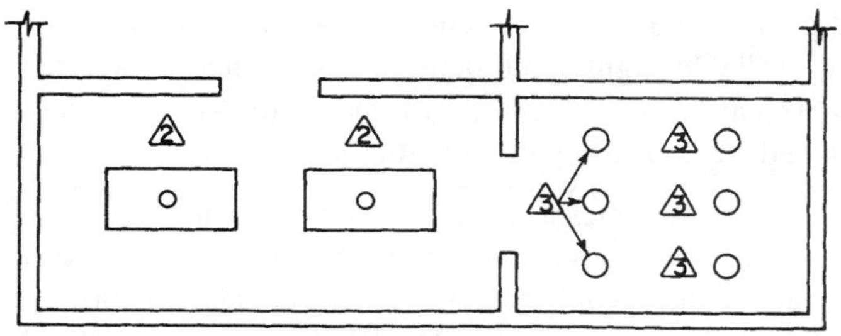

Figure A2-17: *Method of identifying lighting fixtures of different types*

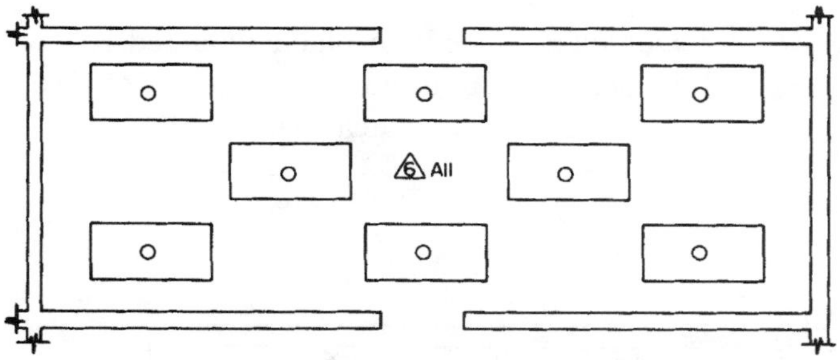
All

Figure A2-18: *Method of identifying several lighting fixtures of one type*

LIGHTING FIXTURE SCHEDULE						
FIXT. Type	Manufacturer's Description	LAMPS No.	LAMPS Type	Volts	Mounting	Remarks

Figure A2-19: *Typical lighting fixture schedule used on an electrical working drawing to identify lighting fixtures*

For example, one consulting engineering firm in Washington, DC, did a large amount of electrical designs for hospitals all over the eastern United States. On one of their jobs, there were over 100 duplex receptacles mounted horizontally in the backsplash of lavatory countertops, while there were over 300 conventional duplex receptacles.

If a draftsman had to letter a note at each of the receptacles located in the backsplash, it is easily seen that much time would be spent on the drawings. On the other hand, a simple symbol with a written explanation in the legend would tell exactly what work was to be done.

One firm recently designed the electrical systems for a group of quick-service restaurants where several of the duplex receptacles were located in the ceiling. If standard symbols had been used, each location would have looked like the following:

 Flush mounted 120V duplex receptacles mounted in the ceiling

Much time was saved by composing a new symbol for the ceiling-mounted receptacles and eliminating the descriptive note at each of the dozen or so receptacles. A note appeared only once in the symbol list as follows:

 Flush-mounted 120V, duplex "twist-lock" receptacle mounted in ceiling with stainless steel plate

The cases are endless, and until a sufficient number of different symbols is available in the standard symbol list, it is highly impractical to use the standard throughout.

If a special outlet is shown on a drawing in only a few locations, then perhaps a descriptive note at each location is best; but if the "special outlet" appears in several locations, then an individual symbol can save much drawing time.

The following are switch and receptacle symbols used in one office on working drawings.

S Single-pole toggle switch mounted 50 inches above finished floor to center of box unless otherwise indicated.

S_2 Two-pole toggle switch mounted 50 inches above finished floor to center of box unless otherwise indicated.

S_3 Three-way switch mounted 50 inches above finished floor to center of box unless otherwise indicated.

S_4 Four-way switch mounted 50 inches above finished floor to center of box unless otherwise indicated.

S_L Low-voltage switch to relay mounted 50 inches above finished floor unless otherwise indicated.

S_D Flush-mounted door switch to control closet light.

S_P Switch with pilot light.

Figure A2-20 illustrates some practical applications of switch symbols used on working drawings. Notice that the single-pole switches are used to control lighting from one point, while the three- and four-way switches, used in combination, control a light or series of lights from two or more points.

The two-pole switch is used to control a series of lights on two separate circuits with only one motion. The switch-pilot light combination is used where it is practical to notice if the item controlled by the switch is energized, such as a light in an attic or closet.

Door switches to control closet lights are quite common in residential construction. When the closet door is opened, the switch button is released and in turn energizes the circuit to the closet light. When the door is closed, the light is de-energized.

Receptacle symbols used on working drawings are numerous. Some consulting engineering firms have used over 50 different symbols for receptacles and power outlets. However, many drawings contain only six different symbols to cover most of the applications. They are as follows:

⊖ Duplex grounded receptacle mounted 18 inches above finished floor to center of box unless otherwise indicated.

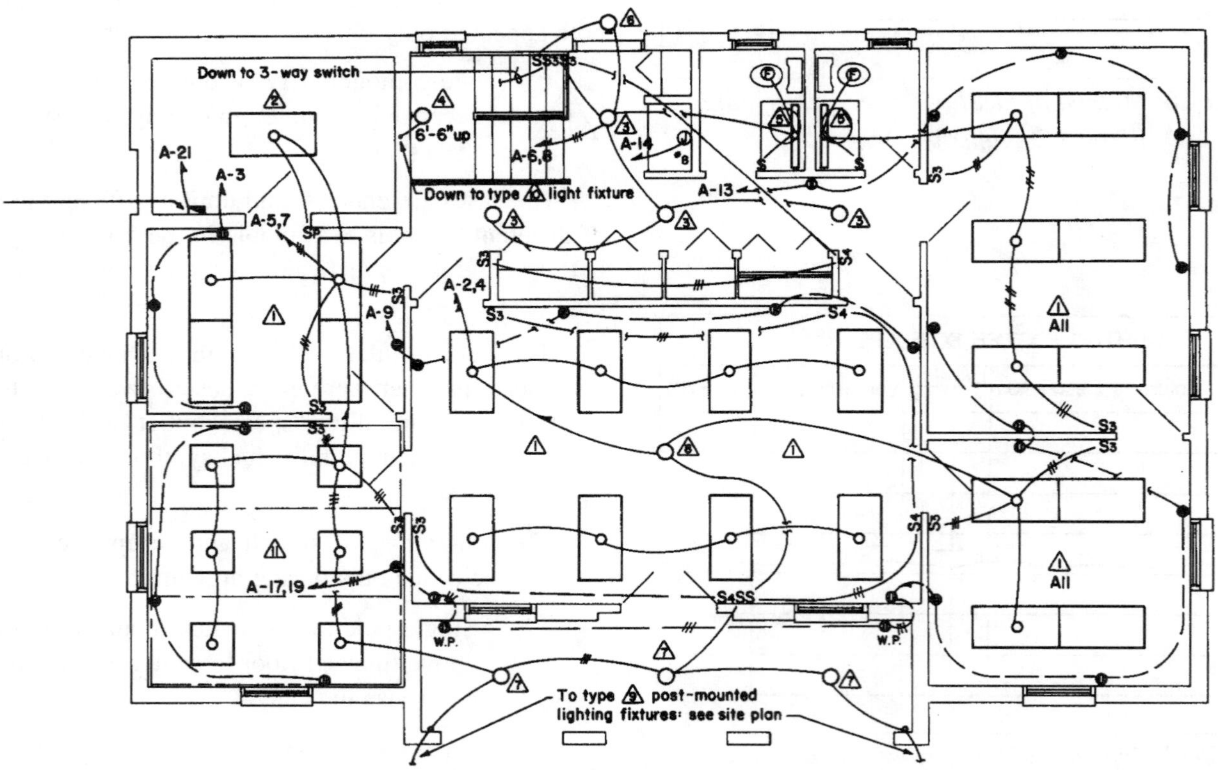
Down to 3-way switch
6'-6" up
Down to type light fixture
A-21
A-3
A-5,7
A-14
A-13
A-6,8
A-2,4
A-9
A-17,19
All
W.P.
To type post-mounted lighting fixtures: see site plan

Figure A2-20: *Practical applications of switch symbols used on working drawings*

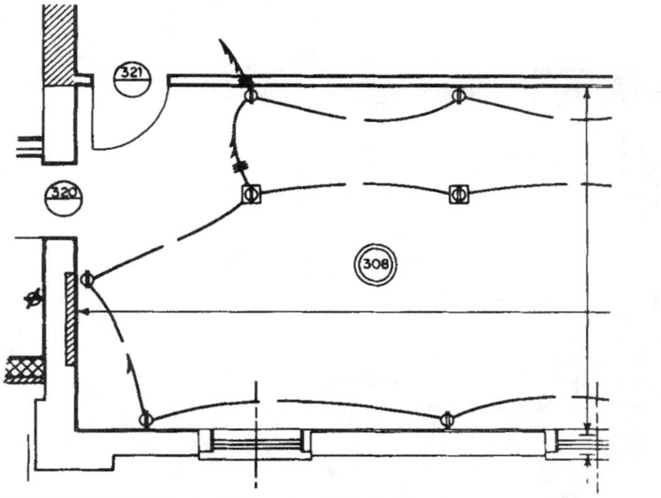
321
320
308

Figure A2-21: *Practical applications of receptacle symbols used on working drawings*

 Duplex grounded receptacle mounted 6 inches above countertop to center of box.

 Split-wired duplex grounded receptacle with top half switched. Mount 18 inches above finished floor unless otherwise indicated.

30 a
3-pole, 3-wire, 240-V receptacle, amperage as indicated.

a
Special outlet or connection, letter indicates types; see legend at end of symbol list.

Floor outlets are indicated on many drawings with a square with the appropriate symbol drawn inside:

 Floor-mounted single receptacle, grounded.

 Floor-mounted duplex receptacle, grounded.

J
Floor-mounted junction box.

Figure A2-21 shows some practical applications of receptacle symbols as used on working drawings.

If other symbols are required to indicate various outlets on working drawings, they may be composed and added to the symbol list or legend with a description of their use. Examples are as follows:

 Duplex grounded receptacle with "Twist-Lock" connection.

 Whatever the need may be.

 Whatever the need may be.

When outlets are located in areas requiring special boxes, covers, etc., they are usually indicated by abbreviations.

Example:

W.P.	Indicates weatherproof cover
E.P.	Indicates explosionproof device, fittings, etc.
EMERG	Indicates outlet on emergency circuit.

Service Equipment

Panelboards, distribution centers, transformers, safety switches, and similar components of the electrical installation are indicated by electrical symbols on floor plans and by a combination of symbols or semi-pictorial drawings in riser diagrams. Some these symbols are as follows:

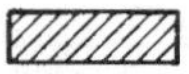 Power panel or main distribution panel

A
Surface-mounted lighting panel; numeral indicates type; see panelboard schedule

A
Flush-mounted lighting panel; numeral indicates type; see panelboard schedule

F
60A
Fusible safety switch; numerals indicate ampere capacity

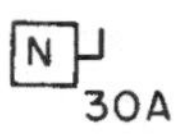
N
30A
Nonfusible safety switch; numerals indicate ampere capacity

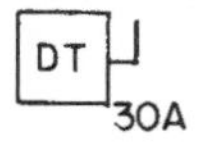
DT
30A
Double-throw switch; numerals indicate ampere capacity

The description of panels and service equipment is usually covered in panelboard schedules such as those in Figures A2-22 and A2-23; at other times a description of the panelboard is covered in the written specifications.

Circuit and feeder symbols have been nearly standardized in that most electrical drawings use a solid line ——— to indicate circuits concealed in ceiling or wall and a broken line — — — for conduit or raceways concealed in floor or ceiling below. Exposed conduit or raceway is also shown with a broken line, but the dashes are shorter than those used for concealed circuits - - -. See Figure A2-24.

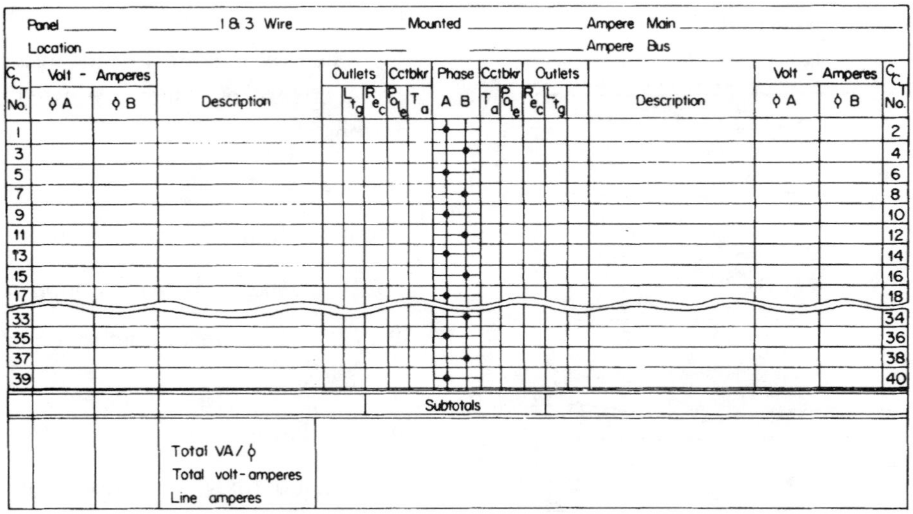
Panel
1 & 3 Wire
Mounted
Ampere Main
Location
Ampere Bus
Volt - Amperes
Description
Outlets
Cctbkr
Phase
A B
Subtotals
Total VA/ϕ
Total volt-amperes
Line amperes

Figure A2-22: *One type of panelboard schedule*

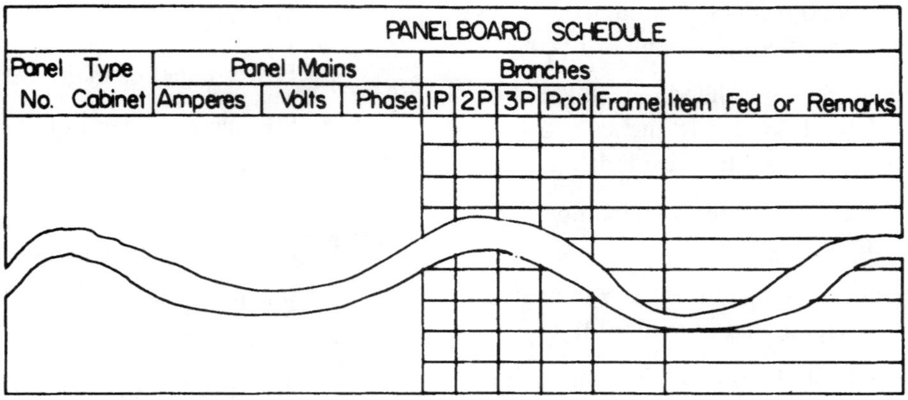
PANELBOARD SCHEDULE
Panel No.
Type Cabinet
Panel Mains
Amperes
Volts
Phase
Branches
1P
2P
3P
Prot
Frame
Item Fed or Remarks

Figure A2-23: *Another type of panelboard schedule*

The variations between electrical designers comes with the method of drawing these lines. Some designers prefer to draw all circuit lines with a straightedge, as shown in Figure A2-25. Others prefer to use a French curve to draw the circuit line; this is illustrated in Figure A2-26. The preference is to use curved lines for all circuits so that they will not be mistaken for building or equipment lines.

Certain letters or numerals may appear in circuit lines. For example, some drawings have used the symbol **—X—X—** to indicate BX or flexible metallic armored cable in electrical wiring systems where both rigid conduit and flexible cable were to be used.

The number of conductors in a conduit or raceway system may be indicated in the panelboard schedule under the appropriate column, but on the other hand, this information is shown on the floor plans along with the circuits. Most workers find this latter method the easiest to follow. For example, one symbol list contains the following symbol to indicate concealed branch circuit wiring:

Branch circuit concealed in ceiling or walls; slash marks indicate number of conductors in run — two conductors are not noted; numerals indicate wire size — No. 12 AWG not noted.

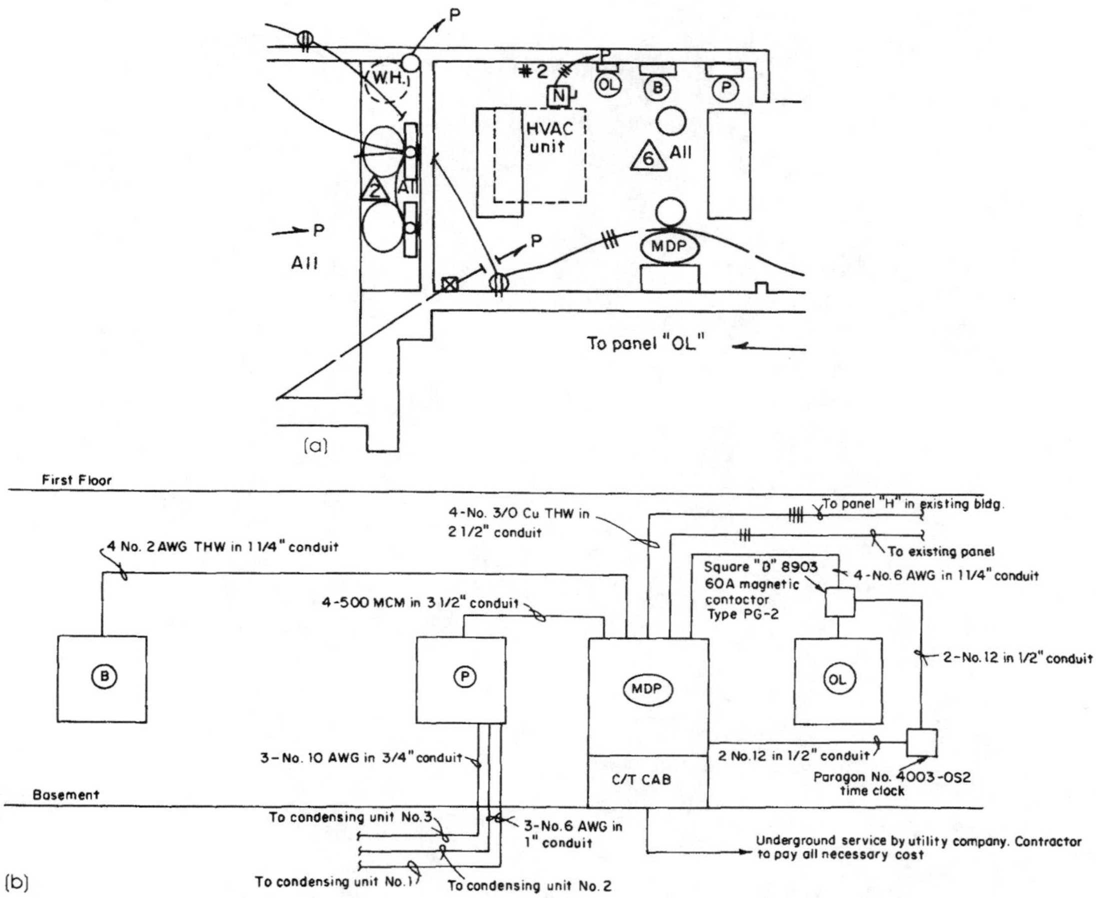
P
W.H.
#2
N
OL
B
P
HVAC unit
6 All
2 All
P
All
P
MDP
To panel "OL"
(a)
First Floor
4-No. 3/0 Cu THW in 2 1/2" conduit
To panel "H" in existing bldg.
4 No. 2AWG THW in 1 1/4" conduit
To existing panel
Square "D" 8903 60A magnetic contactor Type PG-2
4-No.6 AWG in 1 1/4" conduit
4-500 MCM in 3 1/2" conduit
2-No.12 in 1/2" conduit
B
P
MDP
OL
3-No. 10 AWG in 3/4" conduit
2 No.12 in 1/2" conduit
C/T CAB
Paragon No. 4003-OS2 time clock
Basement
To condensing unit No.3
3-No.6 AWG in 1" conduit
Underground service by utility company. Contractor to pay all necessary cost
(b)
To condensing unit No.1
To condensing unit No.2

Figure A2-24: *Method of showing service equipment on working drawings*

The symbol description says that a solid-line circuit with no slash marks or numerals indicates a circuit containing two No. 12 AWG conductors. Three slash marks with no numeral indicates three No. 12 AWG conductors, etc. If we want to indicate four No. 10 AWG conductors in a circuit, the symbol is:

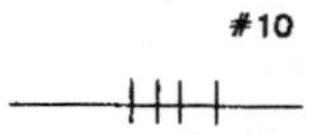
#10

Six slashes through the circuit lines on the drawing would indicate six conductors, etc.

Most electrical drawings use the symbol to indicate home runs to panelboard, with the number of arrowheads indicating the number of circuits in the run and the slash marks indicating the number of conductors in the run. However, since full arrowheads are normally used for call-outs, half-arrowheads are frequently used to indicate branch-circuit home runs to panelboard.

Figure A2-25: *Circuit drawn with straightedge*

Figure A2-26: *Circuit drawn with French curve*

Glossary

AA (Aluminum Association): A manufacturers' association which promotes the use of aluminum.

AAC: All aluminum conductor.

AASC: Aluminum alloy stranded conductors.

Abrasion: The process of rubbing, grinding, or wearing away by friction.

Abrasion resistance: Ability to resist surface wear.

ac (alternating current): 1) A periodic current, the average of which is zero over a period; normally the current reverses after given time intervals and has alternately positive and negative values. 2) The type of electrical current actually produced in a rotating generator (alternator).

Accelerated life tests: Subjecting a product to operating conditions more severe than normal to expedite deterioration to afford some measure of probable life at normal conditions.

Accelerator: 1) A substance that increases the speed of a chemical reaction. 2) Something to increase velocity.

Acceptance test: Made to demonstrate the degree of compliance with specified requirements.

Accepted: Approval for a specific installation or arrangement of equipment or materials.

Accessible (electrical wiring): Capable of being removed or exposed without damaging the building structure or finish, or not permanently closed in by the structure or finish of the building.

Accessible (equipment): Admitting close approach; not guarded by locked doors, elevation, or other effective means.

Accessible, readily: Capable of being reached quickly for operation, renewal, or inspections, without requiring those to whom access is requisite to climb over or remove obstacles or to resort to portable ladders, etc.

ACSR (aluminum, conductor, steel reinforced): A bare composite of aluminum and steel wires, usually aluminum around steel.

Active power: In a 3-phase symmetrical circuit: $\pi = 3vi \cos \theta$; in a 1-phase, 2-wire circuit, $\pi = vi \cos \theta$.

Administrative authority: An organization exercising jurisdiction over the National Electrical Safety Code.

AEIC: Association of Edison Illuminating Companies.

Aggregate: Material mixed with cement and water to produce concrete.

Aging: The irreversible change of material properties after exposure to an environment for an interval of time.

AIA: 1) American Institute of Architects. 2) Aircraft Industries Association.

Air cleaner: Device used for removal of airborne impurities.

Air diffuser: Air distribution outlet designed to direct airflow into desired patterns.

Air entrained concrete: Concrete in which a small amount of air is trapped by addition of a special material to produce greater durability.

Air oven: A lab oven used to heat by convection of hot air.

Al: Aluminum.

Al-Cu: An abbreviation for aluminum and copper, commonly marked on terminals, lugs and other electrical connectors to indicate that the device is suitable for use with either aluminum conductors or copper conductors.

Alive: Energized; having voltage applied.

Alloy: A substance having metallic properties and being composed of elemental metal and one or more chemical elements.

Alternate power source: One or more generator sets, or battery systems where permitted, intended to provide power during the interruption of normal electrical services, or of the public utility electrical service intended to provide power during interruption of service normally provided by the generating facilities on the premises.

Alternating-current module (ac photovoltaic module): A complete, environmentally protected unit consisting of solar cells, optics, inverter, and other components, exclusive of tracker, designed to generate ac power when exposed to sunlight.

Alternator: A device to produce alternating current.

Ambient temperature: Temperature of fluid (usually air) that surrounds an object on all sides.

Ambulatory health care center: A building or part thereof used to provide services or treatment to four or more patients on an outpatient basis. The facility may provide: (1) surgical treatment requiring general anesthesia; or (2) treatment for patients that would render them incapable of taking action for self-preservation under emergency conditions without assistance from others, such as hemodialysis units or freestanding emergency medical units.

Ammeter: An electric meter used to measure current, calibrated in amperes.

Ampacity: The current-carrying capacity of conductors or equipment, expressed in amperes.

Ampere (A): The basic SI unit measuring the quantity of electricity.

Ampere-turn: The product of amperes times the number of turns in a coil.

Amplification: Procedure of expanding the strength of a signal.

Amplifier: 1) A device that enables an input signal to directly control a larger energy flow. 2) The process of increasing the strength of an input.

Amplitude: The maximum value of a wave.

Analog: Pertaining to data from continuously varying physical quantities.

Anesthetizing location: Any area of a health care facility that has been designated to be used for the administration of any flammable or nonflammable inhalation anesthetic agent in the course of examination or treatment including the use of such agents for relative analgesia.

Angle, roll over (overhead): The sum of the vertical angles between the conductor and the horizontal on both sides of the traveler; excessive roll over angles can cause premature splice failures.

Angular velocity: The average time rate of change. of angular position; in electrical circuits = 2 f, and f equals frequency.

Annealing: The process of preventing or removing objectional stresses in materials by controlled cooling from a heated state; measured by tensile strength.

Annealing, bright: Annealing in a protective environment to prevent discoloration of the surface.

Anode: 1) Positive electrode through which current enters a nonmetallic conductor such as an electrolytic cell. 2) The negative pole of a storage battery.

ANSI (American National Standards Institute): An organization that publishes nationally recognized standards.

Antenna: A device for transmission or reception of electromagnetic waves.

Antioxidant: Retards or prevents degradation of materials exposed to oxygen (air) or peroxides.

Appliance: Equipment designed for a particular purpose which utilizes electricity to produce heat, light, mechanical motion, etc., usually complete in itself, generally other than industrial, normally in standard sizes or types.

Approved: 1) Acceptable to the authority having legal enforcement. 2) Per OSHA: A product that has been tested to standards and found suitable for general application, subject to limitations outlined in the nationally recognized testing lab's listing.

Apron: Piece of horizontal wood trim under the sill of the interior casing of a window.

Areaway: Open space below the ground level immediately outside a building. It is enclosed by substantial walls.

Arc: A flow of current across an insulating medium.

Arc furnace: Heats by heavy current flow through the material to be heated.

Arcing time: The time elapsing from the severance of the circuit to the final interruption of current flow.

Arc resistance: The time required for an arc to establish a conductive path in or across a material.

Armature: 1) Rotating machine: the member in which alternating voltage is generated. 2) Electromagnet: the member which is moved by magnetic force.

Armor: Mechanical protector for cables; usually a helical winding of metal tape, formed so that each convolution locks mechanically upon the previous one (interlocked armor); may be a formed metal tube or a helical wrap of wires.

Armored cable (type AC): A fabricated assembly of insulated conductors in a flexible metallic enclosure.

Array: Mechanically integrated assembly of modules or panels with a support structure and foundation, tracker, and other components, as required, to form a direct-current power-producing unit for a solar photovoltaic system.

Arrester: Wire screen secured to the top of an incinerator to confine sparks and other products of burning.

Askarel: A synthetic insulating oil which is nonflammable but very toxic — being replaced by silicone oils.

ASME: American Society of Mechanical Engineers.

ASTM (American Society for Testing and Materials): A group writing standards for testing materials, and specifications for materials.

Asymmetrical: Not identical on both sides of a central line; unsymmetrical.

Attachment plug or cap: The male connector for electrical cords.

Attenuation: A decrease in energy magnitude during transmission.

Audible: Capable of being heard by humans.

Audio amplifier: Electronic equipment that increases the current or voltage (or both) potential of an audio signal intended for use by another piece of equipment.

Automatic: Operating by own mechanism when actuated by some impersonal influence: nonmanual: self-acting.

Automatic transfer equipment: A device to transfer a load from one power source to another, usually from normal to emergency source and back.

Autotransformer: Any transformer where primary and secondary connections are made to a single cell.

Auxiliary: A device or equipment which aids the main device or equipment.

AWG (American Wire Gage): The standard for measuring wires in America.

BX: A nickname for armored cable (wires with a spiral-wound, flexible steel outer jacketing); although used generically, BX is a registered tradename of the General Electric Company.

Backfill: Loose earth placed outside foundation walls for filling and grading.

Back pressure: Pressure in the low side of a refrigerating system; also called suction pressure or low side pressure.

Ballast: A device designed to stabilize current flow.

Balloon framing: System of small house framing; two by fours extending two stories with inch by quarter ledger strips notched into the studs to support the second-story floor beams.

Bank: An installed grouping of a number of units of the same type of electrical equipment; such as "a bank of transformers" or "a bank of capacitors" or a "meter bank," etc.

Bar: A long solid product having one cross-sectional dimension of 0.375 inch or more.

Bare (conductor): Not insulated; nor coated.

Barrier: A partition; such as an insulating board to separate bus bars of opposite voltages.

Base: One of the regions or terminals of a transistor.

Base ambient temperature: The temperature of a cable group when there is no load on any cable of the group or on the duct bank containing the group.

Base load: The minimum load over a period of time.

Battery: A device which changes chemical to electrical energy, used to store electricity.

Beam: A horizontal member of wood, reinforced concrete, steel, or other material used to span the space between posts, columns, girders, or over an opening in a wall.

Bearing plate: Steel plate placed under one end of a beam or truss for load distribution.

Bearing wall: Wall supporting a load other than its own weight.

Bed: Place or material in which stone or brick is laid; horizontal surface of positioned stone; lower surface of brick.

Bedding: A layer of material to act as a cushion or interconnection between two elements of a device, such as the jute layer between the sheath and wire armor in a submarine cable; sometimes incorrectly used to refer to extruded insulation shields.

Belt: The outer protective nonmetallic covering of cable; its jacket.

Belted type cable: A multiple conductor cable having a layer of insulation over the core conductor assembly.

Bench mark: Point of reference from which measurements are made.

Bessel function: A mathematical solution to a differential equation which is used to solve changes in conductor resistance and mutual inductance between conductors with respect to frequency changes due to skin and proximity effects.

Bias vacuum tube: Difference of potential between control grid and cathode. Transistor — difference of potential between base and emitter and base and collector. Magnetic amplifier — level of flux density in the magnetic amplifier core under no-signal condition.

BIL (Basic Impulse Level): A reference impulse insulation strength.

Bimetal strip: Temperature regulating or indicating device that works on the principle that two dissimilar metals with unequal expansion rates, welded together, will bend as temperature changes.

Binder: Material used to hold assembly together.

Birdcage: The undesired unwinding of a stranded cable.

Blister: A defect in metal, on or near the surface, resulting from the expansion of gas in the subsurface zone. Very small blisters may be called "pinheads" or "pepper blisters."

Blocking diode: A diode used to block the reverse flow of current into a photovoltaic source circuit.

Blueprint: See definition of *drawing*.

Boiler: Closed container in which a liquid may be heated and vaporized.

Boiling point: Temperature at which a liquid boils or generates bubbles of vapor when heated.

Bond: A mechanical connection between metallic parts of an electrical system, such as between a neutral wire and a meter enclosure or service conduit to the enclosure for the service equipment with a bonding locknut or bushing; the junction of welded parts; the adhesive for abrasive grains in grinding wheels.

Bonding: The permanent joining of metallic parts to form an electrically conductive path that will insure electrical continuity and the capacity to conduct safely any current likely to be imposed.

Bonding bushing: A special conduit bushing equipped with a conductor terminal to take a bonding jumper; also has a screw or other sharp device to bite into the enclosure wall to bond the conduit to the enclosure without a jumper when there are no concentric knockouts left in the wall of the enclosure.

Bonding jumper: A bare or insulated conductor used to assure the required electrical conductivity between metal parts required to be electrically connected. Frequently used from a bonding bushing to the service equipment enclosure to provide a path around concentric knockouts in an enclosure wall, also used to bond one raceway to another.

Bonding locknut: A threaded locknut for use on the end of a conduit terminal, but a locknut equipped with a screw through its lip. When the locknut is installed, the screw is tightened so its end bites into the wall of the enclosure close to the edge of the knockout.

Border light: A permanently installed overhead strip light.

Braid: An interwoven cylindrical covering of fiber or wire.

Branch circuit: That portion of a wiring system extending beyond the final overcurrent device protecting a circuit.

Branch circuit, appliance: A branch circuit that supplies energy to one or more outlets to which appliances are to be connected, and that has no permanently connected lighting fixtures that are not part of an appliance.

Braze: The joining together of two metal pieces, without melting them, using heat and diffusion of a jointing alloy of capillary thickness.

Breadboard: Laboratory idiom for an experimental circuit.

Breakdown: The abrupt change of resistance from high to low, allowing current flow: an initial rolling or drawing operation.

Breaker strip: Thin strips of material placed between phase conductors and the grounding conductor in flat parallel portable cables; the breaker strips provide extra mechanical and electrical protection.

Breakout: The point at which conductor(s) are taken out of a multiconductor assembly.

Breakout, assembly: An adapter to connect a multipole connector containing two or more branch circuits to multiple individual circuit connectors.

Bridge: A circuit which measures by balancing four impedances through which the same current flows:

Wheatstone — resistance
Kelvin — low resistance
Schering — capacitance, dissipation factor, dielectric constant
Wien — capacitance, dissipation factor

British thermal unit (Btu): Quantity of heat required to raise the temperature of 1 pound of water 1 degree Fahrenheit.

Brittle point: The highest temperature at which a chilled strip of polymer will crack when it is held at one end and impacted at the other end.

Brush: A conductor between the stationary and rotating parts of a machine, usually of carbon.

Buck: Rough wood door frame placed on a wall partition to which the door moldings are attached; completely fabricated steel door frame set in a wall or partition to receive the door.

Buff: To lightly abrade.

Bug: A crimped or bolted type of electrical connector for splicing wires or cables together. Also used as a verb: "bugged." Example: The wires were bugged together.

Bull-cutters: A larger, long-handled tool for cutting the larger sizes of wire and cable, up to MCM sizes.

Bull switch: An externally-operated, wall-mounted safety switch, which may or may not contain overcurrent protection, that is designed for the protection of portable cables and cords.

Buna: A synthetic rubber insulation of styrene-butadiene; was known as GR-S, now as SBR.

Bundled: Cables or conductors that are physically tied, wrapped, taped or otherwise bound together.

Burner: Device in which combustion of fuel takes place.

Bus: The conductor(s) serving as a common connection for two or more circuits.

Busbars: The conductive bars used as the main current supplying elements of panelboards or switchboards; also the conductive bars duct: an assembly of busbars within an enclosure which is designed for ease of installation, have no fixed electrical characteristics, and allowing power to be taken off conveniently, usually without circuit interruption.

Busway: A grounded metal enclosure containing factory-mounted bare or insulated conductors, usually copper or aluminum bars, rods, or tubes.

Butane: Liquid hydrocarbon commonly used as fuel for heating purposes.

Buttress: Projecting structure built against a wall to give it greater strength.

Bypass: Passage at one side of or around a regular passage.

CB: Pronounced "see bee," an expression used to refer to "circuit breaker," taken from the initial letters C and B.

CT: Pronounced "see tee," refers to current transformer, taken from the initial letters C and T.

Cable: An assembly of two or more wires which may be insulated or bare.

Cable, aerial: An assembly of one or more conductors and a supporting messenger.

Cable, armored: A cable having armor (see *armor*).

Cable, belted: A multiconductor cable having a layer of insulation over the assembled insulated conductors.

Cable, bore-hole: The term given vertical riser cables in mines.

Cable clamp: A device used to clamp around a cable to transmit mechanical strain to all elements of the cable.

Cable, coaxial: A cable used for high frequency, consisting of two cylindrical conductors with a common axis separated by a dielectric; normally the outer conductor is operated at ground potential for shielding.

Cable, control: Used to supply voltage (usually ON or OFF).

Cable, duplex: A twisted pair of cables.

Cable, festoon: A cable draped in accordion fashion from sliding or rolling hangers, usually used to feed moving equipment such as bridge cranes.

Cable, hand: A mining cable used to connect equipment to a reel truck.

Cable, parkway: Designed for direct burial with heavy mechanical protection of jute, lead, and steel wires.

Cable, portable: Used to transmit power to mobile equipment.

Cable, power: Used to supply current (power)

Cable, pressure: A cable having a pressurized fluid (gas or oil) as part of the insulation; gas and oil are the most common fluids.

Cable, ribbon: A flat multiconductor cable.

Cable, service drop: The cable from the utility line to the customer's property.

Cable, signal: Used to transmit data.

Cable, spacer: An aerial distribution cable made of covered conductors held by insulated spacers; designed for wooded areas.

Cable, spread room: A room adjacent to a control room to facilitate routing of cables in trays away from the control panels.

Cable, submarine: Designed for crossing under navigable bodies of water; having heavy mechanical protection against anchors, floating debris and moisture.

Cable tray: A rigid structure to support cables: a type of raceway: normally having the appearance of a ladder and open at the top to facilitate changes.

Cable, tray: A multiconductor having a nonmetallic jacket, designed for use in cable trays; (not to be confused with type TC cable for which the jacket must also be flame retardant).

Cable, triplexed d: Helical assembly of 3 insulated conductors and sometimes a bare grounding conductor.

Cable, unit: A cable having pairs of cables stranded into groups (units) of a given quantity, then these groups form the core.

Cable, vertical risers: Cables utilized in circuits of considerable elevation change; usually incorporate additional components for tensile strength.

Cablebus: An assembly of insulated conductors, with fittings and conductor terminations, in a completely enclosed, ventilated protective metal housing.

Cabling: Helically wrapping together of two or more insulated wires.

Calibrate: Compare with a standard.

Calorie: Heat required to raise temperature of 1 gram of water 1 degree centigrade.

Cambric: A fine weave of linen or cotton cloth used as insulation.

Candela (cd): The basic SI unit for luminous intensity: the candela is defined as the luminous intensity of 1/600,000 of a square meter of a blackbody at the temperature of freezing platinum.

Capacitance: The storage of electricity in a capacitor; the opposition to voltage change: units: farad.

Capacitor: An apparatus consisting of two conducting surfaces separated by an insulating material. It stores energy, blocks the flow of direct current, and permits the flow of alternating current to a degree depending on the capacitance and frequency.

Capstan: A rotating drum used to pull cables or ropes by friction; the cables are wrapped around the drum.

Carbon black: A black pigment produced by the incomplete burning of natural gas or oil; used as a filler.

Carbon dioxide: Compound of carbon and oxygen that is sometimes used as a refrigerant.

Cascade: The output of one device connected to the input of another.

Catepuller: Two endless belts which squeeze and pull a cable by friction.

Cathode: 1) The negative electrode through which current leaves a nonmetallic conductor, such as an electrolytic cell. 2) The positive pole of a storage battery. 3) Vacuum tube — the electrode that emits electrons.

Cathode-ray tube: The electronic tube which has a screen upon which a beam of electrons from the cathode can be made to create images; for example; the television picture tube.

Cathodic protection: Reduction or prevention of corrosion by making the metal to be protected the cathode in a direct current circuit.

Cavity wall: Wall built of solid masonry units arranged to provide airspace within the wall.

C-C: Center to center.

CCA (Customer Complaint Analysis): A formal investigation of a cable defect or failure.

CEE (International Commission on Rules for the Approval of Electrical Equipment): Controls the standards for electrical products for sale in Europe; analogous to UL in USA.

Cell line: An assembly of electrically interconnected electrolytic cells supplied by a source of direct-current power.

Cellular metal floor raceways: The hollow spaces of cellular metal floor, that with suitable fittings, may be approved as enclosures for electrical conductors.

Centigrade scale: Temperature scale used in metric system. Freezing point of water is 0 degrees; boiling point is 100 degrees.

CFR (Code of Federal Regulations): The general and permanent rules published in the Federal Register by the executive departments and agencies of the Federal Government. The Code is divided into 50 Titles which represent broad areas; Titles are divided into Chapters which usually bear the name of the issuing agency, e.g. Title 30 — Mineral Resources, Chapter 1 = MESA; Title 29 — Labor, Chapter XVII — OSHA; Title 10 — Energy, Chapter I = NRC.

Cgs: Centimeter, gram, second.

Charge: The quantity of positive or negative ions in or on an object; unit: coulomb.

Charge controller: Equipment that controls dc voltage or dc current (or both) and is used to charge a battery.

Chase: Recess in inner face of masonry wall providing space for pipes and/or ducts.

Check valve: A device that permits fluid flow only in one direction.

Chemical resistance test: Checking performance of materials immersed in different chemicals; loss of strength and dimensional change are measured.

Chimney effect: Tendency of air or gas to rise when heated.

Choke coil: A coil used to limit the flow of alternating current while permitting direct current to pass.

Circuit: A closed path through which current flows from a generator, through various components, and back to the generator.

Circuit breaker: A resettable fuse-like device designed to protect a circuit against overloading.

Circuit foot: One foot of circuit; i.e., if one has a 3-conductor circuit, then each lineal foot of circuit would have 3 circuit feet.

Circular mil: The non-SI unit for measuring the cross-sectional area of a conductor.

CL: Center line.

Class 1E (nuclear power): The safety electrical systems that are essential to emergency reactor shutdown, cooling, and containment.

Class 2 (nuclear power): Items important to reactor operation but not essential to safe shutdown or isolation.

Clearance: The vertical space between a cable and its conduit.

Clearing time: The time from sensing an overcurrent to circuit interruption.

Closing die: A die used to position the individual conductors during cabling.

Coated wire: Wire given a thin coating of another metal such as tin, lead, nickel, etc.; coating by dipping or planting; coating for protection, or to improve its properties.

Coaxial cable: A cable consisting of two conductors concentric with and insulated from each other.

Code: Short for *National Electrical Code*.

Code installation: An installation that conforms to the local code and/or the national code for safe and efficient installation.

Coefficient of expansion: The change in dimension due to change in temperature.

Coefficient of friction: The ratio of the tangential force needed to start or maintain uniform relative motion between two contacting surfaces to the perpendicular force holding them in contact.

Coil: A wire or cable wound in a series of closed loops.

Cold bend: A test to determine cable or wire characteristics at low temperatures.

Cold cathode: A cathode that does not depend on heat for electron emission.

Cold joint: Improper solder connection due to insufficient heat.

Cold welding: Solid-phase welding using pressure without heat.

Cold work: Permanent strain produced by an external force (such as wire drawing) in a metal below its recrystallization temperature.

Collector: The part of a transistor that collects electrons.

Color code: Identifying conductors by the use of color.

Come along: A cable grip (usually of tubular basketweave construction which tightens its grip on the cable as it is pulled) with a pulling "eye" on one end for attaching to a pull-rope for pulling conductors into conduit or other raceway.

Comfort zone: Area on psychrometric chart that shows conditions of temperature, humidity, and sometimes air movement in which most people are comfortable.

Commutator: Device used on electric motors or generators to maintain a unidirectional current.

Comples number: A mathematical expression (a + bi) in which 'a' and 'b' are real, $i^2 = 1$; useful in analyzing certain vectors, especially of electrical fields: j is substituted for i in formulae for electrical circuits — i.e. (a + jb).

Compound fill: An insulation which is poured into place while hot.

Compressibility: A density ratio determined under finite testing conditions.

Compression lug or splice: Installed by compressing the connector onto the strand, hopefully into a cold weld.

Compressor: The pump of a refrigerating mechanism that draws a vacuum or low pressure on the cooling side of a refrigerant cycle and squeezes or compresses the gas into the high pressure or condensing side of the cycle.

Computer: An electronic apparatus: 1) For rapidly solving complex and involved problems, usually mathematical or logical. 2) For storing large amounts of data.

Concealed: Rendered inaccessible by the structure or finish of the building. Wires in concealed raceways are considered concealed, even though they may become accessible by withdrawing them.

Concealed knob and tubing: A wiring method using knobs, tubes, and flexible nonmetallic tubing for the protection and support of single insulated conductors.

Concentricity: The measurement of the center of the conductor with respect to the center of the insulation.

Conductance: The ability of material to carry an electric current.

Conductor: Any substance that allows energy flow through it with the transfer being made by physical contact but excluding net mass flow.

Conductor, bare: Having no covering or insulation whatsoever.

Conductor, covered: A conductor having one or more layers of nonconducting materials that are not recognized as insulation under the *National Electric Code*.

Conductor, insulated: A conductor covered with material recognized as insulation.

Conductor load: The mechanical loads on an aerial conductor — wind, weight, ice, etc.

Conductor, plain: Consisting of only one metal.

Conductor, segmental: Having sections isolated, one from the other and connected in parallel; used to reduce ac resistance.

Conductor, solid: A single wire.

Conductor, stranded: Assembly of several wires, usually twisted or braided.

Conductor stress control: The conducting layer applied to make the conductor a smooth surface in intimate contact with the insulation; formerly called extruded strand shield (ESS).

Conduit: A tubular raceway.

Conduit fill: Amount of cross-sectional area used in a raceway.

Conduit, rigid metal: Conduit made of Schedule 40 pipe, normally 10 foot lengths.

Configuration, cradled: The geometric pattern which cables will take in a conduit when the cables are pulled in parallel and the ratio of the conduit ID to the 1/C cable OD is greater than 3.0.

Configuration, triangular: The geometric pattern which cables will take in a conduit when the cables are triplexed or are pulled in parallel with the ratio of the conduit ID to the 1/C cable OD less than 2.5.

Connection: That part of a circuit which has negligible impedance and which joins components or devices.

Connection, delta: Interconnection of 3 electrical equipment windings in delta (triangular) fashion.

Connection, star: Interconnection of 3 electrical equipment windings in star (wye) fashion.

Connector: A device used to physically and electrically connect two or more conductors.

Connector, pressure: A connector applied using pressure to form a cold weld between the conductor and the connector.

Connector, reducing: Used to join two different size conductors.

Connector strip: A metal raceway containing pendant or flush receptacles.

Constant current: A type of power system in which the same amount of current flows through each utilization equipment, used for simplicity in street lighting circuits.

Constant voltage: The common type of power in which all loads are connected in parallel, different amounts of current flow through each load.

Contact: A device designed for repetitive connections.

Contactor: A type of relay.

Continuity: The state of being whole, unbroken.

Continuous load: In operation three hours or more.

Continuous vulcanization (CV): A system utilizing heat and pressure to vulcanize insulation after extrusion onto wire or cable; the curing tube may be in a horizontal or vertical pole.

Control: Automatic or manual device used to stop, start, and/or regulate flow of gas, liquid, and/or electricity.

Control system: The overall system governing the starting, stopping, direction of motion, acceleration, speed, and retardation of the moving member.

Control, temperature: A thermostatic device that automatically stops and starts a motor, the operation of which is based on temperature changes.

Controller: A device or group of devices that serves to govern in some predetermined manner the electric power delivered to the apparatus to which it is connected.

Convection: The transfer of heat to a fluid by conduction as the fluid moves past the heat source.

Converter: A device that changes electrical energy from one form to another, as from alternating current to direct current.

Cook annealer: An annealer using heavy electrical current through the conductor as the heat source.

Cook buncher: A buncher using controlled diameter and wire position.

Cooling tower: Device that cools water by water evaporation in air. Water is cooled to the wet bulb temperature of air.

Coordination: The selection of system components to prevent the failure of the whole system due to cascading; limiting system failure by activation of the fewest overcurrent devices, hopefully to one.

Copper: A word used by itself to refer to copper conductors. Examples: "A circuit of 500 MCM copper" or "the copper cost of the circuit." It is a good conductor of electricity, easily formed, easily connected to itself and other metals used in electrical wiring.

Copper Development Association: A manufacturer's group to promote the use of copper.

Copper, electrolytic: Copper of high purity, refined by electrolysis, used for electrical conductors.

Copper loss: The energy dissipated in the copper conductors of a circuit, due to heat loss of I^2 R produced by current flow through the conductor. Term is sometimes used to refer to the same type of loss in aluminum circuit conductors.

Cord: A small flexible conductor assembly, usually jacketed

Cord and plug-connected lighting assembly: A lighting assembly consisting of a lighting fixture intended for installation in the wall of a spa, hot tub, or storable pool.

Cord set: A cord having a wiring connector on one or more ends.

Core (cable): The portion of an insulated cable under a protective covering.

Corona: A low energy electrical discharge caused by ionization of a gas by an electric field.

Corrosion: The deterioration of a substance (usually a metal) because of a reaction with its environment.

Coulomb: The derived SI unit for quantity of electricity or electrical charge: One coulomb equals one ampere-second.

Counterbore: A tool that enlarges an already-machined round hole to a certain depth. The pilot of the tool fits in the smaller hole, and the larger part counterbores or makes the end of the hole larger.

Counter emf: The voltage opposing the applied voltage and the current in a coil; caused by a flow of current in the coil; also known as back emf.

Coupling: The means by which signals are transferred from one circuit to another.

Coupon: A piece of metal for testing, of specified size; a piece of metal from which a test specimen may be prepared.

CPE (chlorinated polyethylene): A plastic for jackets.

Crawl space: Shallow space between the first tier of beams and the ground (no basement).

Critical branch: A subsection of the emergency system in a health care facility consisting of feeders and branch circuits supplying energy to task illumination, special power circuits, and selected receptacles serving areas and functions related to patient care, and which are connected to alternate power sources by one or more transfer switches during interruption of the normal power source

Cross head: The mechanism on an extruder where the material is applied; it holds the die, guider and core tube; usually just called "head."

Cross talk: Undesired pickup of signals by a second circuit.

CRT: Cathode Ray Tube.

CU: Copper.

Cure: To change the properties of a polymeric system into a more stable, usable condition by the use of heat, radiation, or reaction with chemical additives.

Current (I): The time rate of flow of electric charges; Unit: ampere.

Current, charging: The current needed to bring the cable up to voltage; determined by capacitance of the cable; after withdrawal of voltage, the charging current returns to the circuit; the charging current will be 90° out of phase with the voltage.

Current density: The current per unit cross-sectional area.

Current-induced: Current in a conductor due to the application of a time-varying electromagnetic field.

Current, leakage: That small amount of current which flows through insulation whenever a voltage is present and heats the insulation because of the insulation's resistance; the leakage current is in phase with the voltage, and is a power loss.

Current limiting: A characteristic of short-circuit protective devices, such as fuses, by which the device operates so fast on high short circuit currents that less than a quarter wave of the alternating cycle is permitted to flow before the circuit is opened, thereby limiting the thermal and magnetic energy to a certain maximum value, regardless of the current available.

Cutout: A fuse holder which may be used to isolate part of a circuit.

Cutout box: A surface mounting enclosure with a cover equipped with swinging doors, used to enclose fuses.

Cut resistance: The ability of a material to withstand mechanical pressure without rupture or becoming ineffective.

Cycle: 1) An interval of space or time in which one set of events or phenomena is completed. 2) A set of operations that are repeated regularly in the same sequence. 3) When a system in a given state goes through a number of different processes and finally returns to its initial state.

Cyclic aging: A test on a closed loop of cable having voltage applied, and induced current applied in cycles to cause expansion and contraction; simulates cable operating in a dry environment.

Damper: Valve for controlling air flow.

Damping: The dissipation of energy with time or distance.

Dead: 1) Not having electrical charge. 2) Not having voltage applied.

Dead-end: A mechanical terminating device on a building or pole to provide support at the end of an overhead electric circuit. A dead-end is also the term used to refer to the last pole in the pole line. The pole at which the electric circuiting is brought down the pole to go underground or to the building served.

Dead-front: A switchboard or panel or other electrical apparatus without "live" (energized) terminals or parts exposed on front where personnel might make contact.

Deadman's switch: A switch necessitating a positive action by the operator to keep the system or equipment running or energized.

Debug: To examine or test a procedure, routine, or equipment for the purpose of detecting and correcting errors especially during start-up.

Defeater: A means to deactivate a safety interlock system.

Deflection: Deviation of the central axis of a beam from normal when the beam is loaded.

Deflection plate: The part of a certain type of electron tube that provides an electrical field to produce deflection of an electron beam.

Demand: 1) The measure of the maximum load of a utility's customer over a short period of time. 2) The load integrated over a specified time interval.

Demand factor: For an electrical system or feeder circuit, this is a ratio of the amount of connected load (in kVA or amperes) which will be operating at the same time to the total amount of connected load on the circuit. An 80% demand factor, for instance, indicates that only 80% of the connected load on a circuit will ever be operating at the same time. Conductor capacity can be based on that amount of load.

Density: Closeness of texture or consistency.

Derating: The intentional reduction of stress/strength ratio in the application of a material; usually for the purpose of reducing the occurrence of a stress-related failure.

Derating factor: A factor used to reduce ampacity when the cable is used in environments other than the standard.

Detection: The process of obtaining the separation of the modulation, component from the received signal.

Device: An item intended to carry, or help carry, but not utilize electrical energy.

Dew point: The temperature at which vapor starts to condense (liquify) from a gas-vapor mixture at constant pressure.

Die: 1) Wire: a metal device having a conical hole which is used to reduce the diameter of wire which is drawn (pulled) through the die or series of dies. 2) Extruder: the fixed part of the mold.

Dielectric: An insulator or a term referring to the insulation between the plates of a capacitor.

Dielectric absorption: The storage of charges within an insulation; evidenced by the decrease of current flow after the application of dc voltage.

Dielectric dispersion: The change in relative capacitance due to change in frequency.

Dielectric heating: The heating of an insulating material by ac induced internal losses; normally frequencies above l0mHz are used.

Dielectric loss: The time rate at which electrical energy is transformed into heat in a dielectric when it is subjected to a changing electric field.

Dielectric phase angle: The phase angle between the sinusoidal ac voltage applied to a dielectric and the component of the current having the same period.

Dielectric strength: The maximum voltage which an insulation can withstand without breaking down; usually expressed as a gradient — vpm (volts per mil).

Diode: A device having two electrodes, the cathode and the plate or anode — used as a rectifier and detector.

Direct current (dc): 1) Electricity which flows only in one direction. 2) The type of electricity produced by a battery and dc generators.

Direction of lay: The lateral direction, designated as left-hand or right-hand, in which the elements of a cable run over the top of the cable as they recede from an observer looking along the axis of the cable.

Disconnect: A switch for disconnecting an electrical circuit or load (motor, transformer, panel) from the conductors which supply power to it, e.g., "He pulled the motor disconnect," meaning he opened the disconnect switch to the motor.

Disconnecting means: A device, a group of devices, or other means whereby the conductors of a circuit can be disconnected from their supply source.

Dispersion: Holding fine particles in suspension throughout a second substance.

Displacement current: An expression for the effective current flow across a capacitor.

Dissipation factor: Energy lost when voltage is applied across an insulation because of capacitance: the cotangent of the phase angle between voltage and current in a reactive component; because the shift is so great, we use the complement (angle) of the angle θ which is used for power factor; dissipation factor = tan = cot θ: is quite sensitive to contamination and deterioration of insulation: also known as power factor (of dielectrics).

Distortion: Unfaithful reproduction of signals.

Distribution panelboard: A single panel or group of panel units designed for assembly in the form of a single panel; including buses and automatic overcurrent devices, and equipped with or without switches for the control of light, heat, or power circuits; designed to be placed in a cabinet or cutout box placed in or against a wall or partition and accessible only from the front.

Distribution, statistical analysis: A statistical method used to analyze data by correlating data to a theoretical curve to a) test validity of data, b) predict performance at conditions different from those used to produce the data: The normal distribution curve is most common.

Diversity factor: The ratio of the sum of load demands to a system demand.

DOAL: Diameter Overall.

DOC: Diameter Over Conductor: note that for cables having a stress control, the diameter over the stress control layer becomes DOC.

DOI: Diameter Over Insulation.

DOJ: Diameter Over Jacket.

Donkey: A motor-driven power machine, on legs, used for threading and/or cutting conduit.

DOS: Diameter Over Insulation Shield.

DOSC: Diameter Over Stress Control.

Double-strength glass: One-eighth inch thick sheet glass (single strength glass is 1/10 inch thick).

Draft indicator: An instrument used to indicate or measure chimney draft or combustion gas movement.

Drain wire: A bonding wire laid parallel to and touching shields.

Drawing: Reducing wire diameter by pulling through dies.

Drawing, block diagram: A simplified drawing of a system showing major items as blocks; normally used to show how the system works and the relationship between major items.

Drawing, line schematic (diagram): Shows how a circuit works.

Drawing, plot or layout: Shows the "floor plan."

Drawing, wiring diagram: Shows how the devices are interconnected.

Drill: A circular tool used for machining a hole.

Drip loop: An intentional sag placed in service entrance conductors where they connect to the utility service drop conductors on overhead services; the drop loop will conduct any rain water to a point lower than the service head, to prevent moisture being forced into the service conductors by hydrostatic pressure and running through the service head into the service conduit or cable.

Drop box: A box containing pendant- or flush-mounted receptacles attached to a multiconductor cable via strain relief, or a multipole connector.

Drum: The part of a cable reel on which the cable is wound.

Dry: Not normally subjected to moisture.

Dry bulb: An instrument with a sensitive element that measures ambient (moving) air temperature.

Dry-niche lighting fixture: A lighting fixture intended for installation in the wall of a pool or fountain, in a niche that is sealed against the entry of pool water.

Drywall: Interior wall construction consisting of plasterboard, wood paneling, or plywood nailed directly to the studs without application of plaster.

Dual extrusion: Extruding two materials simultaneously using two extruders feeding a common cross head.

Duct: A tube or channel through which air is conveyed or moved.

Duct bank: Several underground conduits grouped together.

Ductility: The ability of a material to deform plastically before fracturing.

Dumbbell: A die-cut specimen of uniform thickness used for testing tensile and elongation of materials.

Duty, continuous: A service requirement that demands operation at a substantially constant load for an indefinitely long time.

Duty, intermittent: A service requirement that demands operation for alternate intervals of load and no load, load and rest, or load, no load, and rest.

Duty, periodic: A type of intermittent duty in which the load conditions regularly reoccur.

Duty, short-time: A requirement of service that demands operations at loads and for intervals of time, both which may be subject to wide variation.

Dwarf partition: Partition that ends short of the ceiling.

Dwell: A planned delay in a timed control program.

Dynamic: A state in which one or more quantities exhibit appreciable change within an arbitrarily short time interval.

Dynamometer: A device for measuring power output or power input of a mechanism.

EC: Electrical Conductor of Aluminum.

Ecdentricity: 1) A measure of the entering of an item within a circular area. 2) The percentage ratio of the difference between the maximum and minimum thickness to the minimum thickness of an annular area.

Eddy currents: Circulating currents induced in conducting materials by varying magnetic fields; usually considered undesirable because they represent loss of energy and cause heating.

Edison base: The standard screw base used for ordinary lamps.

EEI: Edison Electric Institute.

Effective temperature: Overall effect on a person of air temperature, humidity, and air movement.

Efficiency: The ratio of the output to the input.

Elasticity: That property of recovering original size and shape after deformation.

Elbow: A short conduit which is bent.

Electric defrosting: Use of electric resistance heating coils to melt ice and frost off evaporators during defrosting.

Electric-discharge lighting: A system of illumination utilizing fluorescent lamps, high-intensity discharge (HID) lamps or neon tubing.

Electric heating: House heating system in which heat from electrical resistance units is used to heat rooms.

Electric vehicle: An automotive-type vehicle for highway use, such as a passenger automobile or bus, primarily powered by an electric motor that draws current from a rechargeable storage battery, fuel cell or another source of electric current. Not included in this category are electric motorcycles, off-road vehicles, industrial transports or equipment, golf carts, or self-propelled electric vehicles.

Electric vehicle charging personnel protective system: A system of personnel protection devices and constructional features that when used together provide protection against electric shock.

Electric vehicle connector: A device that by insertion into an electric vehicle inlet, establishes an electrical connection to the electric vehicle for the purpose of charging and information exchange. This is part of the electric vehicle coupler.

Electric vehicle coupler: A mating electric vehicle inlet and electric vehicle connector set.

Electric vehicle inlet: The device on the electric vehicle into which the electric vehicle connector is inserted for charging and information exchange. For the purposes of the *NEC*, the inlet is considered to be part of the electric vehicle and not part of the electric vehicle supply equipment.

Electric vehicle nonvented storage battery: A hermetically-sealed battery comprised of one or more rechargeable electrochemical cells that has no provision for release of excessive gas pressure, or the addition of water or electrolyte, or for external measurements of electrolyte specific gravity.

Electric vehicle supply equipment: The conductors, including the ungrounded, grounded, and equipment grounding conductors and the electric vehicle connectors, attachment plugs, and all other fittings, devices, power outlets, or apparatuses installed specifically for the purpose of delivering energy from the premises wiring to the electric vehicle.

Electric water valve: Solenoid type (electrically operated) valve used to turn water flow on and off.

Electrical life-support equipment: Equipment whose continuous operation is necessary to maintain a patient's life.

Electrical metallic tubing (type EMT): A listed metallic tubing, of circular cross section, approved for the installation of electrical conductors when joined together with listed fittings.

Electrical nonmetallic tubing (type ENT): A pliable corrugated raceway, of circular cross section, with integral or associated couplings, connectors, and fittings listed for the installation of electrical conductors.

Electrically connected: A connection capable of carrying current, as distinguished from connection through electromagnetic induction.

Electricity: Relating to the flow or presence of charged particles; a fundamental physical force or energy.

Electrocution: Death caused by electrical current through the heart, usually in excess of 50 ma.

Electrode: A conductor through which current transfers to another material.

Electrolysis: The production of chemical changes by the passage of current from an electrode to an electrolyte or vice versa.

Electrolyte: A liquid or solid that conducts electricity by the flow of ions.

Electrolytic cell: A tank or vat, in which electrochemical reactions are caused by applying electrical energy, used for the purpose of refining or producing usable materials.

Electrolytic condenser-capacitor: Plate or surface capable of storing small electrical charges. Common electrolytic condensers are formed by rolling thin sheets of foil between insulating materials. Condenser capacity is expressed in microfarads.

Electromagnet: A device consisting of a ferromagnetic core and a coil that produces appreciable magnetic effects only when an electric current exists in the coil.

Electromotive force (emf) voltage: Electrical force that causes current (free electrons) to flow or move in an electrical circuit. Unit of measurement is the volt.

Electron: The subatomic particle that carries the unit negative charge of electricity.

Electronegative gas: A type of insulating gas used in pressure cables; such as SF6.

Electron emission: The release of electrons from the surface of a material into surrounding space due to heat, light, high voltage, or other causes.

Electronics: The science dealing with the development and application of devices and systems involving the flow of electrons in vacuum, gaseous media, and semiconductors.

Electro-osmosis: The movement of fluids through diaphragms because of electric current.

Electroplating: Depositing a metal in an adherent form upon an object using electrolysis.

Electropneumatic: An electrically controlled pneumatic device.

Electrostatics: Electrical charges at rest in the frame of reference.

Electrotherapy: The use of electricity in treatment of disease.

Electrothermics: Direct transformations of electric and heat energy.

Electrotinning: Depositing tin on an object.

Elevation: Drawing showing the projection of a building on a vertical plane.

Elongation: 1) The fractional increase in length of a material stressed in tension. 2) The amount of stretch of a material in a given length before breaking.

EMA (Electrical Moisture Absorption): A water tank test during which the sample cables are subjected to voltage while the water is maintained at rated temperature; the immersion time is long, with the object being to accelerate failure due to moisture in the insulation; simulates buried cable.

Emergency system: A system of feeders and branch circuits meeting the requirement of the *NEC* Article 700, except as amended by *NEC* Article 517, intended to supply alternate power to a limited number of prescribed functions vital to the protection of life and patient safety, with the capability of automatic restoration of electrical power within 10 seconds of power interruption.

EMI: Electromagnetic interference.

Emitter: The part of a transistor that emits electrons.

Emulsifying agent: A material that increases the stability of an emulsion.

Emulsion: The colloidal suspension of one liquid in another liquid, such as oil in water for lubrication.

Enameled wire: Wire insulated with a thin baked-on varnish enamel, used in coils to allow the maximum number of turns in a given space.

Enclosed: Surrounded by a case that will prevent anyone from accidentally touching live parts.

Energy: The ability to do work; such as heat, light, electrical, mechanical, etc.

Engine: An apparatus which converts heat to mechanical energy.

Environment: 1) The universe within which a system must operate. 2) All the elements over which the designer has no control and that affect a system or its inputs.

EPA (Environmental Protection Agency): The federal regulatory agency responsible for keeping and improving the quality of our living environment — mainly air and water.

EPRI (Electric Power Research Institute): An organization to develop and manage a technology for improving electric power production, distribution and utilization; sponsored by electric utilities.

Equilibrium: Properties are time constant.

Equipment: A general term including material, fittings, devices, appliances, fixtures, apparatus, and the like used as part of, or in connection with, an electrical installation.

Equipment rack: A framework for the support, enclosure, or both, of equipment.

Equipment system: A system of feeders and branch circuits, arranged for delayed, automatic, or manual connection to an alternate power source, and that serves primarily 3-phase power equipment.

Equipotential: Having the same voltage at all points.

Equivalent circuit: An arrangement of circuit elements that has characteristics over a range of interest electrically equivalent to those of a different circuit or device.

Erosion: Destruction by abrasive action of fluids.

Etching: Revealing structural details of a metal surface using chemical or electrolytic action.

ETL: Electrical Testing Laboratory.

Evaporation: A term applied to the changing of a liquid to a gas; heat is absorbed in this process.

Evaporator: Part of a refrigerating mechanism in which the refrigerant vaporizes and absorbs heat.

Excitation losses: Losses in a transformer or electrical machine because of voltage.

Excite: To initiate or develop a magnetic field.

Expansion bolt: Bolt with a casing arranged to wedge the bolt into a masonry wall to provide an anchor.

Expansion joint: Joint between two adjoining concrete members arranged to permit expansion and contraction with temperature changes.

Expansion, thermal: The fractional change in unit length per unit temperature change.

Expansion valve: A device in a refrigerating system that maintains a pressure difference between the high side and low side and is operated by pressure.

Explosion proof: Designed and constructed to withstand an internal explosion without creating an external explosion or fire.

Exposed (as applied to live parts): Live parts that a person could inadvertently touch or approach nearer than a safe distance. This term is applied to parts not suitably guarded, isolated, insulated.

Exposed (as applied to wiring method): Not concealed; externally operable; capable of being operated without exposing the operator to contact with live parts.

Extraction: The transfer of a material from a substance to a liquid in contact with the substance.

Extrude: To form materials to a given cross section by forcing through a die.

Extruder types: a) Strip — uses strips of compound. b) Powder/pellet — uses compound in powder or pellet form.

Eyelet: Something used on printed circuit boards to make reliable connections from one side of the board to the other.

Face: An operation that machines the sides or ends of the piece.

Face of a gear: That portion of the tooth curve above the pitch circle and measured across the rim of the gear from one end of the tooth to the other.

Facsimile: The remote reproduction of graphic material: an exact copy.

Factor of safety: Radio of ultimate strength of material to maximum permissible stress in use.

Fail-safe control: A device that opens a circuit when the sensing element fails to operate.

Failure: Termination of the ability of an item to perform the required function.

Fan: A radial or axial flow device used for moving or producing artificial currents of air.

FAO: This symbol on a mechanical drawing means that the piece is machined or finished all over.

Farad: The basic unit of capacitance: 1 farad equals 1 coulomb per volt.

Fatigue: The weakening or breakdown of a material due to cyclic stress.

Fatigue strength: The maximum stress that can be sustained for a specified number of cycles without failure, the stress being completely reversed within each cycle unless otherwise stated.

Fault: An abnormal connection in a circuit.

Fault, arcing: A fault having high impedance causing arcing.

Fault, bolting: A fault of very low impedance.

Fault, ground: A fault to ground.

Fault-hazard current: The hazard current of a given isolated system, with all devices connected except the line isolation monitor.

Feedback: The process of transferring energy from the output circuit of a device back to its input.

Feeder: A circuit, such as conductors in conduit or a busway run, which carries a large block of power from the service equipment to a sub-feeder panel or a branch circuit panel or to some point at which the block or power is broken down into smaller circuits.

Feeder assembly: The overhead or under-chassis feeder conductors, including the rounding conductor, together with the necessary fittings and equipment or a power supply cord listed for mobile home use, designed for the purpose of delivering energy from the source of the electrical supply to the distribution panelboard within a mobile home.

Ferranti effect: When the voltage is greater than the source voltage in an ac cable or transmission line.

Festoon lighting: A string of outdoor lights that is suspended between two or more points.

Fiber optics: Transmission of energy by light through glass fibers.

Fiddle: A small, hand-operated drill.

Fidelity: The degree to which a system accurately reproduces an input.

Field: The effect produced in surrounding space by an electrically charged object, by electrons in motion, or by a magnet.

Field, electrostatic: The region near a charged object.

Filament: A cathode in the form of a metal wire in an electron tube.

Filler, cable: Materials used to fill voids and spaces in a cable construction: normally to give a smooth outer configuration, and also may serve as flame retardants, etc.

Fillet: The rounded comer or portion that joins two surfaces which are at an angle to each other.

Film: A rectangular product having thickness of 0.010 inch thick or less.

Filter: A porous article through which a gas or liquid is passed to separate out matter in suspension; a circuit or devices that pass one frequency or frequency band while blocking others, or vice versa.

Final: The final inspection of an electrical installation, e.g., "The contractor got the final on the job."

Final tests: Those performed on the completed cable (after manufacturing).

Fines: Fill material such as rocks having 1/8 inch as the largest dimension.

Finish plaster. Final or white coat of plaster.

Fireproof wood: Chemically treated wood; fire-resistive, used where incombustible materials are required.

Fire-rated doors: Doors designed to resist standard fire tests and labeled for identification.

Fire-resistance rating: The time in hours, the material or construction will withstand fire exposure as determined by certain standards.

Fire-shield cable: Material or devices to prevent fire spread between raceways.

Fire-stop: A barrier to prevent fire spread.

Fish: To fish wire or cable means to pull it through conduit, raceway or other confined spaces, like walls or ceilings.

Fish tape: A flexible metal tape for fishing through conduits or other raceway to pull in wires or cables; also made in non-metallic form of "rigid rope" for hand fishing of raceways.

Fitting: An accessory such as a locknut, bushing, or other part of a wiring system that is intended primarily to perform a mechanical rather than an electrical function.

Five hundred thousands: Referring to size of conductors by their MCM rating, e.g., "Two hundred and fifty thousands" is a number of 250 MCM conductors; "Twin three hundred thousands" is two conductors of 300,000 circular mil size.

Flag: A visual indicator for event happenings such as the activation and reclosing of an automatic circuit breaker.

Flame-retardant: 1) Does not support or convey flame. 2) An additive for rubber or plastic that enhances its flame resistance.

Flange: The circular disks on a reel to support the drum and keep the cable on the drum.

Flapper, valve: The type of valve used in refrigeration compressors that allow gaseous refrigerants to flow in only one direction.

Flashover: A momentary electrical interconnection around or over the surface of an insulator.

Flashpoint: A lowest temperature at which a combustible substance ignites in air when exposed to flame.

Flat: Of uniform thickness; eliminates the drops of beams and girders.

Flat cable assemblies (type FC): An assembly of parallel conductors formed integrally with an insulating material web specifically designed for field installation in surface metal raceway.

Flat conductor cable (type FCC): A cable consisting of three or more flat copper conductors placed edge-to-edge, separated and enclosed within an insulating assembly.

Flat wire: A rectangular wire having 0.188 inch thickness or less, 1¼ inch width or less.

Flexible metal conduit: A raceway of circular cross section made of helically-wound, formed, interlocked metal strip.

Flexible metallic tubing: A listed tubing that is circular in cross section, flexible, metallic, and liquidtight without a nonmetallic jacket.

Flexural strength: The strength of a material in bending, expressed as the tensile stress of the outermost fibers of a bent test sample at the instant of failure.

Flitch beam: Built-up beam consisting of a steel plate sandwiched between wood members and bolted.

Floating: Not having a distinct reference level with respect to voltage measurements.

Float valve: Type of valve that is operated by a sphere or pan which floats on a liquid surface and controls the level of liquid.

Flooding: Act of filling a space with a liquid.

Flow meter: Instrument used to measure velocity or volume of fluid movement.

Flux: 1) The rate of flow of energy across or through a surface. 2) A substance used to promote or facilitate soldering or welding by removing surface oxides.

Foamed insulation: Insulation made sponge-like using foaming or blowing agents to create the cells.

Foil: Metal film.

Footing: Structural unit used to distribute loads to the bearing materials.

Footlight: A border light installed on or in a stage.

Forced convection: Movement of fluid by mechanical force such as fans or pumps.

Forming shell: A structure designed to support a wet-niche lighting fixture assembly and intended for mounting in a pool or fountain structure.

Foundation: Composed of footings, piers, foundation walls (basement walls), and any special underground construction necessary to properly support the structure.

FPM: Feet per minute.

FRI: See *VW-1*.

Frequency: The number of complete cycles an alternating electric current, sound wave, or vibrating object undergoes per second.

Friction tape: An insulating tape made of asphalt impregnated cloth; used on 600V cables.

Frost line: Deepest level below grade to which frost penetrates in a geographic area.

Fuel cell: A cell that can continually change chemical energy to electrical energy.

Function: A quantity whose value depends upon the value of another quantity.

Furring: Thin wood, brick, or metal applied to joists, studs, or wall to form a level surface (as for attaching wallboard) or airspace.

Fuse: A protecting device which opens a circuit when the fusible element is severed by heating, due to overcurrent passing through. Rating: voltage, normal current, maximum let-through current, time delay of interruption.

Fuse, dual element: A fuse having two fuse characteristics; the usual combination is having an overcurrent limit and a time delay before activation.

Fuse, nonrenewable or one-time: A fuse which must be replaced after it interrupts a circuit.

Fuse, renewable link: A fuse which may be reused after current interruption by replacing the meltable link.

Fusible plug: A plug or fitting made with a metal of a known low melting temperature; used as a safety device to release pressures in case of fire.

Gain: The ratio of output to input power, voltage, or current, respectively.

Galvanometer: An instrument for indicating or measuring a small electrical current by means of a mechanical motion derived from electromagnetic or dynamic forces.

Garage: A building or portion of a building in which one or more self-propelled vehicles carrying volatile, flammable liquid for fuel or power are kept.

Garden bond: Bond formed by inserting headers at wide intervals.

Gas: Vapor phase or state of a substance.

Gas filled pipe cable: See *Cable, pressure*.

Gate: A device that makes an electronic circuit operable for a short time.

Gauge: 1) Dimension expressed in terms of a system of arbitrary reference numbers; dimensions expressed in decimals are preferred. 2) To measure.

Gem box: The most common rectangular outlet box used to hold wall switches and receptacle outlets installed recessed in walls; made in wide variety on constructions, 2 in. wide by 3 in. high by various depths, without clamps for conduit and with or without clamps for cable (armored for nonmetallic sheathed), single gang boxes which can be ganged together for more than one device.

Generator: 1) A rotating machine to convert from mechanical to electrical energy. 2) Automotive — mechanical to direct current. 3) General — apparatus, equipment, etc., to convert or change energy from one form to another.

Geometric factor: A parameter used and determined solely by the relative dimensions and configuration of the conductors and insulation of a cable.

Girder: A large beam made of wood, steel, or reinforced concrete.

Girt: Heavy timber framed into corner posts as support for the building.

Grade beam: Horizontal, reinforced concrete beam between two supporting piers at or below ground supporting a wall or structure.

Graded insulation: Combining insulations in a manner to improve the electric field distribution across the combination.

Gradient: The rate of change of a variable magnitude.

Gray: The derived SI unit for absorbed radiation dose: one gray equals one joule per kilogram.

Greenfield: Another name used to refer to flexible metal conduit.

Grid: An electrode having one or more openings for the passage of electrons or ions.

Grid leak: A resistor of high ohmic value connected between the control grid and the cathode in a grid-leak capacitor detector circuit and used for automatic biasing.

Grommet: A plastic, metal or rubber doughnut-shaped protector for wires or tubing as they pass through a hole in an object.

Ground: A large conducting body (as the earth) used as a common return for an electric circuit and as an arbitrary zero of potential.

Ground check: A pilot wire in portable cables to monitor the grounding circuit.

Ground coil: A heat exchanger buried in the ground that may be used either as an evaporator or a condenser.

Ground Fault Interrupter (GFI): A protective device that detects abnormal current flowing to ground and then interrupts the circuit.

Grounded conductor: A system or circuit conductor that is intentionally grounded.

Grounded: Connected to earth.

Grounding: The device or conductor connected to ground designed to conduct only in abnormal conditions.

Grounding conductor: A conductor used to connect metal equipment enclosures and/or the system grounded conductor to a grounding electrode, such as the ground wire run to the water pipe at a service; also may be a bare or insulated conductor used to ground motor frames, panel boxes and other metal equipment enclosures used throughout an electrical system. In most conduit systems, the conduit is used as the ground conductor.

Group ambient temperature: The no-load temperature of a cable group with all other cables or ducts loaded.

Grouped: Cables or conductors positioned adjacent to one another but not in continuous contact with each other.

Guard: 1) A conductor situated to conduct interference to its source and prevent its influence upon the desired signal. 2) A mechanical barrier against physical contact.

Guider: The adjustable part of the mold of an extruder.

Gutter: The space provided along the sides and at the top and bottom of enclosures for switches, panels, and other apparatus, to provide for arranging conductors which terminate at the lugs or terminals of the enclosed equipment. Gutter is also used to refer to a rectangular sheet metal enclosure with removable side, used for splicing and tapping wires at distribution centers and motor control layouts.

Guy: A tension wire connected to a tall structure and another fixed object to add strength to the structure.

Half effect: The changing of current density in a conductor due to a magnetic field extraneous to the conductor.

Half hard: A relative measure of conductor temper.

Half lap, joint: Joint formed by cutting away half the thickness of each piece.

Half wave: Rectifying only half of a sinusoidal ac supply.

Hand shake: Requiring mutual events prior to change.

Handhole: A small box in a raceway used to facilitate cable installation into which workmen reach but do not enter.

Handy box: The commonly used, single-gang outlet box used for surface mounting to enclose wall switches or receptacles, on concrete or cinder block construction of industrial and commercial buildings, non-gangable, also made for recessed mounting, also known as "utility boxes."

Hard drawn: A relative measure of temper; drawn to obtain maximum strength.

Hardness: Resistance to plastic deformation usually by deformation: stiffness or temper: resistance to scratching, abrasion or cutting.

Harmonic: An oscillation whose frequency is an integral multiple of the fundamental frequency.

Harness: A group of conductors laced or bundled in a given configuration, usually with many breakouts.

Hat: A special pallet for transporting long rubber strips or coils of wire; the pallets look like a hat.

Hazard current (health care facilities): For a given set of connections in an insulated power system, the total current that would flow through a low impedance if it were connected between either isolated conductor and ground.

Hazardous: Ignitable vapors, dust, or fibers that may cause fire or explosion.

Heat: A fundamental physical force or energy relating to temperature.

Heat dissipation: The flow of heat from a hot body to a cooler body by: 1) convection 2) radiation 3) conduction.

Heat exchanger: A device used to transfer heat from a warm or hot surface to a cold or cooler surface. Evaporators and condensers are heat exchangers.

Heat load: Amount of heat, measured in Btu, that is removed during a period of 24 hours.

Heat pump: A compression cycle system used to supply or remove heat to or from a temperature-controlled space.

Heat sink: A part used to absorb heat from another device.

Heat transfer: Movement of heat from one body or substance to another. Heat may be transferred by radiation, conduction, convection, or a combination of these.

Heat treatment: Heating and cooling a solid metal or alloy to obtain desired properties or conditions; excluding heating for hot work.

Heating equipment: Any equipment used for heating purposes whose heat is generated by induction or dielectric methods.

Heating system: A complete system consisting of components such as heating elements, fastening devices, nonheating circuit wiring, leads temperature controls, safety signs, junction boxes, raceways and fittings.

Heating system (impedance): A system in which heat is generated in a pipeline or vessel wall by causing current to flow through the pipeline or vessel wall by direct connection to an ac voltage source from a dual-winding transformer.

Heating valve: Amount of heat that may be obtained by burning a fuel; usually expressed in Btu per pound or gallon.

Helix: The path followed when winding a wire or strip around a tube at a constant angle.

Henry: The derived SI unit for inductance: one henry equals one weber per ampere.

Hermetic motor: A motor designed to operate within refrigeration fluid.

Hertz: The derived SI unit for frequency: one hertz equals one cycle per second.

Hickey: 1) A conduit bending tool. 2) A box fitting for hanging lighting fixtures.

High-hat: A ceiling recessed incandescent lighting fixture of round cross-section, looking like a man's high hat in the shape of its construction.

High pressure cutout: Electrical control switch operated by the high side pressure that automatically opens an electrical circuit if too high head pressure or condensing pressure is reached.

Hi-pot test: A high-potential test in which equipment insulation is subjected to voltage level higher than that for which it is rated to find any weak spots or deficiencies in the insulation.

Home run: That part of a branch circuit from the panelboard housing the branch circuit fuse or CB and the first junction box at which the branch circuit is spliced to lighting or receptacle devices or to conductors which continue the branch circuit to the next outlet or junction box. The term "home run" is usually reserved to multi-outlet lighting and appliance circuits.

Horsepower: The non-SI unit for power: 1 hp = 1 HP = 746 w (electric) = 9800 w (boiler).

Hot: Energized with electricity.

Hot dip: Coating by dipping into a molten bath.

Hot leg: A circuit conductor which normally operates at a voltage above ground; the phase wires or energized circuit wires other than a grounded neutral wire or grounded phase leg.

Hot junction: That part of the thermoelectric circuit which released heat.

Hot stick: A long insulated stick having a hook at one end which is used to open energized switches, etc.

Hot wire: A resistance wire in an electrical relay that expands when heated and contracts when cooled.

Hub: 1) A fitting to attach threaded conduit to boxes. 2) The central part of a cylinder into which a shaft may be inserted. 3) A reference point used for overhead line layout.

Hum: Interference from ac power, normally of low frequency and audible.

Humidity: Moisture, dampness. Relative humidity is the ratio of the quantity of vapor present in the air to the greatest amount possible at a given temperature.

Hybrid system: A system comprised of multiple power sources.

Hydromassage bathtub: A permanently installed bathtub equipped with a recirculating piping system, pump, and associated equipment, designed so it can accept, circulate and discharge water upon each use.

Hydronic: Type of heating system that circulates a heated fluid, usually water, through baseboard coils. Circulation pump is usually controlled by a thermostat.

Hysteresis: The time lag exhibited by a body in reacting to changes in forces affecting it; an internal friction.

IACS (International Annealed Copper Standard): Refined copper for electrical conductors: 100% conductivity at 20° C.

IBEW: International Brotherhood of Electrical Workers.

IC: Pronounced "eye see." Refers to interrupting capacity of any device required to break current (switch, circuit breaker, fuse, etc.), taken from the initial letters I and C, is the amount of current which the device can interrupt without damage to itself.

ID: Inside diameter.

Identified: Marked to be recognized as grounded.

IEEE: Institute of Electrical and Electronics Engineers.

Ignition transformer: A transformer designed to provide a high voltage current.

Impedance (A): The opposition to current flow in an ac circuit; impedance includes resistance (R), capacitive reactance (xc) and inductive reactance (XL); unit — ohm.

Impedance matching: Matching source and load impedance for optimum energy transfer with minimum distortion.

Impulse: A surge of unidirectional polarity.

Inching: Momentary activation of machinery used for inspection or maintenance.

Incombustible material: Material that will not ignite or actively support combustion in a surrounding temperature of 1200 degrees Fahrenheit during an exposure of five minutes; also, material that will not melt when the temperature of the material is maintained at 900 degrees Fahrenheit for at least five minutes.

Indoor: Not suitable for exposure to the weather.

Inductance: The creation of a voltage due to a time-varying current; the opposition to current change, causing current changes to lag behind voltage changes: Units — henry.

Induction heater: The heating of a conducting material in a varying electromagnetic field due to the material's internal losses.

Induction machine: An asynchronous ac machine to change phase or frequency by converting energy — from electrical to mechanical, then from mechanical to electrical.

Inductor: A device having winding(s) with or without a magnetic core for creating inductance in a circuit.

Infrared lamp: An electrical device that emits infrared rays — invisible rays just beyond red in the visible spectrum.

Infrared radiation: Radiant energy given off by heated bodies which transmits heat and will pass through glass.

In phase: The condition existing when waves pass through their maximum and minimum values of like polarity at the same instant.

Instantaneous value: The value of a variable quantity at a given instant.

Intrinsically safe: Incapable of releasing sufficient electrical or thermal energy under normal or abnormal conditions to cause ignition of a specific hazardous atmospheric mixture in its most ignitable concentration.

Instrument: A device for measuring the value of the quantity under observation.

Insulated: Separated from other conducting surfaces by a substance permanently offering a high resistance to the passage of energy through the substance.

Insulation, electrical: A medium in which it is possible to maintain an electrical field with little supply of energy from additional sources; the energy required to produce the electric field is fully recoverable only in a complete vacuum (the ideal dielectric) when the field or applied voltage is removed: used to a) save space b) enhance safety c) improve appearance.

Insulation, class rating: A temperature rating descriptive of classes of insulations for which various tests are made to distinguish the materials; not related necessarily to operating temperatures.

Insulation fall-in: The filling of strand interstices, especially the inner interstices, which may contribute to connection failures.

Insulation level (cable): The thickness of insulation for circuits having ground fault detectors which interrupt fault currents within 1) 1 minute = 100% level 2) 1 hour = 133% level 3) Over 1 hour = 173% level.

Insulation, thermal: Substance used to retard or slow the flow of heat through a wall or partition.

Integrated circuit: A circuit in which different types of devices such as resistors, capacitors, and transistors are made from a single piece of material and then connected to form a circuit.

Integrated gas spacer cable (type IGS): A factory assembly of one or more conductors, each individually insulated and enclosed in a loose fit, nonmetallic flexible conduit as an integrated gas spacer cable rated 0 through 600 volts.

Integrator: Any device producing an output proportionate to the integral of one variable with respect to a second variable; the second is usually time.

Interactive system: A solar photovoltaic system that operates in parallel with, and may deliver power to, an electrical production and distribution system.

Intercalated tapes: Two or more tapes of different materials helically wound and overlapping on a cable to separate the materials.

Interconnected system: Operating with two or more power systems connected through tie lines.

Interface: 1) A shared boundary. 2) (nuclear power): a junction between Class 1E and other equipment of systems.

Interference: Extraneous signals or power which are undesired.

Interlock: A safety device to insure that a piece of apparatus will not operate until certain conditions have been satisfied.

Intermediate metal conduit (type IMC): A listed steel raceway of circular cross section with integral or associated coupling. It is approved for the installation of electrical conductors and used with listed fittings to provide electrical continuity.

Interpolate: To estimate an intermediate between two values in a sequence.

Interrupting rating: The highest current at rated voltage that a device is intended to interrupt under standard test conditions.

Interrupting time: The sum of the opening time and arcing time of a circuit opening device.

Interstice: The space or void between assembled conductors and within the overall circumference of the assembly.

Inverter: An item which changes dc to ac.

Ion: An electrically charged atom or radical.

Ionization: 1) The process or the result of any process by which a neutral atom or molecule acquires charge. 2) A breakdown that occurs in gaseous parts of an insulation when the dielectric stress exceeds a critical value without initiating a complete breakdown of the insulation system; ionization is harmful to living tissue, and is detectable and measurable; may be evidenced by corona.

Ionization factor: This is the difference between percent dissipation factors at two specified values of electrical stress; the lower of the two stresses is usually so selected that the effect of the ionization on dissipation factor at this stress is negligible.

IPCEA (Insulated Power Cable Engineers Association): The association of cable manufacturing engineers who make nationally recognized specifications and tests for cables.

IR (Insulation resistance): The measurement of the dc resistance of insulating material; can be either volume of surface resistivity; extremely temperature sensitive.

IR drop: The voltage drop across a resistance due to the flow of current through the resistor.

IRK (Insulation dc resistance constant): A system to classify materials according to their resistance on a 1000 foot basis at 15.5°C (60°F).

ISO: International Organization for Standardization who have put together the "SI" units that are now the international standards for measuring.

Isolated: Not readily accessible to persons unless special means for access are used.

Isolated power system: A system comprising an isolating transformer or its equivalent, a line isolation monitor, and its ungrounded circuit conductors.

Isolating: Referring to switches, this means that the switch is not a loadbreak type and must only be opened when no current is flowing in the circuit. This term also refers to transformers (an isolating transformer) used to provide magnetic isolation of one circuit from another, thereby breaking a metallic conductive path between the circuits.

Isolation transformer: A transformer of the multiple-winding type, with the primary and secondary winding physically separated, which inductively couples its secondary windings to the grounded feeder systems that energize its primary windings.

I^2t: Relating to the heating effect of a current (amps-squared) for a specified time (seconds), under specified conditions.

Jack: A plug-in type terminal.

Jacket: A non-metallic polymeric close fitting protective covering over cable insulation; the cable may have one or more conductors.

Jacket conducting: An electrically conducting polymeric covering over an insulation.

Jamming: The wedging of a cable such that it can no longer be moved during installation.

JB: Pronounced "jay bee," refers to any junction box, taken from the initial letters J and B.

Joule: The derived SI unit for energy, work, quantity of heat: one joule equals one newton-metre.

Jumper: A short length of conductor, usually a temporary connection.

Junction: A connection of two or more conductors.

Junction box: Group of electrical terminals housed in a protective box or container.

Jute: A fibrous natural material used as a cable filler or bedding.

ka: KiloAmpere.

kc: Kilocycle, use kiloHertz.

kelvin (K): The basic SI unit of temperature: 1/273.16 of thermodynamic temperature of the triple point of water.

Kilogram (kg): The basic SI unit for mass; an arbitrary unit represented by an artifact kept in Paris, France.

Kilometer: A metric unit of linear measurement equal to 1000 meters.

Kilowatt: Unit of electrical power equal to 1000 watts.

Kilowatt-ft.: The product of load in kilowatts and the circuits distance over which a load is carried in feet; used to compute voltage drop.

Kinetic energy: Energy by virtue of motion.

Kirchoff's Laws: 1) The algebraic sum of the currents at any point in a circuit is zero. 2) The algebraic sum of the product of the cur-rent and the impedance in each conductor in a circuit is equal to the electromotive force in the circuit.

Knockout: A portion of an enclosure designed to be readily removed for installation of a raceway.

KO: Pronounced "kay oh," a knockout, the partially cut opening in boxes, panel cabinets and other enclosures, which can be easily knocked out with a screw driver and hammer to provide a clean hole for connecting conduit, cable or some fittings.

KVA: Kilovolts times Ampere.

LA: Lightning arrestor.

Lamp: A device to convert electrical energy to radiant energy, normally visible light: usually only 10-20% is converted to light.

Laminated core: An assembly of steel sheets for use as an element of magnetic circuit; the assembly has the property of reducing eddy-current losses.

Labeled: Items having trademark of nationally recognized testing lab.

Lagging: The wood covering for a reel.

Lap: The relative position of applied tape edges; "closed butt lap" — tapes just touching: "open butt" or "negative lap"— tapes not touching: "positive lap" or "lap"— tapes overlapping.

Law of charges: Like charges repel, unlike charges attract.

Law of magnetism: Like poles repel, unlike poles attract.

Lay: The axial length of one turn of the helix of any element in a cable.

Lay direction: Direction of helical lay when viewed from the end of the cable.

Lay length: Distance along the axis for one turn of a helical element.

Lead (leed): A short connecting wire brought out from a device or apparatus.

Lead squeeze: The amount of compression of a cable by a lead sheath.

Leading: Applying a lead sheath.

Leakage: Undesirable conduction of current.

Leakage distance: The shortest distance along an insulation surface between conductors.

Leg: A portion of a circuit.

Legend, embossed: Molded letters and numbers in the jacket surface; letters may be raised or embedded.

Lenz' Law: "In all cases the induced current is in such a direction as to oppose the motion which generates it."

Life safety branch: A subsystem of the emergency system consisting of feeders and branch circuits, meeting the requirements of *NEC* Article 700, which are automatically connected to alternate power sources during interruption of the normal power source to provide adequate power needs to ensure safety to patients and personnel.

Lighting outlet: An outlet intended for the direct connection of a lamp holder, lighting fixture, or pendant cord terminating on a lamp holder.

Lightning arrestor: A device designed to protect circuits and apparatus from high transient voltage by diverting the over-voltage to ground.

Limit: A boundary of a controlled variable.

Limit control: Control used to open or close electrical circuits as temperature or pressure limits are reached.

Limiter: A device in which some characteristic of the output is automatically prevented from exceeding in predetermined value.

Line: A circuit between two points: ropes used during overhead construction.

Line, bull: A rope for large loads.

Line, finger: A rope attached to a device on a pole when a device is hung, so further conductor installation can be done from the ground.

Line insulation monitor: A test instrument designed to continually check the balanced and unbalanced impedance from each line of an isolated circuit to ground, and equipped with a built-in test circuit to exercise the alarm without adding to the leakage current hazard.

Line, pilot: A small rope strung first.

Line, tag: A rope to guide devices being hoisted.

Linear: Arranged in a line.

Linearity: When the effect is directly proportional to the cause.

Liquidtight flexible metal conduit: A listed raceway of circular cross section of various types as follows: type LFNC-A; type LFNC-B and type LFNC-C. It is flame resistant, and with fittings, is approved for the installation of electrical conductors.

Lissajous Figure: A special case of an s-y plot in which the signals applied to both axes are sinusoidal functions: useful for determining phase and harmonic relationships.

Listed: Items in a list published by a nationally recognized independent lab that makes periodic tests.

Liter: Metric unit of volume that equals 61.03 cubic inches.

Live: Energized.

Live-front: Any panel or other switching and protection assembly, such as switchboard or motor control center, which has exposed electrically energized parts on its front, presenting the possibility of contact by personnel.

Live load: Any load on a structure other than a dead load; includes the weight of persons occupying the building and free-standing material.

Load: 1) A device that receives power. 2) The power delivered to such a device.

Load-break: Referring to switches or other control devices, this phrase means that the device is capable of safely interrupting load current — to distinguish such devices from other disconnect devices which are not rated for breaking load current and must be opened only after the load current has been broken by some other switching device.

Load center: An assembly of circuit breakers or switches.

Load factor: The ratio of the average to the peak load over a period.

Load losses: Those losses incidental to providing power.

Location, damp: A location subject to a moderate amount of moisture such as some basements, barns, cold-storage, warehouses, and the like.

Location, dry: A location not normally subject to dampness or wetness; a location classified as dry may be temporarily subject to dampness or wetness.

Location, wet: A location subject to saturation with water or other liquids.

Locked rotor: When the circuits of a motor are energized but the rotor is not turning.

Lockout: To keep a circuit locked open.

Logarithm: The exponent that indicates the power to which a number is raised to produce a given number.

Long-time rating (x-ray equipment): A rating based on an operating interval of five minutes or longer.

Looping-in: Avoiding splices by looping wire through device connections instead of cutting the wire.

Loud speaker: Equipment that converts an ac electric signal into an acoustic signal.

Loss: Power expended without doing useful work.

Lug: A device for terminating a conductor to facilitate the mechanical connection.

Lumen: The derived SI unit for luminous flux.

Lus: The derived SI unit for illuminance.

LV: Low voltage.

Machine: An item to transmit and modify force or motion, normally to do work.

Magnet: A body that produces a magnetic field external to itself; magnets attract iron particles.

Magnetic field: 1) A magnetic field is said to exist at a point if a force over and above any electrostatic force is exerted on a moving charge at the point. 2) The force field established by ac through a conductor, especially a coiled conductor.

Magnetic pole: Those portions of the magnet toward which the external magnetic induction appears to converge (south) or diverge (north).

Manhole: A subsurface chamber, large enough for a man, to facilitate cable installation splices, etc., in a duct bank.

Manual: Operated by mechanical force applied directly by personal intervention.

Manufacturing wiring system: A system containing component parts that are assembled in the process of manufacture and cannot be inspected at the building site without damage or destruction to the assembly.

Marker: A tape or colored thread in a cable which identifies the cable manufacturer.

Mass: The property that determines the acceleration the body will have when acted upon by a given force: Unit = qkilogram.

Mat: A concrete base for heavy electrical apparatus, such as transformers, motors, generators, etc., sometimes the term includes the concrete base, a bed of crushed stone around it and an enclosing chain-link fence.

Matrix: A multi-dimensional array of items.

MCM: An expression referring to conductors of sizes from 250 MCM (which stands for Thousand Circular Mils) up to 2000 MCM.

Mean: An intermediate value: arithmetic-sum of values divided by the quantity of the values: the average.

Medium hard: A relative measure of conductor temper.

Medium voltage cable (type MV): A single conductor or multi-conductor solid dielectric insulated cable rated at 2001 volts or higher.

Megger: The term used to identify a test instrument for measuring the insulation resistance of conductors and other electrical equipment; specifically, a megohm (million ohms) meter; but this is a registered trade name of the James Biddle Co.

Megohmmeter: An instrument for measuring extremely high resistance.

Melting time: That time required for an overcurrent to sever a fuse.

Messenger: The supporting member of an aerial cable.

Metal clad (MC): The cable core is enclosed in a flexible metal covering.

Metal-clad switchgear: Switchgear having each power circuit device in its own metal enclosed compartment.

Metal wireway: Sheet metal troughs with hinged or removable covers for housing and protecting electric wires and cables.

Meter: An instrument designed to measure; metric unit of linear measurement equal to 39.37 inches.

Meter pan: A shallow metal enclosure with a round opening, through which a kilowatt hour meter is mounted, as the usual meter for measuring the amount of energy consumed by a particular building or other electrical system.

Mho: Reciprocal of ohm.

MI cable: Mineral insulated, metal sheathed cable.

Mica: A silicate which separates into layers and has high insulation resistance, dielectric strength and heat resistance.

Micrometer (mike): A tool for measuring linear dimensions accurately to 0.001 inch or to 0.01 mm.

Microwave: Radio waves of frequencies above one gigahertz.

Mil: A unit used in measuring the diameter of wire, equal to 0.001 inch (25.4 micrometers).

mm: Millimeter: 1 meter ÷ 1000.

MM: Mining machine.

Mineral insulated metal sheathed cable (type MI): A factory assembly of one or more conductors insulated with a highly compressed refractory mineral insulation and enclosed in a liquidtight and gastight continuous copper or alloy steel sheath.

Minimum average: The specified average insulation or jacket thickness.

Minimum at a point: Specifications that permit the thickness at one point to be less than the average.

Mks: Meter, kilogram, second.

Mobile equipment: Equipment with electric components suitable to be moved only with mechanical aids or which are provided with wheels for movement by a person or powered device.

Modem: Equipment that connects data transmitting/receiving equipment to telephone lines: a word contraction of modulator-demodulator.

Modulation: The varying of a "carrier" wave characteristic by a characteristic of a second "modulating" wave.

Modulus of electricity: The ratio of stress (force) to strain (deformation) in a material that is elastically deformed.

Moisture-repellent: So constructed or treated that moisture will not penetrate.

Moisture-resistance: So constructed or treated that moisture will not readily injure.

Molded case breaker: A circuit breaker enclosed in an insulating housing.

Momentary rating: A rating for x-ray equipment based on an operating interval that does not exceed 5 minutes.

Monitor hazard current: The hard current of a line isolation monitor alone.

Motor: An apparatus to convert from electrical to mechanical energy.

Motor, capacitor: A single-phase induction motor with an auxiliary starting winding connected in series with a condenser for better starting characteristics.

Motor control: Device to start and/or stop a motor at certain temperature or pressure conditions.

Motor control center: A grouping of motor controls such as starters.

Motor effect: Movement of adjacent conductors by magnetic forces due to currents in the conductors.

Mouse: Any weighted line used for dropping down between finished walls to attach to cable to pull the cable up; a type of vertical fishing between walls.

MPT: Male pipe thread.

MPX: Multiplexer.

MTW: Machine tool wire.

Multioutlet assembly: A type of surface or flush raceway designed to hold conductors and attachment plug receptacles.

Multiplex: To interleave or simultaneously transmit two or more messages on a single channel.

Mutual inductance: The condition of voltage in a second conductor because of a change in current in another adjacent conductor.

Mw: Megawatt: 1,000,000 watts.

Mylar®: DuPont trade name for a polyester film whose generic name is oriented polyethylene terephthalate; used for insulation, binding tapes.

N/A: 1) not available 2) not applicable.

National Electrical Code (NEC): A national consensus standard for the installation of electrical systems.

Natural convection: Movement of a fluid or air caused by temperature change.

NBS: National Bureau of Standards.

NC: Normally closed.

Negative: Connected to the negative terminal of a power supply.

NEMA: National Electrical Manufacturers Association.

Neon tubing: Electric-discharge tubing manufactured into shapes that form letters, parts of letters, skeleton tubing, outline lines or other decorative elements or art forms, and that are filled with various inert gases.

Neoprene: An oil resistant synthetic rubber used for jackets; originally a DuPont trade name, now a generic term for polychloroprene.

Network: An aggregation of interconnected conductors consisting of feeders, mains, and services.

Network limiter: A current limiting fuse for protecting a single conductor.

Neutral: The element of a circuit from which other voltages are referenced with respect to magnitude and time displacement in steady state conditions.

Neutral block: The neutral terminal block in a panelboard, meter enclosure, gutter or other enclosure in which circuit conductors are terminated or subdivided.

Neutral wire: A circuit conductor which is common to the other conductors of the circuit, having the same voltage between it and each of the other circuit wires and usually operating grounded; such as the neutral of 3 wire, single-phase, or 3-phase 4-wire wye systems.

Newton: The derived SI unit for force: the force which will give one kilogram mass an acceleration of one meter per second.

NFPA (National Fire Protection Association): An organization to promote the science and improve the methods of fire protection which sponsors various codes, including the National Code.

Nineteen hundred box: A commonly used term to refer to any 2-gang 4-inch square outlet box used for two wiring devices or for one wiring device with a single-gang cover where the number of wires requires this box capacity.

Nipple: A threaded pipe or conduit of less than two feet length.

NO: Normally open.

No-niche lighting fixture: A lighting fixture intended for installation above or below the water line that does not require a niche.

Node: A junction of two or more branches of a network.

Nominal: Relating to a designated size that may vary from the actual.

Nominal rating: The maximum constant load which may be increased for a specified amount for two hours without exceeding temperature limits specified from the previous steady state temperature conditions: usually 25 or 50 percent increase is used.

Nomograph: A chart or diagram with which equations can be solved graphically by placing a straightedge on two known values and reading where the straightedge crosses the scale of the unknown value.

Nonautomatic: Used to describe an action requiring personal intervention for its control.

Noncode installation: A system installed where there are no local, state, or national codes in force.

Nonincendive circuit: A circuit, other than field wiring, in which any arc or thermal effect produced under the intended operating conditions of the equipment, is not capable, under specified test conditions, of igniting the flammable gas-, vapor-, or dust-air mixture.

Nonincendive field wiring: Wiring that enters or leaves an equipment enclosure and, under the normal operating conditions of the equipment, is not capable, due to arcing or thermal effects, of igniting the flammable gas-, vapor-, or dust-air mixture. Normal operation includes opening, shorting, or grounding the field wiring.

Nonmetallic extensions: An assembly of two insulated conductors within a nonmetallic jacket or an extruded thermoplastic covering.

Nonmetallic-sheathed cable (type NMC): A factory assembly of two or more insulated conductors having an outer sheath of moisture-resistant, flame-retardant nonmetallic material.

Nonmetallic wireways: Flame-retardant, nonmetallic troughs with removable covers for housing and protecting electric wires and cables.

Normal charge: The thermal element charge that is part liquid and part gas under all operating conditions.

NPT: National tapered pipe thread.

NR: 1) Nonreturnable reel; a reel designed for one-time use only. 2) Natural rubber.

NSD (neutral supported drop): A type of service cable.

Nylon®: This is the DuPont trade name for polyhexamethylene-adipamide which is the thermoplastic used as insulation and jacketing material.

OC: Overcurrent.

OD: Outside diameter.

OF: Oxygen free.

Offgassing: Percentage of a gas released during combustion.

Ohm: The derived SI unit for electrical resistance or impedance: one ohm equals one volt per ampere.

Ohmmeter: An instrument for measuring resistance in ohms.

Ohm's Law: Mathematical relationship between voltage, current, and resistance in an electric circuit.

OI (Official Interpretation): An interpretation of the *National Electrical Code* made to help resolve a specific problem between an inspector and an installer.

Oil-: The prefix designating the operation of a device submerged in oil to cool or quench or insulate.

Oil-proof: The accumulation of oil or vapors will not prevent safe successful operation.

Oil-tight: Construction preventing the entrance of oil or vapors not under pressure.

OL : Overload.

Open: A circuit which is energized by not allowing useful current to flow.

Opening time: The period between which an activation signal is initiated and switch contacts part.

Optimization: The procedure used in the design of a system to maximize or minimize some performance index.

Oscillation: The variation, usually with time, of the magnitude of a quantity which is alternately greater and smaller than a reference.

Oscillator: A device that produces an alternating or pulsating current or voltage electronically.

Oscillograph: An instrument primarily for producing a graph of rapidly varying electrical quantities.

Oscilloscope: An instrument primarily for making visible rapidly varying electrical quantities: oscilloscopes function similarly to TV sets.

OSHA (Occupational Safety and Health Act): Federal Law #91-596 of 1970 charging all employers engaged in business affecting interstate commerce to be responsible for providing a safe working place: it is administered by the Department of Labor: the OSHA regulations are published in Title 29, Chapter XVII, Part 1910 of the CFR and the Federal Register.

Ought sizes: An expression referring to conductors of sizes No. 1/0, 2/0, 3/0 or 4/0.

Outage: When a component is not available to perform its intended function.

Outdoor: Designed for use out-of-doors.

Outlet: A point on the wiring system at which current is taken to supply utilization equipment.

Outline lighting: An arrangement of incandescent lamps or gaseous tubes to outline and call attention to certain features, such as the shape of a building or the decoration of a window.

Output: 1) The energy delivered by a circuit or device. 2) The terminals for such delivery.

Oven: An enclosure and associated sensors and heaters for maintaining components at a controlled and usually constant temperature.

Overcurrent protection: De-energizing a circuit whenever the current exceeds a predetermined value; the usual devices are fuses, circuit breakers, or magnetic relays.

Overload: Load greater than the load for which the system or mechanism was intended.

Overvoltage (cable): Voltages above normal operating voltages, usually due to: a) switching loads on/off. b) lighting. c) single phasing.

Oxidize: 1) To combine with oxygen. 2) To remove one or more electrons. 3) To dehydrogenate.

Oxygen index: A test to rate flammability of materials in a mixture of oxygen and nitrogen.

Ozone: An active molecule of oxygen which may attack insulation: produced by corona in air.

Pad-mounted: A shortened expression for "pad-mount transformer," which is completely enclosed transformer mounted outdoors on a concrete pad, without need for a surrounding chain-link fence around the metal, box-like transformer enclosure.

Pan: A sheet metal enclosure for a watt-hour meter, commonly called a "meter pan."

Panel: A unit for one or more sections of flat material suitable for mounting electrical devices.

Panelboard: A single panel or group of panel units designed for assembly in the form of a single panel; includes buses and may come with or without switches and/or automatic overcurrent protective devices for the control of light, heat, or power circuits of individual as well as aggregate capacity. It is designed to be placed in a cabinet or cutout box that is in or against a wall or partition and is accessible only from the front.

Paper-lead cable: One having oil-impregnated paper insulation and a lead sheath.

Parallel: Connections of two or more devices between the same two terminals of a circuit.

Parameter: A variable given a constant value for a specific process or purpose.

Pascal: The derived SI unit for pressure or stress: one pascal equals one newton per square meter.

Patch: To connect circuits together temporarily.

Payoff: The equipment to guide the feeding of wire.

PB: Push button; pull box.

PE: 1) Polyethylene 2) Professional engineer.

Peak value: The largest instantaneous value of a variable.

Penciling: The tapering of insulation to relieve electrical stress at a splice or termination.

Period: The minimum interval during which the same characteristics of a periodic phenomenon recur.

Permalloy: An alloy of nickel and iron that is easily magnetized and demagnetized.

Permeability: 1) The passage or diffusion of a vapor, liquid or solid through a barrier without physically or chemically affecting either. 2) The rate of such passage.

Personnel protection system: A system of protection devices and constructional features that when used together provide protection against the electric shock of personnel.

PES: Power Engineering Society of IEEE.

PF: Power factor.

pH: An expression of the degree of acidity or alkalinity of a substance: on the scale of 1-10, acid is under 7, neutral is 7, alkaline is over 7.

Phase: The fractional part "t/p" of the period through which a quantity has advanced, relative to an arbitrary origin.

Phase angle: The measure of the progression of a periodic wave in time or space from a chosen instant or position.

Phase conductor: The conductors other than the neutral.

Phase leg: One of the phase conductors (an ungrounded or "hot" conductor) of a polyphase electrical system.

Phase out: A procedure by which the individual phases of a polyphase circuit or system are identified; such as to "phase out" a 3-phase circuit for a motor in order to identify phase A, phase B and phase C to know how to connect them to the motor to get the correct phase rotation so the motor will rotate in the desired direction.

Phase sequence: The order in which the successive members of a periodic wave set reach their positive maximum values: a) zero phase sequence — no phase shift: b) plus/minus phase sequence — normal phase shift.

Phase shift: The absolute magnitude of the difference between two phase angles.

Phasor quantity: A complex algebraic expression for sinusoidal wave.

Photocell: A device in which the current-voltage characteristic is a function of incident radiation (light).

Photoelectric control: A control sensitive to incident light.

Photoelectricity: A physical action wherein an electrical flow is generated by light waves.

Photon: An elementary quantity of radiant energy (quantum).

Pi (π): The ratio of the circumference of a circle to its diameter.

Pick: The grouping or band of parallel threads in a braid.

Pickle: A solution or process to loosen or remove corrosion products from a metal.

Pickup value: The minimum input that will cause a device to complete a designated action.

Picocoulomb: 10-12 coulombs.

Piezoelectric effect: Some materials become electrically polarized when they are mechanically strained: the direction and magnitude of the polarization depends upon the nature, amount and the direction of the strain: in such materials the reversal is also true in that a strain results from the application of an electric field.

Pigtail: A flexible conductor attached to an apparatus for connection to a circuit.

PILC cable: Paper insulated, lead covered.

Pilot lamp: A lamp that indicates the condition of an associated circuit.

Pilot wire: An auxiliary insulated conductor in a power cable used for control or data.

Pitch diameter: The diameter through the center of a layer in a concentric layup of a cable or strand.

Pitting: Small cavities in a metal surface.

Plasma: A gas made up of charged particles.

Plate: A rectangular product having thickness of 0.25 inch or more.

Plating: Forming an adherent layer of metal on an object.

Plenum: Chamber or space forming a part of an air conditioning system.

Plowing: Burying cable in a split in the earth made by a blade.

Plug: A male connector for insertion into an outlet or jack.

Plugging: Braking an induction motor by reversing the phase sequence of the power to the motor.

Polarity: 1) Distinguishing one conductor or terminal from another. 2) Identifying how devices are to be connected such, as + or −.

Polarization Index: Ratio of insulation resistance measured after 10 minutes to the measure at 1 minute with voltage continuously applied.

Pole: 1) That portion of a device associated exclusively with one electrically separated conducting path of the main circuit of device. 2) A supporting circular column.

Poly: Polyethylene.

Polychloroprene: Generic name for neoprene.

Polyethylene: A thermoplastic insulation having excellent electrical properties, good chemical resistance (useful as jacketing), good mechanical properties with the exception of temperature rating.

Polyphase circuits: ac circuits having two or more interrelated voltages, usually of equal amplitudes, phase differences, and periods, etc: if a neutral conductor exists, the voltages referenced to the neutral are equal in amplitude and phase: the most common version is that of 3-phase, equal in amplitude with phases 120° apart.

Polypropylene: A thermoplastic insulation similar to polyethylene, but with slightly better properties.

Polytetrafluoroethylene (PTFE): A thermally stable (−90 to + 250°C) insulation having good electrical and physical properties even at high frequencies.

Porcelain: Ceramic chinalike coating applied to steel surfaces.

Portable: Designed to be movable from one place to another, not necessarily while in operation.

Portable equipment: Equipment with electric components suitable to be moved by a single person without mechanical aids.

Positive: Connected to the positive terminal of a power supply.

Potential: The difference in voltage between two points of a circuit. Frequently, one is assumed to be ground (zero potential).

Potential energy: Energy of a body or system with respect to the position of the body or the arrangement of the particles of the system.

Potentiometer: An instrument for measuring an unknown voltage or potential difference by balancing it, wholly or in part, by a known potential difference produced by the flow of known currents in a network of circuits of known electrical constants.

Pothead: A terminator for high-voltage circuit conductor to keep moisture out of the insulation and to protect the cable end, along with providing a suitable stress relief cone for shielded-type conductors.

Power: 1) Work per unit of time. 2) The time rate of transferring energy: as an adjective, the word "power" is descriptive of the energy used to perform useful work: pound-feet per second, watts.

Power, active: In a 3-phase symmetrical circuit: $p = 3\ vi \cos \theta$; in a 1-phase, 2 wire circuit, $p = vi \cos \theta$.

Power and control tray cable (type PC): A factory assembly of two or more insulated conductors, with or without associated bare or covered grounding conductors under a nonmetallic sheath, for installation in cable trays, in raceways, or where supported by messenger wire.

Power, apparent: The product of rms volts times rms amperes.

Power element: Sensitive element of a temperature-operated control.

Power factor: Correction coefficient for ac power necessary because of changing current and voltage values.

Power loss (cable): Losses due to internal cable impedance, mainly I^2R: the losses cause heating.

Power pool: A group of power systems operating as an interconnected system.

P-P: Peak to peak.

Precast concrete: Concrete units (such as piles or vaults) cast away from the construction site and set in place.

Premises wiring (system): That interior and exterior wiring, including power, lighting, control and signal circuit wiring together with all of their associated hardware, fittings and wiring devices, both permanently and temporarily installed, that extends from the service point of utility conductors or source of power such as a battery, a solar photovoltaic system, or a generator, transformer, or convector windings, to the outlet(s). Such wiring does not include wiring internal to appliances, fixtures, motors, controllers, motor control centers, and similar equipment.

Premolded: A splice or termination manufactured of polymers, ready for field application.

Pressure: An energy impact on a unit area; force or thrust exerted on a surface.

Pressure motor control: A device that opens and closes an electrical circuit as pressures change.

Primary: Normally referring to the part of a device or equipment connected to the power supply circuit.

Primary control: Device that directly controls operation of heating, air-conditioning, ventilation and similar systems.

Printed circuit: A board having interconnecting wiring printed on its surface and designed for mounting of electronic components.

Process: Path of succession of states through which a system passes.

Program, computer: The ordered listing of sequence of events designed to direct the computer to accomplish a task.

Propagation: The travel of waves through or along a medium.

Property: An observable characteristic.

Protected enclosure: Having all openings protected with screening, etc.

Protector, circuit: An electrical device that will open an electrical circuit if excessive electrical conditions occur.

Prototype: The first full size working model.

Proximity effect: The distortion of current density due to magnetic fields; increased by conductor diameter, close spacing, frequency, and magnetic materials such as steel conduit or beams.

PSI: Pound force per square inch.

PT: Potential transformer.

Pull box: A sheet metal box-like enclosure used in conduit runs, either single conduits or multiple conduits, to facilitate pulling in of cables from point to point in long runs or to provide installation of conduit support bushings needed to support the weight of long riser cables or to provide for turns in multiple-conduit runs.

Pull-down: Localized reduction of conductor diameter by longitudinal stress.

Pulling compound (lubricant): A substance applied to the surface of a cable to reduce the coefficient of friction during installation.

Pulling eye: A device attached to a cable to facilitate field connection of pulling ropes.

Pulsating function: A periodic function whose average value over a period is not zero.

Pulse: A brief excursion of a quantity from normal.

Pumped storage (hydro power): The storage of power by pumping a reservoir full of water during off-peak, then depleting the water to generate when needed.

Puncture: Where breakdown occurs in an insulation.

Purge: To clean.

Push button: A switch activated by buttons.

PVC (polyvinyl chloride): A thermoplastic insulation and jacket compound.

Pyroconductivity: Electric conductivity that develops with changing temperature, and notably upon fusion, in solids that are practically non-conductive at atmospheric temperatures.

Pyrometer: Thermometer that measures the radiation from a heated body.

QA (Quality Assurance): All the planned and systematic actions to provide confidence that a structure, system, or component will perform satisfactorily.

Quadruplexed: Twisting of four conductors together.

Qualified person: A person familiar with construction, operation and hazards.

Quick-: A device that has a high contact speed independent of the operator; example — quick-make or quick-break.

Raceway: Any channel designed expressly for holding wire, cables, or bars and used solely for that purpose.

Rack (cable): A device to support cables.

Radar: A radio detecting and ranging system.

Radial feeder: A feeder connected to a single source.

Radiant energy: Energy traveling in the form of electromagnetic waves.

Radiant heating: Heating system in which warm or hot surfaces are used to radiate heat into the space to be conditioned.

Radiation: The process of emitting radiant energy in the form of waves or particles.

Radius, bending: The radii around which cables are pulled.

Radius, training: The radii to which cables are bent by hand positioning, not while the cables are under tension.

Rail clamp: A device to connect cable to a track rail.

Raintight: So constructed or protected that exposure to a beating rain will not result in the entrance of water.

RAM: Random access memory.

Rated: Indicating the limits of operating characteristics for application under specified conditions.

Rating, temperature (cable): The highest conductor temperature attained in any part of the circuit during a) normal operation b) emergency overload c) shot circuit.

Rating, voltage: The thickness of insulation necessary to confine voltage to a cable conductor after withstanding the rigors of cable installation and normal operating environment.

REA (Rural Electrification Administration): A federally supported program to provide electrical utilities in rural areas.

Reactance: 1) The imaginary part of impedance. 2) The opposition to ac due to capacitance (Xc) and inductance (XL).

Reactor: A device to introduce capacitive or inductive reactance into a circuit.

Real time: The actual time during which a physical process transpires.

Reamer: A finishing tool that finishes a circular hole more accurately than a drill.

Receptacle: A contact device installed at the outlet for the connection of an attachment plug. A single receptacle is a single contact device with no other contact device on the same yoke. A multiple receptacle has two or more contact devices on the same yoke.

Reciprocating: Action in which the motion is back and forth in a straight line.

Recognized component: An item to be used as a sub-component and tested for safety by UL; is UL's trademark for recognized component.

Recorder: A device that makes a permanent record, usually visual, of varying signals.

Rectifiers: Devices used to change alternating current to unidirectional current.

Rectify: To change from ac to dc.

Redraw: Drawing of wire through consecutive dies.

Reducing joint: A splice of two different size conductors.

Redundancy: The use of auxiliary items to perform the same functions for the purpose of improving reliability and safety.

Reel: A drum having flanges on the ends; reels are used for wire/cable storage.

Reference grounding point: The ground bus of the panelboard or isolated power system panel supplying the patient care area in a health care facility.

Reflective insulation: Thin sheets of metal or foil on paper set in the exterior walls of a building to reflect radiant energy.

Refrigerant: Substance used in refrigerating mechanism to absorb heat in an evaporator coil and to release heat in a condenser as the substance goes from a gaseous state back to a liquid state.

Regulation: The maximum amount that a power supply output will change as a result of the specified change in line voltage, output load, temperature, or time.

Reinforced jacket: A cable jacket having reinforcing fiber between layers.

Relative capacitance: The ratio of a material's capacitance to that of a vacuum of the same configuration; will vary with frequency and temperature.

Relay: A device designed to abruptly change a circuit because of a specified control input.

Relay, overcurrent: A relay designed to open a circuit when current in excess of a particular setting flows through the sensor.

Reliability: The probability that a device will function without failure over a specified time period or amount of usage.

Remote-control circuits: The control of a circuit through relays, etc.

Reproducibility: The ability of a system or element to maintain its output or input over a relatively long period of time.

Reservoir, thermal: A body to which and from which heat can be transferred indefinitely without a change in the temperature of the reservoir.

Resin: The polymeric base of all jacketing, insulating, etc. compounds, both rubber and plastic.

Resistance: The opposition in a conductor to current; the real part of impedance.

Resistance furnace: A furnace which heats by the flow current against ohmic resistance internal to the furnace.

Resistance heating element: A specific separate element to generate heat that is applied to, embedded in, or fastened to the pipeline, vessel or surface to be heated.

Resistance, thermal: The opposition to heat flow; for cables it is expressed by degrees centigrade per watt per foot of cable.

Resistance welding: Welding by pressure and heat when the work piece's resistance in an electric circuit produces the heat.

Resistivity: A material characteristic opposing the flow of energy through the material; expressed as a constant for each material: it is affected by temper, temperature, contamination, alloying, coating, etc.

Resistor: A device whose primary purpose is to introduce resistance.

Resistor, bleeder: 1) Used to drain current after a device is deenergized. 2) To improve voltage regulation. 3) To protect against voltage surges.

Resolution: The degree to which nearly equal values of a quantity can be discriminated.

Resonance: In a circuit containing both inductance and capacitance.

Resonating: The maximizing or minimizing of the amplitude or other characteristics provided the maximum or minimum is of interest.

Response: A quantitative expression of the output as a function of the input under conditions that must be explicitly stated.

Restrike: A resumption of current between contacts during an opening operation after an interval of zero current of ¼ cycle at normal frequency or longer.

Reverse lay: Reversing the direction of lay about every five feet during cabling of aerial cable to facilitate field connections.

Reversible process: Can be reversed and leaves no change in system or surroundings.

RF: Radio frequency: 10kGz to GHz.

RFI: Radio frequency interference.

Rheostat: A variable resistor, which can be varied while energized, normally one used in a power circuit.

Rigid metal conduit: A listed metal raceway of circular cross section with integral or associated couplings. Approved for the installation of electrical conductors and used with listed fittings to provide electrical continuity.

Ring-out: 1) A circular section of insulation or jacket. 2) The continuity testing of a conductor.

Ripple: The ac component from a dc power supply arising from sources within the power supply.

Riser: A vertical run of conductors in conduit or busway, for carrying electrical power from one level to another in a building.

Riser valve: Device used to manually control flow of refrigerant in vertical piping.

RMS (Root-mean-square): The square root of the average of the square of the function taken throughout the period.

Rod: The shape of solidified metal convenient for wire drawing, usually 5/16 inch or larger.

ROM: Read only memory.

Romex: General Cable's trade name for type NM cable; but it is used generically by electricians to refer to any nonmetallic sheathed cable.

Rotor: Rotating part of a mechanism.

Rough inspection: The first inspection made of an electrical installation after the conductors, boxes and other equipment have been installed in a building under construction.

Roughing in: The first stage of an electrical installation, when the raceway, cable, wires, boxes and other equipment are installed; that electrical work which must be done before any finishing or cover-up phases of building construction can be undertaken.

Round off: To delete the least significant digits of a numeral and adjust the remaining by given rules.

RPM: Revolutions per minute.

Rubber, chlorosulfonated polyethylene (CP): A synthetic rubber insulation and jacket compound developed by DuPont as Hypalon®.

Rubber, ethylene propylene: A synthetic rubber insulation having excellent electrical properties.

Running board: A device to permit stringing more than one conductor simultaneously.

Safety conductor: A safety sling used during overhead line construction.

Safety control: A device that will stop the refrigerating unit if unsafe pressures and/or temperatures are reached.

Safety factor: The ratio of the maximum stress which something can withstand to the estimated stress which it can withstand.

Safety motor control: Electrical device used to open a circuit if the temperature, pressure, and/or the current flow exceed safe conditions.

Safety plug: Device that will release the contents of a container above normal pressure conditions and before rupture pressures are reached.

Sag: The difference in elevation of a suspended conductor.

Sag, apparent: Sag between two points at 60°F and no wind.

Sag, final: Sag under specified conditions after the conductor has been externally loaded, then the load removed.

Sag, initial: Sag prior to external loading.

Sag, maximum: Sag at midpoint between two supports.

Sag section: Conductor between two snubs.

Sag snub: Where a conductor is held fixed and the other end moved to adjust sag.

Sag, total: Under ice loading.

Sampling: A small quantity taken as a sample for inspection or analysis.

Saturation: The condition existing in a circuit when an increase in the driving signal does not produce any further change in the resultant effect.

Scalar: A quantity (as mass or time) that has a magnitude described by a real number and no direction.

Scan: To examine sequentially, part by part.

Scope: Slang for oscilloscope.

Screen pack: A metal screen used for straining.

SE: Service entrance.

Sealed: Preventing entrance.

Sealed motor compressor: A mechanical compressor consisting of a compressor and a motor, both of which are enclosed in the same sealed housing, with no external shaft or shaft seals, and with the motor operating in the refrigerant atmosphere.

Sealing compound: The material poured into an electrical fitting to seal and prevent the passage of vapors.

Secondary: The second circuit of a device or equipment, which is not normally connected to the supply circuit.

Seebeck Effect: The generation of a voltage by a temperature difference between the junctions in a circuit composed of two homogeneous electrical conductors of dissimilar composition: or in a nonhomogeneous conductor the voltage produced by a temperature gradient in a nonhomogeneous region.

Self inductance: Magnetic field induced in the conductor carrying the current.

Semiconductor: A material that has electrical properties of current flow between a conductor and an insulator.

Sensible heat: Heat that causes a change in temperature of a substance.

Sensor: A material or device that goes through a physical change or an electronic characteristic change as conditions change.

Separable insulated connector: An insulated device to facilitate power cable connections and separations.

Separator: Material used to maintain physical spacing between elements in cables, such as: a layer of tape to prevent jacket sticking to individual conductors.

Sequence controls: Devices that act in series or in time order.

Service: The equipment used to connect to the conductors run from the utility line, including metering, switching and protective devices; also the electric power delivered to the premises, rated in voltage and amperes, such as a "100-amp, 480 volt service."

Service conductors: The supply conductors that extend from the street main or transformers to the service equipment of the premises being supplied.

Service drop: Run of cables from the power company's aerial power lines to the point of connection to a customer's premises.

Service entrance: The point at which power is supplied to a building, including the equipment used for this purpose (service main switch or panel or switchboard, metering devices, overcurrent protective devices, conductors for connecting to the power company's conductors and raceways for such conductors).

Service entrance cable (types SE and USE): A single conductor or multiconductor assembly provided with or without an overall covering, primarily used for services.

Service equipment: The necessary equipment, usually consisting of a circuit breaker(s) or switch(s) and fuses(s) and their accessories, connected to the load end of service conductors to a building or other structure, or an otherwise designated area, and intended to constitute the main control and cutoff of the supply.

Service lateral: The underground service conductors between the street main, including any risers at a pole or other structure or from transformers, and the first point of connection to the service-entrance conductors in a terminal box, meter, or other enclosure with adequate space, inside or outside the building wall. Where there is no terminal box, meter, or other enclosure with adequate space, the point of connection is the entrance point of the service conductors into the building.

Service raceway: The rigid metal conduit, electrical metallic tubing, or other raceway that encloses the service-entrance conductors.

Service valve: A device to be attached to a system that provides an opening for gauges and/or charging lines.

Serving: A layer of helically applied material.

Servomechanism: A feedback control system in which at least one of the system signals represents mechanical motion.

Setting (of circuit breaker): The value of the current at which the circuit breaker is set to trip.

Shaded pole motor: A small dc motor used for light start loads that has no brushes or commutator.

Shall: Mandatory requirement of a Code.

Shaving: Removing about 0.001 inch of metal surface.

Sheath: A metallic close fitting protective covering.

Shield: The conducting barrier against electromagnetic fields.

Shield, braid: A shield of interwoven small wires.

Shield, insulation: An electrically conducting layer to provide a smooth surface in intimate contact with the insulation outer surface; used to eliminate electrostatic charges external to the shield, and to provide a fixed known path to ground.

Shield, tape: The insulation shielding system whose current carrying component is thin metallic tapes, now normally used in conjunction with a conducting layer of tapes or extruded polymer.

Shim: Thin piece of material used to bring members to an even or level bearing.

Shore feeder: From ship to shore feeder.

Short-circuit: An often unintended low-resistance path through which current flows around, rather than through, a component or circuit.

Short cycling: Refrigerating system that starts and stops more frequently than it should.

Shrinkable tubing: A tubing which may be reduced in size by applying heat or solvents.

Shroud: Housing over a condenser or evaporator.

Shunt: A device having appreciable resistance or impedance connected in parallel across other devices or apparatus to divert some of the current: appreciable voltage exists across the shunt and appreciable current may exist in it.

Sidewall load: The normal force exerted on a cable under tension at a bend; quite often called sidewall pressure.

Sign body: A portion of a sign that may provide protection from the weather, but is not an electrical enclosure.

Signal: A detectable physical quantity or impulse (as a voltage, current, or magnetic field strength) by which messages or information can be transmitted.

Signal equipment: Audible and visual equipment such a chimes, gongs, lights, and displays that convey information to the user.

Silicon controlled rectifier (SCR): Electronic semiconductor that contains silicon.

Sill: Horizontal timber forming the lowest member of a wood frame house; lowest member of a window frame.

Sine wave, ac: Wave form of single frequency alternating current; wave whose displacement is the sine of the angle proportional to time or distance.

Single-phase circuit: An ac circuit having one source voltage supplied over two conductors.

Single-phase motor: Electric motor that operates on single-phase alternating current.

Single-phasing: The abnormal operation of a three phase machine when its supply is changed by accidental opening of one conductor.

Skeleton tubing: Neon tubing that is itself the sign or outline lighting and not attached to an enclosure or sign body.

Skin effect: The tendency of current to crowd toward the outer surface of a conductor: increases with conductor diameter and frequency.

Skin effect heating system: A system in which heat is generated on the inner surface of a ferromagnetic envelope embedded in or fastened to the surface to be heated.

Slip: The difference between the speed of a rotating magnetic field and the speed of its rotor.

Slippercoat: A surface lubricant factory applied to a cable to facilitate pulling, and to prevent jacket sticking.

Slot: A channel opening in the stator or rotor of a rotating machine for ventilation and the insertion of windings.

Soap: Slang for pulling compound.

Soffit: Underside of a stair, arch, or cornice.

Soft drawn: 1) A relative measure of the tensile strength of a conductor. 2) Wire which has been annealed to remove the effects of cold working. 3) Drawn to a low tensile.

Solar cell: The direct conversion of electromagnetic radiation into electricity; certain combinations of transparent conducting films separated by thin layers of semiconducting materials.

Solar heat: Heat from visible and invisible energy waves from the sun.

Solar photovoltaic systems: The total components and subsystems that, in combination, convert solar energy into electrical energy suitable for connection to a utilization load.

Solder: To braze with tin alloy.

Solenoid: Electric conductor wound as a helix with a small pitch: coil.

Solidly grounded: No intentional impedance in the grounding circuit.

Solid state: A device, circuit, or system which does not depend upon physical movement of solids, liquids, gases or plasma.

Solid type PI cable: A pressure cable without constant pressure controls.

Solution: 1) Homogenous mixture of two or more components. 2) Solving a problem.

SP: Single pole.

Space heater: A heater for occupied spaces.

Span: A conductor between two consecutive supports.

Spark: A brilliantly luminous flow of electricity of short duration that characterizes an electrical breakdown.

Spark gap: Any short air space between two conductors.

Spark test: A voltage withstand test on a cable while in production with the cable moving: it is a simple way to test long lengths of cable.

SPDT: Single pole double throw.

Specific heat: Ratio of the quantity of heat required to raise the temperature of a body 1 degree to that required to raise the temperature of an equal mass of water 1 degree.

Specs: Abbreviation for the word "specifications," which is the written precise description of the scope and details of an electrical installation and the equipment to be used in the system.

Spectrum: The distribution of the amplitude (and sometimes phase) of the components of the wave as a function of frequency.

Spike: A pulse having great magnitude.

Splice: The electrical and mechanical connection between two pieces of cable.

Splice tube: The movable section of vulcanizing tube at the extruder.

Split fitting: A conduit fitting which may be installed after the wires have been installed.

Split-phase motor: Motor with two stator windings. Winding in use while starting is disconnected by a centrifugal switch after the motor attains speed, then the motor operates on the other winding.

Split system: Refrigeration or air conditioning installation that places the condensing unit outside or remote from the evaporator. It is also applicable to heat pump installations.

Split-wire: A way of wiring a duplex plug outlet (a receptacle) with a 3-wire, 120/240 volt single-phase circuit so that one hot leg and the neutral feeds one of the receptacle outlets and the other hot leg and the common neutral feeds the other receptacle outlet. This gives the capacity of two separate circuits to the duplex receptacle, the split-wired receptacle.

SPST: Single pole single throw.

Spurious response: Any response other than the desired response of an electric transducer or device.

Squirrel cage motor: An induction motor having the primary winding (usually the stator) connected to the power and a current is induced in the secondary cage winding (usually the rotor).

SSR: Solid state relay.

Stand-alone system: A solar photovoltaic system that supplies power independently of an electrical production and distribution network.

Stand lamp (work light): A portable stand that contains a general-purpose lighting fixture or lamp holder with a guard used to provide general illumination on a stage or in an auditorium.

Standard conditions: Temperature of 68 degrees Fahrenheit, pressure of 29.92 inches of mercury, and relative humidity of 30 percent.

Standard deviation: 1) A measure of data from the average. 2) The root mean square of the individual deviations from the average.

Standard reference position: The nonoperated or deenergized condition.

Standing wave: A wave in which, for any component of the field, the ratio of its instantaneous value at one point to that at any other point, does not vary with time.

Standoff: An insulated support.

Starter: 1) An electric controller for accelerating a motor from rest to normal speed and to stop the motor. 2) A device used to start an electric discharge lamp.

Starting relay: An electrical device that connects and/or disconnects the starting winding of an electric motor.

Starting winding: Winding in an electric motor used only during the brief period when the motor is starting.

Static: Interference caused by electrical disturbances in the atmosphere.

Stator: The portion of a rotating machine that includes and supports the stationary active parts.

Steady state: When a characteristic exhibits only negligible change over a long period of time.

Steam: Water in vapor state.

Steam heating: Heating system in which steam from a boiler is conducted to radiators in a space to be heated.

Strand: A group of wires, usually twisted or braided.

Strand, annular: A concentric conductor over a core: used for large conductors (1000 MCM @ 60 Hertz) to make use of skin effect: core may be of rope, or twisted I-beam.

Strand, bunch: A substrand for a rope-lay conductor: the wires in the substrand are stranded simultaneously with the same direction: bunched conductors flex easily and with little stress.

Strand, class: A system to indicate the type of stranding: the postscripts are alpha.

Strand, combination: A concentric strand having the outlet layer of different size: done to provide smoother outer surface: wires are sized with + 5% tolerance from nominal.

Strand, compact: A concentric stranding made to a specified diameter of 8%-10% less than standard by using smaller than normal closing die, and for larger sizes, preshaping the strands for the outer layer(s).

Strand, compressed: The making of a tight stranded conductor by using a small closing die.

Strand, concentric: Having a core surrounded by one or more layers of helically laid wires each of one size, each layer increased by six.

Strand, herringbone lay: When adjacent bunches have opposite direction of lay in a layer of a rope-lay cable.

Strand, regular lay: Rope stranding having left-hand lay within the substrands and right-hand lay for the conductor.

Strand, nonspecular: One having a treated surface to reduce light reflection.

Strand, reverse-lay: A stranding having alternate direction of lay for each layer.

Strand, rope-lay: A conductor having a lay-up of substrands; substrand groups are bunched or concentric.

Strand, sector: A stranded conductor formed into sectors of a circle to reduce the overall diameter of a cable.

Strand, segmental: One having sectors of the stranded conductor formed and insulated one from the other, operated in parallel: used to reduce ac resistance in single conductor cables.

Strand, unilay (unidirectional): Stranding having the same direction of lay for all layers: used to reduce diameter, but is more prone to birdcaging.

Stress: 1) An internal force set up within a body to resist or hold it in equilibrium. 2) The externally applied forces.

Stress-relief cone: Mechanical element to relieve the electrical stress at a shield cable termination; used above 2kV.

Striking: The process of establishing an arc or a spark.

Striking distance: The effective distance between two conductors separated by an insulating fluid such as air.

Stringers: Members supporting the treads and risers of a stair.

Strip: To remove insulation or jacket.

Strip light: A lighting fixture with multiple lamps arranged in a row.

Strut: A compression member other than a column or pedestal.

Studs: Vertically set skeleton members of a partition or wall to which the lath is nailed.

Sub-panel: A panelboard in a residential system which is fed from the service panel; or any panel in any system which is fed from another, or main, panel supplied by a circuit from another panel.

Substation: An assembly of devices and apparatus to monitor, control, transform or modify electrical power.

Superconductors: Materials whose resistance and magnetic permeability are infinitesimal at absolute zero (−273°C).

Supervised circuit: A closed circuit having a current-responsive device to indicate a break or ground.

Surge: 1) A sudden increase in voltage and current. 2) Transient condition.

Surge arrester: A protective device for limiting surge voltages by discharging or bypassing surge current. It also prevents continued flow of follow current while remaining capable of repeating these functions.

Switch: A device for opening and closing or for changing the connection of a circuit.

Switch, ac general-use snap: A general-use snap switch suitable only for use on alternating current circuits and for controlling the following:

- Resistive and inductive loads (including electric discharge lamps) not exceeding the ampere rating at the voltage involved.
- Tungsten-filament lamp loads not exceeding the ampere rating of the switches at the rated voltage.
- Motor loads not exceeding 80 percent of the ampere rating of the switches at the rated voltage.

Switch, ac-dc general-use snap: A type of general-use snap switch suitable for use on either direct or alternating-current circuits and for controlling the following:

- Resistive loads not exceeding the ampere rating at the voltage involved.
- Inductive loads not exceeding one-half the ampere rating at the voltage involved, except that switches having a marked horsepower rating are suitable for controlling motors not exceeding the horsepower rating of the switch at the voltage involved.
- Tungsten-filament lamp loads not exceeding the ampere rating at 125 volts, when marked with the letter T.

Switch, general-use: A switch intended for use in general distribution and branch circuits. It is rated in amperes and is capable of interrupting its rated voltage.

Switch, general-use snap: A type of general-use switch so constructed that it can be installed in flush device boxes or on outlet covers, or otherwise used in conjunction with wiring systems recognized by the *National Electrical Code*.

Switch, isolating: A switch intended for isolating an electrical circuit from the source of power. It has no interrupting rating and is intended to be operated only after the circuit has been opened by some other means.

Switch, knife: A switch in which the circuit is closed by a moving blade engaging contact clips.

Switch-leg: That part of a circuit run from a lighting outlet box where a luminaire or lampholder is installed down to an outlet box which contains the wall switch that turns the light or other load on or off; it is a control leg of the branch circuit.

Switch, motor-circuit: A switch, rated in horsepower, capable of interrupting the maximum operating overload current of a motor having the same horsepower rating as the switch at the rated voltage.

Switchboard: A large single panel, frame, or assembly of panels having switches, overcurrent, and other protective devices, buses, and usually instruments mounted on the face or back or both. Switchboards are generally accessible from the rear and from the front and are not intended to be installed in cabinets.

Symmetrical: Exhibiting symmetry.

Symmetry: The correspondence in size, form and arrangement of parts on opposite sides of a plane or line or point.

Synchronism: When connected ac systems, machines or a combination operate at the same frequency and when the phase angle displacements between voltages in them are constant, or vary about a steady and stable average value.

Synchronous: Simultaneous in action and in time (in phase).

Synchronous machine: A machine in which the average speed of normal operation is exactly proportional to the frequency of the system to which it is connected.

Synchronous speed: The speed of rotation of the magnetic flux produced by linking the primary winding.

Take-off: The procedure by which a listing is made of the numbers and types of electrical components and devices for an installation, taken from the electrical plans, drawings and specs for the job.

Take-up: 1) A device to pull wire or cable. 2) The process of accumulating wire or cable.

Tangent: (Geometry) A line that touches a curve at a point so that it is closer to the curve in the vicinity of the point than any other line drawn through the point: (Trigonometry) In a right triangle it is the ratio of the opposite to the adjacent sides for a given angle.

Tank test: The immersion of a cable in water while making electrical tests; the water is used as a conducting element surrounding the cable.

Tap: A splice connection of a wire to another wire (such as a feeder conductor in an auxiliary gutter) where the smaller conductor runs a short distance (usually only a few feet, but could be up to 25 feet) to supply a panelboard or motor controller or switch. Also called a "tap-off" indicating that energy is being taken from one circuit or piece of equipment to supply another circuit or load; a tool that cuts or machines threads in the side of a round hole.

Tap drill: Drill used to form hole prior to placing threads in hole. The drill is the size of the root diameter of tap threads.

Tape: A relatively narrow, long, thin, flexible fabric, film or mat or combination thereof: helically applied tapes are used for cable insulation, especially at splices: for the first century the primary insulation for cables above 2kV was oil saturated paper tapes.

Target sag: A visual reference used when sagging.

Task illumination: A provision for the minimum lighting required to carry out necessary tasks in a described area, including safe access to supplies and equipment, and access to its exits.

TC: 1) Thermocouple 2) Time constant 3) Timed closing.

TDR: 1) Time delay relay 2) Time domain reflectometer, pulse-echo (radar) testing of cables; signal travels through cable until impedance discontinuity is encountered, then part of signal is reflected back; distance to fault can be estimated. Useful for finding faults, broken shields or conductor.

Technical Appeal Board (TAB) of UL: A group to recommend solutions to technical differences between UL and a UL client.

Telegraphy: Telecommunication by the use of a signal code.

Telemetering: Measurement with the aid of intermediate means that permits interpretation at a distance from the primary detector.

Telephone: The transmission and reception of sound by electronics.

Temper: A measure of the tensile strength of a conductor; indicative of the amount of annealing or cold working done to the conductor.

Temperature, ambient: The temperature of the surrounding medium, such as air around a cable.

Temperature, coefficient of resistance: The unit change in resistance per degree temperature change.

Temperature, emergency: The temperature to which a cable can be operated for a short length of time, with some loss of useful life.

Temperature humidity index: Actual temperature and humidity of a sample of air compared to air at standard conditions.

Temperature, operating: The temperature at which a device is designed or rated for normal operating conditions; for cables: the maximum temperature for the conductor during normal operation.

Temporary: This single word is used to mean either "temporary service" (which a power company will give to provide electric power in a building under construction) or "temporary inspection" (the inspection that a code-enforcing agency will make of a temporary service prior to inspection of the electrical work in a building under construction).

Tensile strength: The greatest longitudinal stress a material — such as a conductor — can withstand before rupture or failure while in service.

Tension, final unloaded conductor: The tension after the conductor has been stretched for an appreciable time by loads simulating ice and wind.

Tension, initial conductor: The tension prior to any external load.

Tension, working: The tension which should be used for a portable cable on a power reel; it should not exceed 10% of the cable breaking strength.

Terminal: A device used for connecting cables.

Termination: 1) The connection of a cable. 2) The preparation of shielded cable for connection.

Testlight: Light provided with test leads that is used to test or probe electrical circuits to determine if they are energized.

Test, proof: Made to demonstrate that the item is in satisfactory condition.

Test, voltage breakdown: a) Step method — applying a multiple of rated voltage to a cable for several minutes, then increasing the applied voltage by 20% for the same period until breakdown. b) Applying a voltage at a specified rate until breakdown.

Test, voltage life: Applying a multiple of rated voltages over a long time period until breakdown: time to failure is the parameter measured.

Test, volume resistivity: Measuring the resistance of a material such as the conducting jacket or conductor stress control.

Therm: Quantity of heat equivalent to 100,000 Btu.

Thermal cutout: An overcurrent protective device containing a heater element in addition to, and affecting, a renewable fusible member that opens the circuit. It is not designed to interrupt short-circuit currents.

Thermal endurance: The relationship between temperature and time of degrading insulation until failure, under specified conditions.

Thermal protector (applied to motors): A protective device that is assembled as an integral part of a motor or motor-compressor and that, when properly applied, protects the motor against dangerous overheating due to overload and failure to start.

Thermal relay (hot wire relay): Electrical control used to actuate a refrigeration system. This system uses a wire to convert electrical energy into heat energy.

Thermally protected (as applied to motors): When the words "thermally protected" appear on the nameplate of a motor or motor-compressor, it means that the motor is provided with a thermal protector.

Thermistor: An electronic device that makes use of the change of resistivity of a semiconductor with change in temperature.

Thermocouple: A device using the Seebeck Effect to measure temperature.

Thermodisk defrost control: Electrical switch with bimetal disk that is controlled by electrical energy.

Thermodynamics: Science that deals with the relationships between heat and mechanical energy and their interconversion.

Thermoelectric generator: A device interaction of a heat flow and the charge carriers in an electric circuit, and that requires, for this process, the existence of a temperature difference in the electric circuit.

Thermoelectric heat pump: A device that transfers thermal energy from one body to another by the direct interaction of an electrical current and the heat flow.

Thermometer: Device for measuring temperatures.

Thermoplastic: Materials which when reheated, will become pliable with no change of physical properties.

Thermoset: Materials which may be molded, but when cured, undergo an irreversible chemical and physical property change.

Thermostat: Device responsive to ambient temperature conditions.

Thermostatic expansion valve: A control valve operated by temperature and pressure within an evaporator coil, which controls the flow of refrigerant.

Thermostatic valve: Valve controlled by thermostatic elements.

Three phase circuit: A polyphase circuit of three interrelated voltages for which the phase difference is 120°: the common form of generated power.

Thumper: A device used to locate faults in a cable by the release of power surges from a capacitor, characterized by the audible noise when the cable breaks down.

Thyratron: A gas-filled triode tube that is used in electronic control circuits.

Timer: Mechanism used to control on and off times of an electrical circuit.

Timer-thermostat: Thermostat control that includes a clock mechanism. Unit automatically controls room temperature and changes it according to the time of day.

Tinned: Having a thin coating of pure tin, or tin alloy; the coating may keep rubber from sticking or be used to enhance connection; coatings increase the resistance of the conductor, and may contribute to corrosion by electrolysis.

Toggle: A device having two stable states: i.e., toggle switch which is used to turn a circuit ON or OFF. A device may also be toggled from slow to fast, etc.

Tolerance: The permissible variation from rated or assigned value.

Toroid: A coil wound in the form of a doughnut; i.e., current transformers.

Torquing: Applying a rotating force and measuring or limiting its value.

Total hazard current: The hazard current of a given isolated system with all devices, including the line isolation monitor, connected.

TPE: Thermal plastic elastomer.

TPR: Thermal plastic rubber.

Tracer: A means of identifying cable.

Transducer: A device by means of which energy can flow from one or more media to another.

Transfer switch: A device for transferring one or more load conductor connections from one power source to another.

Transformer: A static device consisting of winding(s) with or without a tap(s) with or without a magnetic core for introducing mutual coupling by induction between circuits.

Transformer, potential: Designed for use in measuring high voltage: normally the secondary voltage is 120V.

Transformer, power: Designed to transfer electrical power from the primary circuit to the secondary circuit(s) to 1) step-up the secondary voltage at less current or 2) step-down the secondary voltage at more current; with the voltage-current product being constant for either primary or secondary.

Transformer-rectifier: Combination transformer and rectifier in which input in ac may be varied and then rectified into dc.

Transformer, safety isolation: Inserted to provide a nongrounded power supply such that a grounded person accidentally coming in contact with the secondary circuit will not be electrocuted.

Transformer, vault-type: Suitable for occasional submerged operation in water.

Transient: 1) Lasting only a short time; existing briefly; temporary. 2) A temporary component of current existing in a circuit during adjustment to a changed load, different source voltage, or line impulse.

Transistor: An active semiconductor device with three or more terminals.

Transmission: Transfer of electric energy from one location to another through conductors or by radiation or induction fields.

Transmission line: A long electrical circuit.

Transposition: Interchanging position of conductors to neutralize interference.

Traveler: A pulley complete with suspension arm or frame to be attached to overhead line structures during stringing.

Trim: The front cover assembly for a panel, covering all live terminals and the wires in the gutters, but providing openings for the fuse cutouts or circuit breakers mounted in the panel; may include the door for the panel and also a lock.

Triode: A three-electrode electron tube containing an anode, a cathode, and a control electrode.

Triplex: Three cables twisted together.

Trolley wire: Solid conductor designed to resist wear due to rolling or sliding current pickup trolleys.

Trough: Another name for an "auxiliary gutter," which is a sheet metal enclosure of rectangular cross-section, used to supplement wiring spaces at meter centers, distribution centers, switchboards and similar points in wiring systems where splices or taps are made to circuit conductors. The single word "gutter" is also used to refer to this type of enclosure.

Trunk feeder: A feeder connecting two generating stations or a generating station and an important substation.

Tub: An expression sometimes used to refer to large panelboards or control center cabinets, in particular the box-like enclosure without the front trim.

Tube: A hollow long product having uniform wall thickness and uniform cross-section.

Tuning: The adjustment of a circuit or system to secure optimum performance.

Turn: The basic coil element that forms a single conducting loop comprised of one insulated conductor.

Turn ratio: The ratio between the number of turns between windings in a transformer; normally primary to secondary, except for current transformers where it is secondary to primary.

Twist test: A test to grade round material for processibility into conductors.

Two-phase: A polyphase ac circuit having two interrelated voltages.

Two-fer: An adapter cable containing one male plug and two female cord connectors used to connect two loads to one branch circuit.

UHF (ultra high frequency): 300 MHz to 3 GHz.

UL (Underwriters Laboratories): A nationally known laboratory for testing a product's performance with safety to the user being prime consideration: UL is an independent organization, not controlled by any manufacturer: the best known lab for electrical products.

Ultrasonic: Sounds having frequencies higher than 20 KHz: 20 KHz is at the upper limit of human hearing.

Ultrasonic detector: A device that detects the ultrasonic noise such as that produced by corona or leaking gas.

Ultraviolet: Radiant energy within the wave length range 10 to 380 nanometers: invisible, filtered by glass, causes teeth to glow, causes suntan.

Undercurrent: Less than normal operating current.

Underground feeder and branch circuit cable (type UF): Cable of a listed Type UF in sizes No. 14 copper or No. 12 aluminum or copper-clad aluminum through No. 4/0. The conductors must be suitable for branch circuit or feeder usage, and must be the moisture-resistant type, suitable for direct burial in the earth. They must have an overall coating that is flame retardant, and moisture, fungus, and corrosion resistant.

Undervoltage: Less than normal operating voltage.

Ungrounded: Not connected to the earth intentionally.

Universal motor: A motor designed to operate on either ac or dc at about the same speed and output with either.

Urethane foam: Type of insulation that is foamed in between inner and outer walls.

URD (Underground Residential Distribution): A single phase cable usually consisting of an insulated conductor having a bare concentric neutral.

Utilization equipment: Equipment that uses electric energy for mechanical, chemical, heating, lighting, or other useful purposes.

VA: Volts times amps.

Vacuum: Reduction in pressure below atmospheric pressure.

Vacuum pump: Special high efficiency compressor used for creating high vacuums for testing or drying purposes.

Vacuum switch: A switch with contacts in an evacuated enclosure.

Vacuum tube: A sealed glass enclosure having two or more electrodes between which conduction of electricity may take place. Most vacuum tubes have been replaced with solid-state circuitry.

Valve: Device used for controlling fluid flow.

Valve, solenoid: Valve actuated by magnetic action by means of an electrically energized coil.

Vapor: Word usually used to denote vaporized refrigerant rather than gas.

Vapor-safe: Constructed so that it may be operated without hazard to its surroundings in hazardous areas in aircraft.

Vapor, saturated: A vapor condition that will result in condensation into liquid droplets as vapor temperature is reduced.

Vapor-tight: So enclosed that vapor will not enter.

Variable speed drive: A motor having an integral coupling device which permits the output speed of the unit to be easily varied through a continuous range.

VD: Voltage drop.

Vector: A mathematical term expressing magnitude and direction.

Ventilation: Circulation of air, system or means of providing fresh air.

Vernier: An auxiliary scale permitting measurements more precise than the main scale.

Volatile flammable liquid: A flammable liquid having a flash point below 38°C or whose temperature is above its flash point.

Volt: The derived SI unit for voltage: one volt equals one watt per ampere.

Voltage: The electrical property that provides the energy for current flow; the ratio of the work done to the value of the charge moved when a charge is moved between two points against electrical forces.

Voltage, breakdown: The minimum voltage required to break down an insulation's resistance, allowing a current flow through the insulation, normally at a point.

Voltage, contact: A small voltage which is established whenever two conductors of different materials are brought into contact; due to the difference in work functions or the ease with which electrons can cross the surface boundary in the two directions.

Voltage drop: 1) The loss in voltage between the input to a device and the output from a device due to the internal impedance or resistance of the device. 2) The difference in voltage between two points of a circuit.

Voltage, EHV (extra high voltage): 230-765 kV.

Voltage, induced: A voltage produced in a conductor by a change in magnetic flux linking that path.

Voltage rating of a cable: Phase-to-phase ac voltage when energized by a balanced three phase circuit having a solidly grounded neutral.

Voltage regulator: A device to decrease voltage fluctuations to loads.

Voltage, signal: Voltages to 50 V.

Voltage to ground: The voltage between an energized conductor and earth.

Voltage, UHV (ultra-high voltage): 765 + kV.

Voltage divider: A network consisting of impedance elements connected in series to which a voltage is applied and from which one or more voltages can be obtained across any portion of the network.

Voltmeter: An instrument for measuring voltage.

VOM (volt-ohm-multimeter): A commonly used electrical test instrument to test voltage, current, resistance, and continuity. Instruments are available in both dial and the more popular digital meters.

VW-1: A UL rating given single conductor cables as to flame resistant properties, formerly FR-1.

Wall: The thickness of insulation or jacket of cable.

Wall, fire: A dividing wall for the purpose of checking the spread of fire from one part of a building to another.

Waterblocked cable: A multiconductor cable having interstices filled to prevent water flow or wicking.

Waterproof: Moisture will not interfere with operation.

Watertight: So constructed that water will not enter.

Watt: The derived SI unit for power, radiant flux: one watt equals one joule per second.

Watt-hour: The number of watts used in one hour.

Watt-hour meter: A meter which measures and registers the integral, with respect to time, of the active power in a circuit.

Wattmeter: An instrument for measuring the magnitude of the active power in a circuit.

Wave: A disturbance that is a function of time or space or both.

Waveform: The geometrical shape as obtained by displaying a characteristic of the wave as a function of some variable when plotted over one primitive period.

Wavelength: The distance measured along the direction of propagation between two points which are in phase on adjacent waves.

Weatherhead: The conduit fitting at a conduit used to allow conductor entry, but prevent weather entry.

Weatherproof: So constructed or protected that exposure to the weather will not interfere with successful operation.

Weight correction factor: The correction factor necessitated by uneven geometric forces placed on cables in conduits; used in computing pulling tensions and sidewall loads.

Weighting: The artificial adjustment of measurements in order to account for factors different from those factors which would be encountered in normal use.

Weld: The joining of materials by fusion or recrystallization across the interface of those materials using heat or pressure or both, with or without filler material.

Welding cable: Very flexible cable used for leads to the rod holders of arc-welders, usually consisting of size 4/0 flexible copper conductors.

Wet cell battery: A battery having a liquid electrolyte.

Wet locations: Exposed to weather or water spray or buried.

Wet-niche lighting fixture: A lighting fixture intended for installation in a forming shell mounted in a pool or fountain structure where the fixture will be completely surrounded by water.

Winding: An assembly of coils designed to act in consort to produce a magnetic flux field or to link a flux field.

Wire: A slender rod or filament of drawn metal: the term may also refer to insulated wire.

Wire bar: A cast shape which has a square cross section with tapered ends.

Wire, building: That class of wire and cable, usually rated at 600V, which is normally used for interior wire of buildings.

Wire, covered: A wire having a covering not in conformance with *NEC* standards for material or thickness.

Wire, hookup: Insulated wire for low voltage and current in electronic circuits.

Wire, resistance: Wire having appreciable resistance: used in heating applications such as electric toasters, heaters, etc.

WM: Wattmeter.

Work: Force times distance: pound-feet.

Work function: The minimum energy required to remove an electron from the Fermi level of a material into field-free space: Units — electron volts.

Work hardening: Hardening and embrittlement of metal due to cold working.

Xfmr: Transformer.

X-ray: Penetrating short wavelength electromagnetic radiation created by electron bombardment in high voltage apparatus: produce ionization when they strike certain materials.

Index

A

B

C

D

E

F

G

H

I

J

K

L

M

N

O

P

Q

R

S

Contractor's Guide to the Building Code Revised

CONTRACTOR'S GUIDE TO THE BUILDING CODE
JACK M. HAGEMAN
1997 UBC

This new edition was written in collaboration with the International Conference of Building Officials, writers of the code. It explains in plain English exactly what the latest edition of the *Uniform Building Code* requires. Based on the 1997 code, it explains the changes and what they *mean for the builder. Also covers the Uniform Mechanical Code and the Uniform Plumbing Code.* Shows how to design and construct residential and light commercial buildings that'll pass inspection the first time. Suggests how to work with an inspector to minimize construction costs, what common building shortcuts are likely to be cited, and where exceptions may be granted. **320 pages, 8½ x 11, $39.00**

Contractor's Survival Manual

How to survive hard times and succeed during the up cycles. Shows what to do when the bills can't be paid, finding money and buying time, transferring debt, and all the alternatives to bankruptcy. Explains how to build profits, avoid problems in zoning and permits, taxes, time-keeping, and payroll. Unconventional advice on how to invest in inflation, get high appraisals, trade and postpone income, and stay hip-deep in profitable work. **160 pages, 8½ x 11, $22.25**

Drafting House Plans

Here you'll find step-by-step instructions for drawing a complete set of home plans for a one-story house, an addition to an existing house, or a remodeling project. This book shows how to visualize spatial relationships, use architectural scales and symbols, sketch preliminary drawings, develop detailed floor plans and exterior elevations, and prepare a final plot plan. It even includes code-approved joist and rafter spans and how to make sure that drawings meet code requirements.
192 pages, 8½ x 11, $27.50

Electrical Blueprint Reading Revised

Shows how to read and interpret electrical drawings, wiring diagrams, and specifications for constructing electrical systems. Shows how a typical lighting and power layout would appear on a plan, and explains what to do to execute the plan. Describes how to use a panelboard or heating schedule, and includes typical electrical specifications.
208 pages, 8½ x 11, $18.00

Electrician's Exam Preparation Guide

Electrician's Exam Preparation Guide
Certificate of Competence Master Electrician
Newly Revised

Need help in passing the apprentice, journeyman, or master electrician's exam? This is a book of questions and answers based on actual electrician's exams over the last few years. Almost a thousand multiple-choice questions — exactly the type you'll find on the exam — cover every area of electrical installation: electrical drawings, services and systems, transformers, capacitors, distribution equipment, branch circuits, feeders, calculations, measuring and testing, and more. It gives you the correct answer, an explanation, and where to find it in the latest *NEC*. Also tells how to apply for the test, how best to study, and what to expect on examination day.
352 pages, 8½ x 11, $32.00

Estimating & Bidding for Builders & Remodelers w/ CD-ROM

If your computer has a CD-ROM drive, the *CD Estimator* disk enclosed in this book could change forever the way you estimate construction. You get over 2,500 pages from six current cost databases published by Craftsman, plus an estimating program you can master in minutes, plus a 70-minute interactive video on how to use this program, plus an award-winning book. This package is your best bargain for estimating and bidding construction costs. **272 pages, 8½ x 11, $69.50**

Estimating Electrical Construction

Like taking a class in how to estimate materials and labor for residential and commercial electrical construction. Written by an A.S.P.E. National Estimator of the Year, it teaches you how to use labor units, the plan take-off, and the bid summary to make an accurate estimate, how to deal with suppliers, use pricing sheets, and modify labor units. Provides extensive labor unit tables and blank forms for your next electrical job.
272 pages, 8½ x 11, $19.00

How to Succeed With Your Own Construction Business

Everything you need to start your own construction business: setting up the paperwork, finding the work, advertising, using contracts, dealing with lenders, estimating, scheduling, finding and keeping good employees, keeping the books, and coping with success. If you're considering starting your own construction business, all the knowledge, tips, and blank forms you need are here. **336 pages, 8½ x 11, $28.50**

Kitchens & Baths

Practical, detailed solutions to every problem you're likely to approach when constructing or remodeling kitchens and baths. Over 50 articles from the *Journal of Light Construction* plus up-to-date reference material on codes, safety, space planning and construction methods for kitchens and bathrooms. Includes over 400 photos and illustrations. If you're looking for a complete reference on designing, remodeling, or constructing kitchens and baths, you should have this book. **256 pages, 8 x 11, $34.95**

Markup & Profit: A Contractor's Guide

MARKUP & PROFIT
A CONTRACTOR'S GUIDE

In order to succeed in a construction business, you have to be able to price your jobs to cover all labor, material and overhead expenses, and make a decent profit. The problem is knowing what markup to use. You don't want to lose jobs because you charge too much, and you don't want to work for free because you've charged too little. If you know how to calculate markup, you can apply it to your job costs to find the right sales price for your work. This book gives you tried and tested formulas, with step-by-step instructions and easy-to-follow examples, so you can easily figure the markup that's right for your business. Includes a CD-ROM with forms and checklists for your use.
320 pages, 8½ x 11, $32.50

Masonry & Concrete Construction Revised

This is the revised edition of the popular manual, with updated information on everything from on-site preplanning and layout through the construction of footings, foundations, walls, fireplaces and chimneys. There's an added appendix on safety regulations, with all the applicable OSHA sections pulled together into one handy condensed reference. There's new information on concrete, masonry and seismic reinforcement. Plus improved estimating techniques to help you win more construction bids. The emphasis is on integrating new techniques and improved materials with the tried-and-true methods. Includes information on cement and mortar types, mixes, coloring agents and additives, and suggestions on when, where and how to use them; calculating footing and foundation loads, with tables and formulas to use as references; forming materials and forming systems; pouring and reinforcing concrete slabs and flatwork; block and brick wall construction, including seismic requirements; crack control, masonry veneer construction, brick floors and pavements, including design considerations and materials; and cleaning, painting and repairing all types of masonry. **304 pages, 8½ x 11, $28.50**

Plumber's Exam Preparation Guide

Hundreds of questions and answers to help you pass the apprentice, journeyman, or master plumber's exam. Questions are in the style of the actual exam. Gives answers for both the Standard and Uniform plumbing codes. Includes tips on studying for the exam and the best way to prepare yourself for examination day. **320 pages, 8½ x 11, $29.00**

National Construction Estimator

Current building costs for residential, commercial, and industrial construction. Estimated prices for every common building material. Provides manhours, recommended crew, and gives the labor cost for installation. Includes a CD-ROM with an electronic version of the book with *National Estimator*, a stand-alone *Windows*™ estimating program, plus an interactive multimedia video that shows how to use the disk to compile construction cost estimates. **616 pages, 8½ x 11, $47.50. Revised annually**

Residential Wiring to the 1999 *NEC*

Shows how to install rough and finish wiring in new construction, alterations, and additions. Complete instructions on troubleshooting and repairs. Every subject is referenced to the most recent *National Electrical Code*, and there's 22 pages of the most-needed *NEC* tables to help make your wiring pass inspection — the first time. **352 pages, 5½ x 8½, $27.00**

Practical References for Builders

Basic Engineering for Builders

BASIC
ENGINEERING
FOR BUILDERS

If you've ever been stumped by an engineering problem on the job, yet wanted to avoid the expense of hiring a qualified engineer, you should have this book. Here you'll find engineering principles explained in non-technical language and practical methods for applying them on the job. With the help of this book you'll be able to understand engineering functions in the plans and how to meet the requirements, how to get permits issued without the help of an engineer, and anticipate requirements for concrete, steel, wood and masonry. See why you sometimes have to hire an engineer and what you can undertake yourself: surveying, concrete, lumber loads and stresses, steel, masonry, plumbing, and HVAC systems. This book is designed to help the builder save money by understanding engineering principles that you can incorporate into the jobs you bid. **400 pages, 8½ x 11, $34.00**

Blueprint Reading for the Building Trades

How to read and understand construction documents, blueprints, and schedules. Includes layouts of structural, mechanical, HVAC and electrical drawings. Shows how to interpret sectional views, follow diagrams and schematics, and covers common problems with construction specifications. **192 pages, 5½ x 8½, $14.75**

Builder's Guide to Room Additions

How to tackle problems that are unique to additions, such as requirements for basement conversions, reinforcing ceiling joists for second-story conversions, handling problems in attic conversions, what's required for footings, foundations, and slabs, how to design the best bathroom for the space, and much more. Besides actual construction, you'll even find help in designing, planning, and estimating your room addition jobs.
352 pages, 8½ x 11, $27.25

Build Smarter with Alternative Materials

Build Smarter
with
Alternative
Materials
Leon A. Frechette

New building products are coming out almost every week. Some of them may become new standards, as sheetrock replaced lath and plaster some years ago. Others are little more than a gimmick. To write this manual, the author researched hundreds of products that have come on the market in recent years. The ones he describes in this book will do the job better, creating a superior, longer-lasting finished product, and in many cases also save you time and money. Some are made with recycled products — a good selling point with many customers. But most of all, they give you choices, so you can give your customers choices. In this book, you'll find materials for almost all areas of constructing a house, from the ground up. For each product described, you'll learn where you can get it, where to use it, what benefits it provides, any disadvantages, and how to install it — including tips from the author. And to help you price your jobs, each description ends with manhours — for both the first time you install it, and after you've done it a few times. **336 pages, 8½ x 11, $34.75**

Commercial Electrical Wiring

Make the transition from residential to commercial electrical work. Here are wiring methods, spec reading tips, load calculations and everything you need for making the transition to commercial work: commercial construction documents, load calculations, electric services, transformers, overcurrent protection, wiring methods, raceway, boxes and fittings, wiring devices, conductors, electric motors, relays and motor controllers, special occupancies, and safety requirements. This book is written to help any electrician break into the lucrative field of commercial electrical work.
320 pages, 8½ x 11, $27.50

Concrete Construction & Estimating

Explains how to estimate the quantity of labor and materials needed, plan the job, erect fiberglass, steel, or prefabricated forms, install shores and scaffolding, handle the concrete into place, set joints, finish and cure the concrete. Full of practical reference data, cost estimates, and examples.
571 pages, 5½ x 8½, $25.00

Contractor's Growth and Profit Guide

Step-by-step instructions for planning growth and prosperity in a construction contracting or subcontracting company. Explains how to prepare a business plan: select reasonable goals, draft a market expansion plan, make income forecasts and expense budgets, and project cash flow. You'll learn everything that most lenders and investors require, as well as the best way to organize your business. **336 pages, 5½ x 8½, $19.00**

Contracting in All 50 States

CONTRACTING
IN ALL 50 STATES

Every state has its own licensing requirements that you must meet to do business there. These are usually written exams, financial requirements, and letters of reference. This book shows how to get a building, mechanical or specialty contractor's license, qualify for DOT work, and register as an out-of-state corporation, for every state in the U.S. It lists addresses, phone numbers, application fees, requirements, where an exam is required, what's covered on the exam and how much weight each area of construction is given on the exam. You'll find just about everything you need to know in order to apply for your out-of-state license. **416 pages, 8½ x 11, $36.00**

CD Estimator

If your computer has *Windows*™ and a CD-ROM drive, *CD Estimator* puts at your fingertips 85,000 construction costs for new construction, remodeling, renovation & insurance repair, electrical, plumbing, HVAC and painting. You'll also have the *National Estimator* program — a stand-alone estimating program for *Windows*™ that *Remodeling* magazine called a "computer wiz." Quarterly cost updates are available at no charge on the Internet. To help you create professional-looking estimates, the disk includes over 40 construction estimating and bidding forms in a format that's perfect for nearly any word processing or spreadsheet program for *Windows*™. And to top it off, a 70-minute interactive video teaches you how to use this CD-ROM to estimate construction costs. **CD Estimator is $68.50**

Construction Forms & Contracts

125 forms you can copy and use — or load into your computer (from the FREE disk enclosed). Then you can customize the forms to fit your company, fill them out, and print. Loads into Word for *Windows*™, *Lotus 1-2-3*, *WordPerfect*, *Works*, or *Excel* programs. You'll find forms covering accounting, estimating, fieldwork, contracts, and general office. Each form comes with complete instructions on when to use it and how to fill it out. These forms were designed, tested and used by contractors, and will help keep your business organized, profitable and out of legal, accounting and collection troubles. Includes a CD-ROM for *Windows*™ and Mac.
400 pages, 8½ x 11, $41.75

Contractor's Guide to QuickBooks Pro 99

This user-friendly manual walks you through QuickBooks Pro's detailed setup procedure and explains step-by-step how to create a first-rate accounting system. You'll learn in days, rather than weeks, how to use QuickBooks Pro to get your contracting business organized, with simple, fast accounting procedures. On the CD included with the book you'll find a full version of QuickBooks Pro, good for 25 uses, with a QuickBooks Pro file preconfigured for a construction company (you drag it over onto your computer and plug in your own company's data). You'll also get a complete estimating program, including a database, and a job costing program that lets you export your estimates to QuickBooks Pro. It even includes many useful construction forms to use in your business.
312 pages, 8½ x 11, $42.00
Also available: **Contractor's Guide to QuickBooks Pro *version 6*. $39.75**

Contractor's Year-Round Tax Guide Revised

How to set up and run your construction business to minimize taxes: corporate tax strategy and how to use it to your advantage, and what you should be aware of in contracts with others. Covers tax shelters for builders, write-offs and investments that will reduce your taxes, accounting methods that are best for contractors, and what the I.R.S. allows and what it often questions. **192 pages, 8½ x 11, $26.50**

Contractor's Guide to the Building Code Revised

CONTRAC-
TOR'S
GUIDE
TO THE
BUILDING
CODE
1997
UBC

This new edition was written in collaboration with the International Conference of Building Officials, writers of the code. It explains in plain English exactly what the latest edition of the *Uniform Building Code* requires. Based on the 1997 code, it explains the changes and what they *mean for the builder. Also covers the Uniform Mechanical Code and the Uniform Plumbing Code.* Shows how to design and construct residential and light commercial buildings that'll pass inspection the first time. Suggests how to work with an inspector to minimize construction costs, what common building shortcuts are likely to be cited, and where exceptions may be granted. **320 pages, $8^1/_2$ x 11, $39.00**

Contractor's Survival Manual

How to survive hard times and succeed during the up cycles. Shows what to do when the bills can't be paid, finding money and buying time, transferring debt, and all the alternatives to bankruptcy. Explains how to build profits, avoid problems in zoning and permits, taxes, time-keeping, and payroll. Unconventional advice on how to invest in inflation, get high appraisals, trade and postpone income, and stay hip-deep in profitable work. **160 pages, $8^1/_2$ x 11, $22.25**

Drafting House Plans

Here you'll find step-by-step instructions for drawing a complete set of home plans for a one-story house, an addition to an existing house, or a remodeling project. This book shows how to visualize spatial relationships, use architectural scales and symbols, sketch preliminary drawings, develop detailed floor plans and exterior elevations, and prepare a final plot plan. It even includes code-approved joist and rafter spans and how to make sure that drawings meet code requirements.
192 pages, $8^1/_2$ x 11, $27.50

Electrical Blueprint Reading Revised

Shows how to read and interpret electrical drawings, wiring diagrams, and specifications for constructing electrical systems. Shows how a typical lighting and power layout would appear on a plan, and explains what to do to execute the plan. Describes how to use a panelboard or heating schedule, and includes typical electrical specifications.
208 pages, $8^1/_2$ x 11, $18.00

Electrician's Exam Preparation Guide

Electrician's
Exam
Preparation
Guide

Need help in passing the apprentice, journeyman, or master electrician's exam? This is a book of questions and answers based on actual electrician's exams over the last few years. Almost a thousand multiple-choice questions — exactly the type you'll find on the exam — cover every area of electrical installation: electrical drawings, services and systems, transformers, capacitors, distribution equipment, branch circuits, feeders, calculations, measuring and testing, and more. It gives you the correct answer, an explanation, and where to find it in the latest *NEC*. Also tells how to apply for the test, how best to study, and what to expect on examination day.
352 pages, $8^1/_2$ x 11, $32.00

Estimating & Bidding for Builders & Remodelers w/ CD-ROM

If your computer has a CD-ROM drive, the *CD Estimator* disk enclosed in this book could change forever the way you estimate construction. You get over 2,500 pages from six current cost databases published by Craftsman, plus an estimating program you can master in minutes, plus a 70-minute interactive video on how to use this program, plus an award-winning book. This package is your best bargain for estimating and bidding construction costs. **272 pages, $8^1/_2$ x 11, $69.50**

Estimating Electrical Construction

Like taking a class in how to estimate materials and labor for residential and commercial electrical construction. Written by an A.S.P.E. National Estimator of the Year, it teaches you how to use labor units, the plan take-off, and the bid summary to make an accurate estimate, how to deal with suppliers, use pricing sheets, and modify labor units. Provides extensive labor unit tables and blank forms for your next electrical job.
272 pages, $8^1/_2$ x 11, $19.00

How to Succeed With Your Own Construction Business

Everything you need to start your own construction business: setting up the paperwork, finding the work, advertising, using contracts, dealing with lenders, estimating, scheduling, finding and keeping good employees, keeping the books, and coping with success. If you're considering starting your own construction business, all the knowledge, tips, and blank forms you need are here. **336 pages, $8^1/_2$ x 11, $28.50**

Kitchens & Baths

Practical, detailed solutions to every problem you're likely to approach when constructing or remodeling kitchens and baths. Over 50 articles from the *Journal of Light Construction* plus up-to-date reference material on codes, safety, space planning and construction methods for kitchens and bathrooms. Includes over 400 photos and illustrations. If you're looking for a complete reference on designing, remodeling, or constructing kitchens and baths, you should have this book. **256 pages, 8 x 11, $34.95**

Markup & Profit: A Contractor's Guide

MARKUP
&
PROFIT
A CONTRACTOR'S
GUIDE

In order to succeed in a construction business, you have to be able to price your jobs to cover all labor, material and overhead expenses, and make a decent profit. The problem is knowing what markup to use. You don't want to lose jobs because you charge too much, and you don't want to work for free because you've charged too little. If you know how to calculate markup, you can apply it to your job costs to find the right sales price for your work. This book gives you tried and tested formulas, with step-by-step instructions and easy-to-follow examples, so you can easily figure the markup that's right for your business. Includes a CD-ROM with forms and checklists for your use.
320 pages, $8^1/_2$ x 11, $32.50

Masonry & Concrete Construction Revised

This is the revised edition of the popular manual, with updated information on everything from on-site preplanning and layout through the construction of footings, foundations, walls, fireplaces and chimneys. There's an added appendix on safety regulations, with all the applicable OSHA sections pulled together into one handy condensed reference. There's new information on concrete, masonry and seismic reinforcement. Plus improved estimating techniques to help you win more construction bids. The emphasis is on integrating new techniques and improved materials with the tried-and-true methods. Includes information on cement and mortar types, mixes, coloring agents and additives, and suggestions on when, where and how to use them; calculating footing and foundation loads, with tables and formulas to use as references; forming materials and forming systems; pouring and reinforcing concrete slabs and flatwork; block and brick wall construction, including seismic requirements; crack control, masonry veneer construction, brick floors and pavements, including design considerations and materials; and cleaning, painting and repairing all types of masonry. **304 pages, $8^1/_2$ x 11, $28.50**

Plumber's Exam Preparation Guide

Hundreds of questions and answers to help you pass the apprentice, journeyman, or master plumber's exam. Questions are in the style of the actual exam. Gives answers for both the Standard and Uniform plumbing codes. Includes tips on studying for the exam and the best way to prepare yourself for examination day. **320 pages, $8^1/_2$ x 11, $29.00**

National Construction Estimator

Current building costs for residential, commercial, and industrial construction. Estimated prices for every common building material. Provides manhours, recommended crew, and gives the labor cost for installation. Includes a CD-ROM with an electronic version of the book with *National Estimator*, a stand-alone *Windows*™ estimating program, plus an interactive multimedia video that shows how to use the disk to compile construction cost estimates. **616 pages, $8^1/_2$ x 11, $47.50. Revised annually**

Residential Wiring to the 1999 *NEC*

Shows how to install rough and finish wiring in new construction, alterations, and additions. Complete instructions on troubleshooting and repairs. Every subject is referenced to the most recent *National Electrical Code*, and there's 22 pages of the most-needed *NEC* tables to help make your wiring pass inspection — the first time. **352 pages, $5^1/_2$ x $8^1/_2$, $27.00**

Residential Steel Framing Guide

RESIDENTIAL
STEEL FRAMING
CONSTRUCTION GUIDE

Steel is stronger and lighter than wood — straight walls are guaranteed — steel framing will not wrap, shrink, split, swell, bow, or rot. Here you'll find full page schematics and details that show how steel is connected in just about all residential framing work. You won't find lengthy explanations here on how to run your business, or even how to do the work. What you will find are over 150 easy-to-read full-page details on how to construct steel-framed floors, roofs, interior and exterior walls, bridging, blocking, and reinforcing for all residential construction. Also includes recommended fasteners and their applications, and fastening schedules for attaching every type of steel framing member to steel as well as wood.
170 pages, $8^1/_2$ x 11, $38.80

1996 *National Electrical Code* Interpretive Diagrams

Based on the 1996 *NEC*, this new edition visually and graphically explains and interprets the key provisions of the *National Electrical Code* as seen through the eyes of an electrical inspector. All drawings have been carefully prepared and reviewed by inspectors and the Codes and Standards Committee. Here you'll find approved installations of just about every type of electrical part. **270 pages, 5 x 8, $34.95**

Audio: Electrician's Exam Preparation Guide

These tapes are made to order for the busy electrician looking for a better-paying career as a licensed apprentice, journeyman, or master electrician. This two-audiotape set asks you over 150 often-used exam questions, waits for your answer, then gives you the correct answer and an explanation. This is the easiest way to study for the exam.
Two 50-minute audiotapes, $26.50

National Renovation & Insurance Repair Estimator

Current prices in dollars and cents for hard-to-find items needed on most insurance, repair, remodeling, and renovation jobs. All price items include labor, material, and equipment breakouts, plus special charts that tell you exactly how these costs are calculated. Includes a CD-ROM with an electronic version of the book with *National Estimator*, a stand-alone *Windows*™ estimating program, plus an interactive multimedia video that shows how to use the disk to compile construction cost estimates.
568 pages, $8^1/_2$ x 11, $49.50. Revised annually

Vest Pocket Guide to HVAC Electricity

This handy guide will be a constant source of useful information for anyone working with electrical systems for heating, ventilating, refrigeration, and air conditioning. Includes essential tables and diagrams for calculating and installing electrical systems for powering and controlling motors, fans, heating elements, compressors, transformers and every electrical part of an HVAC system. **304 pages, $3^1/_2$ x $5^1/_2$, $18.00**

Plumber's Handbook Revised

PLUMBER'S
HANDBOOK

This new edition shows what will and won't pass inspection in drainage, vent, and waste piping, septic tanks, water supply, graywater recycling systems, pools and spas, fire protection, and gas piping systems. All tables, standards, and specifications are completely up-to-date with recent plumbing code changes. Covers common layouts for residential work, how to size piping, select and hang fixtures, practical recommendations, and trade tips. It's the approved reference for the plumbing contractor's exam in many states. Includes an extensive set of multiple choice questions after each chapter, and in the back of the book, the answers and explanations. Also in the back of the book, a full sample plumber's exam. **352 pages, $8^1/_2$ x 11, $32.00**

Plumbing & HVAC Manhour Estimates

Hundreds of tested and proven manhours for installing just about any plumbing and HVAC component you're likely to use in residential, commercial, and industrial work. You'll find manhours for installing piping systems, specialties, fixtures and accessories, ducting systems, and HVAC equipment. If you estimate the price of plumbing, you shouldn't be without the reliable, proven manhours in this unique book.
224 pages, $5^1/_2$ x $8^1/_2$, $28.25

Renovating & Restyling Older Homes

Any builder can turn a run-down old house into a showcase of perfection — if the customer has unlimited funds to spend. Unfortunately, most customers are on a tight budget. They usually want more improvements than they can afford — and they expect you to deliver. This book shows how to add economical improvements that can increase the property value by two, five or even ten times the cost of the remodel. Sound impossible? Here you'll find the secrets of a builder who has been putting these techniques to work on Victorian and Craftsman-style houses for twenty years. You'll see what to repair, what to replace and what to leave, so you can remodel or restyle older homes for the least amount of money and the greatest increase in value. **416 pages, $8^1/_2$ x 11, $33.50**

National Electrical Estimator

This year's prices for installation of all common electrical work: conduit, wire, boxes, fixtures, switches, outlets, loadcenters, panelboards, raceway, duct, signal systems, and more. Provides material costs, manhours per unit, and total installed cost. Explains what you should know to estimate each part of an electrical system. Includes a CD-ROM with an electronic version of the book with *National Estimator*, a stand-alone *Windows*™ estimating program, plus an interactive multimedia video that shows how to use the disk to compile construction cost estimates.
544 pages, $8^1/_2$ x 11, $47.75. Revised annually

Residential Electrical Estimating

RESIDENTIAL
ELECTRICAL
ESTIMATING
John E. Traister

A fast, accurate pricing system proven on over 1000 residential jobs. Using the manhours provided, combined with material prices from your wholesaler, you quickly work up estimates based on degree of difficulty. These manhours come from a working electrical contractor's records — not some pricing agency. You'll find prices for every type of electrical job you're likely to estimate — from service entrances to ceiling fans.
320 pages, $8^1/_2$ x 11, $29.00

Troubleshooting Guide to Residential Construction

How to solve practically every construction problem - before it happens to you! With this book you'll learn from the mistakes other builders made as they faced 63 typical residential construction problems. Filled with clear photos and drawings that explain how to enhance your reputation as well as your bottom line by avoiding problems that plague most builders. Shows how to avoid, or fix, problems ranging from defective slabs, walls and ceilings, through roofing, plumbing & HVAC, to paint.
304 pages, $8^1/_2$ x 11, $32.50

Wood-Frame House Construction

Step-by-step construction details, from the layout of the outer walls, excavation and formwork, to finish carpentry and painting. Contains all new, clear illustrations and explanations updated for construction in the '90s. Everything you need to know about framing, roofing, siding, interior finishings, floor covering and stairs — your complete book of wood-frame homebuilding. **320 pages, $8^1/_2$ x 11, $25.50. Revised edition**

Craftsman's Illustrated Dictionary of Construction Terms

Almost everything you could possibly want to know about any word or technique in construction. Hundreds of up-to-date construction terms, materials, drawings and pictures with detailed, illustrated articles describing equipment and methods. Terms and techniques are explained or illustrated in vivid detail. Use this valuable reference to check spelling, find clear, concise definitions of construction terms used on plans and construction documents, or learn about little-known tools, equipment, tests and methods used in the building industry. It's all here.
416 pages, $8^1/_2$ x 11, $36.00

Rough Framing Carpentry

If you'd like to make good money working outdoors as a framer, this is the book for you. Here you'll find shortcuts to laying out studs; speed cutting blocks, trimmers and plates by eye; quickly building and blocking rake walls; installing ceiling backing, ceiling joists, and truss joists; cutting and assembling hip trusses and California fills; arches and drop ceilings — all with production line procedures that save you time and help you make more money. Over 100 on-the-job photos of how to do it right and what can go wrong. **304 pages, $8^1/_2$ x 11, $26.50**

Home Inspection Handbook

Every area you need to check in a home inspection — especially in older homes. Twenty complete inspection checklists: building site, foundation and basement, structural, bathrooms, chimneys and flues, ceilings, interior & exterior finishes, electrical, plumbing, HVAC, insects, vermin and decay, and more. Also includes information on starting and running your own home inspection business. **324 pages, $5\frac{1}{2}$ x $8\frac{1}{2}$, $24.95**

National Repair & Remodeling Estimator

The complete pricing guide for dwelling reconstruction costs. Reliable, specific data you can apply on every repair and remodeling job. Up-to-date material costs and labor figures based on thousands of jobs across the country. Provides recommended crew sizes; average production rates; exact material, equipment, and labor costs; a total unit cost and a total price including overhead and profit. Separate listings for high- and low-volume builders, so prices shown are specific for any size business. Estimating tips specific to repair and remodeling work to make your bids complete, realistic, and profitable. Includes a CD-ROM with an electronic version of the book with *National Estimator*, a stand-alone *Windows*™ estimating program, plus an interactive multimedia video that shows how to use the disk to compile construction cost estimates.
312 pages, $8\frac{1}{2}$ x 11, $48.50. Revised annually

Managing the Small Construction Business

Overcome your share of business hassles by learning how 50 small contractors handled similar problems in their businesses. Here you'll learn how they handle bidding, unit pricing, contract clauses, change orders, job-site safety, quality control, overhead and markup, managing subs, scheduling systems, cost-plus contracts, pricing small jobs, insurance repair, finding solutions to conflicts, and much more.
243 pages, $8\frac{1}{2}$ x 11, $27.95

The Contractor's Legal Kit

Stop "eating" the costs of bad designs, hidden conditions, and job surprises. Set ground rules that assign those costs to the rightful party ahead of time. And it's all in plain English, not "legalese." For less than the cost of an hour with a lawyer you'll learn the exclusions to put in your agreements, why your insurance company may pay for your legal defense, how to avoid liability for injuries to your sub and his employees or damages they cause, how to collect on lawsuits you win, and much more. It also includes a FREE computer disk with contracts and forms you can customize for your own use. **352 pages, $8\frac{1}{2}$ x 11, $59.95**

Simplified Guide to Construction Law

Here you'll find easy-to-read, paragraphed-sized samples of how the courts have viewed common areas of disagreement — and litigation — in the building industry. You'll read about legal problems that real builders have faced, and how the court ruled. This book will tell you what you need to know about contracts, torts, fraud, misrepresentation, warranty and strict liability, construction defects, indemnity, insurance, mechanics liens, bonds and bonding, statutes of limitation, arbitration, and more. These are simplified examples that illustrate not necessarily who is right and who is wrong — but who the law has sided with. **298 pages, $5\frac{1}{2}$ x 8-1/2, $29.95**

National Job Cost CD

This new program includes an easy-to-learn estimating program (*National Estimator*), a database of 3,000 lines of manhour data, and Job Cost Wizard, a conversion program that lets you export your estimates into QuickBooks Pro. With this, you can prepare invoices, compare your estimated costs to your actual job costs, and more. Includes an interactive video tutorial that walks you through the estimating and job costing procedure, a 25-use working version of QuickBooks Pro, and complete instructions to help you find where you're making or losing money on your jobs.
National Job Cost CD is $38.50

Craftsman Book Company
6058 Corte del Cedro
P.O. Box 6500
Carlsbad, CA 92018

☎ 24 hour order line
1-800-829-8123
Fax (760) 438-0398

In A Hurry?
We accept phone orders charged to your
❍ Visa, ❍ MasterCard, ❍ Discover or ❍ American Express

Name ____________________

Company ____________________

Address ____________________

City/State/Zip ____________________

❍ This is a residence

Total enclosed ____________________ (In California add 7.25% tax)

We pay shipping when your check covers your order in full.

Card# ____________________

Exp. date ____________ Initials ____________

Tax Deductible: Treasury regulations make these references tax deductible when used in your work. Save the canceled check or charge card statement as your receipt.

Order online http://www.craftsman-book.com
Free on the Internet! Download any of Craftsman's estimating costbooks for a 30-day free trial! http://costbook.com

10-Day Money Back Guarantee

- ❍ 26.50 Audiotape: Electrician's Exam Preparation Guide
- ❍ 34.00 Basic Engineering for Builders
- ❍ 14.75 Blueprint Reading for the Building Trades
- ❍ 34.75 Build Smarter with Alternative Materials
- ❍ 27.25 Builder's Guide to Room Additions
- ❍ 68.50 CD Estimator
- ❍ 27.50 Commercial Electrical Wiring
- ❍ 25.00 Concrete Construction & Estimating
- ❍ 41.75 Construction Forms & Contracts w/CD-ROM for *Windows*™ and Macintosh.
- ❍ 36.00 Contracting in All 50 States
- ❍ 19.00 Contractor's Growth & Profit Guide
- ❍ 42.00 Contractor's Guide to QuickBooks Pro 99
- ❍ 39.75 Contractor's Guide to QuickBooks Pro version 6
- ❍ 39.00 Contractor's Guide to the Building Code Revised
- ❍ 59.95 Contractor's Legal Kit
- ❍ 22.25 Contractor's Survival Manual
- ❍ 26.50 Contractor's Year-Round Tax Guide Revised
- ❍ 36.00 Craftsman's Illustrated Dictionary of Construction Terms
- ❍ 27.50 Drafting House Plans
- ❍ 18.00 Electrical Blueprint Reading Revised
- ❍ 32.00 Electrician's Exam Preparation Guide
- ❍ 69.50 Estimating & Bidding for Builders & Remodelers w/CD-ROM
- ❍ 19.00 Estimating Electrical Construction
- ❍ 24.95 Home Inspection Handbook
- ❍ 28.50 How to Succeed w/Your Own Construction Business
- ❍ 34.95 Kitchens & Baths
- ❍ 27.95 Managing the Small Construction Business
- ❍ 32.50 Markup & Profit: A Contractor's Guide
- ❍ 28.50 Masonry & Concrete Construction Revised
- ❍ 47.50 National Construction Estimator w/FREE *National Estimator* on a CD-ROM.
- ❍ 34.95 1996 *National Electrical Code* Interpretive Diagrams
- ❍ 47.75 National Electrical Estimator w/FREE *National Estimator* on a CD-ROM.
- ❍ 38.50 National Job Cost CD
- ❍ 49.50 National Renovation & Insurance Repair Estimator w/FREE *National Estimator* on a CD-ROM.
- ❍ 48.50 National Repair & Remodeling Estimator w/FREE *National Estimator* on a CD-ROM.
- ❍ 29.00 Plumber's Exam Preparation Guide
- ❍ 32.00 Plumber's Handbook Revised
- ❍ 28.25 Plumbing & HVAC Manhour Estimates
- ❍ 33.50 Renovating & Restyling Older Homes
- ❍ 29.00 Residential Electrical Estimating
- ❍ 38.80 Residential Steel Framing Guide
- ❍ 27.00 Residential Wiring to the 1999 *NEC*
- ❍ 26.50 Rough Framing Carpentry
- ❍ 29.95 A Simplified Guide to Construction Law
- ❍ 32.50 Troubleshooting Guide to Residential Construction
- ❍ 18.00 Vest Pocket Guide to HVAC Electricity
- ❍ 25.50 Wood-Frame House Construction
- ❍ 38.75 Illustrated Guide to the 1999 *National Electrical Code*
- ❍ FREE Full Color Catalog

Prices subject to change without notice

Craftsman Book Company
6058 Corte del Cedro
P.O. Box 6500
Carlsbad, CA 92018

☎ 24 hour order line
1-800-829-8123
Fax (760) 438-0398

Name ____________

Company ____________

Address ____________

City/State/Zip ____________

❍ This is a residence

Total enclosed ____________ (In California add 7.25% tax)

We pay shipping when your check covers your order in full.

In A Hurry?

We accept phone orders charged to your
❍ Visa, ❍ MasterCard, ❍ Discover or ❍ American Express

Card# ____________

Exp. date ____________ Initials ____________

Tax Deductible: Treasury regulations make these references tax deductible when used in your work. Save the canceled check or charge card statement as your receipt.

Order online http://www.craftsman-book.com
Free on the Internet! Download any of Craftsman's estimating costbooks for a 30-day free trial! http://costbook.com

10-Day Money Back Guarantee

❍ 26.50 Audio: Electrician's Exam Preparation Guide
❍ 34.00 Basic Engineering for Builders
❍ 14.75 Blueprint Reading for the Building Trades
❍ 34.75 Build Smarter with Alternative Materials
❍ 27.25 Builder's Guide to Room Additions
❍ 68.50 CD Estimator
❍ 27.50 Commercial Electrical Wiring
❍ 25.00 Concrete Construction & Estimating
❍ 41.75 Construction Forms & Contracts w/CD-ROM for *Windows*™ and Macintosh.
❍ 36.00 Contracting in All 50 States
❍ 19.00 Contractor's Growth & Profit Guide
❍ 42.00 Contractor's Guide to QuickBooks Pro 99
❍ 39.75 Contractor's Guide to QuickBooks Pro version 6
❍ 39.00 Contractor's Guide to the Building Code Rev.
❍ 59.95 Contractor's Legal Kit
❍ 22.25 Contractor's Survival Manual
❍ 26.50 Contractor's Year-Round Tax Guide Rev.
❍ 36.00 Craftsman's Illus. Dictionary of Construction Terms
❍ 27.50 Drafting House Plans
❍ 18.00 Electrical Blueprint Reading Revised
❍ 32.00 Electrician's Exam Preparation Guide
❍ 69.50 Estimating & Bidding for Builders & Remodelers w/CD-ROM
❍ 19.00 Estimating Electrical Construction
❍ 24.95 Home Inspection Handbook
❍ 28.50 How to Succeed w/Your Own Constr. Business
❍ 34.95 Kitchens & Baths
❍ 27.95 Managing the Small Construction Business
❍ 32.50 Markup & Profit: A Contractor's Guide
❍ 28.50 Masonry & Concrete Construction Rev.
❍ 47.50 National Construction Estimator w/FREE *National Estimator* on a CD-ROM.
❍ 34.95 1996 *NEC* Interpretive Diagrams
❍ 47.75 National Electrical Estimator w/FREE *National Estimator* on a CD-ROM.
❍ 38.50 National Job Cost CD
❍ 49.50 National Renovation & Insurance Repair Estimator w/FREE *National Estimator* on a CD-ROM.
❍ 48.50 National Repair & Remodeling Estimator w/FREE *National Estimator*on a CD-ROM.
❍ 29.00 Plumber's Exam Preparation Guide
❍ 32.00 Plumber's Handbook Revised
❍ 28.25 Plumbing & HVAC Manhour Estimates
❍ 33.50 Renovating & Restyling Older Homes
❍ 29.00 Residential Electrical Estimating
❍ 38.80 Residential Steel Framing Guide
❍ 27.00 Residential Wiring to the 1999 *NEC*
❍ 26.50 Rough Framing Carpentry
❍ 29.95 A Simplified Guide to Construction Law
❍ 32.50 Troubleshooting Guide to Residential Construction
❍ 18.00 Vest Pocket Guide to HVAC Electricity
❍ 25.50 Wood-Frame House Construction
❍ 38.75 Illustrated Guide to the 1999 *NEC*
❍ FREE Full Color Catalog

Prices subject to change without notice

Craftsman Book Company
6058 Corte del Cedro
P.O. Box 6500
Carlsbad, CA 92018

☎ 24 hour order line
1-800-829-8123
Fax (760) 438-0398

Name ____________

Company ____________

Address ____________

City/State/Zip ____________

❍ This is a residence

Total enclosed ____________ (In California add 7.25% tax)

We pay shipping when your check covers your order in full.

In A Hurry?

We accept phone orders charged to your
❍ Visa, ❍ MasterCard, ❍ Discover or ❍ American Express

Card# ____________

Exp. date ____________ Initials ____________

Tax Deductible: Treasury regulations make these references tax deductible when used in your work. Save the canceled check or charge card statement as your receipt.

Order online http://www.craftsman-book.com
Free on the Internet! Download any of Craftsman's estimating costbooks for a 30-day free trial! http://costbook.com

10-Day Money Back Guarantee

❍ 26.50 Audio: Electrician's Exam Preparation Guide
❍ 34.00 Basic Engineering for Builders
❍ 14.75 Blueprint Reading for the Building Trades
❍ 34.75 Build Smarter with Alternative Materials
❍ 27.25 Builder's Guide to Room Additions
❍ 68.50 CD Estimator
❍ 27.50 Commercial Electrical Wiring
❍ 25.00 Concrete Construction & Estimating
❍ 41.75 Construction Forms & Contracts w/CD-ROM for *Windows*™ and Macintosh.
❍ 36.00 Contracting in All 50 States
❍ 19.00 Contractor's Growth & Profit Guide
❍ 42.00 Contractor's Guide to QuickBooks Pro 99
❍ 39.75 Contractor's Guide to QuickBooks Pro version 6
❍ 39.00 Contractor's Guide to the Building Code Rev.
❍ 59.95 Contractor's Legal Kit
❍ 22.25 Contractor's Survival Manual
❍ 26.50 Contractor's Year-Round Tax Guide Rev.
❍ 36.00 Craftsman's Illus. Dictionary of Construction Terms
❍ 27.50 Drafting House Plans
❍ 18.00 Electrical Blueprint Reading Revised
❍ 32.00 Electrician's Exam Preparation Guide
❍ 69.50 Estimating & Bidding for Builders & Remodelers w/CD-ROM
❍ 19.00 Estimating Electrical Construction
❍ 24.95 Home Inspection Handbook
❍ 28.50 How to Succeed w/Your Own Constr. Business
❍ 34.95 Kitchens & Baths
❍ 27.95 Managing the Small Construction Business
❍ 32.50 Markup & Profit: A Contractor's Guide
❍ 28.50 Masonry & Concrete Construction Rev.
❍ 47.50 National Construction Estimator w/FREE *National Estimator* on a CD-ROM.
❍ 34.95 1996 *NEC* Interpretive Diagrams
❍ 47.75 National Electrical Estimator w/FREE *National Estimator* on a CD-ROM.
❍ 38.50 National Job Cost CD
❍ 49.50 National Renovation & Insurance Repair Estimator w/FREE *National Estimator* on a CD-ROM.
❍ 48.50 National Repair & Remodeling Estimator w/FREE *National Estimator*on a CD-ROM.
❍ 29.00 Plumber's Exam Preparation Guide
❍ 32.00 Plumber's Handbook Revised
❍ 28.25 Plumbing & HVAC Manhour Estimates
❍ 33.50 Renovating & Restyling Older Homes
❍ 29.00 Residential Electrical Estimating
❍ 38.80 Residential Steel Framing Guide
❍ 27.00 Residential Wiring to the 1999 *NEC*
❍ 26.50 Rough Framing Carpentry
❍ 29.95 A Simplified Guide to Construction Law
❍ 32.50 Troubleshooting Guide to Residential Construction
❍ 18.00 Vest Pocket Guide to HVAC Electricity
❍ 25.50 Wood-Frame House Construction
❍ 38.75 Illustrated Guide to the 1999 *NEC*
❍ FREE Full Color Catalog

Prices subject to change without notice

Mail This Card Today For a Free Full Color Catalog

Over 100 books, annual cost guides and estimating software packages at your fingertips with information that can save you time and money. Here you'll find information on carpentry, contracting, estimating, remodeling, electrical work, and plumbing.

All items come with an unconditional 10-day money-back guarantee. If they don't save you money, mail them back for a full refund.

Order Now! 1 800 829-8123 • 10 Day Money-Back Guarantee
2000 Handbooks for the Professional Builder
Construction References
2000 NATIONAL CONSTRUCTION ESTIMATOR
New 2000 Cost Data! Get it Now!
Craftsman Book Company • 6058 Corte del Cedro • Carlsbad, CA 92009
www.craftsman-book.com

Name ____________

Company ____________

Address ____________

City/State/Zip ____________

Craftsman Book Company / 6058 Corte del Cedro / P.O. Box 6500 / Carlsbad, CA 92018

NO POSTAGE
NECESSARY
IF MAILED
IN THE
UNITED STATES

BUSINESS REPLY MAIL
FIRST CLASS MAIL PERMIT NO. 271 CARLSBAD, CA

POSTAGE WILL BE PAID BY ADDRESSEE

Craftsman Book Company
6058 Corte del Cedro
P.O. Box 6500
Carlsbad, CA 92018-9974

NO POSTAGE
NECESSARY
IF MAILED
IN THE
UNITED STATES

BUSINESS REPLY MAIL
FIRST CLASS MAIL PERMIT NO. 271 CARLSBAD, CA

POSTAGE WILL BE PAID BY ADDRESSEE

Craftsman Book Company
6058 Corte del Cedro
P.O. Box 6500
Carlsbad, CA 92018-9974

NO POSTAGE
NECESSARY
IF MAILED
IN THE
UNITED STATES

BUSINESS REPLY MAIL
FIRST CLASS MAIL PERMIT NO. 271 CARLSBAD, CA

POSTAGE WILL BE PAID BY ADDRESSEE

Craftsman Book Company
6058 Corte del Cedro
P.O. Box 6500
Carlsbad, CA 92018-9974